FRACTURE, FATIGUE, AND WELD RESIDUAL STRESS

presented at
THE 1999 ASME PRESSURE VESSELS AND PIPING CONFERENCE
BOSTON, MASSACHUSETTS
AUGUST 1–5, 1999

sponsored by
THE PRESSURE VESSELS AND PIPING DIVISION, ASME

principal editor
J. PAN
UNIVERSITY OF MICHIGAN

contributing editors

M. T. KIRK
WESTINGHOUSE ELECTRIC COMPANY

G. M. WILKOWSKI
ENGINEERING MECHANICS CORP.

K. K. YOON
FRAMATOME TECHNOLOGIES

P. DONG
BATTELLE

Y. J. CHAO
UNIVERSITY OF SOUTH CAROLINA

THE AMERICAN SOCIETY OF MECHANICAL ENGINEERS
Three Park Avenue / New York, N.Y. 10016

Statement from By-Laws: The Society shall not be responsible for statements or opinions advanced in papers. . . or printed in its publications (7.1.3)

INFORMATION CONTAINED IN THIS WORK HAS BEEN OBTAINED BY THE AMERICAN SOCIETY OF MECHANICAL ENGINEERS FROM SOURCES BELIEVED TO BE RELIABLE. HOWEVER, NEITHER ASME NOR ITS AUTHORS OR EDITORS GUARANTEE THE ACCURACY OR COMPLETENESS OF ANY INFORMATION PUBLISHED IN THIS WORK. NEITHER ASME NOR ITS AUTHORS AND EDITORS SHALL BE RESPONSIBLE FOR ANY ERRORS, OMISSIONS, OR DAMAGES ARISING OUT OF THE USE OF THIS INFORMATION. THE WORK IS PUBLISHED WITH THE UNDERSTANDING THAT ASME AND ITS AUTHORS AND EDITORS ARE SUPPLYING INFORMATION BUT ARE NOT ATTEMPTING TO RENDER ENGINEERING OR OTHER PROFESSIONAL SERVICES. IF SUCH ENGINEERING OR PROFESSIONAL SERVICES ARE REQUIRED, THE ASSISTANCE OF AN APPROPRIATE PROFESSIONAL SHOULD BE SOUGHT.

Authorization to photocopy for internal or personal use is granted to libraries and other users registered with the Copyright Clearance Center (CCC) provided .030¢ per page is paid to CCC, 222 Rosewood Dr., Danvers, MA 01923. Requests for special permission or bulk reproduction should be addressed to the ASME Technical Publishing Department.

ISBN No. 0-7918-1627-3

Copyright © 1999 by
THE AMERICAN SOCIETY OF MECHANICAL ENGINEERS
All Rights Reserved
Printed in U.S.A.

FOREWORD

The 1999 ASME Pressure Vessels and Piping Conference was held in Boston, Massachusetts, August 1 – 5, 1999. This publication contains papers presented at the conference in the areas of fracture, fatigue, and weld residual stress. These papers are organized into four sections:

- Applications of the Master Curve to Assessment of Nuclear Reactor Pressure Vessels
- Fracture Mechanics – Theory and Applications
- Piping Fracture Mechanics
- Weld Residual Stresses and Effects on Fracture/Fatigue

The technical sessions for these papers were sponsored by the Materials and Fabrication Committee and the Design and Analysis Committee of the ASME Pressure Vessels and Piping Division. In accordance with the requirements of ASME, each paper has been subjected to a formal peer review. All of the reviewers' comments have been addressed prior to the preparation of the final manuscripts.

Jwo Pan
The University of Michigan

ACKNOWLEDGMENTS

We would like to thank the authors for their willingness and diligent efforts in preparing the final manuscripts and for sharing their research results. Special thanks and gratitude are extended to the editor's affiliations – The University of Michigan, Westinghouse Electric Company, Framatone Technologies, University of South Carolina, Engineering Mechanics Corporation of Columbus, and Battelle – for their continued support during the preparation of this volume.

We would also like to thank the reviewers for their time and effort, which have contributed to the enhanced quality and clarity of the final papers. We extend our appreciation to the Technical Program Chairman, Dr. A. G. (Jack) Ware, for his patience and careful guidance. Finally, we would like to thank the ASME editorial staff for their cooperation, assistance, and professional work in finalizing this publication.

J. Pan
M. T. Kirk
K. K. Yoon
Y. J. Chao
G. M. Wilkowski
P. Dong

CONTENTS

APPLICATION OF THE MASTER CURVE TO ASSESSMENT OF NUCLEAR REACTOR PRESSURE VESSELS

Introduction
Mark Kirk and Ken Yoon 1

Application of the Master Curve Method to Crack Initiation and Crack Arrest
Kim Wallin 3

Cleavage Fracture in Surface Cracked Plates: Experiments and Numerical Predictions
Xiaosheng Gao, Robert H. Dodds, Robert L. Tregoning, James A. Joyce, and Rick E. Link 11

Statistical Versus Constraint Size Effects on the Transition Regime Toughness: A Preliminary Study of a Pressure Vessel Steel
H. J. Rathbun, G. R. Odette, M. Y. He, G. E. Lucas, and J. W. Sheckherd 17

A Physical Basis for the Master Curve
Marjorie Natishan and Mark Kirk 23

Experimental Validation Work to Prove the Master Curve Concept
D. E. McCabe and M. A. Sokolov 29

On the Use of the Master Curve Based on the Precracked Charpy Specimen
Rachid Chaouadi, Marc Scibetta, Eric van Walle, and Robert Gérard 35

Accuracy in the T_o Determination: Numerical Versus Experimental Results
Carlos A. J. Miranda, Arnaldo H. P. Andrade, John D. Landes, Donald E. McCabe, and Ronald L. Swain 47

Development of the T_o Reference Temperature From Precracked Charpy Specimens
James A. Joyce and Robert L. Tregoning 53

Effect of Loading Rate on Fracture Toughness of Pressure Vessel Steels
K. K. Yoon, W. A. Van Der Sluys, and K. Hour 63

Initial Reference Temperatures and Irradiation Trend Curves for Use With RT_{To}: A Preliminary Assessment
Mark Kirk, Randy Lott, William Server, and Stan Rosinski 71

Application of Master Curve Technology to Estimation of End of License Adjusted Reference Temperature for a Nuclear Power Plant
Randy Lott, Mark Kirk, Charlie Kim, William Server, Chuck Tomes, and James Williams 79

FRACTURE MECHANICS: THEORY AND APPLICATIONS

Introduction
Yuh J. Chao 91

Determination of Fracture Mechanics Characteristics: Initiation or Instability Characteristics as a Basis for Components Assessment
E. Roos and U. Eisele 93

The Effect of Residual Stresses on Crack Extension When Expressed in the K_r–L_r Format
E. Smith 101

Life Estimation of Creep Crack Propagation in Aged Power Plant Pipes Using the Engineering Approach
Shiro Kubo, Masashige Yoshikawa, and Kiyotsugu Ohji 105

Variation of Fracture Toughness With Constraint of PMMA Specimens
Yuh J. Chao, Shu Liu, and Bart J. Broviak 113

Fracture Mechanics Evaluation for and Evolutionary Development of the
Break Preclusion Concept
    E. Roos and K.-H. Herter 121
Mechanical Properties for Fracture Analysis of Mild Steel Storage Tanks
    R. L. Sindelar, P. S. Lam, G. R. Caskey, Jr., and
    L. Y. Woo 129
J-Integral Based Flaw Stability Analysis of Mild Steel Storage Tanks
    P.-S. Lam and R. L. Sindelar 139
Efficient and Reliable Analysis of Structural Collapse by Using a Differential
Quadrature Finite Element Method Along With a Global Secant Relaxation-Based
Accelerated Equilibrium Iteration Procedure
    Chang-New Chen 145

PIPING FRACTURE MECHANICS
Introduction
    G. M. Wilkowski 153
Consideration of Internal Pressure on Pipe Flaw Evaluation Criteria Based
on a New J-Estimation Scheme for Combined Loading
    Naoki Miura and Yeon-Ki Chung 155
The Opening Area Associated With a Crack That is Subjected to a General
Tensile Stress Distribution
    E. Smith 165
Limit Analysis for Pressurized Pipelines With Local Wall-Thinning
    Lianghao Han, Yinpei Wang, and Cengdian Liu 169

WELD RESIDUAL STRESSES AND EFFECTS ON FRACTURE/FATIGUE
Introduction
    Pingsha Dong 177
Residual Stress Simulation Incorporating Weld HAZ Microstructure
    Lubomír Junek, Marek Slovácek, Vladislav Magula, and
    Vladislav Ochodek 179
On the Validation of the Models Related to the Prevision of the HAZ Bahaviour
    Y. Vincent, J. F. Jullien, N. Cavallo, L. Taleb, V. Cano, S. Taheri, and
    Ph. Gilles.. 193
Residual Stress Analysis and Fracture Assessment of Weld Joints in Moment Frames
    Jinmiao Zhang, Pingsha Dong, and Frederick W. Brust 201
Thermo-Mechanical Modeling of Residual Stress and Distortion During
Welding Process
    Yuh-Jin Chao and Xinhai Qi 209
Study of the Material Properties of Thin Pipe Butt Welds (IN C-Mn and Stainless
Steel) on the Welding Residual Stress Distribution by Using Numerical Simulation
    J. J. Janosch and D. Lawrjaniec 215
Welding Residual Stresses With Magnitudes Lower Than the Material Yield Strength
    Kyle Koppenhoefer, William Mohr, and
    J. Robin Gordon 225
A Finite Element Evaluation of Residual Stress in a Thread Form Generated by a
Cold-Rolling Process
    John A. Martin 239
Residual Stresses in Inconel-718 Clad Tungsten Rods Subject to Hot Iso-Static
Pressing
    Ronnie B. Parker and Partha Rangaswamy 255
An Investigation Into the Residual Stresses in an Aluminium 2024 Test Weld
    Robin V. Preston, Simon D. Smith, Hugh R. Shercliff, and
    Philip J. Withers 265
Residual Stress in Transient Zones of a High Vacuum Tube
    Kent K. Leung 279

Influence of Heating Conditions on Temperature Distribution During Local PWHT
   Yukihiko Horii, Jinkichi Tanaka, Masanobu Sato, Hidekazu Murakawa, and Jianhua Wang 287

Determination of Critical Heated Band Width During Local PWHT by Creep Analysis
   Hidekazu Murakawa, Jianhua Wang, Jinkichi Tanaka, Yukihiko Horii, and Masanobu Sato 297

Classical and Emerging Fracture Mechanics Parameters for History Dependent Fracture With Application to Weld Fracture
   Frederick W. Brust 307

Micro- and Macroscopic Fracture Behavior of Heat-Affected Zone in Multi-Pass Welded Cryogenic Steel
   Jae-Il Jang, Young-chul Yang, Woo-sik Kim, and Dongil Kwon 325

Weldment Residual Stresses in a Hydrocarbon Outlet Valve of a Catofin Reactor
   Germán Crespo and Zulay Cassier 335

Influence of the Welding Variables on the Mechanical Properties in Butt Joints for Aluminum 6063-T5
   María Carolina Payares, Minerva Dorta, and Patricia Muñoz-Escalona 339

Author Index 345

APPLICATION OF THE MASTER CURVE TO ASSESSMENT OF NUCLEAR REACTOR PRESSURE VESSELS

Introduction

Mark Kirk
Westinghouse Electric Company
Pittsburgh, Pennsylvania, USA

Ken Yoon
Framatome Technologies
Lynchburg, Virginia, USA

In the early 1970s, ASME adopted an approach to estimate the lower-bound fracture toughness values used to assess the integrity of pressure vessels, including nuclear reactor pressure vessels (RPVs). This approach involves establishing an index temperature, RT_{NDT}, which positions the K_{IC} and K_{IR} curves on the temperature axis. RT_{NDT} is determined through a combination of Charpy V-notch and NDT testing. Also, nuclear RPV surveillance programs currently rely on shifts of CVN transition temperature to estimate the effect of neutron irradiation on RT_{NDT} (see ASTM Standard E185-94 and U.S. NRC Regulatory Guide 1.99, Revision 2). All of these approaches use correlations to fracture toughness, rather than direct measurements of fracture toughness, due to the large specimen size required to measure "valid" linear-elastic fracture toughness values.

Developments since the early 1970s set the scene for fundamental improvements to these correlative techniques. In 1980 Landes and Schaffer noticed a statistical "size" effect for specimens failing by transgranular cleavage. They demonstrated that larger specimens fail at lower toughness values, even when the severe size requirements of linear elastic fracture mechanics are satisfied. Beginning in 1984, Wallin and co-workers from VTT in Finland combined this "weakest link" size effect with micro-mechanical models of cleavage fracture. Wallin proposed a model that accounts successfully for specimen size effects, and provides a means to calculate statistical confidence bounds on cleavage fracture toughness data. These concepts, combined with Wallin's observation (made first in 1984, and reinforced in 1991) that all ferritic steels exhibit the same variation of cleavage fracture toughness with temperature, gave birth to the notion of a "master" transition curve for all ferritic steels.

ASTM recently passed a standard (E1921-97) that describes how to measure the Master Curve index temperature, T_o, based on limited replicate testing. E1921-97 also incorporates a modern understanding of elastic-*plastic* fracture mechanics, and so permits determination of T_o using specimens as small as precracked Charpys. The potential for characterizing the entire fracture mode transition based on direct fracture toughness measurements using specimens already in surveillance programs has sparked considerable interest in Master Curve technology within the commercial nuclear power community. Recently, ASME published Code Case N-629 that permits use of a Master Curve-based index temperature ($RT_{To} \equiv T_o + 35°F$) as an alternative to traditional methods of positioning the ASME K_{IC} and K_{IR} curves.

These recent standardization efforts, and the desire of some electric utilities to use Master Curve technology to support both licensing and plant life extension activities, has spawned research initiatives across a spectrum of topics. The following eleven papers reflect a breadth of work ranging from research to applications. Two papers address the fundamental basis for / assumptions of the Master Curve (see papers by Natishan[1], and Rathbun), four address issues associated with current code implementation of Master Curve technology (see papers by Chaouadi, Joyce, McCabe, and Miranda), three extend the applicability of Master Curve concepts (see papers by Gao, Wallin, and Yoon), and two use the Master Curve to address RPV integrity assessment within a regulatory framework (see papers by Kirk, and Lott). These efforts, and their successful completion, provide critical evidence needed to support future adoption of both ASME codes and Federal Regulations that embrace Master Curve technology.

As session developers, we are indebted to the authors of these papers for their efforts in providing conference attendees with a comprehensive perspective on a topic of considerable current interest. Additionally, we would like to recognize the efforts of many peer reviews, which were critical to assuring the quality, clarity, and technical accuracy of the papers you see here.

---

[1] Papers are referred by the first author's name only.

APPLICATION OF THE MASTER CURVE METHOD TO CRACK INITIATION AND CRACK ARREST

Kim Wallin
VTT Manufacturing Technology
Technical Research Centre of Finland
Espoo, Finland

ABSTRACT

Often, like for operational structures, it is impossible or inappropriate to obtain large material samples for standard fracture toughness determination. This is especially the case with irradiation damage assessment of reactor pressure vessels, but also many other applications have the same restrictions. At VTT a holistic approach by which to determine static and dynamic and crack arrest fracture toughness properties either directly or by correlations from small material samples, has been developed. As a result of the method, often referred to as the Master Curve method, statistically defined fracture toughness estimates, of the material, suitable for structural integrity assessment are obtained. The Master Curve method, developed for brittle fracture initiation estimation, has enabled a consistent analysis of fracture initiation toughness data. Here, the method is briefly presented and shown how to modify it to describe also crack arrest toughness.

INTRODUCTION

Normally, fracture toughness testing standards require the use of comparatively large test specimens to obtain so called valid fracture resistance values. Extreme standards in this respect are the linear-elastic K_{IC} and K_{Ia} standards and the CTOD standard that require elastic behaviour of the test specimen or full section thickness specimens, respectively. Often, like for operational structures, it is impossible or inappropriate to obtain large material samples for standard fracture toughness determination. This is especially the case with irradiation damage assessment of reactor pressure vessels, but also many other applications have the same restrictions. The specimen size requirements are a major obstacle for applying fracture mechanics in structural integrity assessment outside aviation, nuclear and off-shore industries.

At VTT, development work has been in progress for 15 years to develop and validate testing and analysis methods applicable for fracture resistance determination from small material samples. The VTT approach is a holistic approach by which to determine static, dynamic and crack arrest fracture toughness properties either directly or by correlations from small material samples.

Recently, a new testing standard for fracture toughness testing in the transition region (ASTM E 1921-97) has been published. The standard is in a way "first of a kind", since it includes guidelines on how to properly treat the test data for use in structural integrity assessment. No standard, so far, has done this. The standard is based on the VTT master curve approach. Key components in the standard are statistical expressions for describing the data scatter (Wallin, 1984) and for predicting a specimens size (crack front length) effect (Wallin, 1985) and an expression (master curve) for the fracture toughness temperature dependence (Wallin, 1991 and Wallin, 1993). The standard and the approach it is based upon can be considered to represent the state of the art of small specimen brittle initiation fracture toughness characterisation. The key elements of the master curve method are schematically presented in Fig. 1.

The master curve method and the ASTM testing standard allow the use of the elastic-plastic parameter K_{JC}, which enables the use of small specimens. It is possible to characterise the ductile-to-brittle transition region with precracked Charpy specimens as shown in Fig. 2, getting the same as with large K_{IC}-specimens (Sokolov et al., 1997).

Unfortunately no elastic-plastic parameter has been found applicable for crack arrest. Direct determination of K_{Ia} is thus only possible with large specimens behaving in a linear-elastic manner. In the ASME pressure vessel code, reference fracture toughness curves are given both for brittle fracture initiation and crack arrest. Unfortunately, the reference temperature, RTNDT, applied with the ASME reference curves is far from accurate. This is especially the case for brittle fracture initiation toughness. The master curve method enables the determination of a clearly superior reference temperature, T_0, based on the actual initiation fracture toughness, but it does not give any information regarding crack arrest. The ASME code has a fixed relation between the K_{IC} and K_{Ia} reference curves, thus implying a fixed relation between fracture initiation and crack arrest. If this were the general case, the reference temperature T_0 could be used similarly for crack arrest. Unfortunately the relation between fracture initiation and crack arrest is not constant. It can vary considerably from one material to the other (Wallin, 1996).

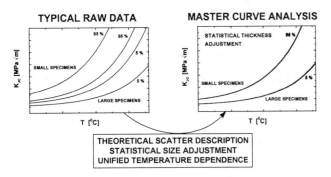

Fig. 1. Principle of the master curve method for brittle fracture initiation toughness.

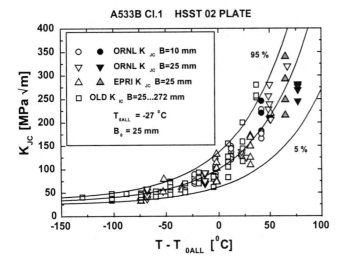

Fig. 2. Comparison between ASTM E 399 K_{IC} and it's elastic plastic equivalent K_{JC} for the ASME reference curve base line material. (Sokolov et al., 1997).

Based on experimental findings (Wallin, 1996), it can be stated that the relation between fracture initiation and crack arrest toughness is more complex than depicted by the ASME pressure vessel code.

In the following, the master curve method is shortly described and the behavior of K_{Ia} is investigated based on a modification of the master curve method.

THE MASTER CURVE METHOD

The method is based on a statistical brittle fracture model (Wallin et al. 1984), which gives for the scatter of fracture toughness (Wallin, 1984)

$$P[K_{IC} \leq K_I] = 1 - \exp\left(-\left[\frac{K_I - K_{min}}{K_0 - K_{min}}\right]^4\right) \quad (1)$$

where $P[K_{IC} \leq K_I]$ is the cumulative failure probability, K_I is the stress intensity factor, K_{min} is the theoretical lower bound of fracture toughness and K_0 is a temperature and specimen size dependent normalization fracture toughness, that corresponds to a 63.2 % cumulative failure probability being approximately $1.1 \cdot \overline{K}_{IC}$ (mean fracture toughness). The model predicts a statistical size effect of the form (Wallin, 1985)

$$K_{B_2} = K_{min} + \left[K_{B_1} - K_{min}\right] \cdot \left(\frac{B_1}{B_2}\right)^{1/4} \quad (2)$$

where B_1 and B_2 correspond to respective specimen thickness (length of crack front).

On the lower shelf of fracture toughness ($K_{IC} \ll 50$ MPa√m) the equations may be inaccurate. The model is based upon the assumption that brittle fracture is primarily initiation controlled, even though it contains a conditional crack propagation criterion, which among others is the cause of the lower bound fracture toughness K_{min}. On the lower shelf, the initiation criterion is no longer dominant, but the fracture is completely propagation controlled. In this case there is no statistical size effect, Eq. (2) and also the toughness distribution differs (not very much) from Eq. (1). In the transition region, where the use of small specimens become valuable, however, Eqs. (1) and (2) are valid.

For structural steels, a "master curve" describing the temperature dependence of fracture toughness has been proposed (Wallin, 1991 and Wallin, 1993).

$$K_0 = 31 + 77 \cdot \exp(0.019 \cdot [T - T_0]) \quad (3)$$

where T_0 is the transition temperature (°C) where the mean fracture toughness, corresponding to a 25 mm thick specimen, is 100 MPa√m and K_0 is 108 MPa√m.

Equation (3) gives an approximate temperature dependence of the fracture toughness for ferritic structural steels and it is comparatively well verified (Wallin, 1998). Keeping the temperature dependence fixed, decreases the effect of possible invalid fracture toughness values upon the transition temperature T_0.

VALIDITY OF K_{JC}

With small specimens, some amount of plasticity is unavoidable. Therefore, it is not possible to determine the linear-elastic K_{IC}, but one is forced to determine it's elastic plastic equivalent K_{JC}. The J-integral based parameter K_{JC} is defined as,

$$K_{JC} = \sqrt{\frac{J_C \cdot E}{(1-\nu)^2}} \quad (4)$$

where E is the modulus of elasticity and ν is the Poisson's ratio.

The validity of K_{JC} has been challenged based upon the assumption that only K_{IC} represents a full plane strain stress state. However, the definition of plane strain for K_{IC} does not in reality correspond to the actual stress state in front of the crack tip. The majority of materials, used for the development of the K_{IC} standard failed by a ductile failure mechanism and the size criteria were selected to correlate the 95 % secant intersection point to 2 % crack growth. Deviations from these criteria cause this correlation to be lost as non-linearities in the elastic compliance are not due solely to crack growth. The secant method cannot differentiate between plasticity and crack growth and therefore size criteria are needed to ensure elastic behavior of the specimen. In some cases the measured 95% secant based K_Q values showed a decreasing trend, with increasing specimen size, whereas in other cases the reverse behaviour was seen. A common feature for K_Q was that the specimen size dependence started to level off with increasing specimen size (an effect that is probably connected to the levelling off of the tearing resistance curve). Thus plane strain was defined as to produce approximately size independent toughness values, not minimum toughness values related to a maximum stress state.

The American Society for Mechanical Engineering (ASME) gives a reference fracture toughness curve based upon K_{IC}. The material constituting the base line results (majority) for the reference curve is an A533B Cl.1 steel plate having the designation HSST 02. The reference curve data base contain 70, HSST 02, K_{IC} values corresponding to specimen thicknesses in the range 25-275 mm. This data set is often referenced to as the *million dollar curve*. When the data is treated by Eqs. (2)-(4), the transition temperature is obtained as $T_0 = -28$ °C (Fig. 2).

Three, elastic plastic K_{JC} data sets, for the same material, have also been determined and are shown in the same figure. One set is part of the Electric Power Research Institute (EPRI) Nuclear Pressure Vessel Steel Data Base and the others are part of the Heavy Section Steel Technology Program performed at Oak Ridge National Laboratory (ORNL) (Sokolov et al. 1997). Two of the data sets are based on 25 mm thick specimens, so that the use of Eq (2) is not necessary. The third data set is based on pre-cracked Charpy-V specimens tested statically. The transition temperatures, Eq (3) for the three data sets are respectively (Fig. 2), $T_0 = -28$°C (EPRI, 25 mm CT), $T_0 = -24$ °C (ORNL, 25 mm CT) and $T_0 = -30$ °C (ORNL, 10 mm CVN_{pc}). These results give strong support to the validity of K_{JC} in respect to K_{IC}. Considering the cost of small specimen fracture toughness testing, the EPRI and ORNL data sets could probably be referred to as *twenty thousand dollar curves*.

In principle, any fracture mechanical parameter is valid as long as it gives a correct description of the stress and strain field in front of the crack tip. Thus, an elastic plastic parameter is just as valid as a linear elastic parameter, as long as it is capable of describing the stresses and strains in the fracture process zone. Small specimens have a smaller measuring capacity than large specimens, but the elastic plastic fracture parameter itself (K_{JC}) is as valid as ASTM E399 K_{IC}.

One major difference between the Master Curve and the ASME K_{IC} reference curve is the curve shape. Since the ASME curve is essentially drawn as a free hand lower bound curve to the HSST-02 K_{IC}-data, it has a steeper rise than the Master Curve. This steepness, which is not visible in the elastic plastic results, has sometimes been used to question the validity of the Master Curve. In order to investigate this discrepancy more closely, the original HSST-02 plate data was re-examined (Witt, 1969, Shabbits et al., 1969 and Mager, 1970). The HSST-02 plate is a 300 mm thick plate from which 25 mm and 50 mm thick specimens were taken from the centre, quarter thickness and surface locations. Larger 100 mm and 150 mm thick specimens were taken from the centre location and large 250 – 300 mm thick specimens were taken as through-thickness specimens. For the *million dollar curve* data set only the valid K_{IC}-results corresponding to the centre and quarter thickness locations were included. The only exception were the large through thickness specimens which also included material close to the plate surface. In the following analysis the K_Q-values have been treated as censored values, denoting that the actual fracture toughness may be higher than recorded. In some cases this may be an erroneous assumption since, failing the ASTM E 399 specimen size requirement does not automatically mean the fracture toughness measured from the specimen includes significant plasticity effects. In order to verify that the K_Q-values are connected with plasticity would require the original load displacement curves and they were not available for the present analysis. Thus, some cases with a large proportion of K_Q-values may lead to too low T_0 estimates. The master curve analysis of the original data is presented in Figs. 3-7.

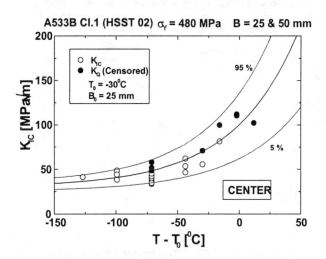

Fig. 3. Original HSST 02 fracture toughness data – 1T & 2T, centre location

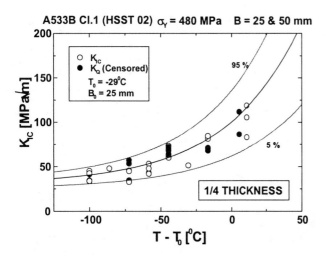

Fig. 4. Original HSST 02 fracture toughness data – 1T & 2T, quarter thickness location

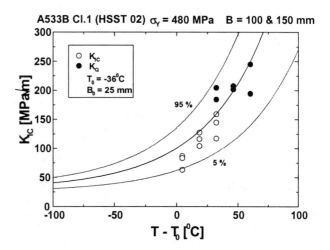

Fig. 6. Original HSST 02 fracture toughness data – 4T & 6T, centre location

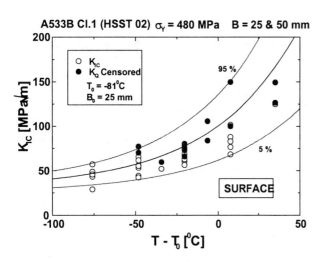

Fig. 5. Original HSST 02 fracture toughness data – 1T & 2T, surface location

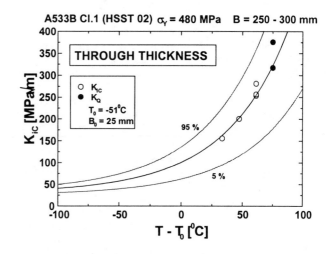

Fig. 7. Original HSST 02 fracture toughness data – 10T - 12T, through thickness location

A striking feature of the data sets is that they seem to follow the master curve temperature dependence quite well (even the large specimens), but there is a big variability in the transition temperatures T_0. This is further examined in Fig. 8. Where the T_0 estimates for each specimen size and location are plotted as a function of specimen location. The individual boxes have the width of the actual specimens and a height corresponding to the ± 1 standard deviation confidence bounds of the T_0 estimate. Presented this way, the variation in T_0 is easily explained. Close to the plate surface, the material has a much higher toughness ($T_0 \approx -80°C$) whereas in the centre region the toughness is clearly lower ($T_0 \approx -30°C$). Since the through thickness specimens include parts of both regions, it is natural for their T_0 value to be in between these values. Thus, the apparent steep rise of the HSST 02 material is not real, but a result of the large specimens including tougher material along their crack fronts. Actually, the HSST 02 plate seem to follow the master curve temperature dependence quite well. This is another indication for that the ASME K_{IC}-reference curve may have a too steep temperature dependence.

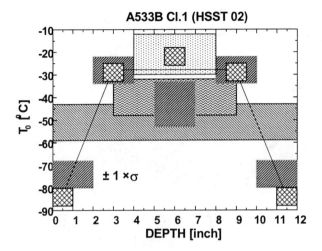

Fig. 8. HSST 02 T_0 estimates as a function of specimen size and thickness location (Individual boxes have the width of the actual specimens and a height corresponding to the +/-1 standard deviation confidence bounds of the T_0 estimate.)

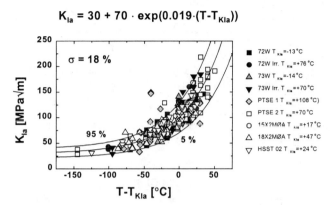

Fig. 9. Scatter and temperature dependence of crack arrest toughness [9].

MASTER CURVE DESCRIPTION OF CRACK ARREST TOUGHNESS

The ASME K_{Ia}-reference curve is based on dynamic fracture toughness and crack arrest data. The ASME K_{Ia}-reference curve was originally drawn as a "free hand" lower limiting curve to one set of K_{Ia} data (HSST 02) and three sets of dynamic K_{Id} data (Marston, 1978). The K_{Id} data was included due to a lack of K_{Ia} data. It appears, from the test results, that dynamic fracture toughness is close to crack arrest, but for a more accurate description it is better to focus only on crack arrest data.

The mechanism of arrest differs from that of initiation. Thus the scatter and size effects are not expected to be the same as for brittle fracture initiation. Mechanistically, arrest occurs when the local crack driving force at the crack tip decreases below the local arrest toughness over a sufficiently large portion of the crack front. A single local arrest is not sufficient to arrest the whole crack front. I.e. the scatter should be more a function of the mean properties of the matrix (and not the local). Therefore the scatter should be less than for initiation and there should not be any statistical size effects in the case of crack arrest.

Nine different sets of K_{Ia} data, including the original K_{Ia} data for the HSST 02 plate used for the construction of the ASME reference curve, were analysed statistically to obtain a better understanding of the K_{Ia}-reference curve. The biggest difference between brittle fracture initiation toughness and crack arrest toughness is that K_{Ia} should not show a statistical size effect. Thus, no size adjustment was performed on the K_{Ia} data.

The analysis of the K_{Ia} results was aimed at finding out whether an unified description of the data was possible. It was decided to try to describe the temperature dependence of the K_{Ia} data by the equation

$$K_{Ia} = 30 + 70 \cdot \exp\{0.019 \cdot (T - T_{KIa})\} \quad (5)$$

where T_{KIa} corresponds to the temperature where the mean K_{Ia} is equal to 100 MPa√m. Equation (5) is of the same form as the standard master curve used for the description of the brittle fracture initiation toughness.

The scatter in K_{Ia} was for simplicity assumed to be log-normal so that the proportional scatter in K_{Ia} is constant. The outcome of the analysis is presented in Fig. 9. The assumed temperature dependence and distribution seem to describe the data quite well. Thus it appears possible to normalise the K_{Ia} transition curve, for these materials, based only on T_{KIa}. Additionally, the same temperature dependence, that is used to describe the brittle fracture initiation toughness, seem to be applicable. The scatter in K_{Ia} appears less than for K_{Jc}. The scatter does not seem to be much material dependent, with the exception of PTSE-1. However, based also on other tests, this material has been found to be macroscopically inhomogeneous (Bryan et al., 1985). It might be considered proper to exclude PTSE-1 from the analysis, but by including it, a slightly more conservative estimate of the K_{Ia} scatter behaviour is obtained. The standard deviation of the total combined data set is $\sigma = 18$ % (compared to $\sigma = 28$ % for brittle initiation toughness).

Based on the success in applying the master curve temperature dependence to K_{Ia} data, it was decided to try to validate the expression further. For the purpose, 54 different data sets containing both static brittle fracture initiation data and crack arrest data were collected from various sources, including some VTT data and much data from literature. The majority of the data refer to pressure vessel steels 15X2MFA, A508 and A533B, but also many welds and other steels are included. The data contain some embrittled (irradiated or heat treated) materials. The materials yield strengths cover a range from 280 MPa...1082 MPa. All data sets were analysed by the master curve expression shown in Fig. 9. The quality of the data sets varied from very few specimens to quite many specimens. Generally, the number of specimens per data set is more than 10.

Overall, the master curve provided a satisfactory description of all data sets. The materials having most data were selected for a further study regarding the validity of the scatter and temperature dependence found in Fig. 9. All the A508 Cl.3, A533B Cl.1 and 15X2MFA and 15X2NMFA data were grouped and plotted, respectively, normalised with the T_{KIa} reference temperature and compared to the assumed temperature dependence and assumed scatter. The resulting figures are presented in Fig. 10 for A508 Cl.3, Fig. 11 for A533B Cl.1 and Fig. 12 for 15X2MFA and 15X2NMFA.

In the case of A508 Cl.3 and 15X2MFA and 15X2NMFA, the assumed scatter and temperature dependence is well verified. Also, in the case of A533B Cl.1, the assumed temperature dependence is well verified, but the scatter at high crack arrest values appear somewhat larger than assumed. These abnormally high crack arrest values refer to non-standard, ESSO type, tests and there is some uncertainty of their validity. Maybe more importantly, the 5 % lower bound curve provides a very good lower bound description of the data for all the materials. Thus it can be concluded that the assumed scatter and temperature dependence appear to provide an adequately

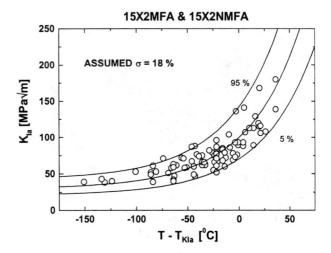

Fig. 12. Temperature dependence and scatter for crack arrest toughness of 15X2MFA and 15X2NMFA.

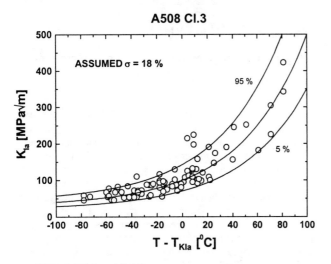

Fig. 10. Temperature dependence and scatter for crack arrest toughness of A508 Cl.3.

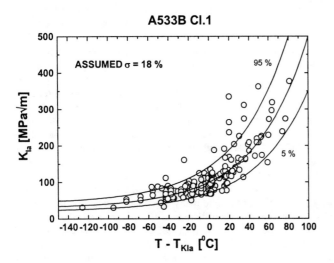

Fig. 11. Temperature dependence and scatter for crack arrest toughness of A533B Cl.1.

accurate universal description of crack arrest toughness for ferritic structural steels. The above conclusion means that if it is possible to correlate T_{KIa} with T_0, it is possible to predict the crack arrest toughness based on static brittle fracture initiation toughness. I.e. small specimens can be used simultaneously for determining K_{JC} and for estimating K_{Ia}.

SUMMARY AND CONCLUSIONS

The Master Curve method, has shortly been described. The method, developed for brittle fracture initiation estimation, enables a consistent analysis of fracture initiation toughness data. Furthermore, it was shown that the method can be easily modified to describe also crack arrest toughness. Overall, it is possible to describe both brittle fracture initiation toughness as well as crack arrest toughness with single temperature parameters T_0 and TK_{Ia}. The relation between these two parameters is, however, not constant, but varies from material to material.

ACKNOWLEDGEMENTS

This work is a part of the Structural Integrity Project (STIN) belonging to the Finnish Research Programme on Nuclear Power Plant Safety (FINNUS), performed at VTT Manufacturing Technology and financed by the Ministry of Trade and Industry in Finland, the Technical Research Centre of Finland (VTT) and the Finnish Centre for Radiation and Nuclear Safety (STUK).

REFERENCES

ASTM E1921-97, 1998, "Standard Test Method for Determination of Reference Temperature, T_0, for Ferritic Steels in the Transition Range," Annual Book of ASTM Standards, Vol 03.01, American Society for Testing and Materials, pp. 1068-1084.

Bryan, R. H., Bass, B. R., Bolt, S. E., Bryson, J. W., Edmonds, D. P., McCulloch, R. W., Merkle, J. G., Nanstad, R. K., Robinson, G. C.,

Thoms, K. R. and Whitman, G. D., 1985, "Pressurized-Thermal-Shock Test of 6-in.-Thick Pressure Vessels. PTSE-1: Investigation of Warm Prestressing and Upper-Shelf Arrest," NUREG/CR-4106, ORNL-6135, Oak Ridge National Laboratory, Oak Ridge, 262 p.

Mager, T. R., 1970, "Heavy Section Steel Technology Program Technical Report No. 10 – Fracture Toughness Characterization Study of A533 Grade B Class 1 Steel," WCAP-7578, Westinghouse Electric Corporation, Pittsburgh, 21 p.

Marston, T. U., (ed.), 1978, "Flaw Evaluation Procedures - Background and Application of ASME Section XI Appendix A," EPRI NP-719-SR, Electric Power Research Institute, Palo Alto, California.

Shabbits, W. O., Pryle, W. H. and Wessel, E. T., 1969, "Heavy Section Steel Technology Program Technical Report No. 6 – Heavy Section Fracture Toughness Properties of A533 Grade B Class 1 Steel Plate and Submerged Arc Weldment," WCAP-7414, Westinghouse Electric Corporation, Pittsburgh, pp. 27-28.

Sokolov, M., Wallin, K.and McCabe D. E., 1997, "Application of Small Specimens to Fracture Mechanics Characterization of Irradiated Pressure Vessel Steels," ASTM STP 1321, J.H. Underwood, B. D. MacDonald and M. R. Mitchell, ed., American Society for Testing and Materials, 17 p.

Wallin, K., Saario, T. and Törrönen, K., 1984, "Statistical Model for Carbide Induced Brittle Fracture in Steel," Metal Science, pp. 13-16.

Wallin, K., 1984, "The Scatter in K_{IC}-Results," Engineering. Fracture. Mechanics, Vol. 19, pp. 1085-1093.

Wallin, K., 1985, "The Size Effect in K_{IC} Results," Engineering. Fracture Mechanics,Vol. 22, pp. 149-163.

Wallin, K., 1991, "Fracture Toughness Transition Curve Shape for Ferritic Structural Steels," Proceedings, Fracture of Engineering Materials & Structures, S. T. Teoh and K. H. Lee, ed., Elsevier Applied Science, pp. 83-88.

Wallin, K., 1993, "Irradiation Damage Effects on the Fracture Toughness Transition Curve Shape for Reactor Pressure Vessel Steels," International Journal of Pressure Vessels and Piping, Vol. 55, pp. 61-79.

Wallin, K., 1996, "Descriptive Characteristic of Charpy-V Fracture Arrest Parameter with respect to Crack Arrest K_{Ia}," ESIS 20, E. van Walle, ed., Mechanical Engineering Publications, pp. 165-176.

Wallin, K., 1998, "Master Curve Analysis of Ductile to Brittle Transition Region Fracture Toughness Round Robin Data - The "EURO" Fracture Toughness Curve," VTT Publications 367, Technical Research Centre of Finland, 58 p.

Witt, F. J., 1969, "Heavy-Section Steel Technology Program Semiannual Progress Report for Period Ending August 31," ORNL-4377, Oak Ridge National Laboratory, Oak Ridge, pp. 69-73.

CLEAVAGE FRACTURE IN SURFACE CRACKED PLATES: EXPERIMENTS AND NUMERICAL PREDICTIONS

Xiaosheng Gao and **Robert H. Dodds**
Department of Civil Engineering
University of Illinois
Urbana, IL 61801
U.S.A.

Robert L. Tregoning
Fatigue and Fracture Branch
Naval Surface Warfare Center
West Bethesda, MD 20817
U.S.A.

James A. Joyce and **Rick E. Link**
Department of Mechanical Engineering
U.S. Naval Academy
Annapolis, MD 21402
U.S.A.

ABSTRACT

This study applies recent advances in probabilistic modeling of cleavage fracture to predict the measured fracture behavior of surface crack plates fabricated from an A515-70 pressure vessel steel. Modifications of the conventional, two-parameter Weibull stress model introduce a non-zero, threshold parameter (σ_{w-min}). The introduction of σ_{w-min} brings numerical predictions of scatter in toughness data into better agreement with experimental measurements. Calibration of this new parameter makes use of the generally accepted, minimum toughness value for ferritic steels adopted in ASTM E-1921 ($K_{min} = 20$ MPa$\sqrt{m}$). The Weibull modulus (m) and scaling parameter (σ_u) are calibrated using a new strategy based on the toughness transferability model, which eliminates the non-uniqueness that arises in calibrations using only small scale yielding toughness data. Joyce, Link and Tregoning recently performed fracture tests on an A515-70 pressure vessel steel. They tested 14 surface crack specimens with essentially identical crack sizes/shapes. Complementary tests performed on deep notch C(T) and shallow notch SE(B) specimens provide the toughness data to calibrate the Weibull stress parameters using the toughness scaling approach. The calibrated model is then applied to predict the measured response of surface crack plates loaded in different combinations of tension and bending. The model predictions accurately capture the measured distributions of fracture toughness values.

INTRODUCTION

Macroscopic values of fracture toughness (J_c or K_{Jc}) measured experimentally over the lower-end of the ductile-to-brittle transition (DBT) range of ferritic steels consistently exhibit a large amount of scatter (for examples, see Wallin 1984; Sorem et al. 1991). In this temperature range, transgranular cleavage triggers the brittle fracture event often in the presence of significant plastic deformation along the crack front. Models to describe the observed scatter generally adopt a weakest link approach consistent with common observations of a single carbide that initiates cleavage along the crack front (see, for example, Wallin, 1984; Wallin et al., 1984; Lin et al., 1986). The volume of highly stressed material along the crack front thus plays a crucial role in driving the fracture process. Elastic-plastic stress fields along the crack front depend strongly on the specimen/component geometry, size, loading mode (e.g. tension vs. bending) and material flow properties. When plastic regions ahead of the crack front interact with nearby (traction free) surfaces of the specimen, the single parameter characterization of the crack front stresses, in terms of the J-integral, breaks down. Under such conditions, a complex (nonlinear) relationship develops between the 3-D crack front stresses, the applied J-value which varies along the crack front, and the corresponding stressed volume of material at the crack front that drives the weakest link mechanism (Moinereau, 1996; Wiesner and Goldthorpe, 1996; Ruggieri et al., 1998; Gao et al., 1998; Gao and Dodds, 1998).

The Weibull stress model originally proposed by the Beremin group (Beremin, 1983) provides a framework to quantify this complex interaction among specimen size, deformation level and material flow properties in a fully 3-D setting. They introduced the scalar Weibull stress (σ_w) as a probabilistic fracture parameter, computed by integrating a weighted value of the maximum principal (tensile) stress over the fracture process zone (i.e., the crack front plastic zone). The Beremin model adopts a two-parameter description for the cumulative failure probability in the form

$$P_f(\sigma_w) = 1 - \exp\left[-\left(\frac{\sigma_w}{\sigma_u}\right)^m\right], \quad (1)$$

with

$$\sigma_w = \left[\frac{1}{V_0}\int_V \sigma_1^m \, dV\right]^{1/m}, \quad (2)$$

where $\overline{V}$ denotes the volume of the cleavage fracture process zone (usually defined as the volume inside the contour of $\sigma_1 \geq \lambda\sigma_0$, σ_0 is the yield stress and $\lambda \approx 2$ is a constant), V_0 is a reference volume ($V_0 = 1$ mm^3 is used in all calculations reported in this paper) and σ_1 is the maximum principal stress acting on material points inside the fracture process zone. Parameters m and σ_u appearing in Eq. (1) denote the Weibull modulus and the scale parameter of the Weibull distribution.

The Weibull stress model defined by Eqs (1–2) represents a pure *weakest link* description of the fracture event. This two-parameter model describes the unconditional cleavage probability that assumes no microcracks arrest (macroscopic cleavage fracture occurs once the critical microcrack experiences propagation). However, the unconditional probability has significant shortcomings to predict cleavage fracture (Anderson et al., 1994; Gao et al., 1998; Gao and Dodds, 1998). First, it implies that a very small K_I (stress intensity factor due to applied load) leads to a finite failure probability, which is not true in reality. Cracks cannot propagate in polycrystalline metals unless sufficient energy exists to break bonds, to drive the crack across grain boundaries and to perform plastic work. Consequently, there must exist a minimum toughness value (K_{min}) below which cracks arrest. K_{min} has an experimentally estimated value of 20 MPa$\sqrt{m}$ for common ferritic steels under SSY conditions, independent of the crack front length. The value of $K_{min} = 20$ MPa$\sqrt{m}$ has been adopted by ASTM E-1921 (1998). Second, the unconditional probability often overestimates the measured scatter of fracture toughness (see Anderson et al., 1994; Gao et al., 1998, Gao and Dodds, 1998 for examples).

Recognition of these problems led some investigators (Bakker and Koers, 1991; Xia and Shih, 1996) to introduce a threshold stress, σ_{th}, directly into the Weibull stress computation of Eq. (2), where σ_{th} is usually taken as a multiple of the uniaxial yield stress. One such proposal for the integrand of Eq. (1) has the form $\{(\sigma_1 - \sigma_{th})/(\sigma_u - \sigma_{th})\}^m$. But rational calibration procedures for σ_{th} remain an open issue. Moreover, the introduction of σ_{th} does not imply the existence of $K_{min} > 0$. To introduce an explicit threshold toughness into the Weibull stress model, we propose a modified form for Eq. (1) given by

$$P_f(\sigma_w) = 1 - \exp\left[-\left(\frac{\sigma_w - \sigma_{w-min}}{\sigma_u - \sigma_{w-min}}\right)^m\right], \quad (3)$$

where σ_{w-min} represents the minimum σ_w-value at which macroscopic cleavage fracture becomes possible. Consistent with the definition of K_{min}, we define σ_{w-min} as the value of σ_w calculated at $K = K_{min}$ in the (plane-strain) SSY model, where the SSY model must have a thickness equal to the configuration of interest for which Eq. (3) is applied. The calibration of σ_{w-min} is straightforward and does not require any additional experimental data. According to Eq. (3), the model to scale toughness values between different specimen sizes, types and loading conditions should be constructed at identical σ_w^* values, where $\sigma_w^* = \sigma_w - \sigma_{w-min}$. Gao et al. (1998) and Gao and Dodds (1998) provide detailed discussions about the three-parameter Weibull stress model and the toughness scaling method based on Weibull stress with $\sigma_{w-min} > 0$.

The applicability of Eq. (3) to predict failure probability for cleavage fracture relies on the calibrated values for Weibull stress parameters. Gao et al. (1998) recently proposed a new calibration procedure based on fracture toughness data measured from two sets of specimens giving rise to different constraint levels at fracture, where m is assumed invariant of constraint at a fixed temperature. By using the toughness scaling model based on the Weibull stress, the calibration process seeks the m-value which corrects the two sets of fracture toughness data to have the same statistical properties under SSY conditions, where fracture toughness data (J_c) follow the Weibull distribution having the form $P_f(J_c) = 1 - \exp[-(J_c/\beta)^2]$ (see Gao et al., 1998 and references therein). Under SSY conditions at fixed values for temperature, loading rate, irradiation, etc., β defines the characteristic toughness for ferritic steels—this observation motivates the calibration procedure to determine m. A relatively small number (6–10) of J_c-values enables accurate estimates for β.

This paper describes applications of these recent developments in modeling cleavage fracture to predict the behavior for various crack configurations of an A515-70 pressure vessel steel, including surface crack specimens loaded by different combinations of tension and bending. We calibrate the three-parameter Weibull stress model using fracture toughness data from deep-notch C(T) specimens and shallow-notch SE(B) specimens, and apply the calibrated model to predict probabilities of cleavage fracture in surface crack specimens. The model predictions agree very well with the experimental data and capture the strong constraint effect on cleavage fracture due to differences in crack geometry and loading mode.

FRACTURE EXPERIMENTS AND NUMERICAL PROCEDURES

Joyce and Link (1996) and Tregoning (1998) recently performed extensive fracture tests on an A515-70 pressure vessel steel in the ductile-to-brittle transition region. The material has a Young's modulus (E) of 200 GPa, Poisson's ratio (ν) of 0.3 and yield stress (σ_0) of 280 MPa at $-7°$C and 300 MPa at $-28°$C. Twelve plane-sided C(T) specimens ($a/W = 0.6$, $B = 25$ mm, $W = 50$ mm) were tested at $-28°$C and twelve plane-sided SE(B) specimens ($a/W = 0.2$, $B = 25$ mm, $W = 50$ mm, $S = 4W$) were tested at $-7°$C. Figure 1 shows the geometries of the surface crack specimens (SC(T)) tested by Joyce and Link. Seven bolt-loaded and seven pin-loaded specimens were tested at $-7°$C. All specimens failed by cleavage without prior ductile tearing. Table 1 lists the J_c-values at fracture, where the J_c-values for SC(T) specimens are centerplane values (at the deepest point on the crack front).

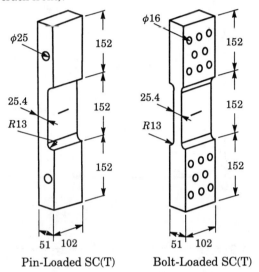

Fig. 1. Geometries of the pin-loaded and bolt-loaded SC(T) specimens. All dimensions are in mm.

Table 1. Measured values of cleavage fracture toughness (J_c) for A515-70 steel (kJ/m^2).

Specimen ID	C(T) ($a/W=0.6$) $-28°$C	SE(B) ($a/W=0.2$) $-7°$C	SC(T) (bolt) $-7°$C	SC(T) (pin) $-7°$C
1	56.7	101.9	107.9	138.7
2	15.5	101.6	79.1	49.3
3	33.6	44.3	116.2	109.1
4	18.0	23.8	107.1	50.5
5	46.6	91.4	84.2	78.0
6	12.5	141.0	110.1	43.8
7	47.2	44.8	101.9	85.8
8	45.5	112.2		
9	49.2	88.6		
10	9.5	184.2		
11	42.0	59.5		
12	18.6	101.2		

Nonlinear analyses were performed on very detailed models of fracture specimens using the three-dimensional, research code, WARP3D (Koppenhoefer et al., 1998). Highly refined 3D finite element meshes are employed for all crack configurations to minimize the errors in Weibull stress calculation caused by large m-values. Figure 2 shows the finite element mesh for the bolt-loaded SC(T) specimen. Symmetry of the geometry and loading condition permits modeling only one-quarter of the specimen. The crack front has a small, initial radius (ϱ_0) to enhance convergence of the finite-strain solutions. The selected value for the initial crack front radius (2.5 μm) does not affect the computed Weibull stress values at deformation (J) levels of interest. The quarter-symmetric 3D model for the bolt-loaded specimen contains 25,650 nodes and 22,800 elements (8-node). Similar levels of mesh refinement are used for the pin-loaded SC(T), the C(T) and the SE(B) specimens.

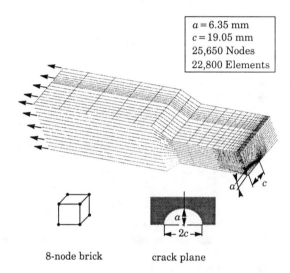

Fig. 2. Quarter-symmetric, finite element mesh for the bolt-loaded SC(T) specimen.

CALIBRATION OF WEIBULL PARAMETERS

Fracture toughness data for the deep-notch C(T) specimen and the shallow-notch SE(B) specimen are used to calibrate the Weibull stress parameters. Because the C(T) specimens and the SE(B) specimens have different test temperatures, toughness values for the C(T) specimens are needed at $-7°$C. Here, we employ the *master curve* approach of ASTM E-1921 (1998) to adjust the C(T) toughness values for the temperature change. The master curve for ferritic steels makes possible the prediction of median fracture toughness ($K_{Jc(med)}$ for 1T thickness) at any temperature (T) in the transition region, provided the reference temperature (T_0) for the material has been determined from SSY fracture toughness data at a single temperature.

The calibration for m is as follows: 1) Assume an m-value and compute the σ_w vs. K_J history for the deep-notch C(T) and the plane-strain, SSY configurations respectively using material flow properties at $-28°$C. Scale the measured toughness values for the C(T) specimens to the SSY configuration. Determine T_0 using the constraint corrected toughness values and estimate K_0 (K_J at 63.2% failure probability) at $-7°C$ (denote as K_0^A) according to ASTM E-1921; 2) Compute the σ_w vs. K_J history for the shallow-notch SE(B) and the plane-strain, SSY configurations, respectively, at $-7°C$. Scale the measured toughness values for the SE(B) specimens to the SSY configuration. Estimate K_0 (at $-7°C$) for the constraint corrected distribution and denote it as K_0^B; 3) Define an error function as $R(m) = (K_0^B - K_0^A)/K_0^A$. If $R(m) \neq 0$, repeat 1) - 2) for additional m-values. The calibrated Weibull modulus makes $R(m) = 0$ within a small tolerance. Figure 3 displays the variation of $R(m)$, with m for this material. $R(m)$ varies strongly with m and there exists one (and only one) calibrated m, $m = 10$, at which $R(m) = 0$.

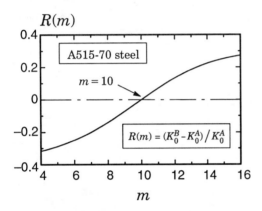

Fig. 3. Variation of $R(m)$, the error of toughness scaling, with m, where $R = 0$ gives the calibrated value for m.

After m is calibrated, σ_{w-min} and σ_u can be easily determined. At $-28°C$, the values of σ_{w-min} and σ_u corresponding to the thickness of the C(T) specimen are 752 MPa and 1441 MPa. At $-7°C$, the values of σ_{w-min} and σ_u corresponding to the thickness of the SE(B) specimen are 702 MPa and 1544 MPa. The value of σ_{w-min} changes between $-7°C$ and $-28°C$ due to the change of yield stress (σ_0) with temperature. The differences in σ_u reflect the combined effects of changes in both σ_0 and fracture toughness (K_0) with temperature.

Figure 4 shows the evolution of cumulative probability for cleavage fracture with increased loading (J) for both C(T) and SE(B) specimens. The solid lines represent predictions of the median fracture probabilities made using the three-parameter Weibull stress model, Eq. (3), with the above calibrated values for the parameters. The symbols indicate median *rank* probabilities for the measured J_c-values. These are computed using $P_i = (i - 0.3)/(N + 0.4)$, where i denotes the rank number and N defines the total number of fracture tests. The calibrated Weibull stress model predicts accurately the *shape* of the median toughness distributions (P_f vs. J) and captures the strong constraint effect on fracture toughness. The calibrated model fits both data sets very well. In Fig. 4, dashed lines indicate 90% confidence limits for the estimates of the experimental rank probabilities. To compute these confidence limits, we assume that the (continuous) P_f-values from Eq. (3) provide the *expected* median rank probabilities for an experimental data set containing the number of measured values (N). Wallin (1989) describes the procedures to compute the 90% confidence limit values.

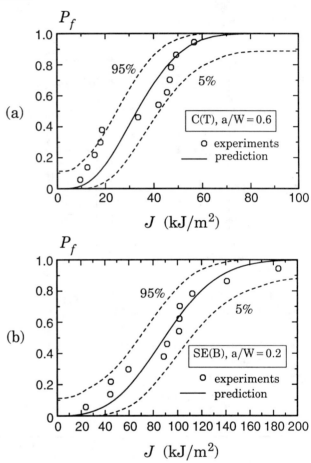

Fig. 4. Comparison of predicted cleavage probabilities (solid lines) with rank probabilities for the measured J_c-values (symbols). The dash lines represent the 90% confidence limits for the median rank probabilities. (a) deep-notch C(T) specimens; (b) shallow-notch SE(B) specimens.

PREDICTION OF CLEAVAGE FRACTURE IN SURFACE CRACK SPECIMENS

This section describes an application of the calibrated three-parameter Weibull stress model to predict the cumulative failure probability for cleavage fracture of the tested surface crack specimens. Figure 5 compares the predicted failure probabilities using the calibrated Weibull stress model ($m = 10$, $\sigma_{w-min} = 739$ MPa and $\sigma_u = 1581$ MPa) with the rank probabilities for measured J_c-values. Here the values of σ_{w-min} and σ_u differ slightly from those for the SE(B) specimen (-7°C). The crack front length of the SC(T) specimens equals 1.67×the crack front length of the SE(B) specimens. By definition, σ_{w-min} equals the value of σ_w calculated at $K = K_{min}$ in the (plane-strain) SSY model with the same crack front length as the actual fracture specimen. Under plane strain, SSY conditions, $\sigma_w \propto B^{1/m}$, where B denotes the thickness of the SSY model, Eq. (2). For surface crack specimens, $\sigma_{w-min} \propto \mathcal{L}^{1/m}$, where $\mathcal{L}$ represents the length of the semi elliptical crack front.

The model predictions capture the measured toughness distributions for both specimens. The pin-loaded specimen has a higher degree of bending load and thus exhibits a higher failure probability at the same J-level compared to the bolt-loaded specimen. The predicted P_f vs. J curve for the pin-loaded specimens *appears* to match the experimental data better than does the predicted curve for the bolt-loaded specimens. However, all experimental data fall within the predicted 90% confidence bounds for both configurations. Figure 5 shows two curves for the predicted failure probabilities of the bolt-loaded specimens. In Eq. (2), the principal stress (σ_1) value appearing in the Weibull stress integral can be assigned the current value at the loading level (J) or the maximum value experienced by the material point during the loading history. Of the four geometric configurations examined in this work, deep-notch C(T)s, shallow-notch SE(B)s and pin/bolt-loaded SC(T)'s, the choice of σ_1 definition makes a difference only for the bolt-loaded, SC(T) configuration as shown. Consequently, the calibration process here to find m, σ_u, σ_{w-min} does not depend on the σ_1 definition. Constraint loss in the bolt-loaded configuration leads to a decrease in near-front stresses under large-scale yielding, and thus use of the maximum σ_1 values earlier in the loading does raise the failure probability. Stresses have smaller values with large-scale yielding but the process zone volume for cleavage ($\overline{V}$) continues to grow with crack front blunting which leads to monotonically increasing failure probabilities. The prediction that includes the history effect provides a slightly better agreement with the experimental data for this very low constraint configuration.

CONCLUDING REMARKS

For the A515-70 pressure vessel steel, the introduction of a non-zero *threshold* value for Weibull stress (σ_{w-min}) in the expression for cumulative failure probability brings numerical predictions of the scatter in fracture toughness data into better agreement with experiments. This reflects an approximate treatment of the conditional probability of propagation in the ductile-to-brittle transition region and is consistent with the experimental observations that there exists a minimum toughness value for cleavage fracture in ferritic steels. Calibration of the threshold Weibull stress makes use of the generally accepted, minimum toughness value for ferritic steels adopted in ASTM E-1921 (1998) ($K_{min} = 20$ MPa$\sqrt{\text{m}}$).

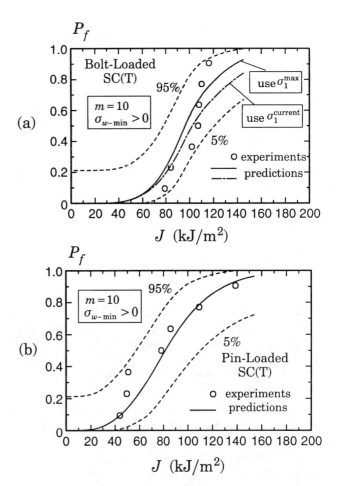

Fig. 5. Comparison of predicted toughness distributions (P_f vs. J) for both bolt-loaded and pin-loaded SC(T) specimens with experimental data, where the solid lines represent model predictions (σ_w is computed using maximum values of σ_1 experienced by the material volume), the symbols represent rank probabilities for the measured J_c-values, and dash lines represent the 90% confidence limits for the median rank probabilities. (a) bolt-loaded specimens; (b) pin-loaded specimens.

The calibration procedure builds upon the *toughness scaling model* between two crack configurations having different constraint levels and exhibits very strong sensitivity to m. It eliminates the recently discovered non-uniqueness that arises in calibrations which use only deep-notch SE(B) or C(T) data. The calibrated Weibull modulus for the material considered in this study is $m = 10$.

The calibrated three-parameter Weibull stress model accurately predicts the *shape* of the toughness distribution (P_f vs. J) for all specimen geometries (deep-notch C(T), shallow-notch SE(B), pin-loaded SC(T) and bolt-loaded SC(T)). The model predictions fit the experimental data sets very well and capture the strong constraint effect on cleavage fracture due to differences in crack geometry and loading mode (bending *vs.* tension).

ACKNOWLEDGEMENTS

This investigation was supported by grants principally from the U.S. Nuclear Regulatory Commission, Office of Regulatory Research and from the Naval Surface Warfare Center, Carderock Division. This work was also supported by National Computational Science Alliance under grant MSS970006N that made available computer resources on the NCSA SGI/CRAY Origin2000.

REFERENCES

American Society for Testing and Materials, 1998, "Test Method for the Determination of Reference Temperature, T_0, for Ferritic Steels in the Transition Range", ASTM E-1921, Philadelphia.

Anderson, T. L., Stienstra, D. and Dodds, R. H., 1994, "A theoretical framework for addressing fracture in the ductile-to-brittle transition region", In *Fracture Mechanics: 24th Volume, ASTM STP 1207* (Edited by J.D. Landes, D.E. McCabe and J. A. Boulet). American Society for Testing and Materials, Philadelphia, pp. 186-214.

Bakker, A. and Koers, R. W. J., 1991, "Prediction of cleavage fracture events in the brittle-ductile transition region of a ferritic steel", In *Defect Assessment in Components - Fundamentals and Applications, ESIS/EG9* (Edited by Blauel and Schwalbe), Mechanical Engineering Publications, London, pp. 613-632.

Beremin, F. M., 1983, "A local criterion for cleavage fracture of a nuclear pressure vessel steel", *Metallurgical Transactions*, Vol. 14A, pp. 2277-2287.

Gao, X., C. Ruggieri, C. and Dodds, R. H., 1998, "Calibration of Weibull stress parameters using fracture toughness data", *International Journal of Fracture*, to appear.

Gao, X. and Dodds, R. H., 1998, "Constraint effects on the ductile-to-brittle transition temperature of ferritic steels: a Weibull stress model", Submitted to *International Journal of Fracture*.

Joyce, J. A. and Link, R. E., 1996, "Ductile-to-brittle transition characterization using surface crack specimens loaded in combined tension and bending", In *Fracture Mechanics: 28th Volume, ASTM STP 1321* (Edited by J.H. Underwood, B.D. Macdonald and M.R. Mitchell). American Society for Testing and Materials, Philadelphia, pp. 243-262.

Koppenhoefer, K., Gullerud, A., Ruggieri, C., Dodds, R. H. and Healy, B., 1998, "WARP3D: Dynamic Nonlinear Analysis of Solids Using a Preconditioned Conjugate Gradient Software Architecture", *Structural Research Series (SRS) 596*, UILU-ENG-94-2017, University of Illinois at Urbana-Champaign.

Lin, T., Evans, A. G. and Ritchie, R. O., 1986, "A statistical model of brittle fracture by transgranular cleavage", *Journal of Mechanics and Physics of Solids*, Vol. 21, pp. 263-277.

Moinereau, D., 1996, "Local approach to fracture applied to reactor pressure vessel: synthesis of a cooperative programme between EF, CEA, Framatome and AEA", *Journal de Physique IV*, Vol. 6, pp. 243-257.

Ruggieri, C., Dodds, R. H. and Wallin, K., 1998, "Constraint effects on reference temperature, T_0, for ferritic steels in the transition region", *Engineering Fracture Mechanics*, Vol. 60, pp. 19-36.

Sorem, W. A., Dodds, R. H. and Rolfe, S. T., 1991, "Effects of crack depth on elastic-plastic fracture toughness", *International Journal of Fracture*, Vol. 47, pp. 105-126.

Tregoning, R., 1998, Unpublished experimental data.

Wallin, K., 1984, "The scatter in K_{Ic} results" *Engineering Fracture Mechanics*, Vol. 19, pp. 1085-1093.

Wallin, K., Saario, T. and Torronen, K., 1984, "Statistical model for carbide induced brittle fracture in steel", *Metal Science*, Vol. 18, pp. 13-16.

Wallin, K., 1989, "Optimized estimation of the Weibull distribution parameters", Research Report 604, Technical Research Center of Finland, Espoo, Finland.

Wiesner, C. S. and Goldthorpe, M. R., 1996, "The effect of temperature and specimen geometry on the parameters of the local approach to cleavage fracture", *Journal de Physique IV*, Vol. 6, pp. 295-304.

Xia, L. and Shih, C. F., 1996, "Ductile crack growth - III. transition to cleavage fracture incorporating statistics", *Journal of the Mechanics and Physics of Solids*, Vol. 44, pp.603-639.

STATISTICAL VERSUS CONSTRAINT SIZE EFFECTS ON THE TRANSITION REGIME TOUGHNESS: A PRELIMINARY STUDY OF A PRESSURE VESSEL STEEL

H. J. Rathbun, G. R. Odette, M. Y. He, G. E. Lucas and J. W. Sheckherd
University of California, Santa Barbara
Santa Barbara, California, USA

ABSTRACT

Cleavage initiation fracture toughness in the transition temperature region manifests a large inherent scatter and an explicit effect of specimen size and geometry. Weakest link statistical models, that have been developed to rationalize the observed scatter, have also been interpreted to attribute size effects in well constrained specimens to the out-of-plane crack front length, B. In this case, the toughness scales as $B^{-1/4}$, consistent with a Weibull slope of 4. While observations of such size scaling in fully constrained specimens supports this concept, the onset of small scale yielding constraint loss, primarily associated with the in-plane ligament dimension, b, may occur under conditions nominally viewed and being fully constrained. Further, the corresponding effects on toughness tend to mimic the putative B scaling, as $b^{-1/4}$. Unfortunately, most previous studies have used specimens with self-similar geometries, involving proportional variations in both B and b. Such confounding makes proper interpretation of statistical (B) versus constraint (b) size scaling effects difficult. Since the assumption of $B^{-1/4}$ scaling is central to the new Master Curve method for characterizing toughness in the transition, it is important to isolate constraint versus statistical effects. Therefore, a quasi single-variable experiment on the HSST-02 reference plate was carried out using pre-cracked (a/W ≈ 0.5), side-grooved (≈20%), three-point bend specimens with both variations in B (2.6 to 43 mm) at a constant b (8.0 mm) and variations in b (1.3 to 8.0 mm) at a constant B (10.2 mm). The test temperature (-61°C) and other test conditions were held constant. The data for this series of experiments showed that variations in B had relatively little effect on the mean toughness (82.1 +/- 3.0 MPa√m) for the thicker specimens (B ≥ 10 mm). However, the mean toughness systematically increased with smaller b, approximately following a $b^{-1/4}$ scaling. Notably, effects of b were observed in regimes nominally associated with high crack front constraint. These preliminary observations suggest that scaling observed in some prior size effects studies may have been caused by constraint loss rather than by statistical crack front length effects. Additional research is being carried out to further explore these questions.

INTRODUCTION AND BACKGROUND

Proper measurement and application of cleavage initiation fracture toughness data for structural steels is important to a wide range of the industrial, energy and civil infrastructures of our economy. The most salient features of the toughness-temperature relation, $K_{Jc}(T)$, in the transition are:

1. The mean toughness varies by approximately an order of magnitude.

2. Individual data are widely scattered around the mean.

3. The mean toughness and scatter often show an explicit effect of specimen size and geometry.

The recently approved ASTM Standard Practice E 1921-97 represents a revolutionary advance in characterizing toughness in the transition, since it permits establishing the entire $K_{Jc}(T)$ curve with measurements on a relatively small number of relatively small specimens [1,2]. The method rests on several key assumptions which appear to be empirically successful, including: a) the shape of the toughness temperature curve has a constant master curve (MC) shape $K_{Jc}(T' = T-T_o)$ that can be indexed on an absolute temperature, T, scale using a reference temperature, T_o, at a reference toughness of 100 MPa√m; b) T_o can be estimated by testing as few as six specimens; c) the scatter in fracture toughness is governed by Weibull weakest link statistics, that can be used to set upper and lower confidence bounds; d) there is a minimum toughness of 20 MPa√m that is dissipated in any cleavage event; e) size effects are primarily due to the crack front length, B, and can be accounted for by a $B^{-1/4}$ scaling law [2,3,4]; and f) constraint effects can be effectively treated by a censoring procedure for $K_{Jc} > (E\sigma_y b/30)^{1/2}$, where σ_y is the yield stress, E is the elastic modulus and b is the length of the unbroken ligament [1]. Note, according to ASTM E 1921-97, for medium strength steels with $\sigma_y \approx 500$ MPa, a precracked Charpy specimen provides fully constrained toughness up to values of about 130 MPa√m.

The broad objective of this work is to carry out a rigorous assessment of the assumptions underlying the MC method. Better understanding of the underlying physics of fracture in the transition region will be particularly important to structural applications that differ significantly from test conditions. This paper reports preliminary results of items e) and f) listed above, related to statistical versus constraint mediated size effects. Data, and specific studies in the literature, can be cited to support both statistical and constraint size effects models [2,5-11]. However, with few exceptions [9-11], these studies have not explicitly separated the single-variable effects of B and b. In the majority of cases, tests have been performed on self-similar specimens, with different absolute sizes, but with the same relative dimensions, i.e. B/b = constant [4]. The major objectives of this preliminary study were to isolate the single-variable effects of B at constant b and b at constant B.

CLEAVAGE MECHANISMS AND MECHANICS

To begin, it is useful to establish a mechanistic framework for evaluating cleavage fracture in the transition. All viable micromechanical models postulate that cleavage occurs when a sufficiently large tensile stress operates over a sufficiently large, microstructurally mediated region in the vicinity of a blunting crack tip. This concept was originally proposed by Ritchie, Knott and Rice (RKR) in terms of a critical stress, σ^*, operating at a critical distance, λ^* [12]. For conditions of small scale yielding (SSY), the distance from the crack tip to a σ^* contour increases with the elastic-plastic energy release rate, J. The critical J_c for cleavage initiation is reached when this distance equals λ^*. The cleavage fracture condition can be described in terms of the equivalent elastic-plastic toughness, $K_{Jc} = (E'J_c)^{1/2}$ where E' is the plane strain modulus. The RKR model has been the basis for many progeny [5,6,7,8,13-24], including a variety of statistical models.

Most statistical models assume that fracture occurs when a sufficiently large volume, V, of material ahead of a blunting crack tip is subjected to sufficiently large local stress. The probability of fracture increases with V, which is the product of the in-plane stressed area, A, times the effective crack front length, B, V = AB. For mean toughness, this can be most simply expressed as a critical stress, σ^*, critical volume, V^*, criteria, based on the assumption that cleavage initiates and propagates from a single weakest link trigger site, causing macroscopic fracture [2,5,25,26]. While subcritical microcleavage events may occur, it is assumed that they do not influence the trigger event for macrocleavage. Statistical variations in toughness are physically associated with variations in σ^*/V^*, such as might arise from size, spatial and orientation distributions of large grain boundary carbides or other trigger sites. Fractographic observations of well-defined single trigger site locations support this concept. Under conditions of SSY, A varies with the applied J^2 or K_J^4, which is consistent with a Weibull slope of 4 and a $B^{-1/4}$ toughness size scaling.

However, other models, including statistical variants, postulate that macroscopic fracture is controlled by a critical stress-critically stressed area criteria, σ^*/A^*, rather than σ^*/V^*. The rationale is that macroscopic cleavage requires the interaction of a number of sympathetic micro-events within the highly stressed process zone and along a crack front. Alternatively, this can be described in terms of both in-plane and out-of-plane critical energy release rate, or resistance curve behavior, involving the collective process zone mechanics. If multiple, interacting events with a spacing along the crack front of order L_c << B control cleavage initiation, statistical toughness dependence on B is weakened or eliminated. Observations of multiple regions of damage prior to macrocleavage based on fractography, confocal microscopy-fracture reconstruction and acoustic emission methods support this concept [27,28]. Further, it is noted that, in σ^*/V^* models, the differential volume element has a high aspect ratio [3]. In the lateral direction, it is order B (>> 1 mm), compared to the much smaller in-plane process zone dimensions (< 1 mm). However, the microstructural distribution of the putative trigger points is expected to be roughly isotropic and evenly spaced in all directions. The physical justification for a high aspect ratio differential volume element is not clear.

If statistically mediated B scaling is not responsible for observed size scaling, then the question remains - what is? In SSY, the tip stress fields of deep (a/W ≈ 0.5) cracks are spatially self-similar, and can be fully characterized by a single loading parameter, J. Thus, in this regime, the fields are independent of the overall specimen dimensions; hence, there is no size effect. However, as the critical J for cleavage increases and/or specimen dimensions decrease below a fully constrained condition, the crack tip stress fields are no longer J-dominated and self-similar. In this regime, the fields explicitly depend on the absolute J, the specimen dimensions and the constitutive properties of the material. Explicitly, within a specified stress, σ, contour, generally expressed as a multiple, R, of σ_y, $R = \sigma/\sigma_y$, A no longer varies with J^2. Indeed, there is a gradual transition to a large scale yielding regime where A depends only weakly on J. The amount of constraint loss depends on the ligament, b, and thickness, B, dimensions. However, for standard specimen geometries with B ≈ 1 to 2b, constraint loss is primarily controlled by b. In this case, the overall loss of constraint can be represented by a non-dimensional parameter, M:

$$M = b\sigma_y / J \qquad (1)$$

Beyond a critical value of M that depends on material flow properties and thickness B, crack front conditions approach those of plane-strain SSY. ASTM E 1921-97 specifies a minimum M of 30 to maintain a condition of high crack front constraint.

Three-dimensional (3-D) finite element methods (FEM) have been used to compute constraint loss effects for a specified specimen configuration and constitutive properties [14,26,29,30]. The σ^*/A^* (or V^*) models provide a convenient and physically motivated method for characterizing the effects of deviations from SSY on toughness. The SSY-J is calculated as the J required to produce the same stressed area, A (or V), as in an actual specimen, for a specified R. Constraint loss is then characterized by the ratio of the actual to SSY-J, J/J_{ssy}. The J/J_{ssy} ratio depends on the constitutive properties and R, as well as $b\sigma_y/J$. The calculations reported here for a B = b four point bend specimen were carried out with the ABAQUS FEM code, using a Ramberg-

Osgood constitutive law, with a strain-hardening exponent of 10 and $E/\sigma_y = 500$ [29]. The ratio K/K_{ssy} is calculated as $(J/J_{ssy})^{1/2}$. Figure 1 shows the ratio of the actual to SSY K as a function of $(M_{ssy}/M)^{1/4}$ for three values of R. Here, M_{ssy} is defined at $K/K_{ssy} \approx 1.01$. While there is curvature for larger $(M_{ssy}/M)^{1/4}$, note the initial scaling is linear. These results indicate that, within a limited regime of initial constraint loss, toughness follows roughly a $b^{-1/4}$ scaling. This regime extends over what is equivalent to a large range of B varying by factors from about 6 to 13. Notably, the M_{ssy} range from about 215 to 400, thus are much larger than the value of 30 nominally associated with SSY.

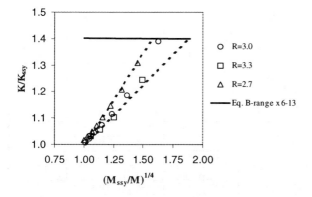

Figure 1. The effects of loss of SSY constraint on toughness as a function of $(M_{ssy}/M)^{1/4}$ as described in text.

EXPERIMENT

The objective of the experiment was to measure the effect of variations in B at constant b and b at constant B on the mean K_{Jc}. It must be emphasized, as is typical of most real world experiments, that attempts to maintain purely single variable conditions were only partially successful. However, this work sought to minimize such effects in the following ways:

1. The geometry and dimensions of all pre-cracked (final ΔK = 15-17 MPa√m) three point bend specimen parameters were approximately constant (except B for the constant b tests, and b for constant B tests), including: span to width ratio (S/W ≈ 3.2), crack length to width ratio (a/W ≈ 0.5) and side grooves (≈ 20%, zero degree flank angle and root radius of 0.15 mm).

2. Six tests were carried out for each specimen configuration to evaluate a mean K_{Jc}. Two additional tests on materials from coupons that showed somewhat higher average toughness, and that failed by ductile tearing in the thinnest specimens were not included in the analysis.

3. All tests were conducted under identical conditions of temperature and loading rate. The test temperature, -61°C, was selected in the elastic-plastic cleavage initiation regime, so that the mean K_{Jc} of about 80 MPa√m would be well within the nominal SSY regime for the variable B constant b tests (M >> 30), while experiencing large scale yielding at the smallest values of b in the constant B tests.

4. All data was analyzed using a common set of procedures.

The 'well-characterized' material stock available for the study consisted of broken halves of HSST Plate-02 drop weight specimens. The initial coupon dimensions, coupled with the desire for wide variations in B and b, imposed limitations that resulted in somewhat atypical specimen configurations, including: a TS orientation; a relatively short S/W ratio; and both lower and higher ratios of B/b than are found in standard test specimens. In order to promote uniform conditions along the crack front, all specimens were side grooved with narrow slots after pre-cracking. The side grooves appear to have had some influence on the character of the cleavage crack propagation in the smallest B specimens that produced fracture surfaces with macroscopic keyhole shaped unbroken ligament features. In any event, the specimens with B = 2.5 and 5 mm (note that nominal values are used in the text for clarity) clearly experienced significant loss of lateral constraint, consistent with the predictions of 3-D FEM calculations [14]. They are included here simply for completeness.

Tests were conducted on an 88 kN MTS servohydraulic frame equipped with a 454 digital controller at a fixed load point displacement rate of about 2.5 μm/s. The load line displacements (Δ) were evaluated from calibrated actuator LVDT. Following testing, the specimens were broken in liquid N_2 and the crack lengths were measured based on 5- (for smaller thickness) or 9-point (for larger thickness) averages. A brief preliminary optical survey of the fracture surfaces was carried out to verify cleavage initiation and to evaluate the macroscopic character of the fracture event.

The elastic plastic toughness was evaluated from load-displacement, P-Δ, curves as

$$J_c = J_{el} + 2A_p / Bb \qquad (2)$$

based on ASTM E 813 [31], where A_p is the plastic work area under the P-Δ curve. The atypical specimen configurations also added some uncertainty about the proper choice of the elastic-plastic η-factor.

Finally, the limited number of 6 tests at each configuration leads to uncertainty in the mean toughness levels of about 12 MPa√m [32]. Clearly, greater test redundancy would be desirable. However, various statistical tests can be applied to the collective data set to mitigate the impact of uncertainties in individual mean toughness evaluations. Indeed, none of the considerations listed above are expected to have a major impact on the tentative conclusions of this study.

RESULTS

The results of the tests are summarized in Table 1 along with details of the specimen configurations. All fractures were by cleavage, preceeded by no or minimal ductile tearing. The E and σ_y are taken as 213 GPa and 535 MPa respectively. Figure 2 plots all the K_{Jq} versus B data for the constant b tests. The large filled circles are the measured mean K_{Jq} and the open circles are the mean constraint corrected toughness, based on 3-D FEM calculations [14,29]. The solid line is the predictions of the size scaling recommended in ASTM E 1921-97 to normalize to a common B = 25 mm:

$$K_{Jc}(B) = \left[K_{Jc(B=25)} - 20\right] \cdot \left(\frac{25}{B}\right)^{1/4} + 20 \quad (3)$$

A least squares fit to the data for B ≥ 10 mm to an equation in the form $K_{Jc}(B) = [K_{Jc(B=25)} - 20](25/B)^p + 20$ yielded p = 0.03 ± 0.02 at the 95% confidence level.

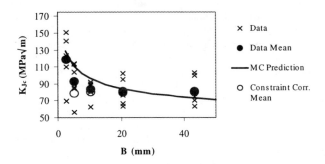

Figure 2. Variation of measured K_{Jc} with B (for b held constant).

Figure 3 plots all the K_{Jq} versus b data for the constant B tests. Again, the large filled circles are the measured mean K_{Jq} and the open circles are the mean constraint corrected toughness. The mean toughness clearly increases with decreasing b at constant B. A least squares best fit to form $K_{Jc}(b) = [K_{Jc(b=8.0)}](8.0/b)^q$ yields a q = 0.22± 0.03 at the 95% confidence level. The constraint correction estimates are consistent with a mean SSY toughness of about 82 MPa√m, except at the smallest b, where somewhat lower toughness values are predicted. This may be due to the fact that the constraint calculations modeled side grooves with much larger flank angles (45°) and root radii (0.5 mm) than in the actual specimens.

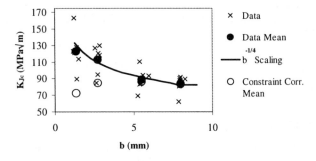

Figure 3. Variation of measured K_{Jc} with b (for B held constant).

Table 1. Specimen Dimensions and Fracture Toughness Data

Specimen ID	B (mm)	b (mm)	W (mm)	S (mm)	K_{Jc} (MPa√m)	M
T02A	43.2	8.0	16.0	50.8	78.1	149.5
T03A	43.1	7.4	15.9	50.8	70.0	172.1
T05A	43.3	7.2	16.0	50.8	63.0	206.7
T08A	43.3	7.4	15.9	50.8	74.8	150.7
T13A	43.3	7.8	15.9	50.8	103.2	83.5
T14A	43.2	8.0	16.2	50.8	99.7	91.7
T02B2	20.5	7.7	15.9	50.8	102.3*	83.8
T03B2	20.5	7.7	15.9	50.8	63.7	216.2
T05B2	20.4	7.8	15.9	50.8	95.3	97.9
T08B2	20.5	7.8	15.9	50.8	81.9	132.5
T13B2	20.4	7.8	15.9	50.8	66.8	199.2
T14B2	20.4	7.9	16.2	50.8	76.9	152.2
T02B1	10.3	7.9	15.9	50.8	90.1	110.9
T03B1	10.3	7.9	15.9	50.8	80.8	137.9
T05B1	10.3	7.8	15.9	50.8	62.1	230.5
T08B1	10.2	8.0	15.9	50.8	92.0	107.7
T13B1	10.2	7.9	15.9	50.8	88.4	115.2
T14B1	10.2	8.3	16.2	50.8	89.7	117.6
T02B3	5.2	8.3	16.0	50.8	55.9	302.7
T03B3	5.1	8.1	15.9	50.8	113.5	71.7
T05B3	5.2	8.2	15.9	50.8	84.2	131.8
T08B3	5.1	8.2	15.9	50.8	103.6	87.1
T13B3	5.2	8.2	15.9	50.8	87.4	122.3
T14B3	5.1	8.6	16.2	50.8	113.7	75.8
T02B4	2.6	8.2	16.0	50.8	69.4	194.0
T03B4	2.7	8.4	16.0	50.8	150.8	42.1
T05B4	2.6	8.4	16.0	50.8	110.5	78.4
T08B4	2.6	8.4	16.0	50.8	140.4	48.6
T13B4	2.5	8.4	15.9	50.8	119.3	67.3
T14B4	2.6	8.7	16.2	50.8	123.2	65.3
T01A	10.3	1.2	2.7	8.6	163.5	5.1
T06C	10.3	1.5	2.7	8.6	113.8	13.2
T06E	10.2	1.3	2.7	8.6	89.9	18.3
T06G	10.2	1.2	2.7	8.6	121.7	9.2
T10B	10.3	1.4	2.7	8.6	127.3	9.8
T10E	10.2	1.2	2.7	8.6	123.0	9.0
T01B	10.2	2.8	6.0	19.2	118.4	22.8
T06A	10.2	2.8	6.0	19.2	130.0	18.9
T06B	10.2	2.7	6.0	19.2	120.3	21.3
T06D	10.2	2.7	5.9	19.2	95.4	33.8
T10A	10.2	2.6	6.0	19.2	84.8	41.2
T10C	10.2	2.5	6.0	19.2	126.7	17.7
T01C	10.2	5.3	11.0	35.2	69.6	124.7
T06F	10.2	5.4	10.9	35.2	83.2	88.9
T06H	10.2	5.4	11.0	35.2	110.8	50.1
T10F	10.2	5.4	11.0	35.2	83.8	87.6
T10G	10.2	5.6	11.0	35.2	94.1	72.1
T10H	10.2	5.9	11.0	35.2	93.2	77.4

* Note: Specimen T02B2 experienced a minor cleavage "pop-in" with calculated toughness of 44.1 MPa√m.

SUMMARY AND FUTURE WORK

The size scaling and constraint limitation assumptions included in ASTM E 1921-97 appear to be inconsistent with the experimental data obtained in this study. Specifically, beyond a characteristic length dictated by lateral constraint loss, toughness appears to be relatively constant and independent of B. In addition, toughness appears to follow roughly a $b^{-1/4}$ scaling, a result that may have been masked by the simultaneous variations in B and b in many previous size effects studies. Finally, there appear to be modest variations in mean toughness for M-factors significantly greater than 30.

However, a number of issues need to be resolved before firm conclusions can be drawn. This will require additional tests over a wide range of b and B, FEM studies of specimens with atypical geometries, and detailed experiments aimed at developing understanding of the underlying micromechanics of cleavage. Such studies are currently underway at the University of California at Santa Barbara, using a recently acquired section of A533B reactor pressure vessel plate.

ACKNOWLEDGEMENTS

This work was supported by U.S. Nuclear Regulatory Commission Contract NRC-04-94-049. The encouragement and support of the late M. Vassilaros is also greatly appreciated. We thank P. Spätig for many helpful suggestions and conversations. The assistance in test and data evaluation provided by T. Huang is gratefully acknowledged.

REFERENCES

1. ASTM E 1921-97, "Standard Test Method for Determination of Reference Temperature, T_0, for Ferritic Steels in the Transition Range," ASTM 1998.

2. Merkle, J.G., Wallin, K., and McCabe, D.E., "Technical Basis for an ASTM Standard on Determining the Reference Temperature, T_0, for Ferritic Steels in the Transition Range," NUREG/CR-5504, 1998.

3. Wallin, K., "The Scatter in K_{Ic}-Results," *Engineering Fracture Mechanics*, Vol. 19, No. 6, pp. 1085-1093, 1984.

4. McCabe, D.E., "A Comparison of Weibull and β_{Ic} Analysis of Transition Range Fracture Toughness Data," NUREG/CR-5788, 1992.

5. Odette, G.R., unpublished research.

6. Wallin, K., Saario, T., and Törrönen, K., "Statistical Model for Carbide Induced Brittle Fracture in Steel," *Metal Science*, 18, pp. 13-16, 1984.

7. Odette, G.R., Edsinger, K., Lucas, G.E., and Donahue, E., "Developing Fracture Assessment Methods for Fusion Reactor Materials with Small Specimens," ASTM STP-1329, American Society for Testing and Materials, 1998.

8. Odette, G.R., "On the Ductile to Brittle Transition in Martensitic Stainless Steels – Mechanisms, Models and Structural Implications," *Journal of Nuclear Materials*, 212-215, pp. 45-51, 1994.

9. Landes, J.D. and McCabe, D.E., "Effect of Section Size on Transition Temperature Behavior of Structural Steels," ASTM STP-833, American Society for Testing and Materials, pp. 378-392, 1984.

10. Brückner-Foit, A., Ehl, W., Munz, D., and Trolldenier, B., "The Size Effect of Microstructural Implications of the Weakest Link Model," *Fatigue Fract. Engng. Mater. Struct.*, Vol. 13, No. 3, pp. 185-200, 1990.

11. Brückner-Foit, A., Munz, D., and Trolldenier, B., "Micromechanical Implications of the Weakest Link Model for the Ductile-Brittle Transition Region," *Defect Assessment in Components – Fundamentals and Applications,* ESIS/EGF9, Mechanical Engineering Publications, London, pp. 477-488, 1991.

12. Ritchie, R.O., Knott, J.F., and Rice, J.R., "On the Relationship Between Critical Tensile Stress and Fracture Toughness in Mild Steel," *J. Mech. Phys. Solids*, 21, pp. 395-410, 1973.

13. Hahn, G.T., "The Influence of Microstructure on Brittle Fracture Toughness," *Metallurgical Transactions A*, 15A, pp. 947-959, 1984.

14. Nevalainen, M., and Dodds, R.H., Jr., "Numerical Investigation of 3-D Constraint Effects on Brittle Fracture in SE(B) and C(T) Specimens," *International Journal of Fracture*, 74, pp. 131-161, 1995.

15. Curry, D.A., Knott, J.F., "The Relationship Between Fracture Toughness and Microstructure in the Cleavage Fracture of Mild Steel," *Metal Science*, 10, pp. 1-6, 1976.

16. Ritchie, R.O., Server, W.L., and Wullaert, R.A., "Critical Fracture Stress and Fracture Strain Models for Prediction of Lower and Upper Shelf Toughness in Nuclear Pressure Vessel Steels," *Metallurgical Transactions A*, 10A, pp. 1557-1570, 1979.

17. Curry, D.A., "Cleavage Micromechanisms of Crack Extension in Steels," *Metal Science*, 14, pp. 319-326, 1980.

18. Beremin, F.M., "A Local Criterion for Cleavage Fracture of a Nuclear Pressure Vessel Steel," *Metallurgical Transactions A*, 14A, pp. 2277-2287, 1983.

19. Curry, D.A. and Knott, J.F., "Effect of Microstructure on Cleavage Fracture Toughness of Quenched and Tempered Steels," *Metal Science*, 13, pp. 341-345, 1979.

20. Curry, D.A., "Comparison Between Two Models of Cleavage Fracture," *Metal Science*, 14, pp. 78-80, 1980.

21. Evans, A.G., "Statistical Aspects of Cleavage Fracture in Steel," *Metallurgical Transactions A*, 14A, pp. 1349-1355, 1983.

22. Lin, T., Evans, A.G. and Ritchie, R.O., "A Statistical Model of Brittle Fracture by Transgranular Cleavage," *J. Mech. Phys. Solids*, 34-5, pp. 477-497, 1986.

23. Lucas, G.E., Yih, H. and Odette, G.R., "Analysis of Cleavage Fracture Behavior in HT-9 with a Statistical Model," *Journal of Nuclear Materials*, 155-157, pp. 673-678, 1988.

24. Odette, G.R., Chao, B.L. and Lucas, G.E., "On Size and Geometry Effects on the Brittle Fracture of Ferritic and Tempered Martensitic Steels," *Journal of Nuclear Materials*, 191-194, pp. 827-830, 1992.

25. Anderson, T.L. and Dodds, R.H., Jr., "Specimen Size Requirements for Fracture Toughness Testing in the Transition Region," *Journal of Testing and Evaluation*, Vol. 19, No. 2, pp. 123-134, 1991.

26. Dodds, R.H., Jr., Anderson, T.L. and Kirk, M.T., "A Framework to Correlate a/W Ratio Effects on Elastic-Plastic Fracture Toughness (J_c)," *International Journal of Fracture*, 48, pp. 1-22, 1991.

27. Edsinger, K., Odette, G.R., Lucas, G.E. and Wirth, B., "The Effect of Constraint on Toughness of a Pressure Vessel Steel," ASTM STP 1270, American Society for Testing and Materials, 1996.

28. Gerberich, W.W., Chen, S.-H., Lee, C.-S., Livne, T., "Brittle Fracture: Weakest Link or Process Zone Control?," *Metallurgical Transactions A*, 18A, pp. 1861-1875, 1987.

29. He, M.Y. and Odette, G.R., "Three Dimensional FEM Assessment of Constraint Loss Effects on Cleavage Initiation Fracture," in preparation.

30. Koppenhoefer, K.C. and Dodds, R.H., Jr., "Loading Rate Effects on Cleavage Fracture of Pre-Cracked CVN Specimens: 3-D Studies," *Engineering Fracture Mechanics*, Vol. 58, No. 3, pp. 249-270, 1997.

31. ASTM E 813-89, "Standard Test Method for J_{Ic}, a Measure of Fracture Toughness," ASTM 1998.

32. McCabe, D.E., "Issues Addressed in the Development of an ASTM Test Method E1921-97," Presented at the International Group – Radiation Damage Mechanisms 8 Open Workshop, Nashville, TN, USA, January 22, 1999.

A PHYSICAL BASIS FOR THE MASTER CURVE

Marjorie Natishan
University of Maryland
College Park, Maryland, USA

Mark Kirk
Westinghouse Electric Company
Pittsburgh, Pennsylvania, USA

ABSTRACT

An abundance of empirical data supports the use of the Wallin Master Curve to describe the transition fracture toughness behavior of pressure vessel steels, including the notion of a material invariant curve shape. However, the lack of fracture toughness data for a considerable proportion of the nuclear fleet suggests that a purely empirical argument cannot validate the Master Curve for all conditions of interest. In this paper we therefore examine a physical basis for the Master Curve. The information presented herein demonstrates that the temperature dependence of the flow curve is controlled only by the lattice structure. Other factors that can vary with steel composition and heat treatment, such as the grain size and the work hardening rate, influence only the athermal component of the flow curve. Consequently, these factors influence the position of the transition curve on the temperature axis (i.e. T_o as determined by ASTM E1921-97), but *not* its shape. This understanding suggests that the breadth of applicability of the Wallin Master Curve to other microstructures is predictable based on the physics of the fracture process.

INTRODUCTION

The Master Curve approach to characterizing fracture toughness transition behavior of pressure vessel steels is based on statistical analysis of empirical data and not on a physics-based understanding of the fracture behavior of these steels. While there is an over-abundance of empirical data that supports the idea of a single curve shape for all pressure vessel steels this cannot replace the need for solid physical modeling. The lack of fracture toughness data for a considerable proportion of the nuclear fleet suggests that a purely empirical argument cannot validate the Master Curve for all conditions of interest. In order to validate the Master Curve approach research must be undertaken that will provide a physical understanding of the fracture behavior of pressure vessel steels. This will provide the basis for defining the limits of the Master Curve as well as for enabling extrapolation of the Master Curve model to other material conditions with only limited testing.

In their 1984 paper, Wallin, Saario, and Törrönen (WST) [1] suggest a link between the micro-mechanics of cleavage fracture and the empirical observation of a "master" toughness transition curve. Using the Griffith equation as the basis for their model they suggest that temperature dependence enters via the plastic work term shown as w_p in equation (1) and (2). Equation (1) shows the dependence of the critical fracture-inducing particle radius on the applied stress, σ, and a combination of the surface energy and the plastic energy absorbed during crack growth, w_p.

$$r = \frac{\pi(\gamma_s + w_p)E}{2(1-v^2)\sigma^2} \qquad (1)$$

The temperature dependence of w_p is given by:

$$\gamma_s + w_p = A + B \cdot \exp[C \cdot T] \qquad (2)$$

This form of the temperature dependence of the plastic work term follows that given for the Master Curve fracture toughness temperature-dependence in ASTM E1921-97 [2]:

$$K_{Jc(median)} = 30 + 70 \cdot \exp[0.019(T-T_o)] \qquad (3)$$

The numeric values in eq. (3) do not depend on the type of steel, suggesting that steel type does not influence the relationship between ($\gamma_s + w_p$) and temperature in eq. (2). Furthermore, the ASTM standard assumes a dispersion of toughness about this median value that follows a Weibull distribution with a fixed slope of 4 for all steels.

Eq. (1), and other mathematical relationships associated with the Master Curve (i.e. predictions of size effects, predictions of data dispersion), model existing fracture toughness data for nuclear RPV steels very well. Empirical validation of the Master Curve approach will be prohibitively expensive (especially for irradiated materials). Establishment of a sound physical basis for the Master Curve approach offers the potential of broad validation of the concept at reasonable costs, and within a time frame of interest to nuclear licensees.

While WST suggest a micromechanical basis for the form of equations (2) and (3), the proposed model will detail the links between dislocation mechanics-based constitutive models of steel behavior and the temperature and strain rate dependence of the Master Curve.

CONSTITUTIVE MODELS OF MATERIAL DEFORMATION BEHAVIOR

Equations have been developed to describe the mechanical behavior of metals taking into account the various strengthening mechanisms present in the alloy, the crystal structure and the dependence of these mechanisms on temperature and strain rate [3-5]. The dislocation-based deformation models describe how various aspects of the microstructure of a material control dislocation motion and how these effects vary with temperature and strain rate. The microstructural characteristics of interest include short-range barriers to dislocation motion provided by the lattice structure itself that effects the atom-to-atom movement required for a dislocation to change position within the lattice and long range barriers. Long range barriers to dislocation motion are defined by the inter-barrier spacing being typically several orders of magnitude greater than the lattice spacing or short-range barriers. These include point defects (solute and vacancies) precipitates (semi-coherent to non-coherent), boundaries (twin, grain, etc.), and other dislocations.

As a stress is applied to a metal dislocations begin to move resulting in plastic deformation. For a dislocation to move it must change position from one equilibrium position in the lattice to another overcoming an energy barrier to do so. The dislocation requires application of a force to overcome this activation barrier and the magnitude of this force is the Peierls-Nabarro stress [6], τ_{PN}. The wavelength of these barriers is equal to the periodicity of the lattice. The moving dislocation will also encounter other barriers such as solute atoms, vacancies, precipitates, inclusions, boundaries and other dislocations. These additional periodic barriers are described by different spacings and lengths.

Thermal energy acts to increase the amplitude of vibration of atoms about their lattice sites. The effect of this thermal energy is to decrease the height of lattice barriers (short range) to dislocation motion decreasing the force required to move the dislocation from one equilibrium position to the next. The effect of strain rate (dislocation velocity) has a similar but opposite effect to that of temperature. As strain rate and thus dislocation velocity is increased there is less time available for the dislocation to overcome the barrier and the effect of thermal energy is decreased resulting in an increased force required for dislocation motion.

Long range obstacles cannot be overcome by thermal energy because no matter how large the amplitude of atomic vibration the height of the energy barrier required to move the dislocation past these large obstacles is orders of magnitude larger. Obstacles are therefore separated into short-range barriers to dislocation motion (thermally activated) and long range barriers (non-thermally activated). The flow stress of a material can then be expressed as:

$$\sigma_f = \sigma_G(\text{structure}) + \sigma^*(T, \dot{\varepsilon}, \text{structure}) \quad (4)$$

The first term is due to the athermal or long-range barriers to dislocation motion while the second term is due to the thermally activated barriers. The short-range barriers include the Peierls-Nabarro stress and dislocation forests. Peierls-Nabarro stresses (lattice friction stresses) are the controlling short range barriers in BCC metals while dislocation forest structures are the controlling short range barriers in FCC and HCP metals. This difference is responsible for the difference in strain rate sensitivity between BCC and FCC metals.

The temperature dependence of the probability that a dislocation will overcome a short-range obstacle is given by:

$$P_B = \exp(\Delta G / kT) \quad (5)$$

where k is Boltzman's constant, T is temperature of interest, and ΔG is the activation energy of a given barrier. The strain rate effect can be shown to have a similar form:

$$\dot{\varepsilon} = \dot{\varepsilon}_o \exp(\Delta G / kT) \quad (6)$$

Solving this equation for ΔG gives

$$\Delta G = kT \cdot \ln\left(\dot{\varepsilon} / \dot{\varepsilon}_o\right) \quad (7)$$

This equation clearly shows that activation energy for dislocation motion decreases with temperature and increases with strain rate.

Using these equations as the basis for their model Zerilli and Armstrong [5] found that overcoming P-N barriers was the principal thermal activation mechanisms in BCC metals. The spacing of these obstacles is equal to the lattice spacing and thus is not affected by plastic strain as are the forest dislocation barriers in FCC metals. Using this information Zerilli and Armstrong developed an expression for the thermal portion of the stress (assuming a constant obstacle spacing for BCC) as:

$$\sigma^* = C_1 \exp(-\beta T) \qquad (8)$$

where β is a parameter that is dependent on strain and strain rate given by:

$$\beta = -C_2 + C_3 \ln \dot{\varepsilon} \qquad (9)$$

Combining this with commonly accepted terms describing strengthening due to athermal barriers (such as hall Petch strengthening, $kd^{-1/2}$, and strain hardening, $C_4\varepsilon^n$) results in a reasonably complete description of the flow behavior of BCC metals given by:

$$\sigma_f = \sigma_G + C_4 \varepsilon^n + kd^{-1/2} + C_1 \cdot \exp\left[-C_2 T + C_3 T \cdot \ln(\dot{\varepsilon})\right] \qquad (10)$$

The form of this equation is consistent with constitutive equations developed by many other researchers accounting for the action of short range and long range barriers on the activation energy for dislocation motion. This equation does not account for dynamic strain aging, solute drag, climb and other interactions between diffusional processes and dislocation velocity. Programs are ongoing to modify constitutive equations to account for these affects.

To understand the physical basis for a single "Master Curve" for all ferritic steels we need to connect the Master Curve equation back through the equation for plastic work dependence on temperature to the constitutive equations that have been developed to describe the plastic deformation behavior of metals.

The derivation of eq. (10) was based on the assumption of dislocation-controlled plastic deformation in BCC metals. The combination of thermal and athermal terms account for the contribution of the various microstructural features to the stress required for dislocation motion. The last component of eq. (10) is the thermal term; it describes the change in flow curve behavior with temperature. This term is derived based on the effects of temperature on the Peierls-Nabarro stresses within the crystal lattice that resist dislocation motion. This contribution to flow stress decreases exponentially with temperature, in apparent contradiction to the exponential increase of both plastic work and fracture toughness with temperature proposed in the Master Curve method.

The athermal terms in the Z-A equation will therefore only effect the temperature at which the fracture toughness transition occurs. Because the thermal term in the Z-A equation describes the variation of flow behavior with temperature for a BCC material, it is this term that is expected to account for the shape of the fracture toughness transition. However, the thermal term describes a decrease in flow stress with increasing temperature. If we assume a constant strain to fracture at all temperatures, this results in a predicted decrease in the area under the true stress vs. true strain curve, and, consequently, a decrease of plastic work with increasing temperature. This trend is counter to the WST model [1], and suggests that an additional parameter must be considered.

So far we have examined the shift in the flow curve predicted with temperature from the Z-A equation. Integration of this equation gives us plastic work per unit volume but we must also consider the variation of the total strain to fracture with temperature to provide the limits of integration to truly capture the variation of plastic work with temperature. As a first approximation we used the uniform strain from a uniaxial tensile test to provide the limit of integration. Using these data to truncate the true stress, true strain curves and integrating gives us a value of plastic work that increases with increasing temperature, similar to the trend used by Wallin et al. [1]. Details of this calculation are described in the following section.

CALCULATION OF A MASTER CURVE

In eq. (2) we showed that the variation of critical-J values with temperature depends only on the variation of plastic work with temperature. In this section we calculate the variation of plastic work with temperature and compare the form of this variation with the Wallin Master Curve.

The plastic work per unit volume to cause fracture is the area under the stress-strain curve. This area depends on the flow curve (i.e. on the variation of stress with strain), and on the strain the material can withstand prior to fracture. The variation with temperature of these two variables therefore determines the variation with temperature of plastic work.

In general, plastic work can be expressed as follows:

$$w_p = \int_0^{\varepsilon_{fail}(T)} \sigma_f(\dot{\varepsilon}, T) d\varepsilon \qquad (11)$$

where $\sigma_f(\dot{\varepsilon},T)$ represents the flow curve and $\varepsilon_{fail}(T)$ represents the plastic strain to fracture. As discussed previously, eq. (10) proposed by Zerilli and Armstrong [5] describe completely the flow curve for BCC materials, with a thermal term determined only by the lattice parameter. Thus, we adopt eq. (10) for $\sigma_f(\dot{\varepsilon},T)$ in eq. (11), and use the thermal coefficients obtained by Zerilli and Armstrong for Armco Iron [5]. C_1, C_2 and C_3 should apply equally well to all to ferritic steels. Conversely, the athermal terms and coefficients ($\sigma_G + kd^{-1/2} + C_4\varepsilon^n$) are expected to depend on the microstructure of the particular steel considered. Furthermore, we held strain rate fixed at 1×10^{-4} / sec. To account for these material dependencies, we adopt the following values:

- ($\sigma_G + kd^{-1/2}$) = 500 MPa (a value selected to provide a variation of yield and ultimate strength with temperature consistent with that reported for RPV steels [7,8], as illustrated in Fig. 1.),
- C_4 as reported by Zerilli and Armstrong, and
- n = 0.08 (a typical value for RPV steels before irradiation).

While Zerilli and Armstrong provide a rigorous theoretical basis for the flow curve, the strain to fracture under the fully triaxial stress state that exists at the tip of a sharp crack is less well defined. Here we use Considère's construction [9] to define the variation with temperature of the uniform strain to failure under uniaxial loading

conditions (see Fig. (2)), and use this value as the limit of integration in eq. (11). Fig. 3 compares the predicted variation of plastic work with temperature to the median 1T Master Curve for weld 72W reported by Nanstad, et al. [7]. While the predicted temperature dependence of the plastic work term does not match the temperature dependence of the Master Curve, Fig. (3) demonstrates an increase of plastic work per unit volume with increasing temperature.

Two refinements to this calculation are expected to improve substantially the theoretical justification for a single Master Curve shape. In this paper we used the uniform strain from a uniaxial tensile test to set the integration limit of eq. (11) because such information is readily available. The uniform strain for a uniaxial test specimen may not provide an adequate representation of strain to fracture for crack-tip stress states. Further work is needed to quantify the combined effects of stress state and temperature on strain to fracture. Secondly, the plastic work values obtained by integrating the Z-A equation are work per unit volume of material. These values must be multiplied by the volume of material that participates in fracture (e.g. the plastic zone size, the process zone size, two grain diameters) to obtain a value of plastic work in better agreement with other values reported in the literature [10, 11].

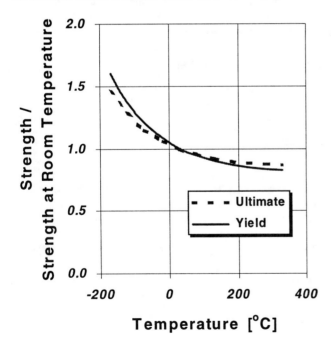

Fig. 1. Variation of yield and ultimate strength with temperature.

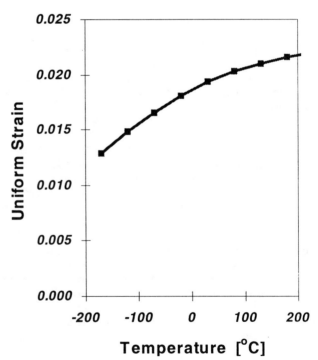

Fig. 2. Variation of the uniform strain under uniaxial loading with temperature.

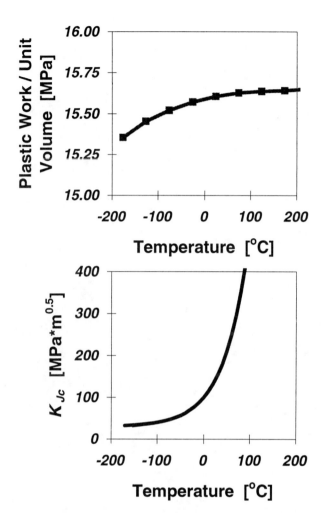

Fig. 3. The temperature dependence of plastic work per unit volume (top) contrasted with the temperature dependence of the Master Curve (bottom).

SUMMARY AND CONCLUSIONS

The information presented herein provides a physical basis for a single "master" toughness transition curve that applies with equal accuracy to all ferritic steels. The temperature dependence of the flow curve is controlled only by the lattice structure. Other factors that can vary with steel composition and heat treatment, such as the grain size and the work hardening rate, influence only the athermal term of the flow curve. Consequently, these factors influence the position of the transition curve on the temperature axis (i.e. T_o as determined by ASTM E1921-97), but *not* its shape. This understanding suggests that the breadth of applicability of Wallin's Master Curve to other microstructures is predictable based on the physics of the fracture process.

In an example calculation, we demonstrate that the plastic work to fracture, which gives the Master Curve its temperature dependence, can be expected to increase with increasing temperature. The temperature dependence of the plastic work to fracture depends on the temperature dependence of two variables: the flow curve and the strain to fracture. The temperature dependence of the flow curve is physically derived by Zerilli and Armstrong and has a sound theoretical basis for ferritic steels. By comparison, the physical basis for the variation with temperature of the strain to fracture requires further attention. Available data from the literature suggests that the temperature dependence may be similar for a variety of steels. Use of literature data in an example calculation predicts the expected increase of plastic work with increasing temperature. Further investigation is, however, needed to establish an appropriate physical basis for the variation of the plastic strain to fracture with temperature, and the effect of stress state on this property.

REFERENCES

[1] Wallin, Saario, and Törrönen, "Statistical Model for Carbide Induced Brittle Fracture in Steel," Metal Science, 18, pp. 13-16, 1984.

[2] ASTM E1921-97, "Test Method for Determination of Reference Temperature, T_o, for Ferritic Steels in the Transition Range," ASTM, 1998.

[3] Johnson, G.R., and W.H. Cook, "A Constitutive Model and Data for Metals Subjected to Large Strains, High Strain Rates and High Temperatures," Proc. 7th Int. Symp. On Ballistics, The Hague, the Netherlands, April 1983, 541-547

[4] Seaman, L., D.R. Curran, and D.A. Shockey, "Computational Models for ductile and Brittle Fracture," J. Appl. Phys., V47, N11 (1976) 4814-26.

[5] Zerilli, F.J. and R.W. Armstrong, "Dislocation-Mechanics-Based Constitutive Relations for Material Dyanmic Calculations," J of Appl. Phys., V61, N5 (1987) 1816-1825

[6] Nabarro, F., Proc. Phys. Soc. 1947, Vol. 59. p 256.

[7] Nanstad, R.K., et al., "Irradiation Effects on Fracture Toughness of Two High-Copper Submerged-Arc Welds, HSSI Series 5., NUREG/CR-5913, US-NRC, 1992.

[8] McGowan, et al., "Characterization of Irradiated Current-Practice Welds and A533 Grade B Class 1 Plate for Nuclear Pressure Vessel Service, NUREG/CR-4880, US-NRC, 1988.

[9] Considère, A., *Ann. Ponts et Chaussées*, Vol. 9, Ser. 6, pp. 574-775, 1885.

[10] Curry, D. and J. Knott, *Metal Science*, V12 (1978) 511-514

[11] Curry, D. and J. Knott, *Metal Science*, V13, (1979) 341-345

EXPERIMENTAL VALIDATION WORK TO PROVE THE MASTER CURVE CONCEPT

D. E. McCabe
M. A. Sokolov
Metals and Ceramics Division
OAK RIDGE NATIONAL LABORATORY
Oak Ridge, TN 37831

ABSTRACT

The introduction of elastic-plastic fracture mechanics and statistical methods into transition temperature characterization of ferritic steels has led to the Master Curve concept. Data scatter, specimen size effects, and a universal transition curve behavior have been identified and explained using a weakest-link statistical concept. This paper presents the experimental evidence to support these findings. However, the modeling that works successfully under most practical conditions does not apply under all scenarios. These limitations are currently being explored. The use of precracked Charpy specimens to produce viable fracture-mechanics data is one of the issues addressed.

INTRODUCTION

Fracture-mechanics analysis methods are currently being used to analyze nuclear reactor vessels under faulted conditions and for pressure/temperature combinations that are permitted during normal start-up and shutdown operations. Perhaps the most significant technical progress has been made during the past few years in improving knowledge of stress distributions during reactor upset transients, allowing more accurate stress-intensity-factor solutions for assumed crack sizes, shapes, and locations. However, the method used to establish fracture-mechanics-based fracture-toughness characterization of the materials being used has not kept pace with the analytical progress. Material fracture-toughness characterization has been handled by a semi-empirical scheme that was devised in the 1970-1972 time-frame (EPRI, 1993). At that time there was very little experience to draw upon, and the fracture-toughness characterization scheme devised was approximate in nature; that is, the methodology could not be expected to be permanent. Some fracture-mechanics concepts developed for use on aerospace materials had been adopted from American Society for Testing and Materials (ASTM) standard method E 399 (ASTM 1982a), using K_{Ic} and adhering to its validity requirements. This approach required huge specimens of reactor pressure vessel (RPV) steels when testing was to be performed in the transition range. Also, data scatter among replicate tests could be substantial, and control of this scatter could not be achieved by constraint condition management of the type that had worked successfully on aerospace materials. The resolution for these difficulties was to develop two working postulates (WRC, 1972). One postulate was that a lower bound curve, drawn beneath a fairly substantial collection of valid K_{Ic} data, could be used as a universal curve. By implication, this curve was expected to underlie the fracture-toughness data for all past and future production RPV steel plates and their weldments. It was also postulated that lower bound K_{Ic} estimates could be suitably indexed with respect to temperature by means of non-fracture-mechanics test methods (ASTM, 1982b; ASME, 1993). This approximate methodology has been in use with only minor modifications for almost 30 years.

New ideas began to emerge starting in about 1980, and there has been continuous progress made since, culminating recently in the Master Curve concept and ASTM standard method E1921-97 (ASTM, 1998). Advanced statistical methods and improved understanding of elastic-plastic test methods have been coupled to define a transition curve that is derived using only fracture-mechanics-based test data. The uncertainties associated with the empirical postulates that had to be employed since the 1970s are eliminated. The reason for excessive data scatter even under controlled constraint conditions can now be explained. Specimen size effect on material fracture toughness is better understood. Consequently, the definition of a transition temperature for a given material can be stated in terms

---

[1] Research sponsored by the Division of Engineering Technology, Office of Nuclear Regulatory Research, U.S. Nuclear Regulatory Commission, under Interagency Agreement DOE 1886-N695-3W with the U.S. Department of Energy under Contract DE-AC05-96OR22464 with Lockheed Martin Energy Research Corporation.

of the more accurate defining temperature, T_o. The key elements of the new technology are as follows.

1. Data scatter is recognized as resulting from randomly distributed cleavage crack-triggering sources contained within the typical microstructure of ferritic steel (Merkle et al., 1998). A three-parameter Weibull statistical model is used to suitably fit observed data scatter. Elastic-plastic fracture-toughness evaluation is expressed in units of a stress-intensity factor, K_{Jc}.

2. The J-Integral at the point of onset of cleavage instability, J_c, is calculated first and converted into its stress-intensity factor equivalent as K_{Jc}. It has been proven that sufficient control of constraint can still be maintained with specimens that are one-fortieth the size required for validity by ASTM E 399.

3. The specimen size effect observed in transition-range testing is quite subtle, and the most accurate modeling of this effect uses a weakest-link assumption, derived from the item 1 observation (Landes and Shaffer, 1980). The recognition of this model enables conversion of K_{Jc} data obtained from specimens of one size to K_{Jc} data for specimens of another size.

4. Use of the above three items has made it possible to observe that most ferritic steels tend to conform to one universal transition curve shape. Hence the existence of a universal "Master Curve" could be demonstrated (Wallin, 1989). Although the Master Curve is based on one selected specimen size, the universality property seems to extend without modification to engineering applications.

The objective of this paper is to present some of the experimental evidence in support of the Master Curve concept and ASTM standard E1921-97.

SPECIMEN SIZE EFFECTS IN THE TRANSITION RANGE

The specimen size effect that has been derived from a weakest-link theory has resulted in the following simple relationship:

$$K_{Jc(1)} = \left(K_{Jc(2)} - 20\right)\left(\frac{B_2}{B_1}\right)^{\frac{1}{4}} + 20, \quad MPa\sqrt{m} \qquad (1)$$

where
B_2 is the thickness of test specimens used to determine toughness $K_{Jc(2)}$,
B_1 is the thickness of prediction.

This method had been developed on the basis of fractographic evidence of cleavage crack triggering sources (Merkle et al., 1998). However, other specimen size effect models that use competing theories had to be considered during the development of standard method E 1921-97. One competing concept is the "local approach," which is in essence a variant of the "RKR" postulate (Beremin, 1983), in which cleavage K_{Jc} is controlled by a critical cleavage stress developing within a certain high-stress location in front of the crack tip. Another competing size effect theory is that the free borders of specimens disrupt the near crack tip stress/strain conditions that would develop normally in infinite bodies. The infinite body condition is referred to as "small-scale yield" (SSY) (Dodds et al., 1997). Finite-element modeling is used to identify and quantify free-border-impacted local crack tip J-integral toughness values. Experimentally determined J-integral values represent a "large-scale yield" (LSY) condition. These LSY-affected toughness values have been presented in parametric form. However, one particular case for the three-point-bend specimen has been mathematically modeled for pressure vessel steels, as follows (Anderson and Dodds, 1991):

$$J_{ssy} = \frac{J_c}{1 + 189\left(\frac{\sigma_{ys} b_o}{J_c}\right)^{-1.31}} \qquad (2)$$

where
b_o is the initial remaining ligament in the specimen,
J_c is the measured J-integral value influenced by LSY conditions,
$K_{Jc} = [J_c E]^{1/2}$.

Equation (2) applies to materials that have a Ramberg-Osgood work-hardening exponent of 10. It is specific to test specimens such as compacts and bend bars that are loaded under dominant bend conditions.

An example comparison of size effect prediction between Eqs. (1) and (2) is shown in Fig. 1. The example uses a postulated $K_{Jc(med)}$ value of 180 MPa$\sqrt{m}$ obtained using 1/2T size specimens. This represents an extreme case to distinctly display the difference in the two methods. Also, a one-to-one proportionality between b_o and B is

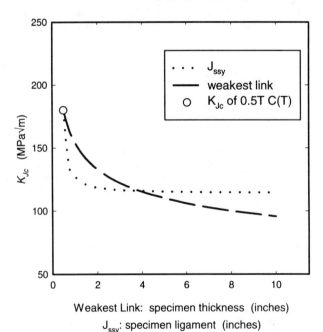

Fig.1. A comparison of the weakest-link and small-scale scale yielding size effect models keyed from a K_{Jc} median toughness value for an 0.5T C(T) specimen.

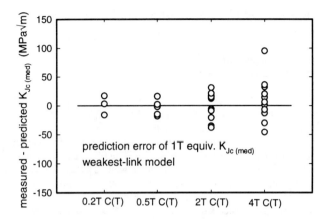

Fig. 2. Prediction error of 1T compact specimen median K_{Jc} using 0.2T, 0.5T, 2T, and 4T specimen K_{Jc} values and the weakest-link size effect model.

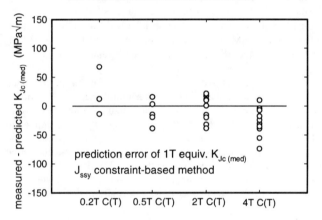

Fig. 3. Prediction error of 1T compact specimen median K_{Jc} using 0.2T, 0.5T, 2T, and 4T specimen K_{Jc} values and the small-scale yield model.

assumed. Figure 1, however, does not show which of the two agrees with experimental evidence. The comparison made with experimental evidence is displayed in Figs. 2 and 3. To prepare these figures, the data were obtained from experiments that had several sizes of specimens tested at one test temperature. A mandatory qualification for acceptable data sets was that at least six 1T specimens had been tested because this would be the reference size for the reference $K_{Jc(med)}$ value of the demonstration. Values of $K_{Jc(med)}$ for other specimen sizes were obtained and were converted into 1T equivalent $K_{Jc(med)}$ values using Eq. (1). These predicted values are plotted in Fig. 2. Figure 3 uses Eq (2) for the same purpose, again keeping in mind that there is a one-to-one proportionality between B and b_o.

The rule of at least six replicate specimens could not be used as qualification for tests at other specimen sizes. In particular, 4T size specimens are seldom ever replicated six times, and as few as two replicate tests had to be acceptable for qualification. This added greatly to the variability. Nevertheless, bias shown by the prediction-error distributions provides sufficient evidence for model evaluation. Bias is not evident in Fig. 2. Figure 3 shows some slight bias that

suggests that there might be a problem with the SSY stress/strain field free surface effect postulate.

Even though the weakest-link size effect model has performed effectively for size effect adjustments to test data, there is a need to be cautious about extending the use of the model beyond its range of applicability. For example, Eq (1) suggests that when the length of exposed crack-front material is extended excessively, it would be possible to reduce the fracture-toughness performance of a material to 20 MPa√m. This is not likely at mid-transition temperatures. McCabe and Merkle (1997) have suggested an idea on how and when this model should be truncated.

Another limitation on the weakest-link model is that the size effect phenomenon will vanish at lower-shelf test temperatures because of a change in the cleavage crack trigger mechanism. The weakest-link mechanism also vanishes as the upper-shelf test temperature is approached. Material toughness on upper shelf is characterized by R-curves, and a great deal of information is on record to show that there is no specimen size effect associated with R-curves (McCabe et al., 1993; McCabe and Ernst, 1983). Hence, the weakest-link effect will gradually vanish as upper shelf is approached. Figures 4 and 5 show why this characteristic of material behavior should not be ignored. Figure 4 has two full R-curves developed on a low upper-shelf material that was tested at a temperature only 25°C (45°F) below the upper shelf. Both tests had to be terminated after full R-curve development.

The purpose of Fig. 4 is to show that when specimens develop slow-stable crack growth prior to cleavage instability, the fracture properties can be affected by the side-groove practice. The data points shown came from tests made at temperatures that were 4, 25, and 50°C (7.2, 45, and 90°F) below the upper-shelf temperature. They represent slow-stable crack growth up to the point of cleavage fracture in some cases and the test termination point K_J values in other cases. Clearly, the specimen side-groove practice can influence K_{Jc} cleavage values when there is significant slow-stable crack growth. This becomes a problem as upper-shelf temperatures are approached.

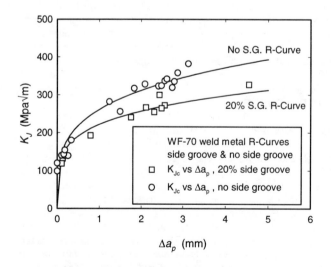

Fig. 4. Upper-shelf R-curves for side-grooved and non-side-grooved low upper-shelf weld metal. Data points are visually measured transition-range stable crack growth, Δa_p, vs. K_{Jc} of K_J values at test termination.

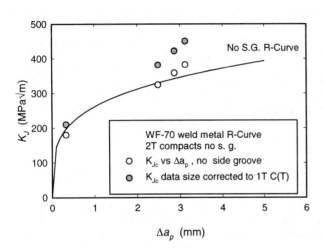

Fig. 5. The effect of size correction to K_{Jc} values on non-side-grooved 2T specimens when the test temperature was within 25°C (45°F) of the upper-shelf temperature.

Of particular interest are the data obtained from four 2T compact specimens that were not side-grooved. Figure 5 shows separately the data from these specimens before and after adjustment to 1T equivalence. Here, at a test temperature just 25°C (45°F) below the upper-shelf temperature, 0°C, the size correction elevates predicted 1T equivalent toughness to a level greater than the stable crack growth resistance capability of this material. Hence, size correction should not have been used. The HSSI 5th irradiation series had also contained evidence of the vanished weakest-link size effect situation in the case of two 8T size compact specimens tested at a temperature close to the upper shelf temperature (Nanstad et al., 1992). Again, specimen size adjustment to 1T equivalence gave excessively high predicted 1T equivalent fracture toughness. Prior slow-stable crack growth up to onset of cleavage fracture was 0.033 and 0.075 in. (0.84 and 1.9 mm) in the two 8T specimens, giving evidence of the proximity to upper-shelf temperature, whereas most of the other test specimens in that program had negligible prior slow-stable crack growth. Side grooving was not used in the 5th irradiation test series.

MASTER CURVE

Even though the concept of universal transition curves has been accepted and used in the ASME code for almost 30 years, the concept of one universal median transition curve applicable to one specimen size (Master Curve) has been received with some skepticism. The ASME lower-bound K_{Ic} curve was established with 11 materials; however, the data from only one of these materials provided the lower values that defined the position and shape of the curve. The ASME lower-bound dynamic K_{Ia} curve was established with only two materials. Figure 6 represents the data available to verify the Master Curve (Merkle et al., 1998). This plot shows median fracture-toughness values for 18 steels, representing weld metals and base metals, in both unirradiated and irradiated conditions. The curve that seems to fit these median K_{Jc} values is the Master Curve as defined in ASTM standard method E1921-97:

$$K_{Jc(med)} = 30 + 70 \cdot \exp[0.019(T - T_o)] \quad \text{MPa}\sqrt{m} \quad (3)$$

where
  T = test temperature,
  T_o = reference temperature.

Figure 7 is shown to compare the two methodologies (Sokolov, 1998). The currently used ASME lower-bound K_{Ic} curve, shown in Fig. 7(a), had utilized only a few of the plotted data points (i.e., the lowest values of K_{Ic}) to arrive at a transition curve shape. Despite the apparent basis of eight materials, the HSST Plate 02 data seem to have controlled the lower-bound curve development. The Master Curve method as applied to setting lower-bound coverage is shown in Fig. 7(b), along with the ASME curve. In the case of the lower bound derived from the Master Curve, all of the data shown were used to arrive at a transition-curve position. For the ASME curve to be close to a lower bound, a reference temperature based on T_o was used. A 35°F (19.4°C) margin had to be added.

In the draft ASME code case for replacing RT_{NDT} with RT_{T0} determined by the Master Curve procedure, RT_{T0} is calculated by adding 35°F (19.4°C) to T_o. A technical basis document pertaining to applications of the Master Curve approach has been published (EPRI, 1998), but exactly how the 35°F (19.4°C) temperature shift was selected is not explained.

Sokolov and Nanstad (1999) examined data from irradiated materials to determine the effect of irradiation damage on the Master Curve shape (see Fig. 8). In this case, each datum is plotted instead of median values as in Fig. 6. Least-squares curve fitting was applied to find the best curve shape. The best fit had a coefficient inside of the Eq. (3) exponential term set at 0.017. When linear regression is used on the Fig. 6 data, the best coefficient is 0.018. In both cases, the differences were not significant from a practical point of view, and alteration of Eq. (3) is not necessary for the irradiated data examined.

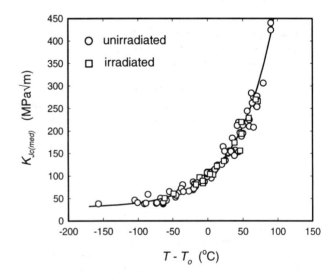

Fig. 6. Currently accumulated median K_{Jc} transition-range data for 18 materials plotted with the ASTM E1921-97 Master Curve.

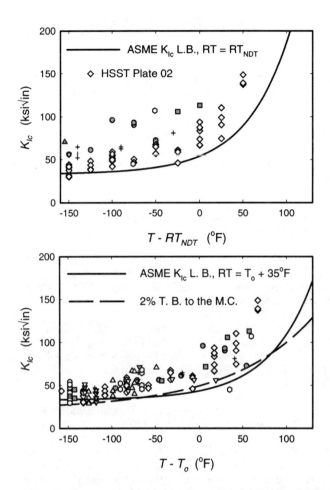

Fig. 7. Comparison of lower bounding of the (top) EPRI K_{Ic} data base values, and (bottom) comparing the ASME method *vs.* the Master Curve method

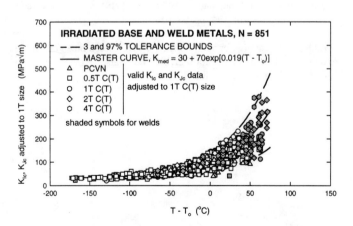

Fig. 8. The irradiation fracture-toughness data size-adjusted to 1T equivalence and normalized to T_o with the Master Curve and 3% and 97% tolerance bounds. E1921-97 valid data only.

DEVELOPMENT OF T_o FROM PRECRACKED CHARPY SPECIMENS

The observation of a weakest-link behavior in steels coupled with elastic-plastic analysis methods has permitted a large reduction in specimen size requirements so that specimens of practical size for laboratory testing can now be used. A practical specimen size for surveillance capsule work is defined as the precracked Charpy specimen. The rationale for this assertion is that in many cases surveillance capsules contain only Charpy and tensile specimens, and only Charpy specimens are currently available for fracture-toughness evaluations. However, the precracked Charpy specimen is of marginal size to be a viable specimen for the development of fracture-mechanics data. The challenge of developing test procedures to produce viable data from such specimens has been undertaken in several projects by various groups. This work is currently in progress, and only preliminary evidence is available at the present time. Figure 9 summarizes the present accumulation by the authors of available information. There is sufficient cause for optimism, but it is premature to conclude that T_o temperatures can be determined from such small specimens without modification to the test procedure. Figure 9 appears to show some bias tendencies because most of the Charpy-generated T_o values appear to be on the high-toughness side of the one-to-one correlation line.

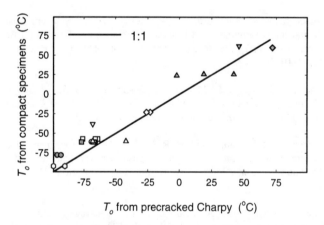

Fig. 9. Comparison of reference temperature, T_o, determination. Compact specimens *vs.* precracked Charpys.

CONCLUSION

Specimen size effects developed in the mid-transition range are most accurately defined by the model derived from weakest-link theory. The mechanism tends to break down at lower-shelf temperatures and at temperatures approaching upper shelf. Specimen size effects tend to vanish as these two conditions are approached.

The Master Curve method of defining ductile-to-brittle behavior of steels has advanced the concept of universal transition curves from speculation to a concept that is supported by theoretical reasoning and proven with ample supporting experimental data. The ASME code has been relying on a universal fracture-toughness curve concept as a working postulate for almost 30 years. The present Master Curve finding suggests that the ASME lower-bound K_{Ic} curve does not have the correct shape. Size effects on fracture toughness and the effect of irradiation damage on transition-curve shape have been unresolved issues for the ASME code curves. The refinement of curve shape definition through introduction of the Master Curve concept has permitted a more accurate evaluation of irradiation damage effects on universal curve shape. For the data base evaluated, the current evidence is that curve shape does not change as a consequence of irradiation.

The precracked Charpy specimen is currently being evaluated as a potentially viable fracture-mechanics specimen. Although there is insufficient evidence for a final conclusion, some bias in T_o prediction appears to be likely.

REFERENCES

American Society for Testing and Materials, 1982a, "Standard Test Method for Plane Strain Fracture Toughness of Metallic Materials," Annual Book of ASTM Standards, Part 10, E399-81.

American Society for Testing and Materials, 1982b, "Standard Method for Conducting Drop-Weight Test to Determine Nil-Ductility Transition Temperature of Ferritic Steels," Annual Book of ASTM Standards, Part 10, E208-81.

American Society for Testing and Materials, 1998, "Standard Test Method for Determination of Reference Temperature, T_o, for Ferritic Steels in the Transition Range," Annual Book of ASTM Standards, Vol. 03.01, E1921-97.

American Society of Mechanical Engineers, 1993, "Fracture Toughness Requirements for Material," ASME Boiler and Pressure Vessel Code, Sect. III, NB 2300, NB-2321.1 & NB-2321.2.

Anderson, T. L., and Dodds, R.H., Jr., 1991, "Specimen Size Requirements for Fracture Toughness Testing in the Ductile Brittle Transition Region," Journal of Testing and Evaluation, Vol. 19, pp. 123-34.

Beremin, F. M., 1983, "A Local Criterion for Cleavage Fracture of a Nuclear Pressure Vessel Steel," Metallurgical Transactions, 14A, pp. 2277-87.

Dodds, R. H., Jr., Ruggieri, C., and Koppenhoefer, K. C., 1997, "3D Effects on Models for Transferability of Cleavage Fracture Toughness," ASTM STP 1321, American Society for Testing and Materials, pp. 179-97.

Electric Power Research Institute, 1993, "White Paper on Reactor Vessel Integrity Requirements for Level A and B Conditions," ASME Section XI Task Group on Reactor Vessel Integrity Requirements Final Report, EPRI TR-100251, Electric Power Research Institute, Palo Alto, CA.

Electric Power Research Institute, 1998, "Application of Master Curve Fracture Toughness Methodology for Ferritic Steels," TR-108390, Electric Power Research Institute, Palo Alto, CA.

Landes, J. D., and Shaffer, D. H., 1980, "Statistical Characterization of Fracture in the Transition Region," ASTM STP 700, American Society for Testing and Materials, pp. 368-82.

McCabe, D. E., and Ernst, H. A., 1983, "A Perspective on R-Curves and Instability Theory," ASTM STP 791, Vol. 1, American Society for Testing and Materials, pp. 561-84.

McCabe, D. E., Landes, J. D., and Ernst, H. A., 1983, "An Evaluation of the J_R-Curve Method for Fracture Toughness Characterization," ASTM STP 803, Vol. II, American Society for Testing and Materials pp. 562-81.

McCabe, D. E., and Merkle, J. G., 1997, "Estimation of Lower-Bound K_{Jc} on Pressure Vessel Steels From Invalid Data," ASTM STP 1321, American Society for Testing and Materials, pp. 199-213.

Merkle, J. G., Wallin, K., and McCabe, D. E., 1998, "Technical Basis for an ASTM Standard on Determining the Reference Temperature, T_o, for Ferritic Steels in the Transition Range," NUREG/CR-5504, U.S. Nuclear Regulatory Commission.

Nanstad, R. K., Haggag, F. M., McCabe, D. E., Iskander, S. K., Bowman, K. O., and Menke, B. H., 1992, "Irradiation Effects on Fracture Toughness of Two High-Copper Submerged-Arc Welds, HSSI Series 5," NUREG/CR 5913, Vol. 1, U. S. Nuclear Regulatory Commission.

Sokolov, M. A., 1998, "Statistical Analysis of the ASME K_{Ic} Data Base," Transactions of the ASME, Vol. 120, pp. 24-27.

Sokolov, M. A., and Nanstad, R. K., 1999, "Comparison of Irradiation-Induced Shifts of K_{Jc} and Charpy Impact Toughness for Reactor Pressure Vessel Steels," Effects of Radiation on Materials: 18th International Symposium, ASTM STP 1325, American Society for Testing and Materials, pp. 167-90.

Wallin, K., 1989, "A Simple Theoretical Charpy V-K_{Ic} Correlation for Irradiation Embrittlement," 1989 PVP/JSME Co-sponsored Conference, PVP Vol. 170, pp. 93-100.

Welding Research Council, 1972, "PVRC Recommendations on Toughness Requirements for Ferritic Materials," ad hoc group on toughness requirements, WRC Bulletin 175.

ON THE USE OF THE MASTER CURVE BASED ON THE PRECRACKED CHARPY SPECIMEN

Rachid Chaouadi, Marc Scibetta, and Eric van Walle
SCK·CEN
Boeretang 200
B-2400 Mol, Belgium

Robert Gérard
Tractebel Energy Engineering
Avenue Ariane, 7
1200 Bruxelles, Belgium

ABSTRACT

Recently, a large worldwide has focused on evaluation of the use of the Master Curve approach to characterize fracture toughness of ferritic steels in the transition regime. This was acknowledged by the recent release of the ASTM Standard Test Method for Determination of Reference Temperature, T_0, for Ferritic Steels in the Transition Range (E1921). The present work aims to investigate the use of the Charpy specimen along with the Master Curve approach to derive the fracture toughness behavior of reactor pressure vessel steels. Therefore, four well characterized and documented reactor pressure vessel steels were selected. A large experimental program to measure fracture toughness with Charpy size specimens was carried out. Four important aspects were investigated:

- the T_0 determination as a function of test temperature;
- the E1921 specimen size requirement (factor M=30);
- the censoring procedure for specimens not satisfying the E1921 size requirements;
- the estimation of the fracture toughness lower bound, and its comparison to the ASME K_{IC} curve.

It is found that within the experimental and statistical uncertainties, the reference temperature T_0 is not affected by the test temperature, even when data are not valid according to E1921 requirements. By application of the censoring procedure, the determination of the reference temperature may lead to non conservative results. Comparison to larger specimen size suggests the use of M=60 rather than 30 to limit the loss of constraint in agreement with finite element calculations. Nevertheless, the differences are not large enough to be statistically significant. The lower bound based on the Master Curve is very close to the experimental lower bound, while the ASME K_{IC} curve tends to be over conservative. Replacing RT_{NDT} by the new index, RT_{T_0}, in the ASME K_{IC} equation reduces this over conservatism.

INTRODUCTION

Characterization of fracture toughness in the transition regime suffers from the size dependence of the results. Indeed, the fracture toughness behavior is affected by two important inter-related phenomena: the statistical size effect and loss of constraint.

SCK•CEN has put an extensive effort in characterizing fracture toughness behavior using Charpy size specimens. The choice of the Charpy geometry to measure fracture toughness is motivated by the following two important points:

- The Charpy specimen is used for monitoring reactor pressure vessel degradation induced by neutron irradiation;
- Reconstitution technology allows re-fabrication of new samples from broken Charpy halves, increasing the number of specimens.

Therefore, various well documented and characterized reactor pressure vessel steels were selected in order to demonstrate the reliability of the methodology. The main geometry is the PreCracked Charpy v-notch (PCCv) specimen (10×10×55 mm). In addition, the 3 Point Bend (3PB) specimen (W=20 mm, B=15 mm, 20% side grooved) and the Compact Tension (CT) specimen were tested for comparison.

The main objective of this work is to investigate the use of the Master Curve, developed by Wallin [1], for fracture toughness characterization of the transition regime. In particular,

- how does the reference temperature, T_0, vary with test temperature?
- is the loss of constraint well evaluated through the factor M=30?
- does the censoring procedure for specimens not satisfying the E1921 [2] size requirements lead to consistent results?
- how does the ASME K_{IC} lower bound curve compare to experimental data and to predicted curves?

Table 1. Chemical composition.

material	C	Si	P	S	Cr	Mn	Ni	Cu	Mo
22NiMoCr37	0.22	0.23	0.006	0.004	0.39	0.88	0.84	0.08	0.51
JRQ	0.18	0.24	0.017	0.004	0.14	1.42	0.84	0.14	0.51
A533B (JSPS)	0.24	0.41	0.028	0.023	0.08	1.52	0.43	0.19	0.49
73W	0.10	0.45	0.005	0.005	0.25	1.56	0.60	0.31	0.58

Table 2. Mechanical properties at ambient temperature.

material	orientation	yield stress σ_y (MPa)	tensile strength σ_u (MPa)	elongation ε_t (%)	reduction of area RA (%)
22NiMoCr37	L	455	609	20	73
JRQ	T	484	625	19	73
A533B (JSPS)	T	460	640	18	58
73W	T; L; S	490	600	22	68

MATERIALS

Four well documented and characterized reactor pressure vessel steels were selected:

- **22NiMoCr37**: This material was extensively investigated, in particular through two round robin exercises related to the local approach to fracture, and to fracture toughness in the ductile to brittle transition regime [3].
- **A533B Cl.1(JSPS)**: This steel was used in the Japanese round robin organized by the Japan Society for the Promotion of Science [4]. It has a low upper shelf energy (70J) as a result of high S and P concentrations.
- **A533B Cl.1(referred as JRQ)**: This IAEA monitor material is well known within the IAEA community. It was also used in a round robin related to fracture toughness [5].
- **73W Weld**: This weld was extensively characterized by ORNL [6], in the un-irradiated as well as the irradiated condition.

The chemical composition and the mechanical properties are reproduced in Table 1 and 2, respectively.

For each steel, at least ten tensile specimens were tested over the temperature range [-200 to +300 °C]. The variation of yield stress, σ_y, with test temperature is fitted with the following equation:

$$\sigma_y = A + B \exp [C\,T]$$

where σ_y is expressed in MPa and the temperature T in °C. The constants A, B and C for each material are given in Table 3. These equations will be used later to determine the maximum K_J capacity of the various specimens. The strain hardening exponent, n, is obtained from the yield to ultimate stress ratio, using the following expression [7]:

$$\frac{\sigma_y}{\sigma_u} = \exp(n) \left[\frac{0.002 + \frac{\sigma_y}{E}}{n} \right]^n$$

Table 3. Fitting constants for the evolution of yield stress with temperature.

material	N	A	B	C
22NiMoCr37	10	409.37	55.50	-0.0118
JRQ	12	443.55	39.26	-0.0127
A533B (JSPS)	17	423.57	40.82	-0.0125
73W	19	466.62	35.73	-0.0129

Notes: N is the number of specimens tested over the temperature range [-200 to +300 °C].

The material is assumed to follow a power law type behavior. The variation of strain hardening exponent with test temperature is fitted with the following polynomial equation:

$$n = A_0 + A_1\,T + A_2\,T^2 + A_3\,T^3 + A_4\,T^4 + A_5\,T^5$$

where the constants A_i are given in Table 4. Figure 1 shows the evolution of the yield stress and the strain hardening exponent as a function of temperature.

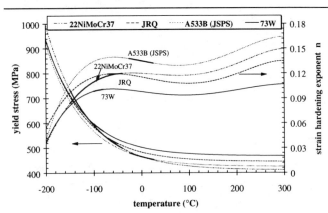

Figure 1. Yield stress and strain hardening coefficient as a function of temperature. Temperature range for K_{Jc} tests **in bold**.

Table 4. Fitting constants for the strain hardening exponent.

material	A_0	$A_1 (\times 10^{-4})$	$A_2 (\times 10^{-7})$	$A_3 (\times 10^{-9})$	$A_4 (\times 10^{-11})$	$A_5 (\times 10^{-14})$
22NiMoCr37	0.120	-0.340	-5.683	7.524	-0.903	-1.623
JRQ	0.116	-1.132	-3.012	7.460	-1.069	-1.419
A533B (JSPS)	0.135	-1.145	-0.477	9.151	-2.510	0.981
73W	0.098	-0.783	-0.257	5.249	-1.797	1.583

Table 5. Impact properties.

material	orientation	$N^{(1)}$	USE (J)	T_{68J} (°C)	T_{41J} (°C)	T_{28J} (°C)	$T_{50\%}$ (°C)	$T_{0.89mm}$ (°C)	RT_{NDT} (°C)
22NiMoCr37	L-S	12	194	-38	-50	-58	-20	-43	-40 [8]
JRQ	T-L	12	215	-4	-19	-30	20	-11	-15 [9]
A533B (JSPS)	T-L	19	72	101	34	16	43	46	$68^{(2)}$
73W	T-L	12	142	-18	-34	-43	-12	-28	-34 [6]

Notes: (1) N is the number of tested specimens.
(2) RT_{NDT} is estimated from the Charpy impact data: RT_{NDT} = max $(T_{68J}; T_{0.89mm})$ - 33°C [10]. Pellini drop weight tests were not performed for A533B (JSPS).

The impact properties derived from Charpy tests are given in Table 5. The transition behavior of the impact energy is shown in Figure 2. This Figure clearly illustrates the differences between the four steels, in particular the upper shelf energy level and the various transition temperatures. In the last column, the Reference Temperature of Nil-Ductility Transition, is also given. The latter is controlled by the Pellini drop weight test except for A533B (JSPS) for which only Charpy impact data were available.

EXPERIMENTAL

Most of the fracture toughness tests are performed on precracked Charpy specimens loaded in three point bending. The tests are carried out in an Instron tensile machine under controlled displacement at a constant crosshead speed of 0.2 mm/min. The specimens are precracked by fatigue up to a "crack length-to-width" ratio close to 0.5. For most specimens, the stress intensity factor during fatigue is maintained below 25 MPa√m. During the last 0.6 mm of precracking, the stress intensity factor is kept below 18 MPa√m. A few samples were precracked above these values but fracture occurs well above the $K_{fatigue}$. Except for the 22NiMoCr37 samples, all specimens are 20% side grooved prior to testing. The specimen is maintained at the test temperature for at least 30 minutes prior to loading. During this period and during testing, the temperature remains well within ±1°C of the nominal value. The specimen is immediately unloaded after unstable crack initiation corresponding to cleavage fracture. The specimens were heat tinted at 290°C for 20 minutes and broken at liquid nitrogen temperature. The initial crack length and the eventual ductile crack extension prior to cleavage are measured on the broken surface using the 9-point method.

The J-integral is calculated from:

$$J = \frac{K^2(1-\nu^2)}{E} + \frac{\eta U_{pl}}{B_n(W-a_0)}$$

where $\eta = 2$ † for SENB geometry and $\eta = 2 + 0.522(1 - a_0/W)$ for the CT specimen. a_0 is the initial crack length, B_n and W are respectively the specimen net thickness and width. U_{pl} is derived from area under the load versus load line displacement curve from which the elastic part is subtracted. For the SENB geometry, the load line displacement (LLD) is derived from the crack mouth opening displacement (CMOD) using the following relation:

$$LLD = \frac{CMOD \quad S/4}{r_p(W-a_0) + a_0 \cos\left[a\tan\left(\frac{CMOD/2}{r_p(W-a_0)+a_0}\right)\right]}$$

which can be approximated by:

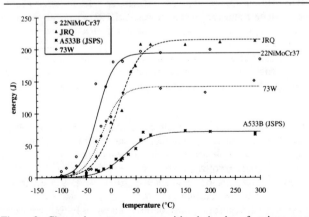

Figure 2. Charpy impact energy transition behavior of various reactor pressure vessel steels. Note that the A533B (JSPS) material exhibits a low upper shelf energy and a high transition temperature.

---

† The E1921 recommends use of $\eta=1.9$. However, as will be indicated later, a better agreement is found with $\eta=2$ when LLD is derived from CMOD.

$$LLD \approx \frac{CMOD \quad S/4}{r_p(W-a_0)+a_0}$$

where S is the span and the rotational factor $r_p=0.44$.

An equivalent elastic plastic stress intensity factor K_{Jc} is then derived from the J-integral using:

$$K_{Jc} = \sqrt{\frac{JE}{1-\nu^2}}$$

where ν is the Poisson ratio ($\nu=0.3$) and E is the elastic modulus (E=210 GPa for 22NiMoCr37 and JRQ, E=206 GPa for A533B (JSPS) and E=207-0.06 T(°C) for 73W [6]). Here, the plane strain rather than plane stress elastic modulus is used. This is more appropriate as plane strain conditions are prevailing at the crack tip. For consistency, all the data reported here are evaluated using this relation (PCCv as well as CT).

3D finite element calculations performed by Nevalainen and Dodds [11] and Koppenhoefer and Dodds [12] suggest the following η values for the deep cracked SENB geometry: $\eta=1.9$ for plain sided, and $\eta=2$ for side grooved samples. By using the above relations for J, LLD and K_{Jc}, and taking $\eta=2$, finite element calculations have shown a very good agreement with K based on the J contour integral. In the range of interest, the fracture toughness is underestimated by less than 2% (conservative side).

Two validity conditions are given in the ASTM E1921 standard [2] in relation to the specimen size:

- the maximum K_J level relative to the specimen size:

$$K_{J\,limit} = \sqrt{\frac{\sigma_y E(W-a_0)}{M}}$$

where, according to E1921, M=30.

- The maximum ductile crack extension:

$$\Delta a_{max}^{ductile} = 0.05(W-a_0)$$

Specimens that do not satisfy these requirements can still be used in a censoring procedure. This is done by replacing the K_{Jc} values by the $K_{J\,limit}$ and calculating K_0 using:

$$K_0 = \left[\sum_{i=1}^{N}\frac{(K_{Jc(i)}-20)^4}{r-(1-\ln(2))}\right]^{1/4} + 20$$

where N is the total number of tests and r the number of 'valid' data.

Due to the limited number of specimens usually tested, a confidence interval should be associated with T_0. In the ASTM E1921 standard, the standard deviation for estimates of T_0 is approximated by $\beta/\sqrt{N}$. For a 95% confidence level, the ΔT_0 temperature deviation is:

$$\Delta T_0 = \frac{\beta}{\sqrt{N}}\alpha = \frac{1.6449 \times \beta}{\sqrt{N}}$$

where β varies between 18 and 22 depending on K_{Jmed} [2]. This relation was verified using experimental data and a Monte Carlo simulation [3].

One of the main advantages of using the Master Curve is that it provides confidence bounds. This is extremely important when safety assessment issues are addressed. For a specific level of probability, the fracture toughness curve is given by:

$$K_{X\%} = C_1 + C_2 \exp[0.019(T-T_0)]$$

where C_1 and C_2 are constants: $C_1=23.5$ and $C_2=24.5$ for X=1% and $C_1=25.4$ and $C_2=37.8$ for X=5%

The best known lower bound curve is probably the K_{IC} ASME curve which is indexed by the RT_{NDT}. This curve is given by [9]:

$$K_{IC} = 36.48 + 22.78 \exp[0.036(T-RT_{NDT})]$$

However, this curve is found too conservative in many cases. Recently, Kirk et al. [13] have assembled a large database of RPV fracture toughness data suggesting that the replacement of RT_{NDT} by a T_0-based index, RT_{T_0} ($RT_{T_0}=T_0+19.4°C$) in the ASME lower bound curve leads to a better agreement with experimental results.

RESULTS

All specimens were analyzed following a similar procedure. Most of the ASTM E1921 testing recommendations were fulfilled as well as the evaluation procedure, in particular for determining the reference temperature T_0. The reference toughness values (K_0) that are given in this paper refer to 1T adjusted values. Additionally, the shape parameter, or Weibull modulus for measuring the scatter is also determined using the maximum likelihood method. However, m was taken equal to 4 for the determination of T_0 and K_0.

The Master Curve approach is then applied to determine the reference temperature following the ASTM E1921 standard procedure. The fracture toughness test data are first normalized to one reference size, namely 1T (25 mm). The use of a small specimen generally results in significant loss of constraint. Data that are outside the validity range can still be used by applying an adequate censoring procedure. Here, four different censoring procedures are used for T_0 determination:

Method 1. all data are taken into account: no censoring is applied;
Method 2. the censoring procedure is applied following the E1921 recommendations: N is the total number of tests, r is the number of valid tests;
Method 3. the censoring procedure is similar to E1921 one except that r is substituted by N;
Method 4. the censoring procedure is similar to E1921 one except that N is substituted by r: only 'valid' data are considered.

For each material, three aspects of the Master Curve approach will be investigated. First, the comparison of the fracture toughness data to the Master Curve and associated probability bounds. Second, as at least two test temperatures were selected, how test temperature affects the reference temperature determination. Finally, the ASME K_{IC} lower bound curve based on the RT_{NDT} is compared to the 5% and 1% probability curves, and to the new proposal for replacing RT_{NDT} by RT_{To} [13].

22NiMoCr37 material

This testing program (see Table 6) was selected for comparison with the European round robin data [14]. Specimens are taken in the L-S orientation in the 1/4 to 3/4 thickness of the plate. All PCCv specimens are plain sided except 6 samples tested at -110°C (20% side grooving). Additionally, two series of nine 0.5T-CT plane sided specimens are tested for comparison.

Table 6. Test matrix: 22NiMoCr37.

test temperature	PCCv	CT
-150	8	
-125	7	
-110	12	
-90	8	9
-60	9	9
-40	8	

All specimens are plain side except 6 PCCv tested at 110°C that were 20% side grooved.

Most of the 22NiMoCr37 specimens do not satisfy the crack front straightness requirement of E1921. This is mainly due to the small size of the specimens and the absence of side grooving. According to E1921, the data that do not satisfy the crack front straightness should be discarded. The requirement is found to be too severe. Indeed, as will be shown later, this does not seem to significantly effect the results.

The observation of the fracture surfaces of small samples in order to determine whether single or multiple initiation occurred is more difficult than with large samples. At -150°C, the fracture surface is flat and no single initiation site could be localized. Between -125 and -91°C, the fracture surface becomes more irregular, but a single initiation site is still difficult to identify. Above -60°C, a single crack initiation site is found.

The results in terms of fracture toughness versus temperature are shown in Figure 3. All data are adjusted to 1 inch (25 mm) thickness. The 1T-adjusted $K_{J\ limit}$ for each geometry is also shown. All specimens failed by cleavage except two specimens tested at -40°C which were interrupted after significant ductile crack extension. A large number of specimens failed above the limit K_J-level according to E1921, necessitating data censoring.

In the test temperature range where loss of constraint is limited, the number of specimens is chosen to comply with the E1921 requirement. Therefore, 8 samples are tested at -150°C while at -125°C, 7 samples are tested. If crack front uniformity is not taken into consideration, all these data can be considered as valid according

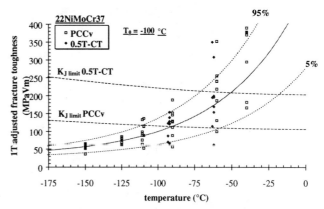

Figure 3. Fracture toughness behavior of 22NiMoCr37 in the transition regime is well described by the Master Curve and associated probability bounds.

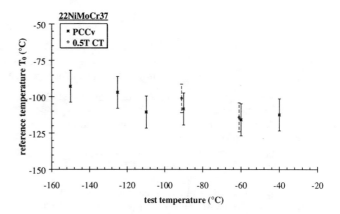

Figure 4. Effect of test temperature on the reference temperature T_0. No significant effect is found.

to E1921 requirements. At -110°C, twelve PCCv samples are tested among which 6 were 20% side grooved. Here, two samples do not satisfy the size requirement. When no censoring is applied, T_0=-110 °C. By censoring according to E1921, T_0 does not change. If r is considered instead of N in the K_0 relation, T_0=-107. Finally, if only the 10 'valid' data points are considered, T_0=-103°C. At higher test temperatures, the number of 'valid' data decreases such that the E1921 requirement on the minimum number of 'valid' data point is not fulfilled anymore. Consequently, the censoring procedure cannot be applied. Nevertheless, T_0 does not vary much in comparison to that determined at lower test temperatures. Figure 4 shows the results of Table 7 including their confidence interval, ΔT_0. It is important to emphasize that the differences observed here are not statistically significant. Globally, the reference temperature T_0 is within the expected confidence limit. The reference temperature based on these data can be estimated as T_0=-100°C (based on specimens tested at -150, -125°C and -110°C). Table 7 summarizes the results of this evaluation of the reference temperature using various data sets. The K_0 and Weibull slope (scatter parameter), m, is also given. For each data set, the number of data satisfying the E1921 size requirements is indicated together with the censoring procedure that is applied.

Table 7. Parameters characterizing fracture toughness of 22NiMoCr37 in the transition regime.

specimen	N	T	T_0	K_0	m	method	remark
	8	-150	-92.7	56.8	5.1	1	non uniform crack front
	7	-125	-97.0	76.0	7.5	1	non uniform crack front
	12	-110	-110.4	108.3	3.6	1	10 PCCv satisfy $K_{J\,limit}$
	12	-110	-107.4	104.0	4.1	3	censoring: r=N=12
PCCv	12	-110	-110.3	108.1	4.1	2	censoring E1921: N=12, r=10
	10	-110	-103.2	98.4	4.1	4	censoring: N=r=10
	8	-90	-109.6	142.2	3.1	1	3 PCCv satisfy $K_{J\,limit}$
	9	-60	-115.8	252.6	3.2	1	1 PCCv satisfies $K_{J\,limit}$
	8	-40	-114.7	348.4	4.5	1	all not valid (2 ductile)
	6	-40	-112.4	334.5	3.7	1	2 ductile discarded
	9	-91	-101.1	123.9	3.8	1	all valid per E1921
	9	-60	-114.2	245.7	2.2	1	7 CT satisfy $K_{J\,limit}$
CT	7	-60	-93.0	174.6	3.5	4	censoring: N=r=7
	9	-60	-96.4	184.3	4.0	3	censoring: r=N=12
	9	-60	-100.1	195.4	4.0	2	censoring E1921: N=9, r=7

Notes: N is the number of specimens, T is the test temperature, T_0 is the reference temperature, K_0 the normalization fracture toughness and m the Weibull modulus.

Comparison with 0.5T CT specimens tested by [15] at -91 and -60°C show that the results closely agree with those of the PCCv samples (see Table 7 and Figure 4). CT samples result in T_0=-101°C. Note that at -60°C, one CT specimen exhibited a very low K_J-value. Unfortunately, the number of available specimens is too limited to perform 12 additional tests as requested by E1921.

As already mentioned, the PCCv specimens are not side grooved. This was deliberately chosen in order to be compared with the test results generated within the European round robin on the same material [3]. As a consequence, most of PCCv samples exhibit a non uniform crack front. According to E1921, these data should be discarded. However, Table 7 shows that they are in very good agreement with CT specimens.

In Figure 5, the measured fracture toughness data are plotted as a function of test temperature. These data are compared to the 5% and 1% curves based on T_0=-100°C and the K_{IC} ASME curves indexed by RT_{NDT} and RT_{To}. If the CT sample that exhibited a very low toughness is ignored, the agreement between experimental data and these lower bound curves is very good. However, the RT_{NDT} based ASME curve is very conservative in comparison to the other curves. Note the excellent agreement between the 1% curve and the RT_{T0} based K_{IC} ASME curve in the [-150 to -75 °C] test temperature range.

JRQ material

This plate material, fabricated in Japan, is the monitor material chosen by the IAEA, in particular within the CRP-III and IV research programs [5,16]. To reduce the material variability, all specimens were taken from the 1/4 and 3/4 thickness. Three geometries are tested beside tensile and impact Charpy samples: PCCv, 3PB, and CT. The specimens are taken in the T-L orientation in the 1/4 to 3/4 thickness of the plate. All specimens are 20% side grooved. Two test temperatures are selected: -70 and -50°C. At least 6 samples were tested per temperature. The test matrix is given in Table 8.

Table 8. Test matrix: JRQ.

test temperature	PCCv	3PB	1/2T-CT
-70	6	6	9
-50	6	6	11

Notes: All specimens are 20% side grooved.
PCCv and CT are taken from the CRP-IV block.
3PB are taken from the ESIS block.

Determination of the Master Curve parameters is performed according to E1921. The results are summarized in Table 9. For the PCCv, the reference temperature, calculated using PCCv tested at -70°C, is T_0=-54°C. For the CT samples, all data satisfy the E1921 requirements. However, at -70°C, one sample exhibited a very high

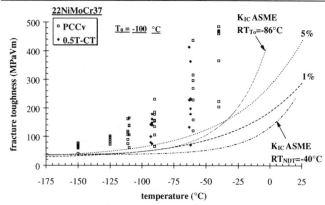

Figure 5. Comparison of fracture toughness data to lower bound curves: ASME K_{IC} versus probability curve.

Table 9. Master curve transition parameters for JRQ material.

geometry	N	T	T_0	K_0	m	method	remark
PCCv	6	-70	-53.8	87.4	2.4	1	valid per E1921
	6	-50	-63.2	129.5	5.7	1	2 PCCv satisfy $K_{J\,limit}$
CT	11	-70	-54.6	88.2	2.9	1	valid per 1921
	10	-70	-41.9	76.0	4.6	1	valid per 1921, outlier discarded
	9	-50	-43.7	99.0	3.5	1	valid per 1921
3PB	6	-70	-50.7	84.2	7.0	1	valid per E1921
	6	-50	-64.1	131.2	3.2	1	valid per E1921

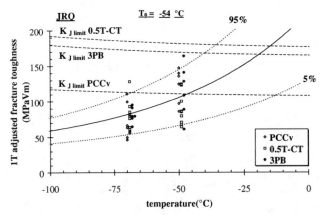

Figure 6. Master curve and associated confidence bounds for JRQ plate material.

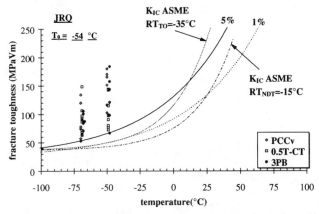

Figure 7. Comparison of fracture toughness data to lower bound curves: ASME K_{IC} versus probability curve.

K_{Jc} value. This data point significantly affects the value of T_0. While T_0=-43°C for specimens tested at -50°C, it varies from T_0=-55 to -42°C depending of whether this data point is taken into consideration or discarded. This data point can be considered an outlier. Indeed, by calculating individual T_0 for each sample, this 'outlier' point lies outside the 2σ confidence interval on average T_0.

Examination of 3PB data show a very good agreement with PCCv data. However, JRQ material is known to be quite inhomogeneous. It should be mentioned that PCCv and CT specimens are taken from the same block (CRP-IV block) while the 3PB specimens are manufactured from another one (ESIS block). Therefore, material inhomogeneity and geometry effects are difficult to separate.

Given the limited number of specimens, it is difficult to attribute the 15°C difference between T_0 determined using PCCv and T_0 from CT samples to the specimen size or to the material inhomogeneity. It will be shown later that the difference most probably results from the specimen size rather than material variability.

In Figure 6, the Master Curve and associated 95 and 5% bounds compared to the 1T adjusted experimental data show a good agreement (T_0 is taken as -54°C).

As for 22NiMoCr37, the lower bound curves are consistent with the experimental data, and the RT_{NDT} based K_{IC} ASME curve is very conservative (see Figure 7). Here also, up to -25°C, the 1% curve and the RT_{T_0}-based curve are identical.

Table 10. Test matrix: A533B (JSPS).

	SCK•CEN	JSPS Round Robin [4]			
test temperature	PCCv	0.5T CT	1T CT	2T CT	4T CT
-120	6	--	--	--	--
-25	10	18	16	2	--
0	20	23	24	6	2
+25	10	12	12	2	1

Number of specimens tested within the JSPS round robin are also indicated. All PCCv samples are 20% side grooved except at 0°C where 10 samples were plain sided.

A533B (JSPS) material

This material, used by the Japan Society for the Promotion of Science [4], is characterized by a high transition temperature and a low upper shelf. This is very interesting as irradiation induces such degradation in RPV steels. Forty PCCv samples of similar orientation (T-L) were manufactured from the broken CT samples and tested in the temperature range of interest [17]. In addition, six samples were tested at -120°C where fracture occurs in the linear elastic regime. All specimens were 20% side grooved except 10 samples tested at 0°C. The test matrix is given in Table 10.

At -120°C and -25°C, all samples fail in a brittle mode, without any ductile extension prior cleavage. At 0°C and +25°C, many samples exhibit a pop-in behavior. At +25°C, most samples exhibit ductile crack growth prior to cleavage. The data analysis in terms of T_0, K_0 and m for each geometry, size and test temperature are gathered in Table 11. Figure 8 shows the 1T-adjusted fracture toughness data as a function of test temperature.

At -120°C, the $K_{J\ med}$ is very low and, consequently, no size correction is applied, as recommended by E1921. According to E1921, these data should not be used, as the uncertainty on T_0 is very high. At -25°C, all samples fail below the $K_{J\ limit}$ except one which is slightly above. The censoring procedures show that E1921 leads to non conservative values in comparison with not censoring the data. At 0°C, among the 20 PCCv specimens, only 7 samples lie below $K_{J\ limit}$. Here, the censoring procedure according to E1921 has no effect on T_0. At +25°C, all PCCv are not valid. On the other hand, most of the CT samples tested within the JSPS round robin comply with the E1921 requirements. The T_0 values determined with various data sets vary within 2 to 14°C, The mean value being T_0=8°C. The PCCv specimens result in T_0=-7°C, in compliance with E1921. Figure 8 shows how the Master Curve and its 95% confidence bounds compare to the experimental data. As with JRQ, a shift of 15°C between the two geometries is observed. Figure 9 shows the effect of test temperature and specimen geometry and size on the T_0 determination.

Figure 10 compares the fracture toughness data to the predictive lower bound curves. The experimental data are well bounded with the various curves, with the K_{IC} ASME curve based on RT_{NDT} being the most conservative.

73W weld material

The 73W weld material is one of the two high-Copper submerged-arc welds thoroughly investigated at ORNL within the 5th irradiation series of HSSI program [6]. Some of the broken samples, un-irradiated as well as irradiated large CT specimens, and a small block for Charpy size manufacturing was provided by ORNL. Reconstitution technology allowed us to machine Charpy specimens from the broken CT specimens. All specimens are taken in the T-L orientation. Details on fabrication and test results can be found in [18, 19]. The PCCv test results analyzed here were not reconstituted. Table 12 summarizes the data analysis. Eight PCCv specimens tested at -100°C lead to T_0=-69°C. At -80°C, of 11 PCCv samples tested, 4 fail above the $K_{J\ limit}$ level. If all data are considered regardless their validity (in terms of $K_{J\ limit}$), T_0 is found equal to -89°C. This results in a 20°C shift in comparison with -100°C data. The application of the censoring procedure according to E1921 slightly affects T_0. The best agreement with -100°C data is obtained when discarding data that are above the $K_{J\ limit}$. When data obtained by ORNL are analyzed in a similar way, the large CT samples lead to T_0=-65°C [19].

According to E1921, PCCv specimens result in T_0=-77°C ((-68.6-84.7)/2). Figure 11 shows that many CT specimens lie outside the 5% bound. However, Figure 12 shows that the 1% bounds fit the experimental data very well. The K_{IC} ASME curve based on RT_{T_0} underestimates the lower bound, while an extra margin remains if RT_{NDT} is used. It is important to emphasize here that T_0 is based on PCCv samples. When T_0 is taken equal to -63°C, a much better agreement to the experimental data is found.

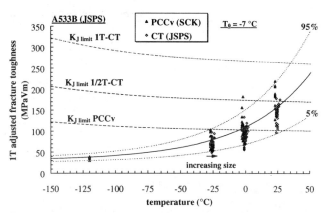

Figure 8. Master curve and associated confidence bounds for A533B (JSPS) material. JSPS round robin data are also included for comparison.

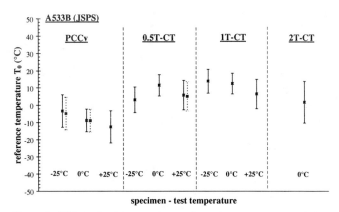

Figure 9. Effect of test temperature and specimen geometry and size on the reference temperature, T_0. Charpy size specimens result in reference temperature of 15°C lower than the CT samples.

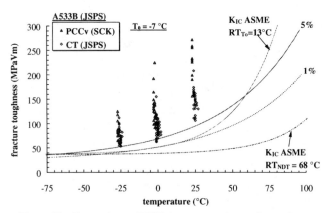

Figure 10. Comparison of 73W fracture toughness data to lower bound curves: ASME K_{IC} versus probability curves.

Table 11. Master curve transition parameters for A533B (JSPS).

specimen	N	T	T_0	K_0	m	method	remark
PCCv	6	-120	20.9	36.2	5.2	1	no size correction, all valid
	10	-25	-3.4	81.8	4.0	1	all
	10	-25	-3.1	81.6	4.0	3	censor N=r=10
	10	-25	-4.8	83.3	4.0	2	censor E1921 N=10, r=9
	9	-25	1.6	77.3	4.4	4	censor N=r=9
	20	0	-8.8	121.6	3.6	1	all
	20	0	7.2	97.8	17.3	3	censor N=r=20
	20	0	-9.0	121.9	17.3	2	censor E1921: N=20, r=7
	7	0	13.2	90.6	9.5	4	censor N=r=7
	10	25	-12.6	187.6	7.1	1	all (all not valid)
0.5T	16	-25	3.1	76.0	4.2	1	valid per E1921
	23	0	11.5	92.6	3.9	1	valid per E1921
	12	25	5.9	141.3	6.3	1	all
	11	25	8.7	135.5	10.6	4	only valid
	12	25	5.1	142.8	7.1	2	censoring per E1921
	12	25	6.4	140.1	7.1	3	
1T	18	-25	14.0	67.6	6.1	1	valid per E1921
	24	0	12.5	91.4	6.9	1	valid per E1921
	12	25	6.5	140.0	7.4	1	valid per E1921
2T	6	0	1.8	105.2	7.8	1	valid per E1921

Table 12. Master curve analysis of 73W weld material. All PCCv specimens are 20% side grooved.

specimen	N	T	r	T_0	K_0	m	method	remark
PCCv	8	-100	--	-68.6	73.2	6.6	1	valid per E1921
	11	-80	--	-89.0	122.0	2.8	1	all data
	7	-80	7	-66.1	89.9	4.2	4	only valid data
	11	-80	10	-77.7	104.4	4.2	3	censoring N=r=11
	11	-80	7	-84.7	114.9	4.2	2	censoring E1921
CT	--	--	--	-65[(1)]			1	CT (ORNL data [6])

Notes: (1) T_0 is determined using the Wallin's multiple temperature method. Note that plane strain elastic modulus is used for J - K_J conversion. (T_0=-61°C when plane stress elastic modulus is used).

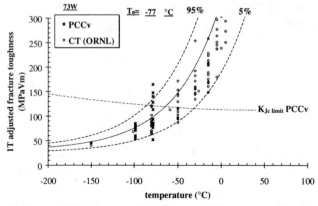

Figure 11. Master curve and associated confidence bounds for the 73W weld material. ORNL data are also included for comparison.

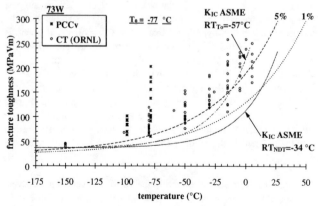

Figure 12. Comparison of 73W fracture toughness data to lower bound curves: ASME K_{IC} versus probability curve.

DISCUSSION

The E1921 standard is found to be a very efficient tool for characterizing fracture toughness in the transition regime. However, with increasing the fracture toughness databank, some minor modifications of the actual procedure will certainly improve the accuracy on T_0 determination. The use of the precracked Charpy geometry is found very promising although T_0 tends to be slightly lower than with large CT specimens.

Given the limited number of specimens, it is extremely difficult to separate the effects of size, geometry and test temperature from material variability. It should be understood here that there is a inherent material variability, but this inhomogeneity can be enhanced when comparing data of specimens taken from different locations.

Within all these uncertainties, it is found that test temperature does not significantly affect the determination of T_0, even when data are not fully satisfying the E1921 requirements. In comparison to large samples, the PCCv samples lead to T_0 values which are 5 to 20°C lower (see Table 13). The ASTM E1921 procedure is used for determining T_0. As a consequence, the censoring procedure as required by the above standard is used. This censoring procedure does not affect the final results: T_0 remains almost unchanged by application of the censoring procedure. Monte Carlo simulation performed by Scibetta [20] clearly indicates the non conservative character of this censoring procedure. A better agreement is found if only valid data are taken into account or if non valid data are replaced by the $K_{J\,limit}$ and N=r.

The E1921 discards specimens that do not fulfill the crack front uniformity (individual crack lengths within ±7% of the average crack length, or within 0.5 mm, whichever is larger). On this basis 90% of the data for 22NiMoCr37, which were plain sided, should not be used for T_0 determination. It is shown that despite the non uniformity of the crack, T_0 agrees very well with the one determined with the large samples. It would be advisable to take into account such data and only mention the crack front non-uniformity in the test report.

A question arises regarding the appropriateness of the M=30 limit factor in E1921: is it too low? Nevalainen and Dodds [11] performed detailed 3D finite element calculations. These calculations suggest M=60 for the material under consideration. Table 13 shows that the agreement is much improved in comparison to CT geometry if M=60 is used instead of M=30. Another very interesting geometry, the circumferentially-cracked round bar (CRB), leads to very similar results (see Table 13). Scibetta [20-21] has shown that loss of constraint appears very early in this geometry. By scaling this geometry to the small scale yielding condition, the results shown in Table 13 are in good agreement with large samples. The size adjustment is applied to the volume rather than to the thickness in order to take loss of constraint into account, the small scale yielding (SSY) condition being the reference state.

In order to estimate the reliability the various lower bound curves, the temperatures corresponding to K=100 MPa√m are compared in Table 14. The K_{IC} ASME curve indexed by the RT_{NDT} is clearly shifted to higher temperature in comparison to a RT_{T_0}-indexed K_{IC} curve, and to a 1% curve. A very good agreement is noticed between these last curves, the 1% curve being more conservative. This is illustrated in Figure 13 where all PCCv data are plotted as a function of kkkkkkk(T-T_0) and compared to RT_{T_0} and 1% curves. In the region of test temperature. In the temperature range of T_0±30°C, both curves are identical. For higher test temperatures, the two curves largely deviate from each other. In particular, above T_0+30°C, more experimental data are required to estimate the reliability of both curves. The current regulation based on RT_{NDT} gives an additional 20 to 40°C extra margin to the actual lower bound.

CONCLUSIONS

This work has shown that the Charpy size specimen combined with the Master Curve approach can reliably characterize fracture toughness in the transition regime. The determination of the reference

Table 13. Influence of the $K_{J\,limit}$.
A better agreement is obtained when increasing M to 60.

M		22NiMoCr37	JRQ	A533B (JSPS)	73W
30	T_0 PCCv	-100	-54	-7	-77
30	T_0 large samples	-101	-43	8	-64
60	T_0 PCCv	-95	-42[(1)]	5	-69
60	T_0 large samples	-101	-43	8	-64
SSY	T_0 CRB samples	-90	-36	14	--

Notes: (1) Due to the limited number of samples, T_0 is determined using the E1921 censoring procedure (N=6, r=4).

Table 14. Comparison between various lower bound curves expressed at the 100MPa√m K_{Jc} level.

$T_{100MPa\sqrt{m}}$	22NiMoCr37	JRQ	A533B (JSPS)	73W
5% curve	-64	-18	29	-41
RT_{T_0} curve	-57	-6	41	-29
1% curve	-45	6	53	-17
K_{IC} ASME curve	-12	13	96	-6

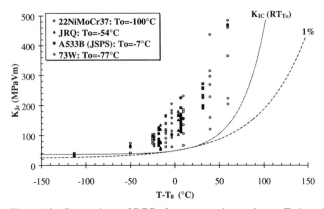

Figure 13. Comparison of PCCv fracture toughness data to T_0-based predictive lower bound curves for the four materials. Very good agreement between experimental and predictive curves is observed. At the upper transition region, the 1%-curve is more conservative than the RT_{T_0}-based curve.

temperature, T_0, according to ASTM E1921 standard by using a limited number of samples, is found independent from test temperature. However, it is important to associate with T_0 a confidence interval that depends on the size of the data set.

With small specimens, the application of the censoring procedure becomes sometimes necessary, in particular when testing at temperatures close to or higher than T_0. It is observed that the censoring procedure as recommended by the ASTM E1921 standard does not affect the results, which can lead sometimes to non conservative results. The T_0 derived from PCCv specimens lie 5 to 20°C below those derived from large samples (mainly CT). Increasing the M factor of the $K_{J\ limit}$ to 60, as suggested by finite element calculations, largely improves the agreement. However, considering the limited experimental effort presented here, further work is needed to confirm this finding.

For safety assessment considerations, the lower bound fracture toughness is of prime importance. It is found that the 1% curve and the RT_{T_0} based ASME curve are very good estimates of the fracture toughness lower bound. Above $T_0+30°C$, more experimental data are required to estimate the reliability of both curves. Comparison to the ASME K_{IC} curve (indexed to RT_{NDT}) indicates that the latter results in an over conservative temperature margin of about 30°C.

ACKNOWLEDGMENTS

The authors are grateful to R.K Nanstad from ORNL for providing the 73W weld material, and to K. Onizawa from JAERI who made all the necessary arrangements with the Japan Steel Works Ltd. to provide the A533B (JSPS) steel. The authors acknowledge the technical support of the LHMA team, in particular L. Van Houdt and A. Pellettieri who performed most of the tests reported here, and R. Mertens for specimen fabrication.

REFERENCES

[1] Wallin K., "Statistical modelling of fracture in the ductile-to-brittle transition region", Defect Assessment in Components - Fundamentals and Applications, ESIS/EGF9, Edited by J.G. Blauel and K.-H. Schwalbe, 1991, Mechanical Engineering Publications, London, pp. 415-445.

[2] ASTM E1921, Test method for the determination of reference temperature, T_0, for ferritic steels in the transition regime, Annual Book of ASTM Standards, Section 3, Vol. 03.01, 1998.

[3] Chaouadi, R., Analysis of fracture toughness behaviour of 22NiMoCr37 steel in the transition regime (SM&T Round Robin), SCK•CEN Report BLG-799, December, 1998.

[4] "Standard Test Method for Fracture Toughness within Ductile-Brittle Transition Range," Standard of the 129th Committee, Japan Society for the Promotion of Science, 1995, (in Japanese).

[5] van Walle E., Chaouadi R., Scibetta M, Puzzolante J.L., Fabry A. and Van de Velde J., Belgian contribution to the IAEA CRP-IV programme on "Assuring Structural Integrity of Reactor Pressure Vessel Steels", SCK•CEN Report BLG-754, October, 1997.

[6] Nanstad R.K., Haggag F.M., McCabe D.E., Iskander S.K., Bowman K.O. and Menke B.H., "Irradiation effects on fracture toughness of two high-copper submerged-arc welds, HSSI series 5", NUREG/CR-5913, ORNL/TM-12156/V1, 1992.

[7] ESIS P2-92, Procedure for determining the fracture toughness behaviour of materials, European Structural Integrity Society, January 1992.

[8] Mudry M. and Di Fant M., A round robin on the measurement of local criteria, Final Report RE 93.319, 1993.

[9] Manufacturing history, initial mechanical properties and radiation damage in the IAEA reference steel JRQ of ASTM A533 type B Class 1 steel plate.

[10] Gérard R., Survey of national regulatory requirements, AMES Report n°4, June, 1995.

[11] Nevalainen M. and Dodds R.H. Jr., Numerical investigation of 3-D constraint effects on brittle fracture in SE(B) and C(T) specimens, UILU-ENG-95-2001, University of Illinois, February 1995.

[12] Koppenhoefer K.C and Dodds R.H. Jr., Loading rate effects on cleavage fracture of precracked CVN specimens: 3-D studies, Engineering Fracture Mechanics, Vol. 58, No. 3, 1997, pp. 249-270.

[13] Kirk M., Lott R., Server W., Hardies R. and Rosinski S., Bias and precision of T_0 values determined using ASTM standard E1921-98 for nuclear pressure vessel steels, Effect of Radiation on Materials,: 19th International Symposium, M.L Hamilton, M. Grossbeck and A.S. Kumar, Eds, American Society for Testing and Materials, 1998.

[14] Chaouadi R. and Scibetta M., On the use of the Master Curve concept and the Charpy size specimen to characterize fracture toughness in the transition regime: Part I: RPV steel with high upper Shelf - 22NiMoCr37 SCK•CEN report, R-3294, December 1998.

[15] Scibetta M, Chaouadi R. and van Walle E., Status report on the use of the CRB for the measurement of fracture toughness of RPV steels, SCK•CEN Report BLG-763, February, 1998.

[16] van Walle E., Fabry A., Puzzolante J.L., Pouleur Y., Verstrepen A., Wannijn J.P. and Van de Velde J., Belgian contribution to the IAEA Phase 3 Coordinated Research Programme on "Optimisation of Reactor Pressure Vessel Surveillance Programmes and Their Analysis", SCK•CEN Report, November, 1993.

[17] Chaouadi R. and Scibetta M., On the use of the master curve concept and the Charpy size specimen to characterize fracture toughness in the transition regime: Part II: RPV steel with low upper shelf - A533B (JSPS), SCK•CEN report, R-3326, March 1999.

[18] Chaouadi R., van Walle E., Fabry A., Scibetta M. and Van de Velde J., Fracture toughness of precracked Charpy specimens of irradiated 73W weld material, Effect of Radiation on Materials,: 19th International Symposium, M.L Hamilton, M. Grossbeck and A.S. Kumar, Eds, American Society for Testing and Materials, 1998.

[19] Chaouadi, R., Fracture toughness measurements in the transition regime using small size samples, Small Specimen Test Techniques, ASTM STP 1329, W.R. Corwin, S.T. Rosinski and E. van Walle, Eds., American Society for Testing and Materials, 1998, pp. 214-237.

[20] Scibetta M., Contribution to the evaluation of the circumferentially-cracked round bar for fracture toughness determination of reactor pressure vessel steels, Ph.D. thesis, May, 1999.

[21] Scibetta M, Chaouadi R. and van Walle E., Fracture toughness evaluation of circumferentially-cracked round bars: theoretical aspects, submitted for publication, 1998.

ACCURACY IN THE T_o DETERMINATION: NUMERICAL VERSUS EXPERIMENTAL RESULTS

Carlos A. J. Miranda and Arnaldo H. P. Andrade
IPEN-CNEN/SP
San Paulo, Brazil

John D. Landes
University of Tennessee
Knoxville, Tennessee, USA

Donald E. McCabe and Ronald L. Swain
Oak Ridge National Laboratory
Oak Ridge, Tennessee, USA

ABSTRACT

The characterization of transition fracture toughness using the Master Curve concept with T_o has been proven to work well for a wide range of ferritic steels. However, the validity of the reference temperature (T_o) determination using the toughness values obtained from small specimens is still a concern among researchers, principally when toughness results are obtained using Charpy specimens. Due to their reduced level of constraint, they should be tested in the lower portion of the transition region, where the uncertainties are great due to a relatively flat Master Curve.

This work presents some experimental toughness results, obtained with an A508 Class 3 steel, using specimens smaller than the unit size. The test matrix contains 59 specimens at four temperatures. Two sets of twelve ½T C(T) specimens each were tested at -100 °C and at -75 °C. These two sets gave almost the same T_o value and are used as reference for the comparison with the toughness results from the other geometries: the "regular" 10mm x 10mm pre-cracked Charpy specimen and SE(B) specimens with a 9mm x 18mm cross-section. The ability of small bend specimens to give reproducible T_o values, the validity of the use of small specimens, and the best temperature range for testing are also discussed.

INTRODUCTION

The Master Curve Method was developed by Wallin (1991a, b) and can be used to describe the entire toughness versus temperature behavior for ferritic steels in the transition fracture region. It is based on the reference temperature (T_o) determination from just one set of toughness values measured in a given temperature (T). A recent study involving many heats of ferritic steels (Kirk and Lott, 1998) shows that the Master Curve fits well for a wide range of ferritic steels in the transition region. This Master Curve method was adopted by the ASTM, standard E1921-97 (1997), to determine the T_o value for ferritic steels in the transition region. By this standard, the minimum number of valid experimental results in the data set, for a given temperature T, should be six. Eq. (1) gives the size criterion for a measured toughness value, expressed as K_{Jc}, to be valid (E and σ_{ys} are the material elastic modulus and its yield stress, b_o is the remaining ligament of the specimen).

$$K_{Jc,\lim it} \leq \sqrt{\frac{E \sigma_{ys} b_o}{M}} \quad (1)$$

The M value of Eq. (1) was set to 30 to limit the small specimen size effect that introduces error between near crack tip (local J) and far-field J that is measured in experiment (Ruggieri et al., 1998). So, indirectly this size criterion imposes an upper limit in the test temperature that depends on the geometry and the material.

All results are referred to a unit size (1T: W = 50.8 mm, B = 25.4 mm) suggesting that this is a recommended test size. A weakest-link thickness correction should be applied to the obtained toughness values when the tested specimens have a different thickness.

The reliability in the reference temperature (T_o) determination using the toughness values obtained from specimens smaller than this unit size is still a concern among researchers (McCabe, 1998), principally those toughness results obtained using Charpy specimens and their sub-sized versions. Due to the reduced level of constraint in these small specimens, and to satisfy eq. (1), they should be tested in the lower portion of the transition region, where the uncertainties are greater due to being in a relatively flat part of the Master Curve.

In a previous work (Miranda and Landes, 1998) a method was shown to determine the confidence level in the T_o determination, as a function of the number of valid results and the test temperature. The work does not take into consideration the geometry from which the toughness values were obtained. From Miranda and Landes (1998) it can be seen that by using 6 specimens, or six valid results, one can get about 90% of confidence when the test temperature (T) is the same of the reference temperature T_o (i.e.: $T-T_o = 0$ °C). For the same test temperature we need about 15 valid results to have 98% of confidence in the T_o determination, i.e.: the obtained T_o falls within an interval of +/- 10 °C of the true T_o.

This work presents some experimental T_o determinations obtained for an A508 Class 3 steel using specimens smaller than the unit size. The validity of the use of small specimens, like the pre-cracked Charpy specimen (B=W=10mm) or bend bars, with B=9mm and W=18mm, and the accuracy of the obtained reference temperature, T_o, are discussed in this paper. The best temperature range for testing small specimens is also discussed.

TEST MATRIX

A heat of an A508 Class 3 steel was used to obtain three plates with dimensions of about 1500mm x 1000mm x 130mm (length, width, and thickness) each. A piece was taken from one of these plates to machine the fracture mechanics specimens used in this work. All of the specimens were machined in the L-T orientation and they were taken from the full thickness of the plate. This includes 0.5T C(T), Charpy size and 0.354T SE(B) specimens with nominal a/W = 0.6, 0.5, and 0.5 respectively. The test matrix (type, number of specimens and test temperature) is presented in Table 1 along with some other information.

A test temperature of -106°C was chosen to allow a direct comparison with toughness values obtained previously for the same heat of this A508 steel (Aquino, 1997). These previous toughness results, using mostly pre-cracked Charpy specimens tested at -106°C and -120°C show that there is no significant influence of the orientation on the toughness. From the measured toughness values presented by Aquino (1997), the average reference temperature value should be around -115°C. These toughness results, including those results obtained with six SE(B) specimens tested a -106°C, will, also, be used in the discussion in this paper.

OBTAINED TOUGHNESS RESULTS

The first group of results was obtained using two sets of 12 ½T C(T) specimens tested at -100°C and -75°C, and three sets of 6 pre-cracked Charpy specimens tested at -106°C, -90°C and -75°C. Some of the Master Curve parameters (K_o, T_o, etc) are presented in Table 1. To obtain these parameters the ASTM E1921-97 standard procedure (ASTM, 1997) was applied. The toughness values, normalized to 1T thickness, and the Master Curve with the associated 5% and 95% of confidence bound curves are presented in Figure 1. Data sets tested at the same temperature are presented with a horizontal shift for clarity purposes.

No stable crack growth was observed in any specimen tested at -106°C, -100°C, and -90°C. Less than 0.1mm of stable crack growth was observed in some of the Charpy specimens tested at -75°C. The average value determined by testing C(T) specimens ($T_o = -92.7$°C) will be considered as the reference temperature value for this material.

From eq. (1), the $K_{Jc,limit}$ for the ½T C(T) and the Charpy specimens are about 200 and 140 MPa√m, respectively. Accordingly, all C(T) results but one obtained at -75°C are valid. This shows that it would be possible to test at a higher temperature using this geometry and still obtain valid results for T_o determination. Due to the exponential nature of the Master Curve this higher test temperature for this geometry should be not far from -70°C, i.e: $(T-T_o)_{max} \approx 22$°C. All Charpy results obtained at -106°C and at -90°C are valid, and just one of six Charpy result obtained at -75°C is valid. This shows that for this type of geometry the highest reasonable test temperature for Charpy specimens is around -90°C, i.e: $(T-T_o)_{max} \approx 0$°C and, so, it is not possible to test Charpys at a temperature much higher than T_o.

The set of Charpy results obtained at -90°C has a median value slightly lower than the median value that comes from the group obtained at -106°C. Despite the fact that all toughness values are valid for T_o determination, this is an unusual behavior, different from the expected one. It is due to a combination of the experimental and statistical uncertainties and the material variability. The influence of these factors is stronger as the number of specimens is reduced.

Due to this fact it was decided to test five remaining Charpy specimens at -90°C. This additional work will be done in a near future. In doing this the confidence in the T_o determination, at -90°C, will increase from about 84% (6 specimens) to about 95% (11 specimens) assuming that all new values will be valid ones for T_o determination (Miranda and Landes, 1998).

The second group of results, using two sets of SE(B) specimens, normalized for 1T thickness, is presented also in Figure 1. The other parameters related with T_o calculation are presented in Table 1. The tests were performed at -106°C (8 specimens) and at -75°C (9 specimens). Again, no stable crack growth was observed in the specimens tested at -106°C, and just one specimen tested at -75°C had a small amount of stable crack growth (<0.1mm).

The $K_{Jc,limit}$ value for these SE(B) specimens is about 190 MPa√m. Therefore all toughness values obtained at -106°C are valid, and just 3 values at -75°C are valid for T_o determination. There are 3 other values very near this $K_{Jc,limit}$ value. This shows the possibility to obtaining the minimum number of valid results at -75°C by increasing the number of test specimens. However, it seems that this temperature is, for this geometry, an upper limit i.e: $(T-T_o)_{max} \approx 20$ °C and, so that it is not possible to test at higher temperatures to obtain T_o with this geometry (SE(B), B=9mm, W=18mm).

Previous Results

In a previous study, conducted by Aquino (1997) using the same heat of material, 34 'regular' pre-cracked Charpy specimens (10mm x 10mm) and 6 SE(B) specimens with B = 9 mm and W = 18 mm were tested for toughness measurements. Figure 2 shows them as corrected to 1T thickness in graphical form. The sets at the same temperature are presented with a shift for clarity purposes. Some other parameters are presented in Table 2 along with the T_o values.

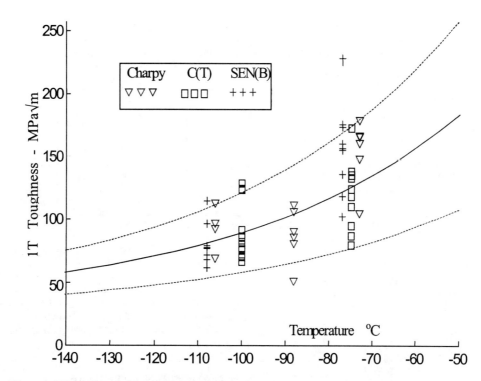

Figure 1: Master Curve with the 5% and 95% bounds and the toughness results (transformed to 1T values)

Table 1: Test matrix and Master Curve parameters

Geometry	Test Temperature	# Specimens / Valid Results	Master Curve Parameters			Average
			$K_{Jc,median}$[1]	K_o[1]	T_o	T_o
	°C	-----	MPa√m		°C	
CT	-75	12 / 11	124.4	134.4	-90.7	**-92.7**
	-100	12 / 12	93.4	100.5	-94.8	
Charpy	-75	6 / 1	165.1	179.1	[-109.6][2]	-----
	-90	6 / 6	89.3	96.0	-81.3	-90.4
	-106	6 / 6	91.8	98.7	-99.4	
SE(B)	-75	9 / 3	172.1	186.7	[-112.2][2]	------
	-106	8 / 8	82.1	88.1	-90.5	-90.5
				Overall T_o average value		-91.2

[1] – after thickness adjustment;   [ ][2] – Not a good / valid value according to E1921-97

In his work Aquino (1997) assumed: σ_{ys} = 550 MPa at -106°C and σ_{ys} = 600 MPa at -120°C for this material. These same values were assumed in the present work. All specimens have nominal a/W = 0.5. No crack growth information was reported. The toughness results obtained were presented in a table format.

Two SE(B) specimens were tested at each orientation (T-L, L-T and S-T). At -106°C two sets of six Charpy specimens each were tested: one set for the T-L orientation and another set for the L-T orientation. At -120°C six Charpy for the T-L direction, six for the S-T direction and 10 Charpy for the L-T direction were tested.

Table 2: Previous results (Aquino, 1997)

Geometry	Test Temperature	# Specimens / Valid Results	Master Curve Parameters			Average T_o
			$K_{Jc,median}$[1]	K_o[1]	T_o	
	°C	-----	MPa√m		°C	
Charpy	-106	12 / 7	117.	126.	-117.	-116.
	-120	22 / 19	94.	101.	-115.	
SE(B)	-106	6 / 4	142.	153.	[-131][2]	------
			Overall T_o average value			-116.

[1] – After thickness adjustment;   [ ][2] – Not a good / valid value according to E1921-97

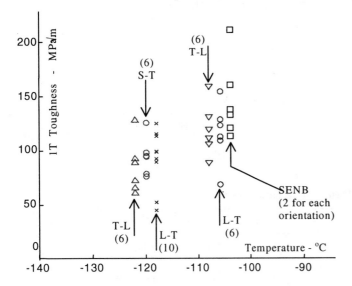

Figure 2: Previous results obtained for Charpy and SE(B) specimens tested at -120°C and at -106°C (Aquino, 1997).

At -106°C there are 7 valid Charpy toughness results from 12 tested and there are 19 valid from 22 tested at -120°C. So at both temperatures there are sufficient toughness values to perform the determination of the reference temperature. Among the six SE(B) specimens tested at -106°C there are only 4 valid toughness values. So the T_o value obtained from the SE(B) results is not a valid value. However T_o was calculated and shown in Table 2 only for comparison and completeness purposes.

The average T_o value that came from these results (-116°C, table 2) is almost 25°C lower than the average value obtained in the present work (-91.2°C, Table 1). One explanation for this difference can be the plate from which the specimens were taken. The reason for this difference is under investigation.

DISCUSSION

For the T_o determination, all ½T C(T) toughness values, but one, were valid. The two sets tested at -100°C and -75°C, with 12 specimens each, gave almost the same T_o value, within +/- 2°C from its average value. The 8 SE(B) specimens (8 valid results), tested at -106°C gave a T_o = -90.5°C. A value very near the obtained one using the C(T) results. The two sets of 6 Charpy results (6 valid) gave the largest T_o scatter: -81.3°C (at -90°C) and -99.4°C (at -106°C). Its average T_o value (-90.4°C) is very near the overall average T_o (-91.2°C).

Maximum Test Temperature

(1) *CT* - Considering this overall average T_o value, or the average T_o value that comes from the ½T C(T), as the reference temperature for this material, the toughness results obtained in this study show that it is possible to have a very reliable reference temperature value by testing the ½T C(T) specimens at temperatures a little higher than -75°C. That is: $(T-T_o)_{max} \approx 22°C$.

(2) *Charpy* - Considering the behavior observed in the toughness results obtained with the Charpy specimens at -90°C, it can be expected that higher toughness values will be obtained when additional Charpy sized specimens are tested at this temperature. For the Charpy geometry the limit temperature seems to be at -90°C, or $(T-T_o)_{max} \approx 0°C$.

(3) *SE(B)* - Using SE(B) specimens with B=9mm and W=18mm the maximum test temperature to obtain valid results for T_o determination seems to be around -75°C, so at $(T-T_o)_{max} \approx 17°C$.

This work does not investigate the lower temperature ($T-T_o < 0$ °C) range. However, from a previous work, using numerical simulation (Miranda and Landes, 1998), it was shown that, due to the flat Master Curve in the lower transition region it is not advisable to test at $(T-T_o) < -25°C$. So, for this material, the best test temperature range to measure toughness for T_o calculation, using small specimens is $-25°C < (T-T_o) < 20°C$ for the ½T C(T) and 0.354T SE(B), and $-25°C < (T-T_o) < 0°C$ for the 'regular' (B=W=10mm) pre-cracked Charpy specimen.

CONCLUSIONS

Some toughness results were measured using sub-sized specimens: ½T C(T), 0.354T SE(B), and pre-cracked Charpy (0.394T, B=W) specimens. The results showed that, with the severe limitations on the upper limit of the test temperature (T) imposed by E1921, it is possible to obtain reliable values of the reference temperature (T_o) using sub-sized specimens. The best test temperature range to measure toughness to obtain the reference temperature for this material was determined.

ACKNOWLEDGEMENTS

The first author would like to thank the support that was given him by the Brazilian research agency CNPq (Conselho Nacional de Desenvolvimento Científico e Tecnológico) through the process 200.681/97-4(nv) and by the Brazilian Nuclear Regulatory Agency CNEN (Comissão Nacional de Energia Nuclear). This work was performed during his split scholarship program, at the University of Tennessee at Knoxville, TN, USA, as part of his PhD program. He would like also to thank Dr. R. Nanstad, Oak Ridge National Laboratory, who arranged for admittance and for use of laboratory equipment, and to R. L. Swain who guided the testing work.

REFERENCES

Wallin, K., 1991, "Fracture Toughness Transition Curve Shape for Ferritic Structural Steels," Proceedings of the Joint FEFG/ICF International Conference on Fracture of Engineering Materials, Singapore, August 6-8, 1991, pp. 83-88.

Wallin, K., 1991, "Statistical Modeling of Fracture in the Ductile-to-Brittle Transition Region," Defects Assessment in Components – Fundamentals and Applications, ESIS/EGF9, Ed. By J. G. Blauel and K. H. Schwalbe, Mechanical Engineering Publications, London, pp. 415-445.

Kirk, M. T., Lott, R., 1998, "Empirical Validation of the Master Curve for Irradiated and Un-Irradiated Reactor Pressure Vessels," Joint ASME/JSME Pressure Vessel and Piping Conference, July 27-30, San Diego, CA, USA.

ASTM, 1997, "Test Method for Determination of the Reference Temperature, T_o, for Ferritic Steels in the Transition Range", ASTM E1921-97, American Society for Testing and Materials.

McCabe, D. E., 1998, "Outstanding Issues for Future Research Activities in Ductile-Brittle Fracture," NSWC/USNA Workshop on Fracture in the Ductile-Brittle Fracture Transition, July 14-15, US Naval Academy, Annapolis, MD.

Miranda, C. A. and Landes, J. D., 1998, "Defining The Confidence Level in The Reference Temperature Determination", submitted to *Engineering Fracture Mechanics*.

Aquino, C. T. E., 1997, "An Approach to the Fracture Toughness Variation in the Ductile-Brittle Transition for Nuclear Pressure Vessel Steels" (in Portuguese), doctoral thesis, IPEN/USP, São Paulo, Brazil.

C. Ruggieri, R. H. Dodds Jr, and K. Wallin, 1998, "Constraint Effects on Reference Temperature, T_o, for Ferritic Steels in the Transition Region," *Engineering Fracture Mechanics*, Vol. 60, No. 1, pp. 19-36.

DEVELOPMENT OF THE T_o REFERENCE TEMPERATURE FROM PRECRACKED CHARPY SPECIMENS

James A. Joyce
United States Naval Academy
Annapolis, MD 21402
Ph: (410) 293-6503
Fax: (410) 293-2591
jaj@usna.navy.mil

Robert L. Tregoning
Naval Surface Warfare Center
9500 MacArthur BLVD
West Bethesda, MD 20817
Ph: (301) 227-5145
Fax: (410) 227-5548
tregoningrl@nswccd.navy.mil

ABSTRACT

The master curve approach specified within the new ASTM E1921 Test Standard is a significant advance in defining an indexing temperature, T_o, the median fracture toughness, and associated failure probability bounds for ferritic steels in the ductile-to-brittle transition regime. An objective in developing this standard test procedure has been that it should, if possible, allow the use of precracked Charpy-size specimens to measure T_o and the associated fracture performance. However, the technical basis document (Merkle et al., 1998) which supports much of the E1921 criteria presents no evidence to justify using the smallest allowable number of Charpy-size specimens to estimate the reference temperature.

Further, during the development of the standard, there was no independent experimental verification that the specified deformation criterion would result in acceptably accurate T_o values. Computational and analytical support of the deformation criterion for precracked Charpy specimens is also dramatically lacking. Recent computational work by Ruggieri et al. (1998) supports a deformation criterion that is almost twice as stringent as the E1921 standard. There is also no historical precedent within current fracture toughness testing standards to support the E1921 relaxation of the deformation criterion.

This work evaluates this criteria experimentally for several materials using experimental data sets which contain sufficient numbers of both precracked Charpy and 1T (or larger) fracture toughness measurements. The data demonstrates that the Charpy-size specimens accurately predict T_o for certain materials, while in other materials, the Charpy results are significantly non-conservative, even when the data fall within the allowable E1921 deformation criterion. This effect may be a function of the degree of crack tip constraint loss, which strongly depends on the material flow properties. An attempt is made to correct for the constraint loss in the Charpy-size specimen and remove any bias in the measured T_o using a simple constant stressed volume cleavage failure criteria (Nevalainen and Dodds, 1995). While this method does decrease the difference between T_o values measured using Charpy and conventional 1T specimens, detailed cleavage initiation models which can be independently calibrated will be required to rigorously account for any constraint loss in Charpy specimens.

INTRODUCTION

The reference temperature, T_o, and "Master Curve" approach defined by the new ASTM E1921-97 Test Standard is a significant advance in defining the median fracture toughness and the associated confidence bounds for ferritic steels in the lower ductile-to-brittle transition regime. The method has been demonstrated (Kirk and Lott, 1998) to be an excellent model for a wide range of pressure vessel steels, irradiated steels, and other structural ferritic steels. The master curve appears to be an applicable model of cleavage initiation at elevated loading rates (Joyce, 1997), and has even been explored as a method to characterize crack arrest toughness (Wallin, 1999).

One objective during the development of the standard test procedure was to allow the measurement of T_o and the median master curve using precracked Charpy specimens. This feature is attractive to the commercial nuclear industry because Charpy specimens are predominant within surveillance capsule programs. However, the deformation criterion (M limit) that results from such an allowance is roughly six times less stringent than E1820-96 requirements. Deformation criteria have been previously standardized using experimental or computational results which demonstrate that the measured toughness results are unaffected by constraint loss up to the prescribed limit (Kirk et al., 1993). However, technical justification for the E1921 size requirement has not been rigorously developed.

A technical support document for E1921-98 has been recently prepared (Merkle et al., 1998), and includes much of the rational behind the requirements of this new standard. A large database of applicable historical testing has also been accumulated which also supports much of the basis for the standard (Kirk and Lott, 1998). However, these historical data sets predominantly consist of 1T or larger fracture toughness specimens. Almost no experimental data sets have been developed which demonstrate the validity of the E1921 method using the minimum allowable number of precracked Charpy-size specimens, especially when the limiting size criterion of E1921 is approached.

Computational and analytical support for the use of Charpy-size specimens near the minimum size requirements of E1921 is also dramatically lacking. Recent computational work by Ruggieri et al. (1998) supports a size criterion which is almost twice as stringent as the E1921 standard. This work shows that the size criteria in other existing fracture toughness standards are too severe when 3D analysis and statistical factors as well as realistic material properties are included in the analysis. However, this work predicts that as the deformation criterion of E1921 is approached, pronounced constraint loss can occur and result in a specimen-size dependent T_o measurement. Constraint loss always causes a less conservative estimation of T_o, although the bias increases as a material's yield strength and strain hardening decreases.

Both in-plane and out-of-plane constraint loss may be important for describing this phenomenon. One of the fundamental tenets of E1921 is that there is an equivalent cleavage fracture probability at all locations along the crack front. This translates to a requirement within E1921 to employ a "statistical size correction" to adjust the fracture toughness data from different thickness test samples to a single, consistent 1T size for evaluation. While this procedure is generally useful for reconciling toughness measurements (Wallin, 1998), it has not been adequately verified for precracked Charpy specimens that fail near the E1921 size criteria. It is possible that out-of-plane constraint loss may contribute by increasing the amount of low constraint volume along the crack front as plasticity grows from the free surface. If this occurs, the likelihood of cleavage initiation is not equivalent throughout the specimen thickness, and would be proportionally less than in larger, predominantly plane-strain fracture toughness specimens.

Ruggieri et al. (1998), and Nevalainen and Dodds (1995) have developed techniques based on 3D finite element analysis. These provide methodologies for assessing and correcting for both in-plane and out-of-plane constraint loss in standard test specimen geometries. One approach (Nevalainen and Dodds, 1995) has been utilized herein to adjust the T_o values measured using precracked Charpy specimens in an effort to eliminate any bias introduced by constraint loss. These adjusted values are compared with T_o measurements on the same materials from conventional 1T fracture toughness specimens where constraint loss is not a concern.

EXPERIMENTAL DETAILS

Material Description

Data from five pressure vessel grade steels is presented and discussed in this paper. Steels include an ASTM A515 Grade 70 steel, an ASTM A508 Class 3 forging steel, and three ASTM A533 Grade B steels. Note that two of the A533B steels were tested in the as-received condition while HSST plate 14 has been heat treated to simulate yield strength elevation due to radiation embrittlement (McAfee et al., 1998). Chemical compositions for the five steels are shown in Table 1, and near-ambient tensile mechanical properties are presented in Table 2. Tensile testing was also conducted on all materials over the range of subsequent fracture toughness testing temperatures. Further information on the other four materials can be found in the references indicated in Table 2.

Table 1: Material Chemical Compositions (wt%)

Element	A515 (GGS)	A533B (HSST-14)	A533B (HEW)	A533B (JSPS)	A508 (Lidbury)
Carbon	0.31	0.22	0.26	0.24	0.16
Manganese	0.95	1.44	1.33	1.52	1.32
Phosphorous	0.021	0.005	0.010	0.028	0.005
Sulfur	0.017	0.003	0.014	0.023	0.002
Silicon	0.20	0.20	0.22	0.41	0.20
Nickel	0.01	0.62	0.58	0.43	0.71
Molybdenum	0.01	0.56	0.52	0.49	0.51
Chromium	0.04	0.06	0.06	0.08	0.16

Table 2: Near-ambient Tensile Mechanical Properties

Material ID	Temp (°C)	σ_{ys} (MPa)	σ_{ult} (MPa)	%Elong (25 mm)	% R.A.
A515 (GGS) [Porr et al., 1995]	21	291	559	25	45
A533B (HSST-14) [McAfee et al., 1998]	40	617	775	16	58
A533B (HEW) [Tregoning and Joyce, 1999]	24	488	644	28	70
A533B (JSPS) [Chaouadi, 1998]	25	461	639	18	59
A508 (Lidbury) [Lidbury and Moskovic, 1993]	22	414	561	NR[1]	NR

[1] Not Reported

Transition Regime Fracture Toughness Testing

HEW A533B, HSST Plate 14 A533B, and GGS A515 Testing. Benchmark tests to determine each material's reference temperature, T_o, were conducted using standard 1T C(T) or SE(B) specimens. All testing was performed according to ASTM E1921 specifications. At least the requisite the requisite minimum number of samples were evaluated at each temperature. Additional testing was conducted for the HEW A533B and GGS A515 materials to increase the confidence in T_o. Lack of material precluded more 1T testing of the HSST Plate 14 A533B alloy.

Small specimen testing using precracked Charpy (W = 10mm, B = 10mm) specimens was conducted at various temperatures below the 1T measured T_o value in an effort to develop E1921-valid data sets. Typically, testing was continued until at least the minimum number of allowable uncensored results was achieved. . However, it was not possible to obtain the requisite number of uncensored results using a reasonable number of specimens for some precracked Charpy data-sets tested near T_o. Additional 1/2T C(T) data is also reported for the HSST Plate 14 A533B material. This data was gathered previously both by Joyce et al. (1998) and Oak Ridge National Laboratory (ORNL) [McAfee et al., 1998].

All small and large specimens were located at the same depth within the original plate thickness. Generally, this ranged between 1/8 to 1/2 of the thickness. The GGS A515 and HEW A553B Charpy and 1T specimens were also extracted from adjacent plate sections. The HSST Plate 14 533B 1/2T C(T) and Charpy specimens tested here and in Joyce et al. (1998) were machined directly from the 1/4 span of several 1T SE(B) specimens. The ORNL 1/2T C(T) specimens did come from a different section of the original plate. More information on the testing conducted on the GGS A515 material can be found in Joyce and Link (1997). The HSST Plate 14 testing is detailed in McAfee et al. (1998) and Joyce et al. (1998), while the HEW A533B testing is more completely summarized in Tregoning and Joyce (1999).

JSPS A533B and A508 Cl3 Forging Testing. Large specimen (> 1T) testing of the JSPS A533B (Chaouadi, 1998) and A508 Cl3 forging materials [11] was performed at several test temperatures. Small specimen testing of the JSPS A533B material was conducted using both 1/2T C(T) and precracked Charpy specimens. Specimens were tested at several temperatures, both above and below T_o (Chaouadi, 1998). The A508 Cl3 alloy was also evaluated above and below T_o using precracked Charpy specimens. All large and small specimen testing at each single temperature generally concurred with ASTM E1921 requirements. The JSPS Charpy specimens were extracted from one of the large 2T C(T) specimens. The relative location of the A508 and remaining JSPS specimens has not been verified, but it is likely that similar care has been taken with respect to sectioning.

Analysis

J_c Evaluation. For the A515, HSST Plate 14 A533B, and HEW A533B materials, the J-integral at cleavage initiation (J_c) was evaluated as per ASTM E1921. The only wrinkle for the SE(B) specimen type is that the plastic component of the J-integral (J_{pl}) was determined from the crack opening displacement (COD) as developed in Kirk and Dodds (1993). The J_{pl} value is therefore defined as

$$J_{pl} = \frac{\eta_{J-C} A_p}{B_N (W - a_o)} \quad (1)$$

where a_o is the initial crack length, W is the specimen width, B_N is the net specimen thickness, and A_p is the plastic area under the load vs. COD curve. The η_{J-C} factor is given by

$$\eta_{J-C} = 3.785 - 3.101(\frac{a_o}{W}) + 2.018(\frac{a_o}{W})^2 \quad (2)$$

The accuracy of this formulation has been numerically verified in Kirk and Dodds (1993). The C(T) specimens utilized the more traditional load-line displacement formulation presented in ASTM E1921. It should be noted that ASTM E1921 does allow J_{pl} to be determined using the COD. However, no equations are presented for this calculation.

Tabulated J_c values was extracted from Lidbury and Moskovic (1993) for the A508 Cl3 forging material and Chaouadi (1998) for the JSPS A533B plate. Actual specimen geometry and initial crack lengths were available in Chaouadi (1998), but nominal values were only provided in Lidbury and Moskovic (1993). Tensile properties over the range of fracture toughness testing temperatures were also obtained. This information was utilized in the subsequent T_o analysis.

Evaluation of T_o. Data for distinct alloys, test temperatures, and specimen geometry was first grouped. The T_o value was then determined for each unique data-set as per E1921 methodology. If the calculated T_o is invalid, the reason has been noted. As prescribed, the censoring limit ($K_{Jc(limit)}$) was based on the material's yield strength (σ_{ys}) at the test temperature. Linear interpolation was used when the measured σ_{ys} and fracture toughness testing temperatures did not coincide.

The data was also analyzed using the multi-temperature maximum likelihood approach described in Wallin (1995). This approach uses data at different temperatures to determine a single estimate of T_o. The E1921 censoring criteria was also used. Merkle, et al (1998) has previously shown that this approach is identical to E1921 for data-sets conducted at a single temperature. The advantage of the multi-temperature approach is that the confidence in the T_o estimate increases with additional data. All the large specimen data (> 1T size) for each material was analyzed for a single T_o estimate using this approach, while the smaller 1/2T C(T) and precracked Charpy data were kept distinct.

EXPERIMENTAL RESULTS

The value for T_o obtained using large specimen (> 1T) data is defined as T_{ols}. This value has been based on all available data to ensure the highest confidence. This data is summarized in Table 3 along with the test temperatures used in the estimate and the total number of uncensored data (r). The T_{ols} values are assumed to represent the best estimate material properties for SSY fracture conditions. These measures should not be affected by constraint loss.

All the as-measured toughness data from the five data sets is illustrated in Fig. 1. Here the data has been normalized by the appropriate T_{ols} shown in Table 3. The 1T Master Curve and associated failure probability bounds have been superimposed on this figure. The relationship between toughness and specimen size is readily apparent in this figure. Data from specimens smaller than 1T lie above the 1T median curve, while larger specimen data tend to fall below this benchmark.

This trend is generally explained by the statistical thickness correction which normalizes the data by equating the highly stressed material volume (Merkle et al., 1998). Data properly normalized to a common 25.4 mm reference thickness should eliminate any size-induced bias unless other contributing effects exist. The raw data was adjusted using the statistical thickness correction as part of the standard T_o analysis. This corrected data is presented in Fig. 2-4 for several of the materials studied. The E1921 Master Curve and the

90% confidence bands have also been superimposed in these figures based on the T_{ols} data presented in Table 3.

Table 3: Large Specimen Reference Temperature (T_{ols}).

Material ID	Spec. Type	Test Temp. (°C)	T_{ols} (°C)	T_{ols}^c (°C)	"r"
A515 (GGS)	1T CT	-70, -28	-9	-4	24
A533B (HSST-14)	1T SEB	-5	-59	-50	7
A533B (HEW)	1T CT	-40	-75	-63	26
A533B (JSPS)	1T CT	-25, 0, 25	15	30	54
A508 (Lidbury)	3.2T SEB & 7.9T SEB	-65, -45, -30, -15, & 9	-103	-93	60

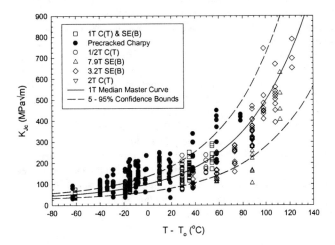

Figure 1: Experimental Data From Five Pressure Vessel Materials Plotted Without Size Correction.

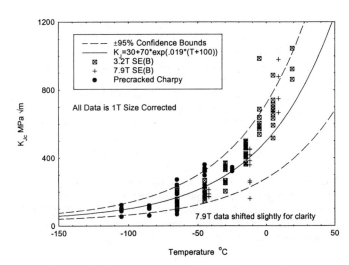

Figure 2: Size-Corrected A508 Cl3 Forging Data

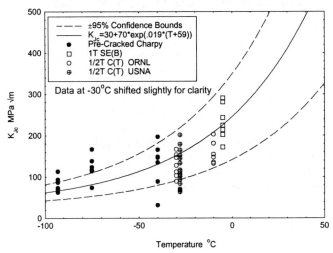

Figure 3: Size-Corrected A533B HSST Plate 14 Data

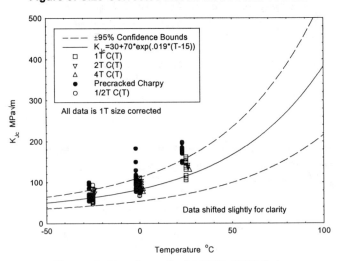

Figure 4: Size-Corrected A533B JSPS Data

For the A508 Class 3 forging material (Fig. 2), the Charpy data agrees very well with the results of the larger specimens after the statistical size correction. The Charpy and larger specimen data for the GGS A515 alloy also shows reasonable agreement, although a preponderance of Charpy results fall above both the median Master Curve and the upper 95% confidence band. However, Charpy results for the three A533B materials (e.g. Figs. 3 & 4) are significantly elevated above the toughness measurements obtained from the larger size specimens.

Tabulated T_o reference temperatures calculated according to E1921 are presented for each individual data set in Table 4. The reference temperatures calculated from the Charpy specimens result in T_o estimates that are generally 10 to 20°C below the estimates obtained from the larger geometries for the A533B steels. If these Charpy size specimens were used to predict the transition behavior of the larger geometries, a non-conservative T_o measure would be obtained.

There are three conceivable explanations that rationalize the generally high toughness measurements (and consequently low T_o values) obtained from the precracked Charpy specimens. These include simple statistical uncertainty in T_o due to the relatively small sampling size, material variability, and inherent measurement bias. The expected uncertainty in T_o due to sample size is illustrated in Fig. 5, which has been obtained from Merkle et al. (1998). These confidence bounds were developed by Monte Carlo simulation using a theoretical Weibull toughness distribution to determine the related T_o distribution. More information is provided in Merkle et al. (1998). Note that ordinate of this figure is normalized by $\sqrt{r}$ to allow direct comparison of data-sets with different sample sizes.

Table 4: Summary of Reference Temperature Results

Material ID	Spec. Type	Test Temp. (°C)	T_o (°C)	r	T_o^c (°C)	M_{min}	M_{avg}
A515 (GGS) n = 5 σ_o/σ_1= 4.1	1T C(T)	-70	-7	12	0	190	539
	1T C(T)	-28	-11	12	-3	105	251
	Charpy[4]	-72	-22	6	-10	41	188
	Charpy	-38	-22	9	-12	12	57
	Charpy	-28	-3	8	0	11	62
	Charpy[4]	-19	-11	3	-9	7	31
A533B (HSST-14) n = 10 σ_o/σ_1= 3.4	1T SE(B)	-5	-59	7	-39	35	59
	1/2T C(T) (ORNL)	-10	-46	4	-17	30	53
	1/2T C(T) (ORNL)	-30	-46	8	-26	45	101
	1/2T C(T)	-30	-47	16	-25	21	99
	Charpy	-93	-76	9	-63	34	80
	Charpy[4]	-75	-89	4	-69	15	35
	Charpy[4]	-40	-58	2	-40	11	97
A533B (HEW) n =10 σ_o/σ_1= 3.4	1T C(T)	-40	-75	26	-64	35	85
	Charpy	-83	-93	10	-74	13	32
A533B (JSPS) n = 10 σ_o/σ_1= 3.4	2T C(T)	0	5	6	16	503	710
	1T C(T)	-25	18	19	35	399	677
	1T C(T)	0	16	24	31	191	310
	1T C(T)	25	9	12	25	102	155
	1/2T C(T)	-25	6	16	27	111	229
	1/2T C(T)	0	14	23	34	56	163
	1/2T C(T)[3]	25	8	12	29	31	57
	Charpy	-25	0	10	20	32	72
	Charpy	0	-6	9	20	10	30
	Charpy[4]	25	---	0	24	8	10
A508 (Lidbury) n = 5 σ_o/σ_1= 4.1	7.9T SE(B)	-15	-95	8	-93	140	336
	3.2T SE(B)	-65	-101	8	-93	168	361
	3.2T SE(B)	-45	-97	11	-89	69	176
	3.2T SE(B)	-30	-100	10	-92	56	89
	3.2T SE(B)	-15	-114[3]	16	-94	20	37
	Charpy[4]	-85	-104	3	-82	14	45
	Charpy[4]	-105	-104	5	-86	24	67

[3] Ductile growth limits exceeded in several of these tests.
[4] Sampling size requirements not met at this temperature

The expected uncertainty figure was developed for the data presented in the following manner. First, the large specimen T_o value for each material (T_{ols}) was taken to be the "true" value for the material. Then the difference between T_o values calculated from the small (1/2C(T) and precracked Charpy) specimens (T_{oss}) and T_{ols} was plotted after normalization by the number of uncensored results.

Where appropriate, the T_{oss} value has been determined using the multitemperature technique. Since the multitemperature technique involves data at different temperatures, the value plotted for $T-T_{ols}$ for each data-set was approximated by the average of all the uncensored results. Each data point on this figure therefore represents a unique material and test specimen geometry.

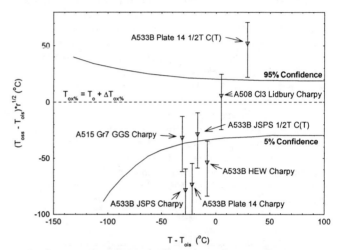

Figure 5: Measurement Uncertainty For Estimations From Finite Sample Sizes

There is uncertainty in the value of T_{ols} since it is determined from a finite sampling size, and not a theoretical distribution. The uncertainty is displayed in Fig. 5 by placing error bars about each data point to reflect the uncertainty in T_{ols}. While the actual uncertainty is a weak function of $T - T_o$ for $T - T_o > -50°C$ (see Fig. 5), a simple estimate was obtained using the confidence bounds at $T -T_o = 100°C$. Additional absolute uncertainty bars could be added to the data to reflect the error in the $T - T_{ols}$ axis, but this is generally small due to the large number of specimens tested, and does not aid in the interpretation of the results.

Figure 5 shows that all the A533B Charpy specimen data falls well below the 5% confidence bound. This implies that it is highly unlikely that the lower T_o measurements obtained from this geometry are due to statistical sampling uncertainty. Further, the A515 Gr 7 material is close to the 5% confidence bound along with the 1/2T C(T) JSPS A533B results. Only the A508 precrack Charpy and HSST plate 14 1/2T C(T) results fall above the large specimen results.

This figure also provides additional indirect evidence that material uncertainty is an unlikely cause for the T_o bias measured using precracked Charpy specimens. Variability due to material uncertainty should be random, and T_{oss} data should lie both above and below T_{ols}. Random scatter is clearly not apparent in Fig. 5. Additionally, great care was taken to ensure consistent specimen location for the HSST Plate 14 A533B, HEW A533B, and A515 testing. It is likely that specimen location was similarly controlled in the Lidbury A508 and JSPS A533B data-sets, although this fact has not been directly verified. This direct and indirect evidence together, seem to rule out material variability as a primary cause for the T_{oss} and T_{ols} differences.

Since neither material variability nor statistical uncertainty appears culpable, the T_{oss} bias in certain materials must result from some inherent feature of the Charpy specimen geometry or the testing conditions. Charpy specimen testing usually requires testing within a tight temperature range. The allowable upper test temperature is defined by the M {M = b σ_{ys}/J_c, where b is the specimen's remaining ligament} = 30 criteria in E1921, and usually is below T_o. The lower test temperature is governed by the onset of lower shelf fracture where the E1921 weakest link modeling breaks down. ASTM E1921 limits the lower test temperature to approximately $T_o - 50°C$.

The result of this tight temperature range is that data from precracked Charpy specimens often approaches the E1921 M criteria more closely than data from larger specimen types. The effect of the deformation level at fracture can be studied by considering $T_{oss} - T_{ols}$ as a function of the mean deformation level, M_{avg}, of each data set (Fig. 6). M_{avg} was determined from the uncensored toughness values of each single temperature data-set (Table 4). For this figure, T_{oss} is simply the single temperature 1/2T C(T) and precracked Charpy T_o data summarized in Table 4. Also shown on Fig. 6 are the 85% confidence bounds provided in E1921 for a data set consisting of six specimens.

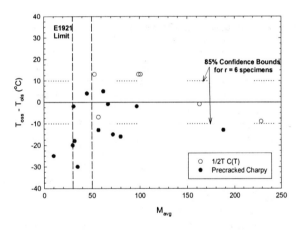

Figure 6: Effect of Deformation Level on T_o Values.

Clearly, for $M_{avg} > 100$, there appears to be good agreement between T_{oss} and T_{ols}. However, for $50 < M_{avg} < 100$, there are several instances where the T_{oss} value is significantly less than T_{ols}. As M_{avg} decreases below 50, the data show an increasing bias in T_o. It should be noted that all but one of the data points below the lower confidence bound were measured on A533B material.

DISCUSSION

Constraint Correction

If the bias in T_o is related to the deformation level at fracture, it is possible that constraint loss may be the cause. Presently, ASTM E1921, and the corresponding technical support document, stipulate that T_o is transferable with an accuracy of 20°C using the minimum allowable sample size and the M = 30 deformation criteria. However, this stipulation does not seem to be supported by the data assembled here. There is currently no provision in E1921 for constraint correction, although some studies (Ruggieri et al., 1998, Nevalainen and Dodds, 1995) have implied that some correction may be necessary near the existing deformation criteria. If constraint loss is the source for the apparent bias in T_o measurements from precracked Charpy specimens, it should be possible to reduce or eliminate this bias by "correcting" these measurements for constraint loss. In this initial study, a simple approach developed by Nevalainen and Dodds (1995) was utilized in an attempt to eliminate constraint differences between data sets from different specimen types.

This study conducted detailed three-dimensional finite element analysis of standard specimen geometries to examine the evolution of the local crack tip stress field perpendicular to the crack plane (σ_1) during loading. The study also considered the effect of material flow properties on these fields. Material flow was modeled using a two-part uniaxial constitutive model of the form:

$$\frac{\varepsilon}{\varepsilon_o} = \frac{\sigma}{\sigma_o} \quad \varepsilon \leq \varepsilon_o; \quad \frac{\varepsilon}{\varepsilon_o} = \left(\frac{\sigma}{\sigma_o}\right)^n \quad \varepsilon > \varepsilon_o \quad (3)$$

where ε_o and σ_o define limits for the initial linear portion of the stress strain curve. The numerical analysis investigated two strain hardening exponents (n = 10 and n = 5), and assumed that $E/\sigma_o = 500$ and Poisson's ratio = 0.3. In this analysis, σ_o is roughly equivalent to the 0.2% offset yield stress as illustrated in Table 2. All future discussion and calculations utilize σ_{ys} in place of σ_o as appropriate.

Nevalainen and Dodds (1995) assume that the volume of "critically stressed material" ahead of the crack tip governs fracture. Then, specimen toughness, J_c, is related to the material toughness under idealized, three-dimensional small scale yielding (SSY) deformation, $\bar{J}_o$, by requiring equivalency of the "critically stressed volume" between the two deformation states. This study also attempted to separate in-plane and out-of-plane constraint loss using the following formulation:

$$\bar{J}_o = J_o(J_c)(B_{eff}/B_{nom})^{1/2} \quad (4)$$

where J_o is the in-plane small scale yielding (SSY) value at J_c, B_{eff} is the effective thickness of the test geometry, B_{nom} is a nominal specimen thickness without side grooves, and $\bar{J}_o$ is the three-dimensional SSY toughness at J_c.

The J_c to J_o relation defines in-plane constraint loss. It is conceptually similar to the 2D toughness scaling model presented earlier by Anderson and Dodds based on equivalency of critically stressed areas (Anderson and Dodds, 1991). A series of curves are numerically developed which relate remote loading parameters J_c to J_o by equating the area of an assumed normalized stress contour level (σ_1/σ_o).

A compilation of these curves for an n = 10, $E/\sigma_o = 500$ material is provided in Fig. 7 for the specimen types studied. The thickness dependent SSY toughness, J_o, for a given experimentally measured ($J_c = J_{avg}$) value can be determined utilizing these curves. However, as seen in this figure, the relationship between J_c and J_o is a strong function of the particular stress contour which is assumed to govern fracture.

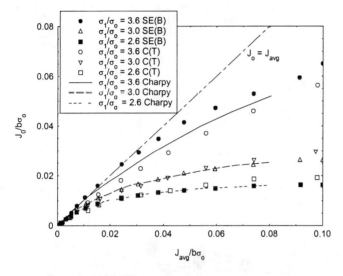

Figure 7: Toughness Scaling Relationships for Test Specimen Geometries

It is interesting to recognize that for a particular choice of strain hardening and stress contour ratio, σ_1/σ_o, the in-plane toughness scaling relationships of these three specimen types are predicted to be almost equivalent. This part of the constraint correction provides the greatest portion of the reference temperature (T_o) shift. However, the amount of the shift is almost solely a function of the "M" value at fracture and not the specimen geometry.

The $(B_{eff}/B_{nom})^{1/2}$ quantity has been developed by Nevalainen and Dodds to correct for out-of-plane constraint loss as predicted by the 3D computational analysis. This quantity is the equivalent fraction of the specimen thickness which is subjected to the same high stress conditions as the center-plane of the specimen. This ratio is a function of the deformation level at failure (Fig. 8), although the predicted effect saturates by $J_{avg}/(b\sigma_{ys}) > 0.02$ (M < 50).

The $(B_{eff}/B_{nom})^{1/2}$ component is geometry dependent, although it predicts less through thickness constraint loss for the Charpy {W/B = 1, SE(B)} geometry than for the standard SE(B) or C(T) specimen geometries.). This quantity is also generally dependent on the strain hardening exponent, but it is not strongly dependent on the stress contour ratio σ_1/σ_o. Regardless, this term provides relatively little reference temperature shift since it saturates quickly and only corrects K_{Jo} by the 1/4 power in the E1921 T_o analysis.

Constraint Corrected Data Analysis

The constraint correction analysis was implemented here by fitting simple functions to the Nevalainen and Dodds contours, and using these relationships in a spread sheet to evaluate the constraint corrected SSY toughness (J_o) from the J_c values for each specimen. The in-plane toughness scaling correction component was obtained by fitting a 3rd order polynomial to each toughness scaling contour.

The magnitude of the constraint correction is governed by the σ_1/σ_o contour chosen. In this feasibility study, the specific σ_1/σ_o ratio (see Table 4) was chosen to provide only a moderate constraint correction, even at high deformation levels ($J_o/J_c < 2$ for M $\approx$ 10) for all materials. Extreme constraint correction using low stress ratios was avoided.

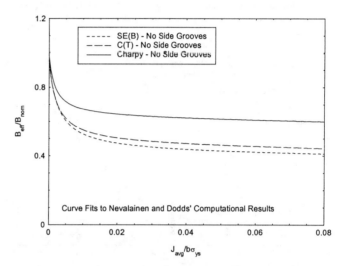

Figure 8 B_{eff}/B_{nom} Relationship for Charpy, SE(B), and C(T) Specimens.

Also, the same stress contour was applied for like materials. All A533B materials were corrected using the n = 10 relationships, while the A515 and A508 alloys were analyzed using the n = 5 relationships. The A533B, and A515 strain hardening exponents are close to these modeled values, while the A508 material actually falls between n = 5 and 10. The hardening exponent and stress contour values used for each material are summarized in Table 4. It should be emphasized that this simple approach was followed simply to examine the relative magnitude of the correction.

Similarly, the Nevalainen and Dodds results were fit and used to determine B_{eff} for each specimen as a function of $J_c/b\sigma_{ys}$. The fit used the following form:

$$B_{eff}/B_{nom} = \frac{\left[L(J_c/b\sigma_{ys})^2 + M(J_c/b\sigma_{ys}) + N\right]}{O + (J_c/b\sigma_{ys})} \quad (5)$$

where L, M, N, and O are fitting coefficients. No stress contour choice was necessary for the out-of-plane constraint correction component.

The procedure used to evaluate the constraint corrected reference temperature, T_o^c, is summarized as follows. The measured J_c values are corrected for in-plane constraint $\{J_o(J_c)\}$. Then, this value is converted to a 1T equivalent K_{Jo} values using the standard E1921 procedure, except that B_{eff} is used instead of B_{nom} for statistical thickness correction. The corrected Weibull scale parameter K_o^c and T_o^c were then calculated as per E1921, except that no additional censoring was performed. Censoring is directly accounted for in the constraint correction process.

Constraint Corrected Results

The application of the constraint correction analysis results in an overall improvement in the correspondence between the data obtained from the various specimen sizes and the master curve and confidence bounds, as shown for each material in Figs. 9 - 13. The constraint corrected results are also summarized (as T_o^c) in Table 3 for the large specimen, multi-temperature results, and Table 4 for the small specimens and individual data sets.

large and small specimen T_o values remain within 10°C. The corrected data is still well predicted by the large specimen failure probability bounds (Fig. 9). For the GGS A515 alloy, the E1921 T_{oss} results may be slightly biased ($\approx$ –5 to -6°C). Constraint correction reduces this small bias (Table 4). Further, the Charpy data is positioned better within the 1T Master Curve failure probability bounds (Fig. 10).

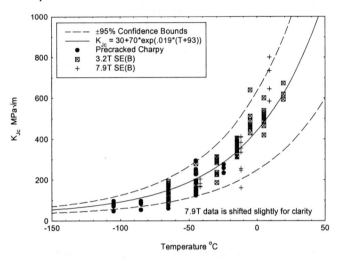

Figure 9: Constraint Corrected A508 Data.

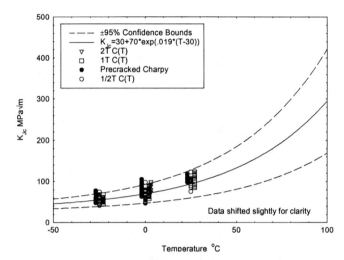

Figure 11: Constraint Correct A533B (JSPS) Data

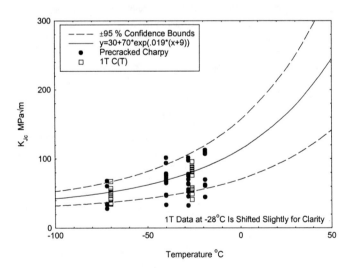

Figure 10: Constraint Corrected A515B (GGS) Data

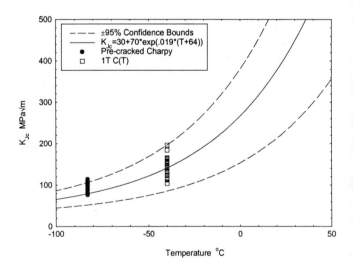

Figure 12: Constraint Corrected A533B (HEW) Data

For the A508 forging, the T_o values from large and small specimen measurement generally corresponded well prior to constraint correction. Constraint correction did not dramatically alter the large specimen data except for the 3.2T SE(B) data set at –15°C. This data set agreed much more closely with the other results after constraint correction. The T_o^c results for the Charpy specimens are shifted to a greater extent than the large specimen data, although the

For the JSPS A533B plate results, the initial E1921 T_o Charpy predictions were non-conservative by 15 to 20°C compared to measurements using 1/2T, 1T, and 2T C(T) specimens. After constraint correction, the average T_o^c of the Charpy specimens (21°C) falls in the midst of the average T_o^c measurements of the 1/2T (30 °C), 1T (30°C), and 2T (16°C) C(T) specimens. The corrected results for this material (Fig. 11) illustrate that the Charpy specimen data remains elevated in the large-specimen confidence bands. However, there is a closer correspondence between the Charpy and C(T) data than before constraint correction.

The E1921 T_o Charpy predictions for the HEW A533B material are similarly improved by constraint correction. The initial Charpy results were non-conservative by 17°C with respect to the 1T C(T) results. After correction, the Charpy specimen T_o^c results are only non-conservative by 9 °C. The corrected data for this material (Fig. 12) again appears to be elevated with respect to the 1T results. However, the correspondence is again better than the E1921 results.

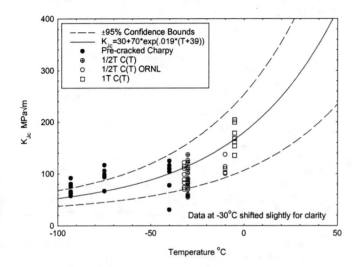

Figure 13: **Constraint Corrected A533B (HSST Plate 14) Data**

Unlike these other examples, constraint correction offers only slight improvement to the A533B HSST Plate 14 data-set, at least for the stress contour parameters chosen here. The 1/2T C(T) specimens tend to fall below the median toughness master curve (Fig. 13), while the Charpy specimens remain concentrated on the high side of the confidence band. The T_o value for the 1T testing is -59°C. The mean of the Charpy T_o results is approximately –66°C which is reasonably close to the 1T SE(B) result. There was also a large number (32 total) of 1/2T C(T) tests conducted, and these T_o results are all consistently around -45°C (Table 4).

While significant material variability from intense carbide has been shown to exist for HSST Plate 14 (Joyce et al., 1998), these specimens, as described earlier, were predominantly located within one section of the original plate. Therefore, material variability does not likely explain the apparent differences between the SE(B) and C(T) specimen results, both before and after constraint correction. It is interesting to note that the 1T SE(B) tests were conducted at a test temperature well above T_o, and the deformation at fracture is high (35 < M < 98) for this data set. This is the only 1T data set studied here that fails within this high deformation range. It is therefore possible that there may be fundamental differences between SE(B) and C(T) specimens that are not solely explained by constraint loss. This possibility will be examined in future work.

The difference between the small specimen and large specimen constraint corrected data is summarized as a function of M_{avg} in Fig. 14. In general, constraint correction does appear to remove most of the bias associated with the precracked Charpy T_o measurements. Only the HSST Plate 14 data, as mentioned earlier, remains outside the confidence bands. It appears as if the 1/2T C(T) data has been over corrected while the Charpy data has been slightly under corrected. As mentioned earlier there appear to be distinct differences between the C(T) and SE(B) results for this material.

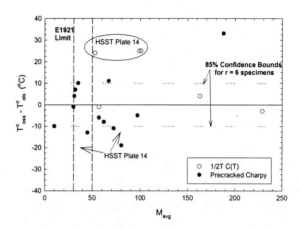

Figure 14: **Small and Large Specimen Difference After Constraint Correction.**

The remaining Charpy and 1/2T C(T) data sets with $M_{avg} > 100$ are generally consistent and are affected little by the constraint correction process as should be expected. For $M_{avg} < 100$, the precracked Charpy results are more evenly scattered about the T_{ols}^c value, and there is no discernible trend with M_{avg} as in Fig. 12. Constraint correction therefore seems like a promising approach for rectifying large and small specimen T_o measurements. However, as seen in Fig. 7, the choice of the critical stress contour level that governs cleavage critically defines the amount of correction for a given M_{avg} value. A more rigorous method for defining these parameters will be necessary to fully verify the correctness of this approach.

CONCLUSIONS

Reference temperatures were obtained as per ASTM E1921 for five pressure vessel steels using precracked Charpy specimens. The results were generally non-conservative when compared to T_o values from larger, standard fracture toughness specimens. The degree of non-conservatism is a function of the material and the general deformation level at fracture for the data set. While a more stringent deformation criterion would eliminate this bias, it would also greatly restrict the use of Charpy size surveillance specimens. It appears that using reference temperature results based solely on precracked Charpy specimens may require that some type of "Charpy temperature shift" be applied if the data will be used to determine the applicable reference temperature for the large-scale structure.

An alternative suggestion is to use constraint correction analyses to adjust the cleavage fracture data to a common small scale yielding basis, and then calculate the reference temperature from the constraint corrected data. This process was applied in this paper to the five materials studied. The corrected results exhibited an overall improvement in the correlation between reference temperatures predicted by the Charpy and larger size specimens. The constraint correction procedure adjusts the fracture toughness for both an in-plane fracture toughness scaling component and an out-of-plane B_{eff} component. The in-plane component is largely a function of the deformation level at fracture, material flow properties, and the governing stress contour. The out-of-plane component is primarily a function of the specimen geometry. Further work, however, is required to remove the rather arbitrary choice of stress ratio contour used in the present implementation of the constraint correction procedure.

REFERENCES

Anderson, T.L. and Dodds, R.H., Jr. "Specimen Size Requirements for Fracture Toughness Testing in the Transition Region," Journal of Testing and Evaluation, JTEVA, Vol. 19, No. 2, March 1991, pp. 123-134.

Chaouadi, R., "Fracture Toughness Measurements in the Transition Regime Using Small Size Samples," Small Specimen Test Techniques, ASTM STP 1329, W.R. Corwin, S.T. Rosinski, and E. van Walle, Eds., American Society for Testing and Materials, 1998.

Gao, X., Dodds, R.H., Jr., Tregoning, R.L., Link, R.E., and Joyce, J.A., "A Weibull Stress Model to Predict Cleavage Fracture in Plates Containing Surface Cracks," submitted to Fatigue & Fracture of Engineering Materials & Structures, 1999.

Joyce, J.A., "On The Utilization of High Rate Charpy Test Results and the Master Curve to Obtain Accurate Lower Bound Toughness Predictions in the Ductile-to-Brittle Transition," Small Specimen Test Techniques, ASTM STP1329, W.R. Corwin, S.T. Rosinski, and E.van Walle, Eds., American Society for Testing and Materials, pp. 253-273, 1997.

Joyce, J.A., and Link, R.E., "Ductile-to-Brittle Transition Characterization Using Surface Crack Specimens Loaded in Combined Tension and Bending," Fatigue and Fracture Mechanics: 28th Volume, ASTM STP 1321, J.H. Underwood, B.D. MacDonald, and M.R. Mitchell, Eds., American Society for Testing and Materials, 1997.

Joyce, J.A., Tregoning, R.L., and Zhang, X.J., "Application of Master Curve Technology to Biaxial and Shallow Crack Fracture Data for A533B Steels," 19th International Symposium on Effects of Radiation on Materials, ASTM E10, Seattle WA, June 16-18, 1998.

Kirk, M.T., and Dodds, R.H., Jr., "J and CTOD Estimation Equations for Shallow Cracks in Single Edge Notch Bend Specimens," NUREG/CR-5969, U.S. Nuclear Regulatory Commission, Washington, DC, July 1993.

Kirk, M.T., Koppenhoefer, K.C., and Shih, C.F., "Effect of Constraint on Specimen Dimensions Needed to Obtain Structurally Relevant Toughness Measures," Constraint Effects in Fracture, ASTM STP 1171, E.M. Hackett, et al., Eds., American Society for Testing and Materials, pp. 2-20, 1993.

Kirk, M., and Lott, R., "Empirical Validation of the Master Curve for Irradiated and Unirradiated Reactor Pressure Vessel Steels," ASME/JSME Pressure Vessel and Piping Conference, San Diego, CA., July 1998.

Lidbury, D., and Moskovic, R., "Assessment of the Ductile-to-Brittle Transition Toughness Behavior of an A508 Class 3 PWR Pressure Vessel Steel by a Statistical Approach," Pressure Vessel Integrity, PVP, Vol 250, The American Society of Mechanical Engineers, pp. 283-294, 1993.

Merkle, J.G., Wallin, K., and McCabe, D.E., "Technical Basis for an ASTM Standard on Determining the Characterization Temperature, T_o, for Ferritic Steels in the Transition Range," NUREG/CR-5504, U.S. Nuclear Regulatory Commission, Washington, DC, November 1998.

McAfee, W.J., Bass, B.R., and Bryson, J.W., "Development of a Methodology for the Assessment of Shallow-Flaw Fracture in Nuclear Reactor Pressure Vessels: Generation of Biaxial Shallow-Flaw Fracture Toughness Data," ORNL/NRC/LTR-97/4, Oak Ridge National Laboratory, July 1998.

Nevalainen, M., and Dodds, R.H., Jr., "Numerical Investigation of 3-D Constraint Effects on Brittle Fracture in SE(B) and C(T) Specimens," International Journal of Fracture, Vol. 74, pp. 131-161, 1995.

Porr, W.C., Jr., Link, R.E., Waskey, J.P., and Dodds, R.H., Jr., "Experimental Application of Methodologies to Quantify the Effect of Constraint on J_c for a 3-D Flaw Geometry," Fracture Mechanics: Volume 26, ASTM STP 1256, W.G. Reuter, J.H. Underwood, and J.C. Newman, Eds., American Society for Testing and Materials, pp. 2-20, 1995.

Ruggieri, C., Dodds, R.H., Jr., and Wallin, K., "Constraint effects on reference temperature, T_o, for ferritic steels in the transition region," Engineering Fracture Mechanics, Vol. 60, No. 1, pp. 19-36, 1998.

Tregoning, R.L. and Joyce, J.A. "T_o Estimation from Small Specimens," submitted to Engineering Fracture Mechanics, 1999. [6]

Wallin, K., "Validity of Small Specimen Fracture Toughness Estimates Neglecting Constraint Corrections," Constraint Effects in Fracture Theory and Applications: Second Volume, Mark Kirk and Ad Bakker, Eds., American Society for Testing and Materials, 1995.

Wallin, K, "Application of the Master Curve Method to Crack Initiation and Crack Arrest," submitted to ASME 199 PVP Conference, Boston, MA, 1-5 August, 1999.

Wallin, K., "Master Curve Analysis of Ductile to Brittle Transition Region Fracture Toughness Round Robin Data: The "Euro" Fracture Toughness Curve," VTT Manufacturing Technology, 367, 1998.

PVP-Vol. 393, Fracture, Fatigue and Weld Residual Stress
ASME 1999

EFFECT OF LOADING RATE ON FRACTURE TOUGHNESS OF PRESSURE VESSEL STEELS

K. K. Yoon
Framatome Technologies

W. A. Van Der Sluys
McDermott Technology

K. Hour
BWX Technologies

ABSTRACT

The master curve method has recently been developed to determine fracture toughness in the brittle-to-ductile transition range. This method was successfully applied to numerous fracture toughness data sets of pressure vessel steels. Joyce (1997) applied this method to high loading rate fracture toughness data for SA-515 steel and showed the applicability of this approach to dynamic fracture toughness data. In order to investigate the shift in fracture toughness from static to dynamic data, B&W Owners Group tested five weld materials typically used in reactor vessel fabrication in both static and dynamic loading. The results were analyzed using ASTM Standard E 1921 (ASTM, 1998).

This paper presents the data and the resulting reference temperature shifts in the master curves from static to high loading rate fracture toughness data. This shift in the toughness curve with the loading rate selected in this test program and from the literature is compared with the shift between K_{Ic} and K_{Ia} curves in ASME Boiler and Pressure Vessel Code. In addition, data from the B&W Owners Group test of IAEA JRQ material and dynamic fracture toughness data from the Pressure Vessel Research Council (PVRC) database (Van Der Sluys, 1998) are also presented. It is concluded that the master curve shift due to loading rate can be addressed with the shift between the current ASME Code K_{Ic} and K_{Ia} curves.

INTRODUCTION

Recently published ASTM Standard E 1921 provides acceptable practices for conducting fracture toughness tests in the brittle-to-ductile transition temperature regime and the use of Weibull statistical analysis methods to compile and interpret the test data. The resulting median fracture toughness, $K_{Jc(med)}$, yields the reference temperature T_0, which is the temperature at which the median fracture toughness has a value of 100 MPa$\sqrt{m}$, thus defining the master curve. This provides a means to determine a material-specific fracture toughness curve for the material of interest. The master curve can then be applied to reactor vessel integrity analysis. This method has been successfully applied to numerous fracture toughness data sets of pressure vessel steels.

The proposed approach for implementation of the master curve method to the ASME Boiler and Pressure Vessel Code utilizes a method for directly indexing the K_{Ic} curve based on an empirically derived relationship between the master curve and the K_{Ic} curve. The approach described here is intended to achieve the short-term objective of PVRC Task Group on Master Curve Toughness defining an alternative to RT_{NDT} for indexing fracture toughness. This approach permits use of fracture toughness properties based on the Master Curve methodology without the need to redefine the lower bound toughness curves, K_{Ic} and K_{Ia}, that are currently defined in the ASME Code. Since the Master Curve is based on static, crack initiation toughness properties, the relationship is established with the ASME Code K_{Ic} curve developed as a lower bound representation of initiation toughness, whereas the K_{Ia} curve is based on dynamic and crack arrest toughness. The advantage in this approach is that it permits the direct measurement of the indexing parameter T_0 from fracture toughness tests rather than the indirect parameter RT_{NDT} from Charpy and drop-weight tests.

Joyce (1997) recently demonstrated that the master curve method is equally applicable to dynamically tested fracture data and that the reference temperature obtained from dynamic fracture toughness data exhibits a systematic shift in T_0 from that of quasi-static test data. Wallin (1997) also analyzed dynamic fracture toughness data using the master curve method and calculated the T_0 shifts from static to dynamic data in terms of loading rates. When the loading rate anticipated in the nuclear power plant applications are considered, the predicted ΔT_0 is bounded by the shift between the K_{Ic} and K_{Ia} curves.

FRACTURE TOUGHNESS TEST PROCEDURES

Precracked Charpy specimens made of Linde-80 weld metals were tested in the transition temperature range according to ASTM Standard E 1921-97 (ASTM, 1998). This standard provides acceptable practices for conducting fracture toughness tests in the transition temperature regime and the use of Weibull statistical analysis methods to compile and interpret the test data. This provides a means to determine a material-specific fracture toughness curve for a given material. The master curve can then be applied to reactor vessel integrity analysis. B&W Owners Group has successfully performed this type of testing on both unirradiated and irradiated Charpy specimens using quasi-static loading rate (Hour and Yoon, 1997).

Pre-cracking of the non-side-grooved Charpy specimens was conducted on a MTS servo-hydraulic machine in compliance with ASTM Standard E 1921-97. The fracture tests were performed in displacement control mode. A double cantilever beam clip gage was mounted onto the knife edges to monitor the crack mouth opening displacement (CMOD). Force and displacement were monitored continuously through loading and unloading cycles with a load cell and an LVDT interfaced to a personal computer. The maximum possible actuator speed of this test machine was approximately 1.12 inch/sec. Depending on the nature of specimen failure, the time to failure ranged from 20 mili-seconds to several seconds. All tests were performed in an ATS split-type furnace designed to handle both elevated temperature and sub-zero temperature testing. Temperature was determined by placing a Type K thermocouple wire onto the specimen surface. A soak time of at least 20 minutes, after reaching test temperature, was used to ensure uniform temperature distribution in the specimen. Temperature control was provided by automated control of a solenoid valve that regulates the flow rate of liquid nitrogen into the furnace. Test temperatures were controlled to within ± 4 °F of a selected test temperature.

HIGH LOADING RATE TEST DATA AND ANALYSIS

All tests were conducted using the maximum actuator speed of the test machine. Data were collected continuously during the test. Since some of the tests were completed within 20 milliseconds, a high data collection rate was used to ensure that a sufficient number of data points were collected for data analysis purpose. A spreadsheet was constructed to perform the calculation, using the imported time, force, load-line displacement, and CMOD from the raw data file. The spreadsheet then was used to calculate J_c, K_{Jc}, time to failure, and dK_I/dt according to E 1921. Stress intensity factor rate (dK_I/dt) was calculated by the following equation:

$$dK_I/dt = K_{Jc} /(\text{time to failure})$$

Depending on the failure mechanism of the specimen (thus the time to failure), the time rate of change in stress intensity factor ranged from 879 to 3626 MPa$\sqrt{m}$/sec. (800 to 3300 ksi$\sqrt{in}$/sec.).

The test temperature for each group of pre-cracked Charpy specimens depends on the material's fracture toughness property, and is very important to obtain valid K_{Jc} data points. For each material set, a scoping test was performed to establish an appropriate test temperature. Guided by the results of the scoping tests, seven groups of Linde 80 weld specimens were tested in three-point bending. A total of 53 K_{Jc} values were obtained, ranging from 67 to 117 MPa$\sqrt{m}$.

Through careful test temperature selections, all 53 K_{Jc} data are valid. Therefore, the upper limit K_{Jc} was met and no censoring was necessary in accordance with ASTM Standard E 1921. The K_{Jc} uppper limit is defined as

$$K_{Jc(limit)} = (Eb_0\sigma_{ys}/30)^{0.5}$$

where E = elastic modulus
b_0 = initial remaining ligament
σ_{ys} = yield strength

The yield strength was obtained from tensile tests. The lower limit for this program is 45 MPa$\sqrt{m}$. and all data were well above the lower toughness limit. The optical crack length was measured according to the standard. No crack growth was observed on any Charpy size specimens. The crack lengths determined from these measurements were used in the data analysis.

The median toughness $K_{Jc(med)}$ and reference temperature, T_0, for each set of material were calculated according to ASTM Standard E 1921-97 and the results are shown in Table 1.

Applicability Of Master Curve Method To Dynamic Fracture Data

Dynamic fracture toughness data are K_{Jc} data obtained by a higher rate of loading. Joyce (1997) and Wallin (1997) demonstrated that the Master Curve method is equally applicable to a high loading rate fracture test data. The results from the B&W Owners Group dynamic tests also confirmed that the Master Curve method is applicable to the Linde 80 weld data. A Weibull and a K_{Jc}–temperature plots are shown in Figures 1 and 2 for SA-1526 weld. Similar plots are applicable to other Linde 80 welds tested in this program. When these toughness plots are indexed to the reference temperature, T_0, the results are as shown in Fig. 3. This shows that all the data points are within 5%/95% tolerance bound.

Table 1. $K_{Jc(med)}$ and Reference Temperature for Linde-80 Weld Metals

Material ID	$K_{Jc(med)}$ MPa√m	Dynamic Test T_0, °C	Static Test[+] T_0, °C
SA-1526	81.8	-71.4	N/A
WF-182-1	69.7	-58.2	-108
WF-193	75.9	-62.2	-112
WF-25	76.8	-79.9	-92
WF-25*	80.8	-75.9	N/A
WF-70	73.5	-81.6	-77
WF-70*	78.1	-86.9	-77

[+] T_0 for static test data are from the earlier B&W Owners Group programs.
*Material from a different plant.

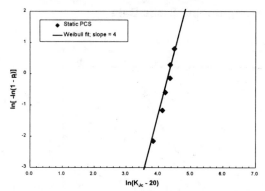

Figure 1. Weibull Plot of WF-182-1 High Loading Rate Data

In Fig. 4, all the dynamic data in the PVRC database (Van Der Sluys et al., 1998) are presented, indexed by T_0. This database includes old data and precracked Charpy bend test data. In this figure, all precracked Charpy test data that exceed the $K_{Jc(limit)}$ were excluded. This plot is significant in illustrating the applicability of the Master Curve method to dynamic fracture toughness data.

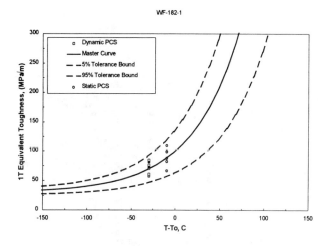

Figure 2. Master Curve and K_{Jc} Data - WF-182-1

REFERENCE TEMPERATURE COMPARISON

T_0 Shift Due To High Loading Rate

Joyce demonstrated that there is a systematic shift in T_0 as a function of loading rate as shown in Fig. 5. The B&W Owners Group dynamic data for five Linde 80 (seven groups) welds are also plotted in Fig. 5. These data show a similar trend as the SA 515 data. Both static and dynamic fracture toughness of these Linde 80 welds are higher than the SA 515 steel, however, T_0 shifts due to higher loading rate are approximately equal to the SA 515 data. Another dynamic fracture toughness data set is reported by Yoon and Hour (1999). This set is for the IAEA round-robin test material JRQ which is a SA-533 plate. The T_0 shift due to high loading rate for JRQ material also exhibits a similar trend observed in SA 515 as shown in Fig.5.

A similar comparison is shown in Fig. 6 where the Japanese Welding Society dynamic fracture toughness data from the PVRC database are plotted with SA 515 data. These data also follow the same trend observed in Fig. 5 and exhibit a similar T_0 shift behavior from.

Following Wallin's suggestion (1997), Figures 5 and 6 can be combined into one T_0 shift (ΔT_0) versus loading rate plot shown in Fig. 7. With the exception of WF-70 and WF-25, all the industry data exhibit similar trends as the SA 515 does. The WF-70 and WF-25 welds show less shift between static and dynamic data, however, they are conservatively bounded by the best fit equation to the SA 515 data.

The shift in T_0 between static and dynamic fracture toughness data can be modeled according to a best-fit equation through the SA 515 data. That is;

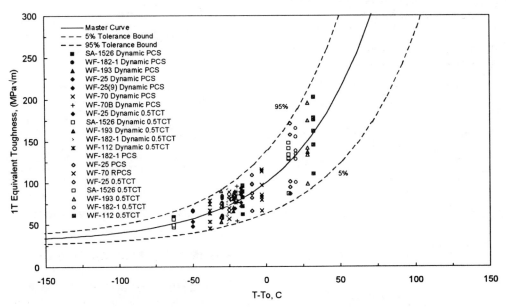

Figure 3. All High Loading Rate Data and Corresponding Static Data

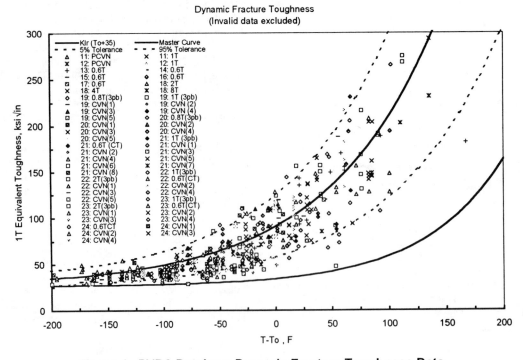

Figure 4. PVRC Database Dynamic Fracture Toughness Data

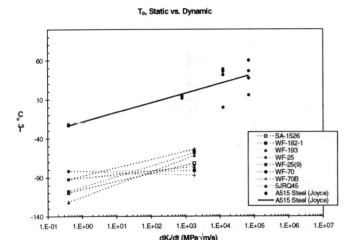

Figure 5. SA 515 Data and B&WOG Linde 80 Weld Data

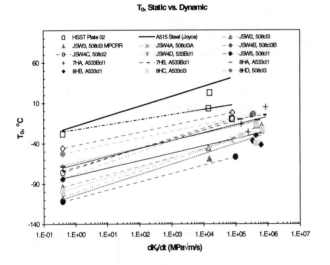

Figure 6. T_0 - PVRC Database Dynamic Fracture Toughness Data

$$\Delta T_0 = 5.338 \ln(dK_I/dt) - 17.11, \; ^0C$$

where dK_I/dt is loading rate in MPa√m/second. The weld metals WF-70 and WF-25 show a lesser shift. The reasons for the lesser shift for these metals are under study at this point, however, these welds are conservatively bounded by the above shift equation.

EFFECT OF T_0 SHIFT TO ASME CODE TOUGHNESS CURVES

Wallin (1997) elected to use the loading rate, dK_I/dt (MPa√m/sec.) instead of more conventional strain-rate for dynamic test data analysis, since the strain-rate in a fracture test specimen is somewhat controversial parameter. He established shift, ΔT_0, based on a definition of qausi-static loading rate of 1 MPa√m/sec.

Since ΔT_0 is a function of loading rate, it is interesting to determine what range of loading rates are anticipated for the nuclear power plant applications. Most high rate of loading rate can be found in the pressurized thermal shock (PTS) scenario transients, however, the loading rate for the severest PTS event is rather small. One of the high loading rate cases is shown in Fig. 7. Which is the case of Main Steam Line break transient listed in Selby (1985). The applied K_I-time history curves are shown in Fig. 7 for crack depths ranging from 10% to 95% through-wall. The highest loading rate is found in the 85% through wall crack depth curve (Curve P in Fig. 7) and is estimated to be only 1.3 MPa√m/sec. This is a quasi-static loading rate. For more realistic crack depths of 50% through-wall (Curve L in Fig. 7) or less, the loading rate is less than 0.4 MPa√m/sec., which corresponds to a typical loading rate for static tests.

To make a conservative assessment, it is assumed that the maximum loading rate anticipated in the PTS events is 10 MPa√m/sec. From Fig. 8, the predicted ΔT_0 is approximately 22^0C. This shift is less than the shift between the Code K_{Ic} and K_{Ia} curves, which is about 52.9^0F (30^0C) as shown in Fig. 9. Therefore, an equivalent RT_{NDT} value determined from a statically determined T_0 can be used in the K_{Ia} curve that will account for the shift due to the maximum loading rate anticipated in PTS transients, which have very low probability of occurrence.

CONCLUSIONS

1. High loading rate fracture toughness tests were performed on unirradiated pre-cracked Charpy specimens in the transition temperature range in accordance with ASTM Standard E 1921. The Master Curve method is applicable to the dynamic fracture toughness test data.
2. The shift in T_0 between static and dynamic fracture toughness data can be modeled according to a best-fit equation through the SA 515 data.
3. This equation is applicable to other industry data shown in this paper with the exception of the B&W Owners Group data WF-70 and WF-25. These weld materials exhibit less shift, however, these welds are conservatively bounded by the above shift equation.
4. The loading rates in the PTS transients are quasi-static.

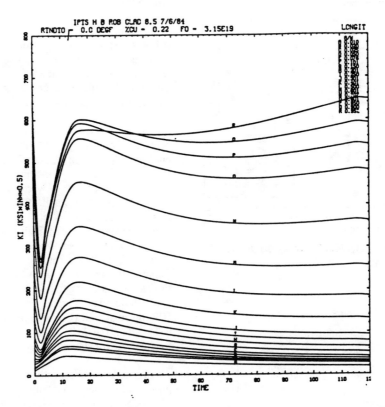

Figure 7. Transient 8.5: K_{Ic} vs. Time (minutes) for Various Depths in Wall

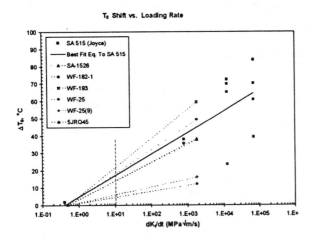

Figure 8. ΔT_0 vs. Loading Rate

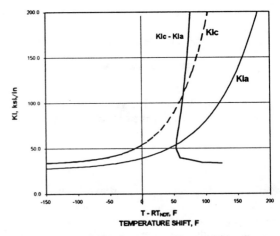

Figure 9. Shift between K_{Ic} and K_{IR} Curves

ACKNOWLEDGEMENT

Authors wish to acknowledge the support provided by the B&W Owners Group for this work.

REFERENCES

ASTM, 1998, Standard E 1921-97, "Standard Test Method for the Determination of Reference Temperature, T_0, for Ferritic Steels in the Transition Range," <u>1998 Annual Book of ASTM Standards</u>, Vol. 03.01, American Society for Testing and Materials, West Conshohocken, PA.

Hour K. Y. and Yoon K. K., 1997, "Fracture Toughness Test on Pre-cracked Charpy Specimens in the Transition Range for Linde 80 Weld Metals," <u>Small Specimens Test Technique</u>, <u>ASTM STP 1329</u>, W.R. Corwin, S.T. Rosinski, and E. Van Walle, Eds., American Society for Testing and Materials, 1997.

Joyce, J. A., 1997, "On the Utilization of High Rate Charpy Test Results and the Master Curve to Obtain Accurate Lower Bound Toughness Predictions in the Ductile-to-Brittle Transition, Small Specimen Test techniques," <u>Small Specimens Test Technique</u>, <u>ASTM STP 1329</u>, W.R. Corwin, S.T. Rosinski, and E. Van Walle. Eds., ASTM

Selby, D. L., et al., 1985, "PTS Evaluation of the H. B. Robinson Unit 2 Nuclear Power Plant," Oak Ridge National Laboratory, NUREG/CR-4183, September 1985.

Van Der Sluys, W. A., Yoon, K. K., Killian, D. E., Hall, J. B., 1998, "Fracture Toughness of Ferritic Steels and ASTM Reference Temperature (T_0)," BAW-2318, Framatome Technologies.

Yoon, K. K. and Hour, K., 1999, "Dynamic Fracture Toughness and Master Curve Method Analysis of IAEA JRQ Material," Structural Mechanics in Reactor Technology Conference, Seoul, Korea.

Wallin, K., 1997, "Loading Rate Effect on the Master Curve T_0," VTT Manufacturing Technology, Espoo, Finland.

INITIAL REFERENCE TEMPERATURES AND IRRADIATION TREND CURVES FOR USE WITH RT_{To}, A PRELIMINARY ASSESSMENT

Mark Kirk and Randy Lott
Westinghouse Electric Company
Pittsburgh, Pennsylvania, USA

William Server
ATI Consulting
Pinehurst, North Carolina, USA

Stan Rosinski
Electric Power Research Institute
Charlotte, North Carolina, USA

ABSTRACT

The "Master Curve," as initially proposed by Wallin and co-workers, has emerged as a robust characterization of the transition fracture toughness of ferritic steels. The American Society for Testing and Materials (ASTM) recently published a standard method for determining the Master Curve index temperature (T_o) based on limited fracture toughness testing at one temperature using specimens as small as precracked Charpys. The potential for characterizing the entire fracture mode transition based on limited testing of specimens already in surveillance programs has sparked considerable interest within the commercial nuclear power community. Recently, ASME published Code Case N-629 that permits use of a Master Curve-based reference temperature ($RT_{To} \equiv T_o + 35°F$) as an alternative to traditional methods of positioning the ASME K_{IC} curve. In this paper we provide information supportive of RT_{To} application to nuclear RPV assessment on a generic basis. Specifically, we develop generic unirradiated initial reference temperature (IRT) values for use with RT_{To}. These values suggest that currently accepted generic IRT values (-56°F for CE welds, and -5°F for Linde 80 welds) are overly conservative, with the degree of over-conservatism dependent upon flux type. Furthermore, we demonstrate that the NRC Regulatory Guide 1.99 Rev. 2 fluence function characterizes irradiation-induced shifts of the fracture toughness transition curve as well as it does shifts of the Charpy transition curve. These findings are based on a fracture toughness database for RPV steels.

INTRODUCTION AND OBJECTIVE

To maintain their operating licenses, nuclear plant operators must demonstrate that the effects of irradiation embrittlement do not compromise the safe operation of their reactor pressure vessel (RPV). The variation of the RPV steel's fracture toughness with temperature provides a key input to these analyses. Currently, the ASME K_{Ic} curve provides this toughness characterization. The K_{Ic} curve is a lower bound curve that is indexed to a reference temperature (i.e. RT_{NDT}) defined in Section III of the ASME Boiler and Pressure Vessel Code, and determined by testing nil-ductility temperature specimens per ASTM E208, and by testing Charpy-V notch (CVN) specimens as per ASTM E23.

In current applications, RT_{NDT} establishes the position of the K_{Ic} curve on the temperature axis. A two-step process establishes the appropriate RT_{NDT} for an irradiated material, and thus the appropriate variation of K_{Ic} with temperature for use in a RPV assessment. First, RT_{NDT} is determined in the unirradiated condition for the limiting material in the RPV (i.e. the part of the RPV thought to have the lowest toughness). Throughout the operational life of the vessel, tests of CVN specimens removed from surveillance capsules quantify the shift of the 41 J (30 ft-lb) CVN transition temperature produced by irradiation. This shift is added, along with margins to account for uncertainties, to the RT_{NDT} for the unirradiated material to estimate the RT_{NDT} of the irradiated material [RG1.99].

This approach to estimating fracture toughness after irradiation relies on correlations to fracture toughness rather than on direct measurements. The large specimen size required to measure "valid" linear-elastic fracture toughness values necessitated this approach. This correlative approach typically under-predicts fracture toughness, sometimes to a considerable degree. A perception of low toughness can lead to premature plant closure because of regulatory limits on pressurized thermal shock [10CFR50.61] Furthermore, low toughness limits the permissible pressure-temperature envelope for routine heat-up and cool-down operations. This may cause performance of these operations in a manner that increases overall plant risk due to the increased probability for pump trips, and due to the increased time spent in transient, *vs.* steady state, conditions.

Technological advancements in the last 25 years set the scene for a more direct approach to the characterization of irradiation effects on the fracture toughness of RPV steels. The "Master Curve," as initially proposed by Wallin and co-workers [Wallin 1984], has emerged as a robust characterization of the transition fracture toughness of ferritic steels. The Master Curve describes the variation of fracture toughness with temperature in fracture mode transition based on the results of fracture toughness tests. Fracture toughness values are used to estimate an index temperature T_o, which positions the Master Curve along the temperature axis.

ASTM recently published a standard method for determining T_o based on 6-10 fracture toughness tests conducted at a single temperature [ASTM E1921]. The specimen size requirements of E1921-97 permit determination of valid toughness (K_{Jc}) values using specimens as small as precracked Charpys. Charpy specimens are included in all nuclear surveillance programs as per the requirements of 10CFR 50 Appendix H and the guidance of ASTM E185-94. The potential for characterizing the entire fracture mode transition based on limited testing of specimens already in surveillance has sparked considerable interest within the nuclear power community. These developments mark the first opportunity to know the fracture toughness of an operating reactor vessel based on direct measurements without resort to the use of empirical correlations since the initiation of commercial nuclear power generation at Shippingport Pennsylvania in 1957.

A number of recent investigations have empirically validated the premises that underlie the Master Curve, and demonstrated that it applies equally well to RPV steels both before and after irradiation [Wallin 1993, Sokolov 1996, Kirk 1998]. Efforts to provide a physical rationale for why a Master Curve should exist for such a wide strength and chemical composition range also appear promising [Natishan 1998]. Finally, basis documents concerning both Master Curve technology and the ASTM testing standard have been published [EPRI 1998, Merkle 1998]. This background work has led to recent passage of ASME Code Case N-629 that permits use of a Master Curve-based reference temperature, RT_{To}, as an alternative to RT_{NDT} for indexing the ASME K_{IC} curve [ASME N629]. ASME Code Case N-629 defines RT_{To} as follows:

$$RT_{To} \equiv T_o + 35°F \qquad (1)$$

A companion paper in this Proceedings provides an example application of RT_{To} to assessment of the adjusted reference temperature at end of license (*EOL*) for one commercial nuclear plant [Lott 1999]. In this paper we provide information that supports RT_{To} application to nuclear RPV assessment on a more generic basis.

CURRENT PRACTICE FOR *ART* DETERMINATION

RPV integrity analysis requires an evaluation of the fracture toughness of the irradiated steel. Determination of RT_{NDT} for the unirradiated steel (the *IRT*, or initial reference temperature) is the first step in this evaluation. The *IRT* is modified by adding (a) the shift in Charpy transition temperature at 41J (30 ft-lbs) produced by irradiation (ΔRT_{NDT}), and (b) a margin term (*M*) to account for uncertainties [RG1.99]. This calculation produces a reference temperature adjusted for the effects of irradiation, or *ART*. Reg. Guide 1.99 (Rev. 2) defines the *ART* as follows:

$$ART = IRT + \Delta RT_{NDT} + M \qquad (2)$$

where

$$\Delta RT_{NDT} = (CF)(\phi t)^{(0.28-0.1\log \phi t)}$$

CF = the Chemistry Factor, either fit to surveillance Charpy data or estimated from chemical composition
ϕt = fluence / 10^{19} n/cm^2
$M = \sqrt{\sigma_I^2 + \sigma_\Delta^2}$
σ_I = the uncertainty in *IRT*
σ_Δ = the uncertainty in ΔRT_{NDT}.

Currently, plant (or material) specific measurements of both *IRT* and ΔRT_{NDT} are preferred, but generic values are used when direct measurement is not possible. Generic values of *IRT* and ΔRT_{NDT} are estimated as follows:

Generic Values of *IRT*: When direct measurements of *IRT* based on *NDT* and Charpy tests are not available, generic mean values are assigned to welds based on flux type [SECY-82-465, 10CFR50.61]. The SECY document established a generic *IRT* of -56°F for all Combustion Engineering (CE) welds (Linde flux types 0091, 0124, and 1092). This -56°F was the mean of 82 unirradiated RT_{NDT} values (49 of Linde 0091, 20 of Linde 0124, and 13 unidentified). Similarly, the SECY document established a generic *IRT* of 0°F (later reduced to -5°F) for Babcock and Wilcox (B&W) welds (Linde 80 flux) based on 10 unirradiated RT_{NDT} values. No generic *IRT* values are recognized currently for plate or forging materials.

Generic Values of ΔRT_{NDT}: When the surveillance program includes the limiting material, Chemistry Factor (*CF*) is determined by fitting the Charpy transition temperature shift caused by irradiation at 41J to the Reg. Guide 1.99 fluence function as follows:

$$CF = \frac{\sum_{i=1}^{n}\left[A_i(\phi t_i)^{(0.28-0.1\log\phi t_i)}\right]}{\sum_{i=1}^{n}(\phi t_i)^{(0.56-0.2\log\phi t_i)}} \quad (3)$$

where A_i are the measured values of ΔRT_{NDT}, ϕt_i is the fluence (/10^{19} n/cm^2) for each ΔRT_{NDT}, and n (the number of measured ΔRT_{NDT} values) must be at least 2. In situations where the limiting material is not in surveillance, CF may be estimated via a correlation with copper content, nickel content, and product form (plate or weld) [RG1.99, Randall 1987].

In this paper we focus on development of generic IRT values for use with RT_{To}, and we demonstrate that form of the Reg Guide 1.99 (Rev. 2) fluence function applies as well to ΔRT_{To} as it does to ΔRT_{NDT} values. Questions regarding the appropriate uncertainty factors (i.e. σ_I and σ_Δ) for use with RT_{To}, while important to the development of a complete methodology, are reserved as a topic for future study.

DATABASE

A database of fracture toughness values for ASTM A533B, ASTM A508, and ASTM A302 steels (and their weldments) is currently being assembled from the literature. Table 1 summarizes how the data now available divide between the different product forms and irradiation conditions. In addition to fracture toughness data, the database also includes other indications of transition shift (i.e. ΔT_{NDT}, ΔRT_{NDT}, and ΔCV_{30}), chemical composition data, and strength data. References which have contributed significantly to this database are listed at the end of this paper, or in a previous paper by these authors [Kirk 1998].

The T_o values reported here are calculated using the maximum likelihood method proposed by Wallin [1995] to make best use of the available toughness data. T_o is determined iteratively using the following equation:

$$\sum_{i=1}^{n}\left[\frac{\delta_i \cdot \exp\{c(T_i-T_o)\}}{a-K_{min}+b\cdot\exp\{c(T_i-T_o)\}}\right] - \sum_{i=1}^{n}\left[\frac{(K_{Jc}^i-K_{min})^4 \cdot \exp\{c(T_i-T_o)\}}{\|a-K_{min}+b\cdot\exp\{c(T_i-T_o)\}\|^5}\right] = 0 \quad (4)$$

where
- n is the number of toughness specimens tested
- T_i is the test temperature
- K_{Jc} is the lower of the measured K_{Jc} value and the E1921-97 $K_{Jc(limit)}$ value. These K_{Jc} values are converted to 1T equivalence.

- a = 28.179 ksi√in
- b = 69.993 ksi√in
- c = 0.0106
- K_{min} = 18.18 ksi√in

- δ_i is 1 if E1921-97 size requirements are met
  is 0 if E1921-97 size requirements are not met

In no case has a T_o value been calculated from less than 6 K_{Jc} values that satisfy the E1921-97 validity criteria. Other studies have demonstrated that T_o values calculated by the maximum likelihood technique agree well with T_o values calculated as per ASTM E1921-97 protocols. In cases where data only exists at one temperature, eq. (4) and E1921-97 produce identical results.

Table 1. The number of K_{Jc} Values in the Database.

Designation	Unirradiated	Irradiated
Linde 0091	71	77
Linde 0124	178	160
Linde 1092	141	22
Linde 80	117	189
A302B	55	135
A302B Modified	6	0
A533B Cl. 1	664	126
A508 Cl. 2	72	4
A508 Cl. 3	427	0
Weld Subtotal	507	448
Plate Subtotal	725	261
Forging Subtotal	499	4
Total	1731	713
Welds	29%	63%
Plates	42%	37%
Forgings	29%	1%

INITIAL UNIRRADIATED REFERENCE TEMPERATURE, IRT

Figure 1 summarizes available results pertinent to determination of generic values of RT_{To} for different weld flux types and ASTM material types. A total of 66 individual RT_{To} values are available currently, each calculated from independent K_{Jc} data sets. If K_{Jc} data sets for the same material are available from multiple sources, these data sets are combined together and input to eq. (4) to obtain a single T_o value. Here we define "same material" as follows: same heat number for plates and forgings, same weld wire heat number and flux type for welds. The upper part of Figure 1 provides the individual RT_{To} values while the lower part of the figure compares the estimated mean and 95% confidence bounds on the mean with the currently accepted generic IRT values for CE and B&W welds (no generic IRT values are currently recognized for plates and forgings). Comparison is made between currently accepted generic values and mean

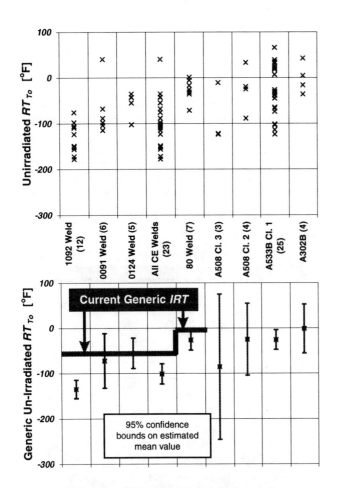

Fig. 1. Values of RT_{To} before irradiation for a variety of RPV steels and their weldments (numbers in parentheses are the number of RT_{To} values).

RT_{To} values of because current generic IRTs were established from experimental data as mean values [SECY-82-465]. Furthermore, the IRT is clearly intended to be a best estimate value in eq. (2), with the uncertainty in IRT accounted for by the separate σ_i term. Some of the data groupings are clearly too small to make a realistic estimate of the mean value, these we present for information only. However, the "All CE Welds" and "Linde 80" data sets both contain a sufficient number of RT_{To} observations for accurate determination of a mean value. Comparison of the best estimate mean values of RT_{To} to the currently accepted generic IRT values [SECY-82-465, 10CFR50.61] suggests that both CE welds and Linde 80 welds are assessed conservatively, by 45°F and 22°F, respectively. Furthermore, the generic IRT of Linde 1092 welds is assessed even more conservatively (by 79°F based on currently available data).

These data provide an initial indication of generic IRT values for RT_{To}-based reactor vessel integrity assessments.

Furthermore, available evidence suggests that currently accepted generic IRT values are overly conservative, with the degree of over-conservatism dependent upon flux type. These material-to-material differences indicate that current technology may not enforce a uniform safety margin across all operating commercial nuclear power plants.

IRRADIATION TREND CURVES

Reg. Guide 1.99 (Rev. 2) includes the following fluence function, which describes the increase in transition temperature produced by irradiation:

$$\Delta RT_{NDT} = (CF)(\phi t)^{(0.28 - 0.1 \log \phi t)} \quad (5)$$

This functional form was established by curve-fitting a database of 179 41J Charpy shift values [Randall 1987]. Current efforts within ASTM Committee E10.02 focus on updating eq. (5) to include the more substantial empirical database developed since the introduction of eq. (5), and modification of the fluence function to reflect the current best understanding of the physical mechanisms that cause irradiation damage [Eason 1998].

Lack of sufficient ΔRT_{To} data to permit fitting Eason's new fluence functions well, and limited differences between eq. (4) and Eason's new proposal motivates our focus on the appropriateness of applying a Reg. Guide 1.99 (Rev. 2)-type fluence function to ΔRT_{To} data, i.e.:

$$\Delta RT_{To} = (CF)(\phi t)^{(0.28 - 0.1 \log \phi t)} \quad (6)$$

Irradiation shifts in both Charpy and fracture toughness transitions are controlled to a large extent by increases of material flow strength produced by irradiation. It is therefore reasonable to assume that a fluence function that describes shifts in the Charpy transition temperature might also model well shifts in fracture toughness transition temperature.

All data sets having at least one unirradiated RT_{To} value and two irradiated RT_{To} values were used (four Linde 80 welds, three Linde 1091 welds, three Linde 0124 welds, and six A533B plates). This practice for data consideration is consistent with Reg. Guide 1.99 (Rev. 2) requirements. Chemistry Factor (CF) values were calculated using eq. (3) where the A_i are now the measured values of ΔRT_{To}. Figures 2-5 illustrate the four data sets having the largest number of individual ΔRT_{To} observations. The figure includes three graphs for each data set:

A. ΔRT_{To} vs. ϕt
B. ΔRT_{To} vs. $(\phi t)^{[0.28 - 0.10*\log(\phi t)]}$ (a linearized version of Graph A)

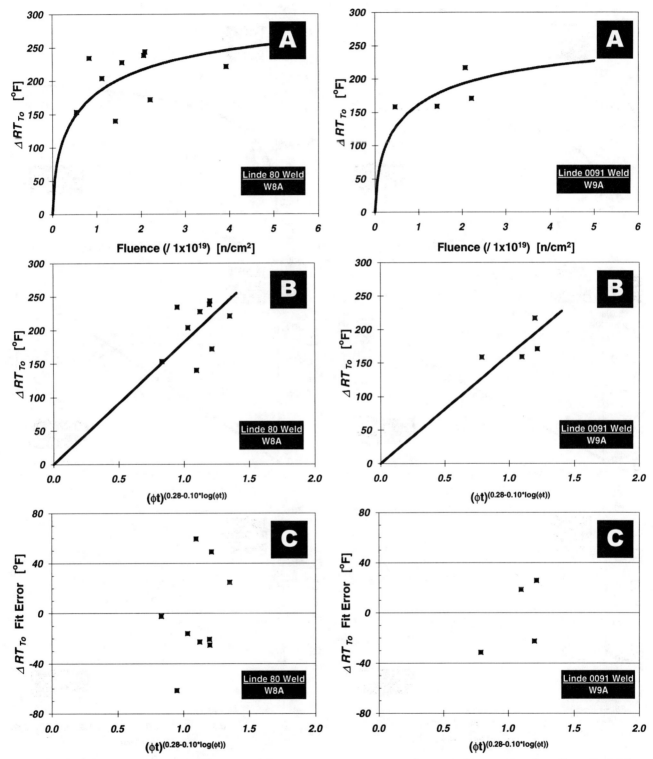

Fig. 2. Variation of ΔRT_{To} with fluence, and comparison with the Reg. Guide 1.99 (Rev. 2) fluence function for a Linde 80 weld [Hawthorne and Hiser 1988, Hawthorne and Hiser 1990(a)].

Fig. 3. Variation of ΔRT_{To} with fluence, and comparison with the Reg. Guide 1.99 (Rev. 2) fluence function for a Linde 0091 weld [Hawthorne and Hiser 1990(a)].

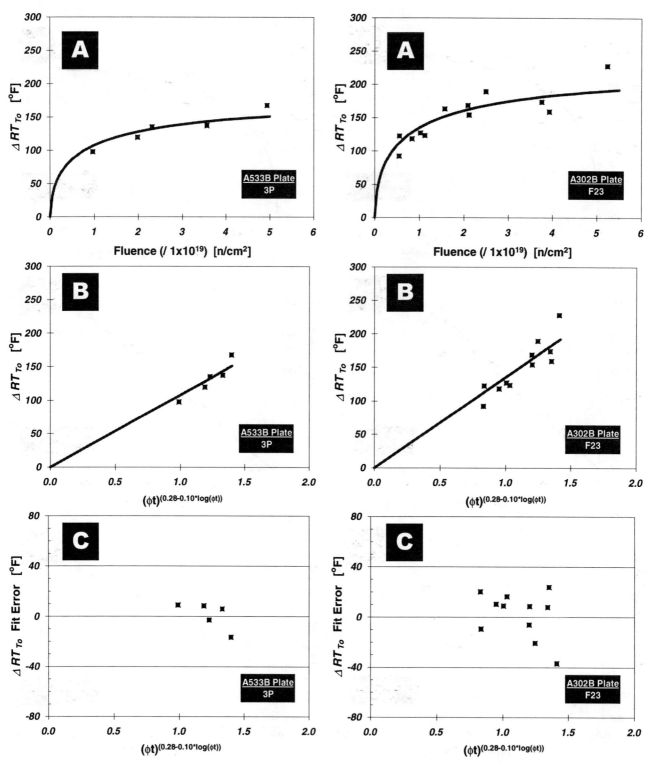

Fig. 4. Variation of ΔRT_{To} with fluence, and comparison with the Reg. Guide 1.99 (Rev. 2) fluence function for a A533B plate [Hawthorne et al. 1984].

Fig. 5. Variation of ΔRT_{To} with fluence, and comparison with the Reg. Guide 1.99 (Rev. 2) fluence function for a A302B plate [Hawthorne et al. 1984].

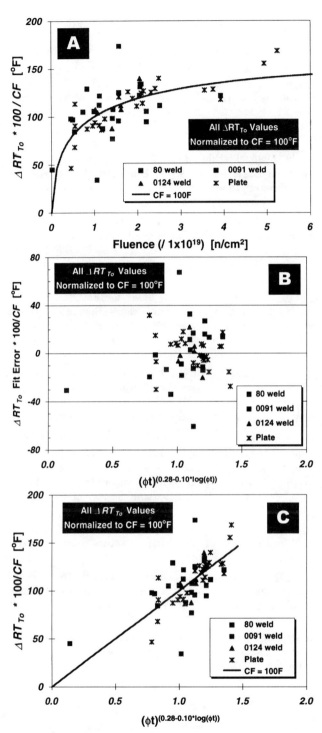

Fig. 6. Variation of ΔRT_{To} with fluence, and comparison with the Reg. Guide 1.99 (Rev. 2) fluence function for all available irradiated data.

C. $\{\Delta RT_{To}|_{measured} - \Delta RT_{To}|_{fit}\}$ vs. $(\phi t)^{[0.28-0.10*\log(\phi t)]}$ (a residual plot for Graph B)

The "C" graphs for these four data sets show that the residuals exhibit only random fluctuation about zero, suggesting that the Reg. Guide 1.99 (Rev. 2) fluence function provides a reasonable fit to ΔRT_{To} data. Figure 6 collects together all 16 data sets by normalizing the ΔRT_{To} values to a common CF of 100°F. As with the four individual data sets, this plot demonstrates that the Reg. Guide 1.99 (Rev. 2) fluence function provides a reasonable fit to ΔRT_{To} data.

SUMMARY AND CONCLUSIONS

In this paper we examine the possibility of establishing generic IRT values and fluence function forms for use in assessments of nuclear reactor vessel integrity based on Master Curve technology. Available empirical evidence supports the following conclusions:

- ✓ The currently accepted value of –56°F for the generic IRT of all Combustion Engineering welds is conservative by at least 45°F. Furthermore, Linde 1092 welds are assessed even more conservatively (by 79°F) relative to the current generic value.
- ✓ The -5°F generic IRT for Babcock and Wilcox / Linde 80 welds is conservative by at least 22°F.
- ✓ The Reg. Guide 1.99 (Rev. 2) fluence function provides a reasonable fit to all available data for irradiation effects on RT_{To}.

Additional data would of course provide valuable reinforcement of these findings. Nevertheless, this study suggests that concerns raised regarding the difficulty of applying Master Curve technology to integrity assessment of nuclear RPVs on a generic basis may not be technically defensible.

ACKNOWLEDGEMENTS

We would like to thank Dr. Mikhail Sokolov of the Oak Ridge National Laboratory for his kind provision of data in electronic form.

REFERENCES

10 CFR 50.61, Nuclear Regulatory Commission 10 CFR Part 50, "Fracture Toughness Requirements for Light Water Reactor Pressure Vessels."

ASME N629, ASME Code Case N-629, "Use of Fracture Toughness Test Data to Establish Reference Temperature for Pressure Retaining Materials for Section XI."

ASTM E23-96, "Standard test Methods for Notched Bar Impact Testing of Metallic Materials," ASTM, 1998.

ASTM E208-95a, "Standard Test Method for Conducting Drop-Weight Test to Determine Nil-Ductility Transition Temperature of Ferritic Steels," ASTM 1998.

ASTM E1921-97, "Test Method for Determination of Reference Temperature, T_o, for Ferritic Steels in the Transition Range," ASTM, 1998.

ASTM E185-94, "Standard Practice for Conducting Surveillance Tests for Light-Water Cooled Nuclear Power Reactor Vessels," ASTM, 1998.

Byrne 1999, "Summary of Master Curve Test Results," private communication.

Eason 1998, Eason, E.D., Wright, J.E., and Odette, G.R., "Improved Embrittlement Correlations for Reactor Pressure Vessel Steels," NUREG/CR-6551, 1998.

EPRI 1998, "Application of Master Curve Fracture Toughness Methodology for Ferritic Steels," EPRI-TR-108390, 1998.

Hawthorne, et al. 1984, Hawthorne, J.R., Menke, B.H., Hiser, A.L., "Notch Ductility and Fracture Toughness Degradation of A302-B and A533-B Reference Plates from PSF Simulated Surveillance and Through-Wall Irradiation Capsules," NUREG/CR-3295 (Vol. 1), USNRC, 1984.

Hawthorne and Hiser 1988, Hawthorne, J.R., Hiser, A.L., "Experimental Assessments of Gundremmingen RPV Archive Material for Fluence Rate Effects Studies," NUREG-CR/5201, US-NRC, 1988.

Hawthorne and Hiser 1990(a), Hawthorne, J.R., Hiser, A.L., "Investigations of Irradiation-Anneal-Reirradiation (IAR) Properties Trends of RPV Welds: Phase 2 Final Report," NUREG-CR/5492, US-NRC, 1990.

Hawthorne and Hiser 1990(b), Hawthorne, J.R., Hiser, A.L., "Influence of Fluence Rate on Radiation-Induced Mechanical Property Changes in RPV Steels," NUREG/CR-5493, US-NRC, 1990.

Hawthorne, et al. 1992, Hawthorne, J.R., Menke, B.H., Loss, F.J., Watson, H.E., Hiser, A.L., and Gray, R.A., "Evaluation and Prediction of Neutron Embrittlement in Reactor Pressure Vessel Materials," EPRI NP-2782, EPRI, 1992

Hiser 1989, Hiser, A.L., "Post-Irradiation Fracture Toughness Characterization of Four Lab-Melt Plates," NUREG/CR-5216 (Rev. 1), US-NRC, 1989.

Kirk 1998, Kirk, M., Lott, R., Kim, C., and Server, W., "Empirical Validation Of The Master Curve For Irradiated And Unirradiated Reactor Pressure Vessel Steels," Presented at the 1998 ASME/JSME Pressure Vessel and Piping Symposium, July 26-30, 1998, San Diego, California, USA.

Lott 1999, Lott, R., Kirk, M., Kim, C., Server, W., Tomes, C., Williams, J., "Application of Master Curve Technology to Estimation of End Of License Adjusted Reference Temperature for a Nuclear Power Plant," Presented at the 1999 ASME Pressure Vessel and Piping Symposium, August 2-5, 1999, Boston, Massachusetts, USA.

McGowan, et al. 1988, McGowan, et al., "Characterization of Irradiated Current-Practice Welds and A533 Grade B Class 1 Plate for Nuclear Pressure Vessel Service, NUREG/CR-4880, U.S. Nuclear Regulatory Commission, 1988.

Merkle 1998, Merkle, J.G., Wallin, K., McCabe, D.E., "Technical Basis for an ASTM Standard for Determining the Reference Temperature, T_o, for Ferritic Steels in the Transition Range," NUREG/CR-5504, 1998.

Natishan 1998, Natishan, M. and Kirk, M., "A Micromechanical Evaluation of the Master Curve," *Fatigue and Fracture Mechanics, 30th Volume, ASTM STP-1360*, K. Jerina and P. Paris, Eds., American Society for Testing and Materials, 1998.

Randall 1987, Randall, P.N., "Basis for Revision 2 of the U.S. Nuclear Regulatory Commission's Regulatory Guide 1.99," *Radiation Embrittlement in Nuclear Pressure Vessel Steels: An International Review (Second Volume), ASTM STP-909*, L.E. Steele, Ed., 1987.

RG1.99, Regulatory Guide 1.99, Revision 2, *Radiation Embrittlement of Reactor Vessel Materials, U.S. Nuclear Regulatory Commission*, May 1988.

SECY-82-465, SECY-82-465, Enclosure A, "NRC Staff Evaluation of Pressurized Thermal Shock," November 1982.

Sokolov 1996, Sokolov, M.A., and Nanstad, R.K., "Comparison of Irradiation Induced Shifts of K_{Jc} and Charpy Impact Toughness for Reactor Pressure Vessel Steels," ASTM STP-1325, American Society of testing and Materials, 1996.

Wallin 1984, Wallin, Saario, and Törrönen, "Statistical Model for Carbide Induced Brittle Fracture in Steel," *Metal Science*, 18, pp. 13-16, 1984.

Wallin 1993, Wallin, K., "Irradiation Damage Effects on the Fracture Toughness Transition Curve Shape for Reactor Vessel Steels," *Int. J. Pres. Ves. & Piping*, 55, pp. 61-79, 1993.

Wallin 1995, Wallin, K., "Re-Evaluation of the TSE Results Based on the Statistical Size Effect," VTT Manufacturing Technology, 1995.

APPLICATION OF MASTER CURVE TECHNOLOGY TO ESTIMATION OF END OF LICENSE ADJUSTED REFERENCE TEMPERATURE FOR A NUCLEAR POWER PLANT

Randy Lott, Mark Kirk, and Charlie Kim
Westinghouse Electric Company
Pittsburgh, Pennsylvania, USA

William Server
ATI Consulting
Pinehurst, North Carolina, USA

Chuck Tomes
Wisconsin Public Service Corporation
Green Bay, Wisconsin, USA

James Williams
NASA, Ames Research Center
Moffett Field, California, USA

ABSTRACT

ASME Code Case N-629 recognizes a Master Curve-based reference temperature, RT_{T_o}, as an alternative estimate of the fracture toughness reference temperature RT_{NDT}. Here we describe procedures that enable application of this technologically superior estimate of the fracture toughness transition temperature to the assessment of commercial nuclear reactor pressure vessels. Margin terms that are functionally equivalent to the margins applied to RT_{NDT} in current regulations are developed. These margin terms are used to calculate an irradiation adjusted reference temperature, or ART, applicable to reactor pressure vessel integrity analysis. Multiple practical situations are considered, including the following: RT_{T_o} measured for the unirradiated material, RT_{T_o} measured on a sample irradiated to end of license (EOL) fluence, and RT_{T_o} measured on a sample irradiated to a fluence other than that at EOL. In each case, procedures for determining a Master Curve-based ART value that meets the intent of the procedures outlined in US-NRC Regulatory Guide 1.99 Rev. 2 and 10 CFR 50.61 are provided. These procedures are illustrated by considering data from the Kewaunee Nuclear Power Station reactor vessel surveillance program.

INTRODUCTION AND OBJECTIVE

The ASME code employs a lower bound fracture toughness transition curve in the analysis of vessel integrity. Currently, the code supplies a lower bound structural curve for static loading rate applications (i.e. the K_{IC} curve). The K_{IC} curve was determined by constructing a reference temperature that defined a lower bound curve for all existing fracture toughness data. The reference temperature selected was RT_{NDT}, an amalgam of the nil-ductility transition temperature (NDT) and the Charpy V-notch transition temperature. In setting the K_{IC} curve, it was assumed that the measured values of fracture toughness were size independent, making RT_{NDT} the only parameter required to determine the bounding structural curve.

Master curve technology [Wallin, 1984(a), 1984(b), 1985, 1993, 1995] provides an alternative means of constructing a lower bound structural curve using a single set of fracture toughness measurements from a specific material. The fact that the Master Curve is determined from measurements of the property of interest (fracture toughness) as opposed to the current approach which determines RT_{NDT} based on other manifestations of the ductile-to-brittle transition (dynamically loaded specimens having either brittle starter notches, or blunt notches) makes this new technology extremely attractive. The only parameter required to establish the Master Curve is the reference temperature, T_o. T_o is determined by performing a series of K_{Jc} fracture toughness tests on precracked specimens, which may be as small as a standard Charpy bar. The Master Curve offers two additional advantages: direct measurement of the fracture toughness transition curves in materials with limited availability of archival materials, and measurement of the transition curve for irradiated materials. Measurements of RT_{NDT} in these cases are impractical or impossible, a fact which necessitates the current correlative approach.

The objective of this effort is to integrate results obtained using the Master Curve procedure into the existing reactor pressure vessel (RPV) evaluation methodology [10CFR50, RG1.99(R2)]. ASME Code Case N-629 [ASME N629] allows use of a Master Curve based reference temperature, RT_{T_o}, as an alternative to RT_{NDT} when establishing the ASME K_{IC}

curve. Although there is no direct correlation between RT_{To} and RT_{NDT} [Sokolov, 1998], either reference temperature can be used to predict a reliable lower bound toughness curve [Kirk, et al., 1998]. In this sense they are functionally equivalent. However, within the integrated family of codes and regulations that govern the operation of a nuclear pressure vessel, the value of RT_{NDT} is often used without direct reference to the K_{IC} curve. For instance, 10CFR50.61 defines the pressurized thermal shock (PTS) screening criteria in terms of maximum allowed values of RT_{NDT} adjusted to account for the effects of irradiation (this adjusted value is called RT_{PTS} or the adjusted reference temperature, ART). These screening criteria were based on a probabilistic analysis that described the temperature dependence of fracture toughness in terms of the ART. While the relationship between the reference temperature and the fracture toughness data is inherent in this analysis, it is not immediately apparent. The normal pressure-temperature operating limits for the reactor are also directly related to the ART. Again, the fracture toughness reference curve is inherent in the calculation but not obvious to the user. Implicit in acceptance of RT_{To} as an alternative means of indexing the fracture toughness reference curve is the assumption that RT_{To} could also be used as a direct replacement for RT_{NDT} in these various RPV integrity analysis.

CURRENT PRACTICE

RPV integrity analysis requires an evaluation of the fracture toughness of the irradiated material. The determination of RT_{NDT} in unirradiated material is the first step in this evaluation. The evaluation is then accomplished by calculating the adjusted reference temperature, ART, which is used as an reference temperature for the K_{IC} and K_{IR} curves. ART is determined by taking the sum of the unirradiated RT_{NDT} value, the Charpy transition temperature shift at 41J (30 ft-lbs) (ΔRT_{NDT}), and a margin term, M:

$$ART = RT_{NDT} + \Delta RT_{NDT} + M \tag{1}$$

where $M = \sqrt{\sigma_I^2 + \sigma_\Delta^2}$, and σ_I and σ_Δ are estimates of the uncertainty in RT_{NDT} and ΔRT_{NDT}, respectively [RG1.99(R2)]. The ART is intended to serve as a conservative estimate of the irradiated value of RT_{NDT}. Regulatory Guide 1.99 Rev. 2 defines the "margin" as *"the quantity, to be added to obtain conservative, upper-bound values of adjusted reference temperature for the calculation required by Appendix G to 10 CFR Part 50"*. Any procedure that replaces RT_{NDT} with RT_{To} in the estimation of ART must maintain the intent and function of the irradiation shift and margin terms.

Nearly every step in the procedures for reactor vessel integrity analysis contains some adjustment to assure a conservative calculation. The margin included in the ART definition is only one contribution to the overall conservatism of the process. To maintain clarity and relevance in the analysis, it is important to only include conservatisms related to the determination of the reference temperature in the definition of ART. For this reason, the margin term used in the current definition of ART includes only the uncertainty in the determination of the unirradiated RT_{NDT} value, σ_I, and the uncertainty in the determination of the Charpy transition temperature shift, σ_Δ.

Current practice recognizes several different situations that require different methods of determining the adjusted reference temperature. These methods are illustrated schematically in Fig. 1. Each path through Fig. 1, represents a different set of assumptions about the credibility of the data and, consequently, the appropriate margins. Four distinct paths are outlined in Table 1. Data from the Kewaunee Nuclear Power Station surveillance weld (Linde 1092 flux, weld wire heat 1P3571) is included in Table 1 to demonstrate the effects of these assumptions. While the best estimates of the irradiated RT_{NDT} values range from 190°F to 203°F, the associated ART values, which represent the best estimate plus a margin term, span a much broader range: 231°F - 256°F. These methods accommodate the amount of available data on the vessel materials. The significance of the margin terms is immediately obvious. In particular, current practice prefers measurements of RT_{NDT}, but allows generic values for classes of materials when measurements. Similarly, current practice prefers measurement of Charpy shifts, but allows the use of predictive equations when an associated margin term, with a unique combination of

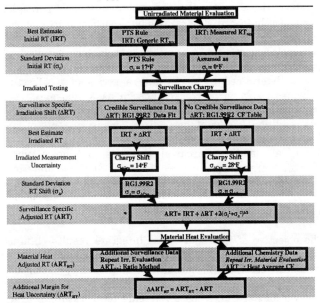

Fig. 1. Paths for estimation of the ART using current technology.

Table 1. *ART* Determination for the Kewaunee Weld and Vessel.

Method	Best Estimate of Initial RT_{NDT} (IRT), °F	Standard Deviation for IRT (σ_I), °F	Kewaunee Surveillance Estimate of Shift (ΔRT), °F	Standard Deviation for ΔRT (σ_Δ), °F	Best Estimate of Irradiated Value, °F	Total Margin (M) °F	Adjusted Reference Temperature (ART) °F	Heat Uncertainty Adjustment (ΔART_{HT}) °F	Heat Adjusted Ref. Temp. (ART_{HT}) °F
1. Current Technology Measured IRT; No Credible CVN Data	Measured Value, $RT_{NDT}=-50$	Assumed 0	RG1.99R2, CF Table 246	RG1.99R2 28	IRT+ΔRT 196	$2(\sigma_I^2+\sigma_\Delta^2)^{1/2}$ 56	IRT+ΔRT+M 252	Ind. Mean Chemistry 36	288
2. Current Technology; Generic IRT; No Credible CVN Data	PTS Rule $RT_{NDT}=-56$	PTS Rule 17	RG1.99R2, CF Table 246	RG1.99R2 28	IRT+ΔRT 190	$2(\sigma_I^2+\sigma_\Delta^2)^{1/2}$ 66	IRT+ΔRT+M 256	Ind. Mean Chemistry 36	292
3. Current Technology; Measured IRT; Credible CVN Data	Measured Value, $RT_{NDT}=-50$	Assumed 0	RG1.99R2, Data Fit 253	RG1.99R2 14	IRT+ΔRT 203	$2(\sigma_I^2+\sigma_\Delta^2)^{1/2}$ 28	IRT+ΔRT+M 231	Ratio Adj. 36	267
4. Current Technology; Generic IRT; Credible CVN Data	PTS Rule $RT_{NDT}=-56$	PTS Rule 17	RG1.99R2, Data Fit 253	RG1.99R2 14	IRT+ΔRT 197	$2(\sigma_I^2+\sigma_\Delta^2)^{1/2}$ 44	IRT+ΔRT+M 241	Ratio Adj. 36	277
5a. Master Curve; Unirradiated T_o; Credible CVN Data	Unirradiated T_o +35°F $RT_{To}=-109$	ASTM $\beta/\sqrt{n}$ 7	RG1.99R2, Data Fit 253	RG1.99R2 & T_o-CVN 30	$RT_{NDT(U)}$+ΔRT_{NDT} 144	$2(\sigma_I^2+\sigma_\Delta^2)^{1/2}$ 62	IRT_{To}+ΔRT_{NDT}+M 206	Ratio Adj. 36	242
5b. Master Curve; Unirradiated T_o; No Credible CVN Data	Unirradiated T_o +35°F $RT_{To}=-109$	ASTM $\beta/\sqrt{n}$ 7	RG1.99R2, CF Table 246	RG1.99R2 & T_o-CVN 39	$RT_{NDT(U)}$+ΔRT_{NDT} 137	$2(\sigma_I^2+\sigma_\Delta^2)^{1/2}$ 79	IRT_{To}+ΔRT_{NDT}+M 216	Ind. Mean Chemistry 36	252
6. Master Curve; Irradiated T_o	NA	NA	NA	ASTM $\sigma_{To}=\beta/\sqrt{n}$ 8	Irradiated T_o +35°F 183	$2\sigma_{To}$ 16	$RT_{To(irr)}$ + M 199	MY meas. w/ Ratio Adj. 35	234
7. Master Curve Shift; Measured $RT_{NDT(U)}$; Irr. T_o-Unirr. T_o	Unirradiated T_o +35°F $RT_{To}=-109$	ASTM $\beta/\sqrt{n}$ 7	Data Fit, CF = 222 292	Similar to RG1.99R2 14	RT_{To}+ΔRT_{To} 183	$2(\sigma_I^2+\sigma_\Delta^2)^{1/2}$ 31	IRT+ΔRT_{To}+M 214	MY meas. w/ Ratio Adj. 35	249

uncertainties (see Table 1). The margin terms reflect both the uncertainty in the measurement of the transition temperature and the uncertainty associated with use of generic data. In general, the largest penalties are applied to situations where material specific data is not available.

The number of possible paths to estimate *ART* has been significantly increased by recent regulatory actions that require the consideration of all sources of data on a material (generally defined as a given weld flux and weld wire heat). In Fig. 1, this effect has been recognized by including an additional requirement for material heat evaluation. The impact of the material heat evaluation can best be understood by segregating these effects into a separate term, ΔART_{HT}, which includes adjustments made to the calculational procedure based on surveillance data and/or chemistry data for the same heat of material taken from sources outside the Kewaunee surveillance program. The results of this evaluation for the Kewaunee vessel weld are supplied in the last two columns of Table 1. For the Kewaunee vessel weld, the Maine Yankee weld is the primary source of additional data.

ALTERNATIVE APPROACHES BASED ON FRACTURE TOUGHNESS TESTING, THE DEFINITION OF RT_{TO}

The Master Curve provides an alternative reference toughness curve based on direct measurements of fracture toughness. T_o indexes the variation of fracture toughness with temperature. While both T_o and RT_{NDT} provide a reference temperature for fracture toughness transition curves, they correspond to very different reference toughnesses. RT_{NDT} corresponds to a toughness of 42 MPa√m on a generic lower bound curve while T_o corresponds to a toughness of 100 MPa√m on a material specific median toughness curve. ASME Code Case N-629 allows adoption of a T_o-based reference temperature (termed RT_{To}) as an alternative to RT_{NDT} [ASME N629] for establishing the ASME K_{IC} curve. The following relationship has been proposed:

$$RT_{To} \equiv T_o + 35° F \qquad (2)$$

The 35°F shift accomplishes the translation from the median curve to a bounding curve, and matches the 42 MPa√m reference toughness value on the K_{IC} curve. The objective of using RT_{To} as an alternative to RT_{NDT} requires some linkage to historical safety margins. To maintain this linkage, the 35°F shift was selected such that the K_{IC} curve, when indexed to RT_{To}, bounds available fracture toughness data in a manner functionally equivalent to the manner in which the K_{IC} curve indexed to RT_{NDT} bounds available fracture toughness data. The basis for the selection of the 35°F shift is contained in the definition of the term "functionally equivalent." The technical basis document for ASME Code Case N-629 contains a detailed discussion of the relationship between RT_{To} and RT_{NDT} [EPRI, 1998]. A brief summary of the major considerations is provided in this paper.

The spacing between the measured toughness values and the resulting reference curve is a measure of the implicit margin on toughness contained in the definition of the reference temperature. As illustrated in Fig. 2, implicit margins on toughness vary considerably for the current RT_{NDT}-based methodology. Therefore, it is not possible to define RT_{To} such that it maintains *the* margin implicit to a RT_{NDT} indexed K_{IC} curve because no single, unique, margin exists currently. Criteria for judging the functional equivalence of the RT_{To} to RT_{NDT} therefore must define both the data set to be analyzed, and an appropriate acceptance level. These criteria must also encompass the ambiguity in the margin inherent in the existing approach. Possible criteria for establishing functional equivalence are as follows:

1. The resulting K_{IC} curve should provide an absolute bound to all data except for the lower shelf data that was never bounded by existing technology (i.e. a RT_{NDT} indexed K_{IC} curve).
2. The implicit margin on toughness must exceed the minimum acceptable margin in the current, RT_{NDT}-based, approach.

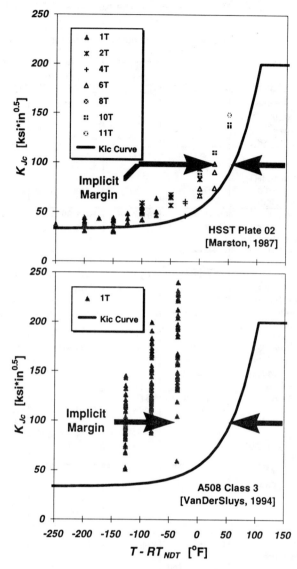

Fig. 2. Comparison of inconsistent manner in which RT_{NDT} positions two different heats of RPV steel relative to the K_{Ic} curve (top: small implicit margin; bottom: large implicit margin).

3. The implicit margin on toughness for the original K_{IC} database should be maintained.

The first criterion imposes a level of conservatism that far exceeds the conservatism implicit in current practice. Moreover, this level of conservatism becomes more extreme as time goes on and more data is collected. The first criteria is therefore an inappropriate basis for establishing functional a equivalence of RT_{To} to RT_{NDT}. Conversely, the second and third criteria provide a basis for functional equivalence that is stable over time, and is consistent with margins implicit in current practice (i.e. a K_{IC} curve referenced to RT_{NDT}). These

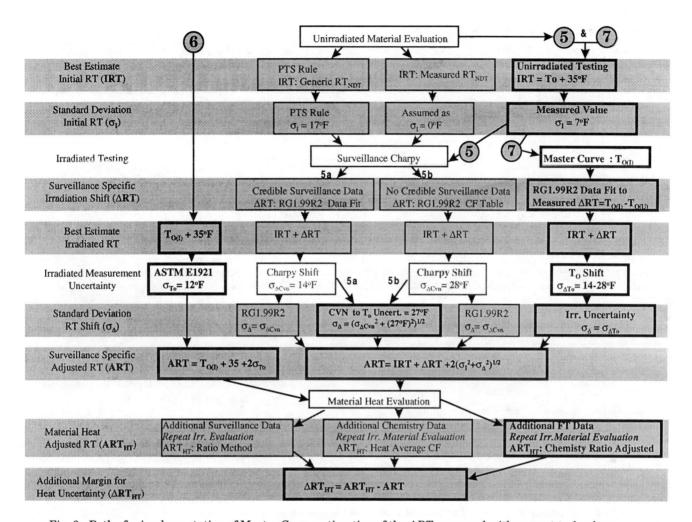

Fig. 3. Paths for implementation of Master Curve estimation of the *ART* compared with current technology.

latter criteria both support the 35°F shift value of ASME Code Case N-629 [Kirk, et al., 1998].

ALTERNATIVE DETERMINATIONS OF THE ADJUSTED REFERENCE TEMPERATURE

Procedures for using RT_{T_o} to calculate *ART* should parallel current practice. Several alternative paths for determining the *ART* are illustrated in Fig. 3. These paths indicate how RT_{T_o} measurements on both unirradiated and irradiated materials can be integrated into an *ART* evaluation in a way that satisfies the intent of Reg. Guide 1.99 (Rev. 2). Shifts, uncertainties, and margins are added to measured RT_{T_o} values as appropriate, always with the aim of satisfying the intent of Reg. Guide 1.99 (Rev. 2). These procedures were developed in support of the Kewaunee Nuclear Power Plant. This analysis for Kewaunee is included here to illustrate the process of calculating ART from RT_{T_o}. Descriptions of the margins adopted in each path, and a basis for these margins, are provided in the following sections. T_o values for the Kewaunee weld (Linde 1092 flux, heat 1P3571) are summarized in Table 2.

Direct Measurements of Fracture Toughness Before Irradiation (Paths 5a & 5b)

Paths 5a and 5b in Fig. 3 illustrate how measurements of T_o in the unirradiated condition (see also methods 5a and 5b in Table 1) can be used in the determination of *ART*. This is the most straightforward application of ASME Code Case N-629. The T_o value measured in the unirradiated condition for the Kewaunee surveillance weld was -144°F. The corresponding best estimate of the initial reference temperature, RT_{T_o}, is therefore -109°F (= T_o + 35°F). The margin term associated with this initial reference temperature is taken as the uncertainty in the T_o determination as defined in ASTM E1921-97 (σ_I =6.5°F). This approach is more conservative than the equivalent RT_{NDT} based approach, where the uncertainty in measured

Table 2. T_o data for Linde 1092 Weld Heat 1P3571.

Fluence [n/cm^2/10^{19}]	Specimen Geometry	T_o Calculation Ref.	# of Valid Tests	# of Tests above K_{Jc} Limit	Test Temp. [°F]	T_o [°F]	$\sigma = (\beta/N^{0.5})$ [°F]
Kewaunee Surveillance Weld							
0	1/2T C(T)	E1921	7	0	-187	-129	12.8
0	PC-CVN	E1921	8	0	-200	-149	12.0
0	Reconstituted PC-CVN	E1921	7	0	-200	-154	12.8
0	All of the Above	ML, Wallin95	22	0	various	-144	6.5
3.36	Reconstituted PC-CVN	E1921	8	1	136	136	10.8
3.36	Reconstituted PC-CVN	N/A	3	0	59	N/A	N/A
3.36	1XWOL	N/A	2	0	136	N/A	N/A
3.36	All of the Above	ML, Wallin95	13	1	various	148	8.2
Maine Yankee Surveillance Weld							
0	Reconstituted PC-CVN	E1921	7	0	-200	-158	12.8
6.11	Reconstituted PC-CVN	E1921	7	1	210	232	11.5

values of the unirradiated reference temperature is defined as zero.

If only unirradiated values of T_o are available, Charpy data must be used to evaluate the irradiation induced shift. Therefore, beyond determination of the initial reference temperature and σ_I, paths 5a and 5b follow the same route as the current Reg. Guide 1.99 Rev. 2 procedures. There is, however, one necessary modification. Paths 5a and 5b combine measured fracture toughness values with a Charpy-based irradiation shift, something outside the scope of Reg. Guide 1.99 (Rev. 2). This combination of a toughness-based initial reference temperature with a Charpy-based shift may produce additional uncertainty in the shift term.

Studies conducted at the Oak Ridge National Laboratory indicate that, while there is a correlation between irradiation shifts as measured by Charpy and measured by toughness specimens, significant variability may be expected on a material by material basis [Rosseel, 1998]. For welds, Rossell reports a 1:1 correlation between Charpy transition temperature shift and fracture toughness transition temperature shift; the 95% confidence bands for this correlation are ±54°F. This uncertainty corresponds to a standard deviation of 27°F. When using an un-irradiated T_o and combining it with a Charpy shift, the shift uncertainty (σ_Δ) is therefore increased as follows:

$$\sigma_\Delta = \sqrt{\sigma_{\Delta CVN}^2 + (27)^2} \qquad (3)$$

where

$\sigma_{\Delta CVN}$ = 14°F for a credible surveillance program, or 28°F for a non-credible surveillance program (both as per Reg. Guide 1.99 (Rev. 2)).

This approach to estimating shift uncertainty when measured RT_{T_o} values are combined with Charpy shifts is highly conservative. Eq. (3) reflects the assumption Charpy shifts are known without error, so *all* of the uncertainty in the relation between Charpy shifts and fracture toughness shifts occurs on the fracture toughness side of the relationship. This assumption is obviously incorrect. It is therefore expected that the 27°F factor in eq. (3) can be reduced without any safety impact as more information becomes available.

Further adjustment of the *ART* is made to account for heat to heat variability. This added uncertainty applies only to the shift term. Therefore, well established heat-average chemistry and ratioing procedures [RG 1.99(R2)] can be applied to the Path 5a and 5b analysis because conventional Charpy-based shifts are used. Table 1 accumulates all of these terms for both paths 5a (credible surveillance, *ART* = 242°F) and 5b (non-credible surveillance, *ART* = 252°F). Because the initial reference temperature determined by Master Curve testing of the unirradiated material is lower than the conventional measurement, the best estimate reference temperature is also lower than that estimated by current practice. However, the additional uncertainty associated with the process of combining fracture toughness measurements with Charpy shifts consumes a portion of this advantage.

Direct Measurements of Irradiated Fracture Toughness at the Fluence of Interest (Path 6)

Master Curve technology enables direct estimation of a reference temperature (RT_{T_o}) for irradiated materials by testing a set of specimens as per E1921-97 that have been irradiated to the fluence of interest. This process eliminates both unirradiated testing to determine the initial reference temperature and σ_I (Step 1), and the Charpy shift term (Step 2). The only contribution to the margin term in this case is the uncertainty in the determination of T_o. This approach is prudent and safe because it includes an explicit margin for measurement uncertainty larger than the margin currently applied to account for uncertainty in a measured initial reference temperature. This margin is never explicitly added in current practice. Instead it is implicit in the relationship between the bounding K_{Ic} curve and the RT_{NDT} index temperature.

From Table 2, T_o for the irradiated Kewaunee surveillance weld is +148°F. Therefore RT_{T_o} in the irradiated condition (i.e. the best estimate of the RT_{NDT}) is 183°F (T_o + 35°F). The Kewaunee specific ART value (+199°F) contains an extra 16°F (2σ) margin term to account for measurement uncertainty. This margin is not required for current technology when direct measurements of RT_{NDT} are used. Adjustments to this procedure for heat uncertainty are discussed separately in a subsequent section.

Determination of Fracture Toughness Shift at Fluences having no Measured Fracture Toughness (Path 7)

The determination of the reference temperature for irradiated materials in cases where available specimens have not been irradiated to the neutron fluence of interest requires use of an irradiation trend curve (see Path 7 in Fig. 3 and Table 1). Path 7 represents the most general application of Master Curve technology. As an example, this path could be used to consider license extension because specimens irradiated to 60 EFPY are not typically available.

The currently accepted curve for predicting the increase of reference temperature due to irradiation is as follows [RG1.99(R2)]:

$$\Delta RT_{NDT} = (CF)\phi t^{(0.28-0.1\log\phi t)} \quad (4)$$

where CF is the chemistry factor and ϕt is the fluence. Although this trend curve was based solely on Charpy data, it is routinely used to shift fracture toughness reference curves. Here we adopt the form of eq. (4) to predict irradiation induced shifts in RT_{T_o}. The data in Fig. 4 substantiate our assumption that the eq. (4) fluence function

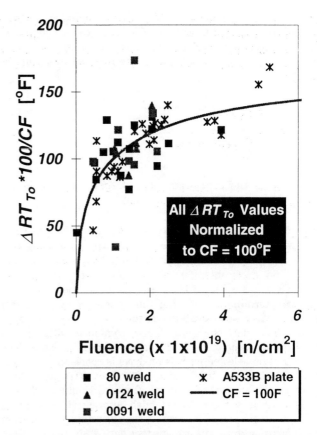

Fig. 4. Comparison of irradiation induced irradiation shifts with the Reg. Guide 1.99 fluence function for a number of RPV steels and weldments.

predicts shifts in RT_{T_o} to an accuracy consistent with established margin terms (M in eq. (1)). These data were drawn from a number of NRC and EPRI publications [Hawthorne, et al., 1984, 1988, 1990(a), and 1990(b); Hawthorne, et al., 1992; Hiser, 1989; McGowan, et al., 1988]. T_o values were calculated using the maximum likelihood method [Wallin 95] and converted to RT_{T_o} using eq. (2). Chemistry factor (CF) was calculated by applying the method of Reg. Guide 1.99 (Rev. 2) to ΔRT_{T_o} values instead of 41 J Charpy shifts. All available materials having at least one unirradiated and two irradiated measurements of T_o are shown in Fig. 4 (four Linde 80 welds, three Linde 1091 welds, three Linde 0124 welds, and six A533B plates).

To apply eq. (4), a determination of the irradiation induced shift in the fracture toughness reference temperature (RT_{T_o}) is required. The measured shift in fracture toughness for the Kewaunee surveillance weld is 292°F at 3.34x10¹⁹n/cm² (292°F = 148°F$_{IRRADIATED}$ - -144°F$_{UN-IRRADIATED}$). The chemistry factor is the CF value that forces the trend curve through this measured value.

If more than one irradiated RT_{To} value is available, a fitting procedure is needed to estimate CF. A procedure similar to that currently used to determine material specific chemistry factors from Charpy surveillance data [RG1.99(R2)] could be used, i.e.:

$$CF = \frac{\sum_{i=1}^{n}\left[A_i x \phi t_i^{(0.28-0.1\log\phi t_i)}\right]}{\sum_{i=1}^{n} \phi t_i^{(0.56-0.2\log\phi t_i)}} \quad (5)$$

where A_i is the measured value of ΔRT_{To} and ϕt_i is the fluence for each ΔRT_{To} determination.

The PTS Rule requires at least two ΔRT_{NDT} measurements to determine a chemistry factor from Charpy data using this formula. While a similar number of ΔRT_{To} measurements would be preferable, for the purposes of this analysis chemistry factors have been calculated on the basis of a single determination. Using this procedure, a chemistry factor of 222°F was calculated for the Kewaunee surveillance weld. This new chemistry factor, combined with eq. (4), predicts the trend curve for Kewaunee shown in Fig. 5. The margin term selected for this analysis is consistent with the use of the Regulatory Guide 1.99 (Rev. 2) trend curve. Because the amount of available fracture toughness data is limited, it is difficult to establish appropriate credibility criteria for the data. Therefore, the uncertainty term from path 5b ($\sigma_D = 28°F$) was assumed for path 7. Note that if the fluence of interest in the analysis is fluence of the measurement, this methodology will reduces to path 6. In this case, there would be a relatively large penalty for selecting path 7 over path 6. However, as the distance between the fluence of interest and the fluence of measurement grows, the amount of extrapolation required increases and the use of the larger uncertainty term seems more reasonable.

Heat to Heat Uncertainty (Surrogate Materials)

As a general rule, the material properties used in the design and analysis of any engineering structure are determined by destructive testing. In particular, determination of the tensile and fracture toughness properties of a structural steel requires destructive tests such as those described in ASTM Standards E8-96, E23-96, E208-95, and E1921-97. In every instance where materials property data is applied in design, the material samples tested to generate the property data stand as a surrogate for the materials used to fabricate the structure. It is critical to the application of the data that the tested material is an appropriate surrogate. Thus, the surrogates question is in fact one concerning material sampling strategies.

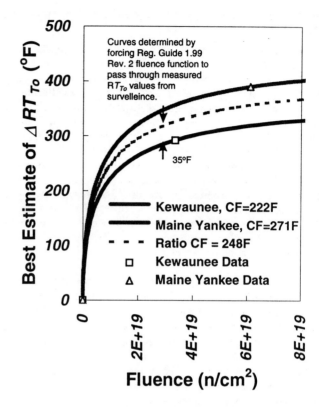

Fig. 5. Variation of RT_{To} with neutron fluence for Kewaunee and Maine Yankee materials.

Several recent observations indicating that the variability in material properties may be larger than originally believed have led to regulatory concerns about the material sampling strategy used in RPV surveillance programs. These observations have led to the adoption of the ratioing and heat average chemistry procedures [RG 1.99(R2)]. Traditionally, these procedures apply to the evaluation of Charpy shifts. In Table 1, the effects of these procedures are accumulated in the ΔART_{HT} term. For the Kewaunee surveillance weld, the ΔART_{HT} term includes data from the Maine Yankee weld. The analysis of the heat uncertainty in the Charpy data is summarized in Table 1. Both the industry mean chemistry approach and the ratio procedure produce additional adjustments for heat chemistry of 36°F. This adjustment applies to all of the Charpy based methods for determining ART (i.e. paths 1-4). This 36°F value also applies to paths 5a and 5b, because the shift term for these paths is based on Charpy data.

An alternative method of determining the heat uncertainty is required for applications which use irradiated T_o measurements in lieu of Charpy shifts (i.e. Paths 6 and 7). For the Kewaunee surveillance weld the heat uncertainty can be estimated because T_o measurements for the Maine Yankee surveillance weld are also available. The value of T_o

for the irradiated Maine Yankee surveillance weld from Table 2 is 232°F at 6.11×10^{19} n/cm². An unirradiated T_o value of -158°F was measured for the Maine Yankee weld by Framatome Technologies on behalf of the Combustion Engineering Owner's Group [Hoffman, 1999]. The irradiation induced shift in T_o was therefore 393°F at 6.11×10^{19} n/cm². This corresponds to a chemistry factor of 271°F (as compared to 222°F for the Kewaunee surveillance weld). The trend curve for the Maine Yankee surveillance weld is compared to the trend curve for the Kewaunee surveillance weld in Fig. 5. The industry average composition for weld heat 1P3571 is intermediate to the two surveillance welds (chemistry factor ratio = 0.536). It would therefore be expected that the heat adjusted trend curve should fall midway between the two surveillance welds. This intermediate trend curve can be constructed by applying the ratio procedure to the respective chemistry factors, which results in a value for CF of 248°F. The trend curve produced for this heat average chemistry factor is also illustrated in Fig. 5. At the projected Kewaunee EOL fluence of 3.36×10^{19} n/cm², the difference between the Kewaunee surveillance weld and the 1P3571 industry average trend curve is 35°F. This value of ΔART_{HT} corresponds closely to the 36°F value found by applying a similar analysis to the Charpy data. The close agreement between these two estimates of ΔART_{HT} indicates the concerns about material variability and surrogate materials are independent of the Master Curve. Therefore it is possible to use the technically superior Master Curve approach and maintain the desired level of margin to accommodate material uncertainty.

Comparison of Methods

Table 1 and Fig. 6 compare the eight different methods for estimating the irradiated fracture toughness transition temperature for the Kewaunee surveillance weld. The most interesting comparisons between methods is in the best estimate of the ART (sixth column in Table 1), and the best estimate of the ART adjusted for uncertainties in initial properties, in shift value, and in heat variability (last column in Table 1).

Paths 1 through 4 are Reg. Guide 1.99 (Rev. 2) assessments of the Kewaunee RPV, which rely on RT_{NDT} and Charpy data. These methods produce best estimates of the irradiated reference temperature that only vary over a limited range (from 190°F to 203°F). However, when the margin terms and heat adjustment terms are included, the corresponding ART values are approximately 80°F higher and the range of values increases by a factor of 2 (from 267°F to 292°F).

The Master Curve test procedure is employed in Methods 5a, 5b, 6 and 7. In all four cases, the best estimate of the irradiated reference temperature is lower than estimates based on RT_{NDT} and Charpy. The lowest best estimates of irradiated reference temperature were obtained in paths

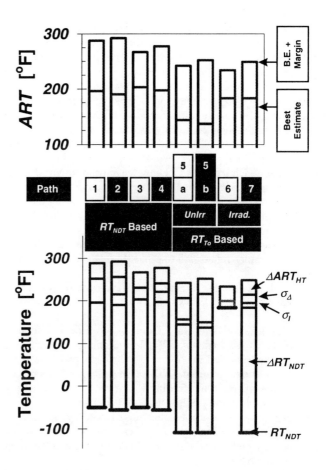

Fig. 6. Comparison of ART values calculated by various techniques.

5a (144°F) and 5b (137°F). These estimates combined the unirradiated RT_{To} measurements with shifts based on Charpy data. The unirradiated RT_{To} value for the Kewaunee weld (-109°F) was significantly lower than the unirradiated RT_{NDT} value (-50°F). Conversely, direct measurement of the ART (Path 6) produces a higher value (199°F). When the margins are added in, the ART value for the direct measurement of irradiated toughness (path 7) is the lowest. This reversal occurs because large margin has been imposed on Paths 5a and 5b to account conservatively for uncertainties arising from the use of Charpy data to shift unirradiated T_o values.

CONCLUSIONS

The Master Curve index temperature (T_o), and the related ASME-defined reference temperature (RT_{To}) for the K_{IC} and K_{IR} curves, can be used to estimate a reference temperature adjusted for the effects of irradiation embrittlement (i.e. the ART). This ART estimate is technologically superior to conventional estimates of ART because T_o is determined by directly measuring fracture toughness, rather than the current practice of inferring it from other manifestations of fracture mode transition. In this paper we described methodologies whereby T_o measurements are combined with margin terms to estimate the *ART*. These margin terms account for uncertainties associated with reference temperature measurement, with the magnitude of the irradiation shift, and with heat-to-heat variability. The proposed methodologies satisfy the intent of US-NRC Reg. Guide 1.99 (Rev. 2), which describes the current methodology for *ART* estimation from nil-ductility temperature, Charpy, and chemistry measurements. The following three new methodologies are proposed herein:

Estimation of *ART* based on T_o measurements made on
1. unirradiated samples, or
2. samples irradiated to the EOL fluence, or
3. samples irradiated to other than the EOL fluence.

Key findings made during the development of these methodologies are as follows:

- ✓ The form of the Reg. Guide 1.99 (Rev. 2) fluence function applies as well to prediction of irradiation induced shifts in T_o as it has to prediction of irradiation induced shifts in Charpy transition temperature. This finding is critical to the success of *ART* estimation methodologies #1 and #3.
- ✓ Surrogate materials concerns caused by sample to sample chemistry variations can be addressed using established ratioing and heat average chemistry procedures.
- ✓ The proposed ART estimation methodologies are at least as conservative, if not more so, than Reg. Guide 1.99 (Rev. 2) procedures. Increases in conservatism relative to current practice occur in two areas:
  - ✓ The margin added to account for uncertainty in a measured value of reference temperature (typically between 6°F and 14°F) is tied to the statistical uncertainty in measured T_o values described in ASTM Standard E1921-97. By comparison, the margin currently added to account for uncertainty in a measured value of reference temperature is 0°F. This additional margin is included in all three *ART* estimation methodologies.
  - ✓ ART estimation methodology #1 combines To values measured before irradiation with Charpy shifts to account for irradiation embrittlement. Our proposed estimate of shift uncertainty for this methodology reflects the rather improbable assumption Charpy shifts are known without error, so all of the uncertainty in the relation between Charpy shifts and fracture toughness shifts occurs on the fracture toughness side of the relationship. This assumption is obviously incorrect. It is therefore expected that the 27°F factor added to compensate for use of a "mixed" methodology can be reduced without any safety impact as more information becomes available.

ACKNOWLEDGEMENTS

The work described in this paper was funded jointly by the Westinghouse Owners' Group, and by the Wisconsin Public Service Corporation.

REFERENCES

10 CFR 50, Nuclear Regulatory Commission 10 CFR Part 50, "Fracture Toughness Requirements for Light Water Reactor Pressure Vessels."

ASME N629, ASME Code Case N-629, "Use of Fracture Toughness Test Data to Establish Reference Temperature for Pressure Retaining Materials for Section XI.

ASTM E1921-97, "Test Method for Determination of Reference Temperature, T_o, for Ferritic Steels in the Transition Range," ASTM, 1997.

EPRI, 1998, "Application of Master Curve Fracture Toughness Methodology for Ferritic Steels," EPRI-TR-108390, 1998.

Hawthorne, et al., 1984, Hawthorne, J.R., Menke, B.H., Hiser, A.L., "Notch Ductility and Fracture Toughness Degradation of A302-B and A533-B Reference Plates from PSF Simulated Surveillance and Through-Wall Irradiation Capsules," NUREG/CR-3295 (Vol. 1), USNRC, 1984.

Hawthorne and Hiser, 1988, Hawthorne, J.R., Hiser, A.L., "Experimental Assessments of Gundremmingen RPV Archive Material for Fluence Rate Effects Studies," NUREG-CR/5201, US-NRC, 1988.

Hawthorne and Hiser, 1990(a), Hawthorne, J.R., Hiser, A.L., "Investigations of Irradiation-Anneal-Reirradiation (IAR) Properties Trends of RPV Welds: Phase 2 Final Report," NUREG-CR/5492, US-NRC, 1990.

Hawthorne and Hiser, 1990(b), Hawthorne, J.R., Hiser, A.L., "Influence of Fluence Rate on Radiation-Induced Mechanical Property Changes in RPV Steels," NUREG/CR-5493, US-NRC, 1990.

Hawthorne, et al., 1992, Hawthorne, J.R., Menke, B.H., Loss, F.J., Watson, H.E., Hiser, A.L., and Gray, R.A., "Evaluation and Prediction of Neutron Embrittlement in Reactor Pressure Vessel Materials," EPRI NP-2782, EPRI, 1992

Hiser, 1989, Hiser, A.L., "Post-Irradiation Fracture Toughness Characterization of Four Lab-Melt Plates," NUREG/CR-5216 (Rev. 1), US-NRC, 1989.

Hoffman, 1999, Private communication.

Kirk, et al., 1998, Kirk, M., Lott, R., Server, W., Hardies, R., and Rosinski, S., "Bias and Precision of T_o Values Determined using ASTM Standard E1921-97 for Reactor Pressure Vessel Steels," *19th International Symposium on the Effects of Irradiation on Materials, ASTM STP-1366*, M.L.

Hamilton, S. Rosinski, M. Grossbeck, and A. Kumar, Eds., American Society for Testing and Materials, 1998.

Marston, 1978, Marston, T.U., "Flaw Evaluation Procedures, Background and Application of ASME Section XI Appendix A," EPRI Report NP-719-SR, Electric Power Research Institute, 1978.

McGowan, et al., 1988, McGowan, et al., "Characterization of Irradiated Current-Practice Welds and A533 Grade B Class 1 Plate for Nuclear Pressure Vessel Service, NUREG/CR-4880, U.S. Nuclear Regulatory Commission, 1988.

Rossell, 1997, Rosseel, T.M., "Heavy-Section Steel Technology Irradiation Program, Semiannual Progress Report for October 1996 - March 1997," NUREG/CR-5913, U.S. Nuclear Regulatory Commission, 1997.

RG 1.99(R2), Regulatory Guide 1.99, Revision 2, *Radiation Embrittlement of Reactor Vessel Materials, U.S. Nuclear Regulatory Commission*, May 1988.

Sokolov, 1998, Sokolov. M.A., "Statistical Analysis of the ASME K_{Ic} Database," *Journal of Pressure Vessel Technology*, Vol. 120, February 1998.

VanDerSluys and Miglin, 1994, VanDerSluys, W.A. and Miglin, M.T., "Results of MPC/JSPS Cooperative Testing Program in the Ductile-to-Brittle Transition Region," *Fracture Mechanics: 24th Volume, ASTM STP 1207*, J.D. Landes, D.E. McCabe, and J.A.M. Boulet, Eds., American Society for Testing and Materials, pp. 308-324, 1994.

Wallin 1984(a), Wallin, Saario, and Törrönen, "Statistical Model for Carbide Induced Brittle Fracture in Steel," *Metal Science*, 18, pp. 13-16, 1984.

Wallin 1984(b), Wallin, K., "The Scatter in K_{Ic} Results," *Engineering Fracture Mechanics*, 19(6), pp. 1085-1093, 1984.

Wallin 1985, Wallin, K., "The Size Effect in K_{Ic} Results," *Engineering Fracture Mechanics*, 22, pp. 149-163, 1985.

Wallin 1993, Wallin, K., "Irradiation Damage Effects on the Fracture Toughness Transition Curve Shape for Reactor Vessel Steels," *Int. J. Pres. Ves. & Piping*, 55, pp. 61-79, 1993.

Wallin 1995, Wallin, K., "Re-Evaluation of the TSE Results Based on the Statistical Size Effect," VTT Manufacturing Technology, 1995.

FRACTURE MECHANICS - THEORY AND APPLICATIONS

Introduction

Yuh J. Chao
Depart of Mechanical Engineering
University of South Carolina
Columbia, SC 29208

Fracture mechanics has been extensively used in the structural integrity assessment of pressure vessels and piping. As such, it has been an active theme in the ASME Pressure Vessels and Pining Conference for many years. In this year's program, there are eight papers in this subject and they are presented in two sessions. The topics covered include fundamental development in fracture mechanics, applications to nuclear plant components, and determination of material fracture properties.

The first paper, by Roos and Eisele, reviews the current status of fracture assessment using various parameters and advocating the use of Ji as the proper parameter for fracture initiation in the ductile-brittle transition regime.

The second paper, by Smith, discusses the effect of residual stress on failure analysis when the applied load is small compared to the limit load.

The third paper, by Kubo, Yoshikawa and Ohji, proposes an engineering approach for creep crack growth analysis based on the J* integral and its application to aged power plant pipes. The proposed engineering approach appears to work for the material studied.

The fourth paper, by Chao, Liu and Broviak, discusses the constraint effect in brittle fracture when the deformation mode is elastic. Through theoretical consideration and test data it shows that the fracture toughness of PMMA varies with the specimen geometry and this variation is attributed to the constraint of the specimens.

The fifth paper, by Roos and Herter, reviews a fracture mechanics based procedure for carrying out "break preclusion" analysis for piping systems.

The sixth paper, by Sindelar, Lam, Caskey, and Woo, compiles the material fracture properties for A285 Grade B carbon steels and discusses their potential applications to the structural integrity of nuclear waste storage tanks. The follow-up paper, by Lam and Sindelar as the seventh paper, uses the material properties for A285 steels in the flaw stability analysis for a postulated flaw in a waste tank.

The last paper, by Chen, deals with the computation aspect in the elastic-plastic regime using the finite element methods.

DETERMINATION OF FRACTURE MECHANICS CHARACTERISTICS: INITIATION OR INSTABILITY CHARACTERISTICS AS A BASIS FOR COMPONENTS ASSESSMENT

E. Roos, U. Eisele
Staatliche Materialprüfungsanstalt (MPA)
University of Stuttgart, Germany

Abstract

For components assessments in the ductile fracture regime it is acknowledged that the analysis should consist of two steps, the first covers the question of initiation loading, the second covers questions of crack extension and instability. It is also acknowledged, that for crack growth and instability analysis a second parameter characterizing the state of stress triaxiality has to be taken into consideration.

In the brittle fracture regime usually no crack growth has to be considered, the analysis of the components loadability can be based on a single parameter, e.g. valid values of fracture toughness K_{Ic}.

In the transition range of ferritic steels the situation is not that clear. It can be seen from experimental investigations, that the failure process also includes ductile crack initiation, ductile crack growth and cleavage instability. The extent of ductile crack growth developing between ductile initiation and cleavage instability depends on several influencing factors, e.g. temperature, specimen size and crack length. However, testing recommendations and standards recommend to determine the instability values at the cleavage fracture event and use these values as material characteristics, the determination of initiation values is usually not recommended.

In this contribution the transferability of fracture mechanics initiation and instability values to components is considered for the transition range of fracture toughness.

Introduction

It is acknowledged that for many component assessments supplementary to the fracture mechanics parameters K_I or J a second parameter characterising the multiaxiality of the stress state has to be taken into consideration.

The necessity to work with such a second parameter depends on several boundary conditions, e.g.
- do we need to analyze only crack initiation?
- do we need to consider instability including prior crack growth?
- which of several available material characteristic parameters are used?

While the situation is quite clear in the pure cleavage fracture range, it can be seen sometimes in the ductile to brittle transition, that the different stages of the failure process are not considered in all required details.

Failure process in fracture mechanics specimen and precracked components

The failure process in ferritic structural steels is influenced by several interacting mechanisms. The most important influencing factors are
- temperature
- „basic ductility behavior" of the material
- geometry of the specimen/component
- geometry of the crack (crack length, crack tip sharpness, crack shape)
- loading situation:
  (tension/bending/pressure
  load controlled/displacement controlled)

The mechanical behavior of ferritic steels shows a clear dependency of temperature. At low temperatures slip processes in the microstructure are restrained, which in general yield macroscopic ductile behavior. At low temperatures a macroscopic brittle behavior (linear load-displacement behavior) of the specimen/component can be observed up to fracture load as a consequence. Pure cleavage fracture is usually found on the fracture surface of such specimens.

With increasing temperature more and more microscopic slip processes can happen and lead to a macroscopic ductile behavior (non linear load displacement behavior) of the specimen/structure.

The extent of „ductility" is strongly affected by the influencing factors mentioned above. The transition from brittle to ductile behavior is fluent. Following the failure process of precracked fracture mechanics specimen, a macroscopic purely linear elastic behavior can be seen at very low temperatures. With increasing test temperatures an increasing deviation from the linear elastic behavior can be observed.

Looking at the tip of a crack in a fracture mechanics specimen, the onset of plastic flow processes at the crack tip takes place at relatively low loading levels, where macroscopically (F-COD) still linear elastic specimen behavior can be observed.

This holds especially for large specimens with dimensions clearly above those that are usually used, 1T- or 2T- Compact Tension specimens, where the relation is small between plastic zone size and zone with elastic material behavior. This means in other words, that at a given temperature in the „transition regime" the failure process in small specimen may be dominated by plastic processes even in macroscopically measurable parameters (Load-CMOD), while in large specimens these macroscopic parameters still show a nearly linear elastic behavior.

It is well known, that the plastic zone size and the stress triaxiality have an influence on each other. This means, that the small specimen, which is dominated by plastic processes, may show a slow but stable, more or less limited ductile crack growth after exceeding a certain critical value of crack tip loading (crack initiation), whereas the large specimen fails by sudden (cleavage) fracture after exceeding this value. When the fracture event of this specimen is seen only as a „material parameter" and the prior specimen behavior is disregarded, an influence of stress triaxiality is already included in this „material parameter". As a consequence, a direct transferability of these parameters to components is no longer given. This is why transferable material characteristics have to be defined very carefully in order not to mix up varios effects.

Determination of fracture mechanics characteristic values

Linear-elastic specimen behavior

To characterize the macroscopical linear-elastic specimen behavior up to fracture usually K_{Ic}-values are used. Validity criteria are given in several national and international standards, e.g. ASTM E 399, ESIS P2-92, ISO 12737, and BS 7448. Most important criteria reflect the macroscopically measurable load-displacement behavior.

In order to provide a technically acceptable procedure, deviations from the pure linear elastic behavior are allowed up to a maximum of 5% of the specimen compliance, fig.1.

Fracture mechanics specimens, which fulfill these criteria, show K_{Ic}-values that are nearly independent of the specimen size.

In the literature, e.g. ASTM E 1921, equations, derived from the weakest link theory are discussed, in order to predict size effects (effects of crack length) on fracture toughness, e.g.:

$$K_{I,B_2} = K_{I,B_1} \left(\frac{B_1}{B_2} \right)^{1/4}$$

where $K_{I,B2}$ is the fracture toughness of a specimen with thickness B_2, calculated from the measured fracture toughness $K_{I,B1}$ of a specimen with thickness B_1. However, the applicability of these formulae on structural steels is still discussed and should be limited - if at all - until further clarification on the pure linear elastic specimen behavior.

ESIS P2-92: Determining the Fracture Behaviour of Materials

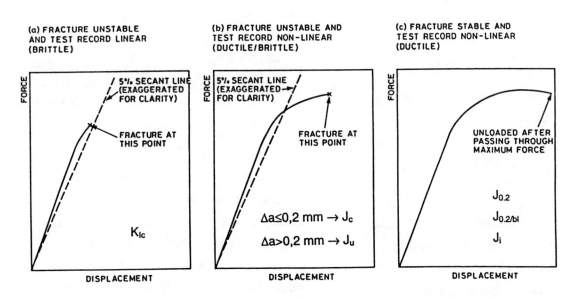

Figure 1: Criteria for the determination of Fracture Toughness values according to ESIS P2-92

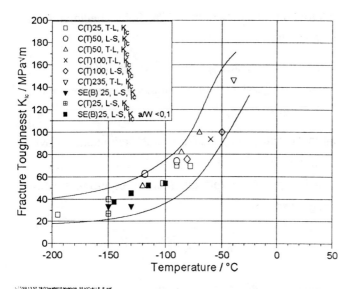

Figure 2: K_{Ic}-values of specimens of varios types, sizes and a/W-ratios

Figure 2 shows valid K_{Ic}-values (according to ASTM E 399) determined by C(T)-specimens with a/W ratio in the range of 0.5 with a thickness between 25 mm and 235 mm. Also shown in this diagram are results of SE(B) specimens with a/W-ratio according to the standard (ASTM E 399) as well as SE(B) specimens with "shallow crack". All specimens (also the shallow crack specimens) failed within the "5%-non-linearity-criterion" of ASTM E 399. It can be seen, that all results are located within a temperature-dependent scatterband without showing any influence of specimen type or crack length.

Seen from the technical point of view it can be said, that transferability of K_{Ic}-values is given within the limitations mentioned above without further corrections concerning specimen size and/or stress triaxiality.

Elastic-plastic specimen behavior

For the quantitative characterization in the upper shelf level of fracture toughness usually the parameter J-Integral is used. Many technical rules, recommendations and guidelines exist describing testing and evaluation procedures for the determination of fracture toughness values at crack initiation, e.g. J_{Ic} (ASTM E 813), $J_{0.2/t.l}$ (ESIS P1-92, DVM 002), and $J_{0.2}$ (ESIS P1-92), fig. 3.

The most important fact when using these and similar values is, that the definition of a stable crack growth is included, and, thus these values are partly remarkably

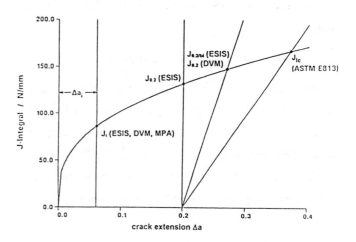

Figure 3: Definition of varios initiation values

higher than the effective initiation values J_i, cf. fig. 3. The determination of J_i either by potential drop technique or by means of the „stretch zone width", as described by Roos et. al. (1988) is also recommended by some testing procedures, e.g. ESIS P1-92, DVM 002, JSME 001. It is remarkable that in ESIS P1-92 the „stretch zone width method" to determine J_i is „considered to be the most accurate method for measuring J close to the onset of crack initiation", whereas $J_{0.2/bl}$-values are looked upon as to provide an „engineering definition of initiation".

By the fact that, e.g. $J_{0.2/bl}$-values or J_{Ic}-values correspond to a more or less relevant extent of slow stable crack growth, these values are influenced by the degree of stress triaxiality, which has a strong impact on the crack growth characteristics of the specimen. This is proven by the following literature (Eisele et al., 1991, 1992, Roos et al. 1993) and can be avoided by using the effective initiation values J_i, which are not affected by stress triaxiality, fig. 4.

It can also be seen as a consequence of this fact, that no real instability values can be defined for ductile fracture as a material characteristic parameter. Attempts to define such a or similar instability values were realized several years ago (e.g. J_{50}, Paris, 1981), but they were not successful and, therefore, were not continued.

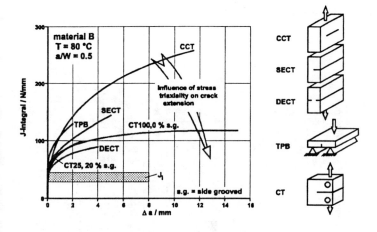

Figure 4: Independence of J_i of stress triaxiality

Fracture mechanics material characterization in the „transition range"

Several suggestions are available for the characterization of fracture toughness in the transition range (ESIS P2-92, BS 7448, ASTM E 1921). These documents show very similar rules.

The first decisive criterion, which parameter has to be determined, is given by the load-displacement behavior of the specimen fig. 5 and fig. 1.

Mainly depending on the extent of plastification (besides other criteria) it needs to be decided, whether a pure load consideration (K_{Ic}, ignoring small extents of plastification)

or an energy consideration (J, taking plastification into account) has to be performed.

Should a J-analysis be necessary, the fracture surface of the specimen needs to be considered as an additional

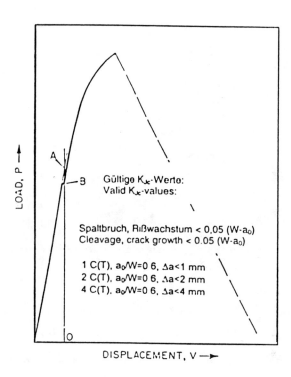

Figure 5: Validity Criteria for K_{Jc} as of ASTM E 1921

criterion. Depending on the extent of ductile crack growth (0 mm < Δa < 0,2 mm, respectively 0,2 mm < Δa < 0,5 mm (ESIS) or Δa < 0,05 (W-a_o) (ASTM)) before the onset of cleavage, a „classification" of the J value obtained has to be made. These values include a stable crack growth which may occur before the onset of cleavage, and is strongly dependent on the specimen size and the a/W-ratio. Thus these values characterize the fracture of the special specimen and not characteristics of a material.

An example for the consequences can be seen in fig. 6, where K_{Ic}, K_{Jc} and K_{Ju} values (according to ESIS P2) of specimens of different sizes for a 10 MnMoNi 5 5 shape welded material are shown.

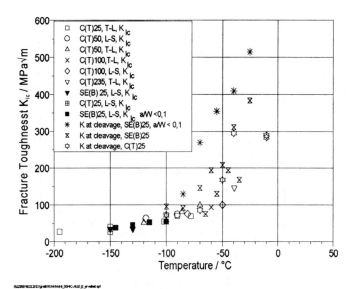

Figure 6: K_{Ic}, K_{Jc} and K_{Ju} values (according to ESIS P2) of specimens of different sizes and a/W-ratios

The most interesting point of this diagram can be seen at a temperature of -40 °C. At this temperature a large compact tension specimen with a thickness of 235 mm gives a fully valid K_{Ic}-value of ~145 MPa√m. Smaller standard specimens (CT25, TPB25) give K_{Jc} and K_{Ju}-values (K at cleavage) of up to 300 MPa√m, specimens with shallow crack reach even more than 400 MPa√m. It can also be seen, that the K_{Jc}, resp. K_{Ju}-values are higher than the valid K_{Ic}-value in the whole temperature range. It needs to be pointed out, that this is not a scatter in fracture toughness but due to a comparison of values with different meaning.

In order to avoid a misleading interpretation of the specimen behavior, we do not have to concentrate on the constraint dependent fracture of the specimen K_{Jc} or K_{Ju}, but we have to evaluate the constraint independent initiation of the specimen. This can be done by applying the J_i-evaluation procedure (by means of stretch zone width measurements) - as it is established for the upper shelf region - for all specimen and load histories above the validity limits of pure K_{Ic}-behavior (Sun et. al. 1992). The K_{IJ}-values (formally determined by J_i) of the different specimen sizes and a/W-ratios are in the same scatterband together with valid K_{Ic}-values along the whole temperature range, fig. 7.

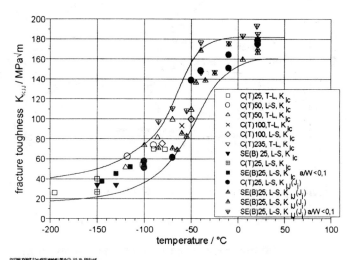

Figure 7: Initiation toughness values of specimens of varios types, sizes and a/W ratios

These results show no influence of specimen size on initiation. This is clearly confirmed in fig. 7, that at -40 °C the K_{Ic}-value of the CT235 specimen is nearly identical with the K_{IJ}-value (determined by J_i) of smaller specimens.

These experimental results show, that by means of the J_i-values, it is possible to characterize the material behavior in the whole temperature range (also in the upper shelf) by using this single parameter, as schematically shown in fig 8.

This means, that for the analysis of the initiation loading of cracked specimens or components based on J_i-, the multiaxiality of the stress state has no influence and does not need to be taken into account.

However, these results show that immediately after exceeding the J_i-loading level, crack growth due to the multiaxiality of the stress state has decisive influence on crack growth, even for very small extents of crack growth. Such small extents of crack growth may be included for example in the J_c- or J_u-values. This is the reason why these values are not transferable and thus are not looked upon as material characteristics.

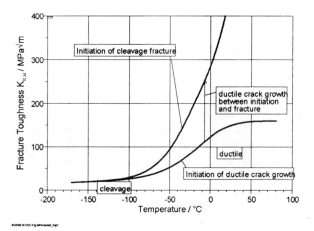

Figure 8: Increasing differences between initiation and instability values depending on the temperature (multiaxiality of the stress state)

Conclusions

As mentioned above, the fracture process of cracked specimens or components is influenced by several interacting parameters.

It can be seen from experimental, theoretical, analytical and numerical studies, that the triaxiality of the stress state is one of the most important influencing factors.

From the results available at the present, the failure process in precracked specimens and components can be roughly divided into two stages.

The first stage is the increase of loading up to the initiation level. This loading level has to be characterized and quantified by the parameter J_i, which is independent of specimen type and specimen size according to the present knowledge. This parameter can be transferred from laboratory specimens to components. The initiation loading of components can be assessed reliably on the basis of this parameter. It should be noted, that by such an analysis no decision can be made, whether the component will fail in case of further increase of loading by sudden fracture (cleavage or ductile fracture), or allow a more or less limited increase of loading accompanied by a more or less stable crack growth.

If loading levels above this initiation level have to be analyzed, the triaxiality of the stress state has to be included in the analysis. It is important to note, that this already includes the loadings corresponding to J_c, J_u, J_{Ic} and comparable values, as mentioned above.

This is why a safety analysis should be performed for the initiation loading on the basis of the J_i-values. The crack growth behavior has to be considered additionally as well as the stress triaxiality (as recommended by Clausmeyer et al. (1991) for the assessment of extended loading.

References

ASTM E 399-90:
Standard Test Method for Plane-Strain Fracture Toughness of Metallic Materials. Annual Book of ASTM Standards, Vol. 03.01.1995

ASTM E 813-89:
Standard Test Method for J_{ic}, a Measure of Fracture Toughness. Annual Book of ASTM Standards, Vol. 03.01., 1995

ASTM E 1921-97
Test Method for the Determination of Reference Temperature, T_o, for Ferritic Steels in the Transition Range, Annual Book of ASTM Standards, Vol. 03.01., 1998

BS 7448: Part 1: 1991
Method for Determination of K_{lc}, Critical CTOD and Critical J values of metallic materials, British Standard Fracture Mechanics Toughness Tests, Part 1, 1991

Clausmeyer, H., K. Kussmaul und E. Roos:
Influence of stress state on the failure behavior of cracked components made of steel.
Appl. Mech. Rev. Vol 44, No 2, February 1991, pp. 77-92
DVM Merkblatt 002:

DVM 002
Ermittlung von Rißinitiierungswerten und Rißwiderstandskurven beiAnwendung des J-Integrals.
DVM Merkblatt 002, Deutscher Verband für Materialprüfung e. V., Juni 1987

Eisele, U., Roos, E.:
Evaluation of Different Fracture-Mechanical J-Integral Initiation Values with Regard to their Usability in the Safety Assessement of Components.
Nuclear Engineering Design 130, pp. 237-247, 1991

Eisele, U., Roos, E., Seidenfuß, M., Silcher, H.:
Determination of J-Integral Based Crack Resistance Curves and Initiation Values for the Assessment of Cracked Large Scale Specimens.
ASTM STP 1131, H.A. Ernst, A. Saxena and D.L. Mc Dowell Eds., ASTM Philadelphia, 1992, pp. 37-59

ESIS P1-92
ESIS Recommendations for Determining the Fracture Resistance of Ductile Materials,
European Structural Integrity Society-ESIS, January 1992

ESIS P2-92
ESIS Procedure for Determining the Fracture Behavior of Materials, European Structural Integrity Society-ESIS, January 1992

ISO 12737
Metallic Materials - Determination of plane-strain fracture toughness. International Standard, 1996-11-15

JSME S001
Standard Method of Test for Elastic-Plastic Fracture Toughness J_{lc}, JSME Standard S 001-1981

Paris, P.C.:
A method of application of elastic plastic fracture mechanics to nuclear vessel analysis. Second int. Symp. On Elastic-Plastic Fracture Mechanics, Philadelphia, October 1981

Roos, E. and U. Eisele:
Determination of Material Characteristic Values in Elastic Plastic Fracture Mechanics by Means of J-Integral Crack Resistance Curves. Journal of Testing and Evaluation, JTEVA, Vol. 16, No. 1, Jan. 1988, pp. 1-11

Roos, E., Eisele, U., Silcher, H.:
Effect of Stress State on the Ductile Fracture Behavior of Large Scale Specimens. ASTM STP 1171,
E.M. Hackett, K.H. Schwalbe and R.H. Dodds, Eds.
ASTM Philadelphia, 1993, pp. 41-63

Sun, G., Eisele, U., Roos, E.:
Ermittlung zähbruchmechanischer Werkstoffkennwerte an Drei-Punkt-Biegeproben im Übergangsbereich der Werkstoffzähigkeit. Materialwissenschaften und Werkstofftechnik, 23, 1992, 250-259

THE EFFECT OF RESIDUAL STRESSES ON CRACK EXTENSION WHEN EXPRESSED IN THE K_r - L_r FORMAT

E. Smith

Manchester University - UMIST Materials Science Centre,
Grosvenor Street, Manchester, M1 7HS, United Kingdom.

ABSTRACT

Normalised stress intensity (K_r) versus normalised limit load (L_r) failure assessment diagrams feature prominently in integrity assessments, concerned with the effects of cracks in engineering structures. The geometry dependence of the failure assessment curve, when expressed in the K_r-L_r format, has been highlighted in recent work. Against this background, the present paper focuses on the effect of residual stresses on the failure assessment curve for the small L_r regime, with the residual stresses being incorporated within the K_r parameter, but not the L_r parameter, as in the R6 procedure. The residual stresses are expressed in terms of a power law variation, and the reduction in the magnitude of K_r, for the case where $L_r \to 0$, is quantified for such a variation. The results highlight the extent to which the K_r-L_r failure assessment curve can lie below the no residual stress curve, and this could be of importance with a low toughness material.

INTRODUCTION

Failure assessment diagrams now feature prominently in integrity assessments of engineering structures containing cracks. The basis of these diagrams is the use of two dimensionless parameters: $K_r = K/K_{IC}$ and $L_r = P/P_L$ (or σ/σ_L). K_{IC} is the material's fracture toughness, K is the linear elastic stress intensity factor due to the applied load (stress): P (σ), while P_L (σ_L) is the limit load (stress). If K_r and L_r are known for a particular structure containing a crack, a judgement as to whether or not crack extension is likely to occur is made on the basis of the position of the K_r-L_r point relative to an appropriate failure assessment curve $K_r = f(L_r)$.

Use of this approach is simplified if the K_r - L_r failure assessment curve is assumed to be geometry independent, as is the case if Options 1 or 2 of the R6 procedure [1] are used. However, prompted by Bloom's work [2], it has become apparent over the last few years that the K_r - L_r failure assessment curve can show a marked geometry dependence, and it has been shown that this is particularly the case where a crack is in a pronounced stress gradient, for example when a crack is at the root of a blunt stress concentration. It is against this background, and as part of a research programme that is focused on this geometry dependence issue, that the present paper addresses the effect of residual stresses on the failure assessment curve, and extends the work described in the author's paper [6] at the 1998 PVP Conference. The residual stresses are incorporated in the K_r parameter, but not the L_r parameter, as in the R6 procedure [1]. They are expressed in terms of a general power law variation (only the linear variation was considered in the author's earlier paper [6]), and the reduction in the magnitude of K_r is quantified for such a variation for the case where $L_r \to 0$, i.e. the intercept of the K_r - L_r failure assessment curve with the K_r axis is quantified. It is assumed that the residual stresses are sufficiently small that there is limited plastic deformation at a crack tip, for this allows the equivalent elastic crack concept [7, 8] to be used. Thus, using the Dugdale-Bilby-Cottrell-Swinden (DBCS) strip yield representation [9, 10] of plastic deformation, it is possible to obtain analytical results, which allow us to clearly see how material and geometrical parameters associated with the residual stress distribution can affect the way and extent to which the K_r-L_r failure assessment curve can lie below the no residual stress curve, for which $K_r = 1$ when $L_r \to 0$; this reduction could be of importance with a low toughness material.

GENERAL THEORETICAL BACKGROUND

With the DBCS strip yield representation of plastic deformation, the material is assumed to be non-work-hardening with σ_y, the material's tensile yield stress, being the tensile stress within a strip yield zone at a crack tip. Crack extension is presumed to occur when the relative displacement υ_T of the crack faces at a crack tip attains a critical value υ_C. The situation where the plastic deformation is limited, i.e. the strip yield zone is small in relation to other characteristic lengths of a configuration, can be considered by use of the equivalent elastic crack procedure [7, 8]. The Mode I stress intensity factor K due to the residual stresses can be written in the following general form:

$$K = \sigma_R h^{1/2} S(\lambda) \quad (1)$$

where h is a characteristic dimension related to the scale of the residual stress distribution, a is the crack depth and $\lambda = $ a/h. With the equivalent elastic crack procedure, the J integral [11] can be written in the form

$$J = \frac{1}{E_O}\left[K(a + \Delta a_E)\right]^2 \quad (2)$$

where Δa_E is the elastically equivalent size of strip yield zone, and is the distance between the initial crack tip and the tip of the equivalent elastic crack. $E_O = E/(1-\nu^2)$ where E is Young's modulus and ν is Poisson's ratio, since we are applying the strip yield model to the plane strain situation. The right-hand side of relation (2) can be expanded to the first two terms, to give

$$J = \frac{K^2}{E_O}\left[1 + \frac{2K'}{K} \cdot \frac{\Delta a_E}{h}\right] \quad (3)$$

where K given by relation (1), and $K' = dK/d\lambda$ are defined with regard to the initial crack tip. Now J, as given by relation (3), is equal to the area W_F under the stress (p)-relative displacement (υ) curve appropriate to the strip yield zone, up to a value of υ that is equal to the displacement υ_T at the crack tip, i.e.

$$\frac{K^2}{E_O}\left[1 + \frac{2K'}{K} \cdot \frac{\Delta a_E}{h}\right] = W_F = \int_0^{\upsilon_T} p(\upsilon)d\upsilon \quad (4)$$

At the onset of crack extension, and with the DBCS strip yield representation, p is constant within the strip yield zone and has a value σ_y, while $\upsilon_T = \upsilon_C$ is the critical value of υ_T at which crack extension occurs. Furthermore [8], we can then equate Δa_E with the elastically equivalent size $R_{E\infty}$ of zone that is associated with a semi-infinite crack in a remotely loaded infinite solid, i.e. $\pi E_O \upsilon_C/24\sigma_y$ [12]. Relation (4) then gives the K value appropriate to the onset of crack extension as

$$K = \left[\frac{E_O \sigma_y \upsilon_C}{1 + \frac{2K'}{K}\frac{R_{E\infty}}{h}}\right]^{1/2} \quad (5)$$

Since $K R_{E\infty}/Kh$ is presumed to be small or otherwise the present analysis is invalid (the step from relation (2) to (3) and the argument preceding relation (5) would be invalid), and because the fracture toughness K_{IC} of the material is equal to $[E_0\sigma_Y\upsilon_C]^{1/2}$ with the strip yield representation, then relations (1) and (5) give, to the first two terms on the right hand side,

$$K_r = \frac{K}{K_{IC}} = 1 - \frac{\pi\sigma_R^2}{24\sigma_y^2}SS' \qquad (6)$$

and this is the value of K_r when $L_r \to 0$ for the residual stress distribution.

APPLICATION OF THE GENERAL THEORY

We now apply the general theory developed in the preceding section. Thus let us consider the situation where there is a two-dimensional crack of length $2a$ in a large (infinite) plate that is subjected to a distribution of tensile residual stress $\sigma(x)$ which, in the crack's absence, is described by the relation

$$\sigma(x) = \sigma_R \sum_{n=0}^{\infty} A_n \left(\frac{|x|}{h}\right)^n \qquad (7)$$

with n having positive integer values and with x being measured from the crack centre along the crack plane. The crack tip stress intensity K due to this distribution is given by the expression [13]

$$K = \sigma_R \sum_{n=0}^{\infty} A_n a^{1/2} \frac{\Gamma\left(\frac{n}{2}+\frac{1}{2}\right)}{\Gamma\left(\frac{n}{2}+1\right)}\left(\frac{a}{h}\right)^n \qquad (8)$$

whereupon relations (1) and (8) give, with $\lambda = a/h$

$$S(\lambda) = \sum_{n=0}^{\infty} A_n \lambda^{n+1/2} \frac{\Gamma\left(\frac{n}{2}+\frac{1}{2}\right)}{\Gamma\left(\frac{n}{2}+1\right)} \qquad (9)$$

The value of K_r when $L_r \to 0$ is then given by substituting for $S(\lambda)$ in relation (6).

Let us now consider some special cases. For the case where $\sigma(x) = \sigma_R$, $A_0 = 1$ and all the other A_n are equal to zero, whereupon

$$K_r = 1 - \frac{\pi^2\sigma_R^2}{48\sigma_y^2} \qquad (10)$$

For the case where $\sigma(x) = \sigma_R[1-(|x|/h)]$, $A_0 = 1$, $A_1 = -1$ and all the other A_n are equal to zero, whereupon

$$K_r = 1 - \frac{\pi^2\sigma_R^2}{48\sigma_y^2}\left(1-\frac{2\lambda}{\pi}\right)\left(1-\frac{6\lambda}{\pi}\right) \qquad (11)$$

in accord with the result obtained in the author's paper [6] at the 1998 PVP Conference. For the case where $\sigma(x) = \sigma_R[1-(x/h)^2]$, $A_0 = 1$, $A_2 = -1$ and all the other A_n are equal to zero, whereupon

$$K_r = 1 - \frac{\pi^2\sigma_R^2}{48\sigma_y^2}\left(1-\frac{\lambda^2}{2}\right)\left(1-\frac{5\lambda^2}{2}\right) \qquad (12)$$

DISCUSSION

The present paper has been concerned with the effect of residual stresses on the form of the failure assessment curve when this is represented in the K_r - L_r format, with attention being focused on the case where $L_r \to 0$, and the analysis proceeding from the basis that there is limited plastic deformation at a crack tip, i.e. the residual stresses are not unduly large in relation to the yield stress. For a general stress distribution, simulated by a power law variation, the paper has quantified the extent to which the K_r - L_r failure curve can lie below the no residual stress curve, for which $K_r = 1$ when $L_r \to 0$, and specific examples have been considered in the preceding section. Linear and quadratic stress distributions have been considered, in both cases, the stress decaying on moving away from the crack centre. The analyses are probably most realistic for situations where the crack size a is small compared with the scale h of the residual stress distribution which might for example arise as a consequence of welding. Here it has been shown that K_r is less than unity when $\lambda = a/h$ is less than $\pi/6$ for a linear distribution and when $\lambda = a/h$ is less than $(2/5)^{1/2}$ for a quadratic distribution. The reductions are greatest with a flat residual stress distribution, and will be greater than those determined in the present paper if there is extensive plastic deformation at a crack tip, i.e. if the residual stresses are large in relation to the yield stress.

Though the results are clearly of practical concern with a material of low toughness where it is possible for crack extension to occur at small crack sizes, this is unlikely to be the case with ductile materials. However, the results obtained in this paper do show, once again, that there are problems in considering secondary stresses within the K_r - L_r format, as indeed has been recognised in the R6 procedure [1].

REFERENCES

1. I. Milne, R.A. Ainsworth, A.R. Dowling and A.T. Stewart, Int. Jnl. Press. Vess. and Piping, 32 (1988) 2.

2. J.M. Bloom, PVP - Vol. 287 ASME (1994) 147.

3. Q. Hong, T. Ying and S. Liankiu, Int. Jnl. Press. Vess. and Piping, 57 (1994) 201.

4. E. Smith, PVP - Vol. 324 ASME (1996) 143.

5. E. Smith, PVP - Vol. 346 ASME (1997) 3.

6. E. Smith, PVP - Vol. 373 ASME (1998) 517.

7. G.R. Irwin, Ninth Int. Congress App. Mech., Vol XIII, Paper 101 (II), University of Brussels, (1957) 245.

8. J. Planas and M. Elices, Int. Jnl. Fracture, 51 (1991) 139.

9. D.S. Dugdale, Jnl. Mechs. Phys. Solids, 8 (1960) 100.

10. B.A. Bilby, A.H. Cottrell and K.H. Swinden, Proc. Roy. Soc. A272 (1963) 304.

11. J.R. Rice, in Fracture, ed. H. Liebowitz, Vol. 2, Academic Press, New York, (1968) 191.

12. T.M. Edmunds and J.R. Willis, Jnl. Mechs. Phys. Solids, 25 (1977) 423.

13. H. Tada, P.C. Paris and G.R. Irwin, "The Stress Analysis of Cracks Handbook", Del Research Corporation, Hellertown, Pa, USA (1973).

Life Estimation of Creep Crack Propagation in Aged Power Plant Pipes Using the Engineering Approach

Shiro Kubo and **Masashige Yoshikawa**
Osaka University
Department of Mechanical Engineering and Systems
2-1, Yamadaoka, Suita, Osaka, 565-0871 Japan
Phone: +81-6-6879-7304   Fax: +81-6-6879-7305
E-mail: kubo@mech.eng.osaka-u.ac.jp

Kiyotsugu Ohji
Ryukoku University
Faculty of Science and Engineering
1-5, Yokotani, Oe-cho, Seta, Ohtsu, 520-2194 Japan
Phone: +81-77-543-7411   Fax: +81-77-543-7457
E-mail: ohji@rins.ryukoku.ac.jp

ABSTRACT

Estimation of creep lives was made for a $2\frac{1}{4}$Cr-1Mo steel used in a fossil power plant for about 120,000 hrs. It was found that the creep crack propagation rate da/dt can be correlated with the modified J-integral J^* (C^* integral). The engineering approach, which combined J-integral analyses for fully plastic materials with the creep characteristic, was applied to estimate the value of the modified J-integral. It was assured that the creep crack propagation behavior in the C(T) specimen can be reconstructed using the estimated J^* value and $da/dt - J^*$ characteristics. The engineering approach was then applied to the estimation of creep propagation behavior of an axial crack and a circumferential crack in straight pipes of the power plant. The estimated creep crack propagation lives were large enough for relatively long cracks.

NOMENCLATURE

- a : crack length
- a_0 : initial crack length
- b : ligament length
- E : Young's modulus
- da/dt : creep crack propagation rate
- h_1 : hardening function
- J^* : modified J-integral (C^* integral)
- J_{el} : elastic J-integral
- J^*_{LSC} : J^* under large scale creep
- J^*_{SSC} : J^* under small scale creep
- K_{I} : stress intensity factor
- n : creep exponent
- P : applied load
- P_0 : reference load
- t : time
- t_{pl} : time for plasticity correction
- α : material constant
- $\dot{\delta}$: load–point displacement rate
- $\dot{\epsilon}$: creep strain rate
- ϵ_0 : material constant
- $\dot{\epsilon}_0$: material constant
- η : parameter for tension component correction
- σ : applied stress
- σ_0 : material constant
- σ_{net} : net section stress
- σ_{p} : proof stress

INTRODUCTION

The number of fossil power plants, whose accumulated operation times have exceeded 100,000 hours and which are still operating in Japan, U.S. and Europe, is increasing. Due to cost cuts for power generation the assessment and control of safety of existing fossil power plants are necessary. For the safety assessment of fossil power plants, it is required to estimate the residual creep lives and fatigue lives.

In the present paper estimation of creep lives was made for a $2\frac{1}{4}$Cr-1Mo steel used in a fossil power plant for about 120,000 hrs. The creep crack growth rates were characterized using the modified J-integral (C^* integral), J^*. The engineering approach, which combined J-integral analyses for fully plastic materials with the creep characteristic, was applied to estimate the value of the modified J-integral, J^*. The engineering approach was then applied to the estimation of creep propagation behavior of postulated cracks in pipes of the power plant.

CHARACTERIZATION OF MATERIAL PROPERTIES

Material

The material investigated is a $2\frac{1}{4}$Cr-1Mo steel, which has been used in elbow parts of high-temperature reheat steam generator pipes for about 120,000 hours. The temperature of the steam was 841 (K). The location of material cut-out is shown in Fig. 1.

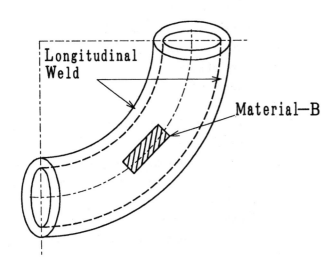

Fig. 1 Location of material cut-out from aged pipes

Creep Properties

Creep tests were conducted for the aged material at 841(K). In the elbows the stress in the circumferential direction is higher than that in the longitudinal direction. To investigate the effect of the direction on creep properties creep testes were conducted for specimens whose longitudinal direction is coincident with the longitudinal direction or the circumferential direction of the elbow pipes. The test results under the longitudinal and circumferential directions are denoted hereafter by characters B3V and B3C, respectively.

Figure 2 shows examples of creep curves under the applied stress of (120MPa). It can be seen from the figure that the transient creep region is negligible and the tertiary crrep is predominant in the creep deformation.

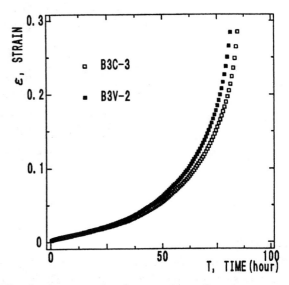

Fig. 2 An example of comparison of creep curves under loads in longitudinal and circumferential directions

The minimum creep rate was determined from creep curves. The creep rate showed its minimum in the initial part of creep curves. Figure 3 show the relationship between the minimum creep rate and the applied stress. The effect of loading direction on the relation is not significant. Although there is a scatter in the data due to ill-definedness of the minimum creep rate, the relation can be approximated by the following Norton type law:

$$\dot{\epsilon} = \alpha \dot{\epsilon}_0 (\sigma/\sigma_0)^n = 9.58 \times 10^{-20} \sigma^{7.56} \qquad (1)$$

($\dot{\epsilon}$: in 1/h, σ : in MPa),

where $\alpha, \dot{\epsilon}_0$ and σ_0 are material constants and n denotes the creep exponent.

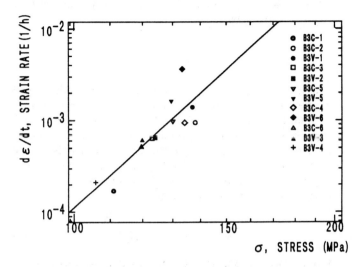

Fig. 3 Relation between minimum creep rate $\dot{\varepsilon}$ and applied stress σ

Creep Crack Propagation Characteristics

Creep crack propagation tests were conducted at 841(K). Figure 4 show the geometry of the C(T) specimen with side-grooves used in the study. To investigated the effect of the direction on creep crack propagation, the loading direction was selected to be coincident with the circumferential or the longitudinal direction of the pipe. For the specimens loaded in the circumferential directions, which are denoted by B2C in the following, the crack is expected to propagate in the longitudinal direction. For the specimens loaded in the longitudinal directions, which are denoted by B2L in the following, the crack is expected to propagate in the circumferential direction. The crack length was monitored by dual probe method, which used a reference electric potential reading together with an electric potential reading near and across the crack.

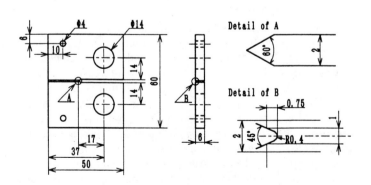

Fig. 4 Geometry of C(T) specimen used for creep crack propagation tests (dimensions in mm)

The modified J-integral J^* (Ohji et al., 1974, 1976, 1988), which was equivalent to the C^* integral (Landes et al., 1976) was evaluated using the load-point displacement rate $\dot{\delta}$. Kubo (1975) and Ohji et al. (1978, 1988) and Shih (1983) proposed the following equation for the evaluation of J^* of deep-cracked bend specimens:

$$J^* = \frac{2n}{n+1}\sigma_{\text{net}}\dot{\delta}, \qquad (2)$$

where σ_{net} denotes the net section stress evaluated at side-groove root. By referring the tension component correction of the J-integral estimation proposed by Landes et al. (1979) and Clarke et al. (1979), Kino et al. (1986) modified the equation as follows:

$$J^* = \frac{2n}{n+1}\frac{1+\eta}{1+\eta^2}\sigma_{\text{net}}\dot{\delta}, \qquad (3)$$

where η is defined as:

$$\eta = \sqrt{\left(\frac{2a}{b}\right)^2 + 2\left(\frac{2a}{b}\right) + 2} - 2\left(\frac{2a}{b}\right),$$

where a and b denote the crack length and the ligament length, respectively. Equation (3) was used in the present study.

The creep crack propagation rate da/dt was correlated with the stress intensity factor K_{I}, the net section stress σ_{net} and the modified J-integral J^*. Figure 5 shows an correlation of da/dt with J^*. The creep crack growth rates along weld bead are shown with characters C3 for comparison purpose. As have been found for many materials (Ohji et al., 1988), da/dt was correlated well with J^* in Fig. 5, but not with K_{I} and σ_{net}. The effect of loading direction, and therefore the effect of the crack propagation direction, is not seen in the figure.

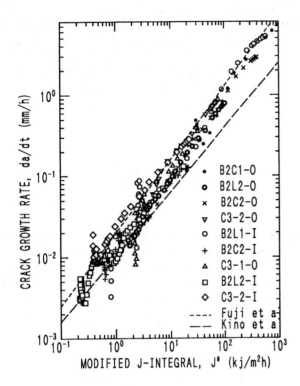

Fig. 5 Correlation of creep crack propagation rate da/dt with the modified J-integral J^*

The dotted line and the long dashed line in the figure show the $da/dt - J^*$ relationship obtained for $2\frac{1}{4}$Cr-1Mo steels by Fuji et al. (1992) and Kino et al. (1986), respectively. The Fuji's and Kino's equations are expressed as,

$$da/dt = 2.0 \times 10^{-2} J^{*0.904}, \quad (4)$$

$$da/dt = 1.0 \times 10^{-2} J^{*0.8}, \quad (5)$$

$(da/dt : \text{in mm/h}, J^* : \text{in kJ/m}^2\text{h})$

respectively. Equations (4) and (5) seem to give an upper bound and an lower bound of the crack propagation rates. Equation (4), which gives a conservative estimation, is used in the present study.

ESTIMATION OF CREEP CRACK PROPAGATION BEHAVIOR USING THE ENGINEERING APPROACH

Estimation of Modified J-Integral under Large Scale Creep

In the engineering approach proposed by Kumar (1981) the J-integral (Rice, 1968) is evaluated for fully-plastic power law stress-strain relation expressed as:

$$\epsilon = \alpha\epsilon_0(\sigma/\sigma_0)^n, \quad (6)$$

where ϵ_0 and n are material constants. The J-integral for Eq. (6) is given by

$$J = \alpha\epsilon_0\sigma_0 b h_1 \left(\frac{P}{P_0}\right)^{n+1}, \quad (7)$$

where P and P_0 are applied load and a reference load, and h_1 denotes a hardening function which depends on geometry and exponent n. The value of h_1 is tabulated for certain combinations of specimen geometry and the exponent n. Under large scale creep deformations the strain rate is given by Eq. (1). By applying the Hoff's analogy (Hoff, 1954) to the J-integral, the modified J-integral J^* under the large scale creep is given by:

$$J^*_{\text{LSC}} = \alpha\dot{\epsilon}_0\sigma_0 b h_1 \left(\frac{P}{P_0}\right)^{n+1} \quad (8)$$

Estimation of Modified J-Integral under Small Scale creep and Large Scale Creep

It has been found by the present authors (Ohji et al., 1979, 1980, 1988; Kubo, 1982, 1990, 1998) and Riedel and Rice (1980) that under small-scale creep, in which creep dominated region is small compared to specimen dimensions, the modified J-integral in the vicinity of the crack tip is much higher than that under the large scale creep. To take the effect of small scale creep into account the following equation was proposed.

$$J^* = J^*_{\text{SSC}} + J^*_{\text{LSC}} = \frac{J_{\text{el}}}{(n+1)(t+t_{\text{pl}})} + J^*_{\text{LSC}}, \quad (9)$$

where J_{el} is the elastic J-integral. To take the effect of plastic strain t_{pl} was introduced (Ohji et al., 1979, 1980), which is evaluated using proof stress σ_p and Young's modulus E together with creep parameters as:

$$t_{\text{pl}} = \frac{\sigma_p^{1-n}}{(n+1)E\alpha\dot{\epsilon}_0\sigma_0^{-n}} \quad (10)$$

Estimation of Creep Crack Propagation Behavior in C(T) Specimen

To examine the applicability of the engineering approach it was applied to the crack propagation in the C(T) specimen. The modified J-integral J^* under large scale creep was evaluated using Eq. (8). The value of hardening function h_1 for a given combination of geometry and n was determined by applying linear interpolation to tabulated values for certain combination of geometry and n given by Kumar (1981). The modified J-integral J^* taking account of small scale creep and large scale creep was evaluated using Eq. (9). These estimation of J^* was made for the plane strain and plane stress conditions. The creep crack propagation rate da/dt was estimated by substituting the values of J^* thus obtained into Eq. (4) describing the creep crack propagation characteristics.

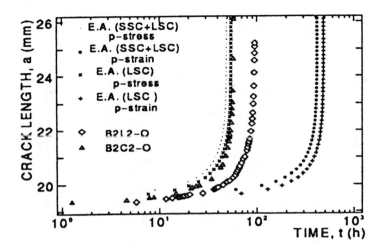

Fig. 6 Comparison of estimated and experimental creep crack propagation curves

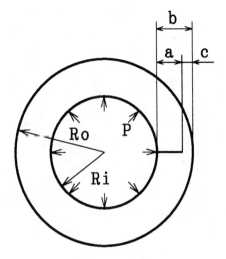

(a) Axial crack

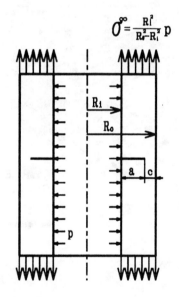

(b) Circumferential crack

Fig. 7 Inner cracks in a pipe

Figure 6 shows an example of comparison between estimated and experimental creep crack propagation curves. From the figure it is seen that the effect of small scale creep is not significant. The creep crack is predicted to propagate faster under the plane stress conditions than under the plane strain conditions. The experimental creep crack propagation curves were reconstructed by using the engineering approach under plane stress conditions, indicating the applicability of the engineering approach. This may be due to the fact that the thickness of the C(T) specimen is not large enough to induce high constraint in the thickness direction and the plane stress condition is prevailing in the specimen.

Estimation of Propagation Behavior of Inner Crack in Pipes

The engineering approach was applied to the estimation of creep crack propagation in straight pipes, for which the tabulated values of the hardening function h_1 is given by Kumar (1981). As shown in Figs. 7 (a) and (b) an axial crack and a circumferential crack with a constant length in radial direction were considered. The estimation for these cracks gives conservative ones for axial cracks and circumferential cracks with nonconstant lengths. The estimation was made for a pipe with outer radius of 660.4 (mm), inner radius of 598.4 (mm), and internal pressure p of 3.04 (MPa). The value of hardening function h_1 for a given combination of geometry and n was determined by interpolating the tabulated values again.

The estimated crack propagation curves using the J^* value under the large scale and that under the small scale and the large scale creep for the initial crack length a_0 of 0.5, 1, 2, 3 and 5 (mm) is shown in Fig. 8. It is seen that the effect of small scale creep is small. From the comparison between the axial and circumferential cracks with the same initial lengths a_0 the axial crack is easier to propagate than the circumferential crack.

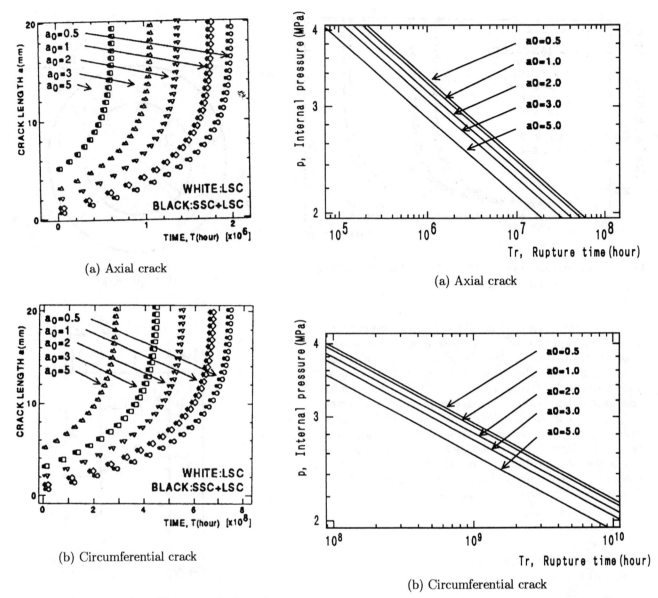

Fig. 8 Estimated propagation of inner cracks in a pipe

Fig. 9 Estimated relationship between rupture time and internal pressure for various initial lengths a_0 of axial crack and circumferential crack in a pipe

Figure 9 (a) and (b) show the estimated relationship between rupture time and internal pressure for various initial lengths a_0 of axial crack and circumferential crack, respectively. From the figure the rupture time is large enough even for relatively large crack when the internal pressure is less than 3 (MPa).

CONCLUSIONS

For the safety assessment of fossil power plants, the estimation of residual creep lives was made for a $2\frac{1}{4}$Cr-1Mo steel used in a fossil power plant for about 120,000 hrs. Creep tests were conducted to obtain creep characteristics of the aged steel. Creep crack propagation experiments were also conducted using the C(T) specimen. It was found that the creep crack propagation rate da/dt can be correlated well with the modified J-integral J^* (C^* integral), as it has been found for many other materials. The engineering approach was applied to estimate the value of the modified J-integral J^*. The effect of small scale creep was taken into consideration. It was assured that the creep crack propagation behavior in the C(T) specimen can be reconstructed using the estimated J^* value under the plane stress condition and $da/dt - J^*$ characteristics. The engineering approach was then applied to the estimation of creep propagation behavior of postulated axial and circumferential cracks in a straight pipe of the power plant. The estimated creep crack propagation lives were large enough for relatively long cracks.

REFERENCES

Clarke, G.A., and Landes, J.D., 1979, "Evaluation of the J Integral for the Compact Specimen", J. Test. Eval., Vol. 7, pp. 264-269.

Fuji, A., 1992, Tetsu-to-Hagane, (in Japanese), Vol. 48, p. 1729.

Hoff, N.J., 1954, "Approximate Analysis of Structures in the Presence of Moderately Large Creep Deformations", Quart. Appl. Math., Vol. 12, pp. 49-55.

Kino, H., Nakashima, K., Yamanouchi, H., and Yamashita, Y., 1986, "Creep Crack Growth Behavior in $2\frac{1}{4}$Cr-1Mo Steel", J. Soc. Mater. Sci., Japan, (in Japanese), Vol. 35, pp. 254-259.

Kubo, S., 1975, "Mechanical Analysis of Creep Crack Propagation", Doctral Thesis, Osaka University.

Kubo, S., 1981, "Combined Effect of Creep Recovery-Hardening and Elastic Straining on the Stress and Strain-Rate Fields Near a Crack Tip in Creeping Materials", Brown University Report, MRL E-133, pp. 1-35.

Kubo, S., and Ohji, K., 1990, "Small-Scale and Large-Scale Creep Predictions of Notch Stress and Strain under Creep Conditions", Creep in Structures, Zyczkowski, M. ed., Springer, pp. 363-370.

Kubo, S., Kinugasa, H., and Ohji, K., 1998, "Estimation of Stress Concentration at Shallow Notches under Creep Conditions Based on the Small-Scale and Large-Scale Creep Concepts", PVP-Vol.365, ASME, pp. 381-389.

Kumar, V., 1991, "An Engineering Approach for Elastic-Plastic Fracture Analysis", EPRI NP-1931.

Landes, J.D., and Begley, J.A., 1976, "A Fracture Mechanics Approach to Creep Crack Growth", ASTM STP 590, pp. 128-148.

Landes, J.D., Walker, H., and Clarke, G.A., 1979, "Evaluation of Estimation Procedures Used in J-Integral Testing", ASTM STP 668, pp. 266-287.

Ohji, K, Ogura, K., and Kubo, S., 1974, "An Application of the J- Integral to Creep Crack Problems", Preprint of Japan. Soc. Mech. Engrs. 740-11, pp. 207-209

Ohji, K., Ogura, K., and Kubo, S., 1976, "Mechanics of Creep Crack Propagation under Longitudinal Shear and Its Application", Trans. Japan. Soc. Mech. Engrs., Vol.42, pp. 350-357.

Ohji, K., Ogura, K., and Kubo, S., 1978, "Estimates of J-Integral in the General Yielding Range and Its Application to Creep Crack Problems", Trans. Japan. Soc. Mech. Engrs., Vol.44, pp. 1831-1837.

Ohji, K., Ogura, K., and Kubo, S., 1979, "Stress Field and Modified J-Integral Near a Crack Tip Under Condition of Confined Creep Deformation", Preprint of Japan. Soc. Mech. Engrs. 790-13, pp. 18-20.

Ohji, K., Ogura, K., and Kubo, S., 1980, "Stress Field and Modified J-Integral Near a Crack Tip Under Condition of Confined Creep Deformation", J. Soc. Mater. Sci., Japan, (in Japanese), Vol. 29, pp. 465-471.

Ohji, K., and Kubo, S., 1988, "Fracture Mechanics Evaluation of Crack Growth Behavior under Creep and Creep-Fatigue Conditions", Current Japanese Materials Research - Vol. 3 High Temperature Creep-Fatigue, Elsevier, pp. 91-113.

Rice, J.R., 1968, "A Path Independent Integral and the Approximate Analysis of Strain Concentration by Notches and Cracks", J. Appl. Mech., Vol. 35, pp. 379-386.

Riedel, H., and Rice, J.R., 1980, "Tensile Cracks in Creeping Solids", ASTM STP 700, pp. 112-130.

Shih, T.T., 1983, "A Simplified Test Method for Determining the Low Rate Creep Crack Growth Data", ASTM STP 791 Vol. II, pp. II-232-II247.

VARIATION OF FRACTURE TOUGHNESS WITH CONSTRAINT OF PMMA SPECIMENS

Yuh J. Chao, Shu Liu and Bart J. Broviak
Department of Mechanical Engineering, University of South Carolina
Columbia, SC 29208, USA
Tel: (803)777-5869, Fax : (803)777-0106
e-mail : chao@sc.edu

ABSTRACT

The effect of constraint on fracture under predominantly elastic deformation and under Mode I loading conditions is discussed. Williams' asymptotic series solutions, including higher order terms, for the stress and strain fields at a crack tip are applied to the interpretation of fracture of brittle materials. Fracture tests using PMMA were performed on specimens covering a broad range of constraint levels obtained through different geometry of the specimens. It is demonstrated that the apparent fracture toughness of the material varies with the specimen geometry or the constraint level. Furthermore, this variation can be interpreted by using T or A_3 as the constraint parameter where T and A_3 are the amplitudes of the second and the third term in the Williams' series solution, respectively. The implications of this constraint effect to the CTOD fracture criterion and the energy release rate are discussed.

1. INTRODUCTION

Constraint effect in fracture is qualitatively referred to as the dependence of fracture toughness upon the specimen thickness, in-plane geometry, and loading configuration. For brittle fracture in elastic-plastic deformation, it is now recognized that the single fracture parameter J is only valid under special circumstance when a high degree of constraint at the crack-tip exists. In general, the critical J at fracture changes according to the shape and size of the crack configuration. Recent analytical, numerical and experimental studies (e.g. Chao, et al. 1994, Yang, et al. 1993, Hancock, 1992, O'Dow and Shih, 1992) have attempted to interpret this fracture behavior using both J and a second parameter, A_2, T or Q, where J represents the loading level and the second parameter a constraint level.

For brittle fracture in predominantly elastic deformation mode, current fracture assessment methodology relies on a critical stress intensity factor, K_{IC} or plane-strain fracture toughness, which is assumed to be a material property and represents a material's resistance to fracture under plane-strain conditions. Recent studies by Chao and Zhang (1997), and Richardson and Goree (1993), however, have shown that the apparent fracture toughness K_C varies with specimen geometry and loading configuration for brittle materials. While this constraint effect is similar to those for ductile materials, the trend for the dependence of the fracture toughness K_C on the constraint is opposite to it for ductile materials. In this paper, K_C is used as the critical stress intensity factor or the apparent fracture toughness for a cracked specimen, and K_{IC} is a subset of K_C when the specimen geometry and size requirements in ASTM E399 are met.

The work by Chao and Zhang (1997) is based on (1) crack tip mechanics fields in a brittle material that can be characterized by the Williams' series solutions, (2) fracture event controlled by a critical stress or critical strain in front of the crack tip, and (3) the levels of the critical stress/strain that are characterized by more than one term in the Williams' series solutions. Under certain constraint levels the stress and strain at a critical distance ahead of the crack tip for brittle materials cannot be characterized by a single parameter K alone. Therefore, if two terms in the Williams' series are used, the fracture event may be characterized by K and another parameter, where K represents the loading level and the second parameter the constraint level. The second parameter depends on whether the fracture is stress or strain controlled.

The key argument in the work of Chao and Zhang (1997) is that the controlling stress/strain for fracture is at a *finite* distance r_c, in contrast to r→0, from the crack tip. Therefore, the stress level may not always be characterized by the first term in the Williams' solution, or the K-stress. For a given material, if the K-dominant zone at a crack tip encompasses the fracture process zone the actual stress at r_c is close to the K-stress. Thus, using a critical stress σ_c as the fracture criterion is equivalent to using a critical K, such as K_{IC}, since these two are related by the first term in the Williams' equation. On the other hand, if the K-dominant zone is smaller than r_c, the stress at $r = r_c$ cannot be completely characterized by the K-stress alone. Under these conditions, using K to characterize the fracture event is not sufficient. Furthermore, the size of the K-dominant zone depends upon the specimen geometry, size, crack length, and loading configuration. Thus, the critical stress intensity factor for a given cracked specimen becomes a function of these factors. This behavior can be generally referred to as the constraint effect in brittle fracture.

In this paper, we further examine this constraint effect in the fracture of brittle materials. Governing equations for both stress and strain in the vicinity of a crack tip are revisited. Fracture tests using Polymethyl Methacrylate (PMMA, Acrylic or Plexiglas) plate were performed to examine the brittle fracture. Compact tension (CT), single edge notch in tension (SEN) and double cantilever beam (DCB) specimens with various crack lengths are used in order to cover a broad range of constraint levels. It is demonstrated that the critical stress intensity factor or apparent fracture toughness of PMMA varies with the specimen geometry or the constraint level. This variation can be interpreted by using T or A_3 as the constraint parameter where T and A_3 are the amplitudes of the second and the third term in the Williams' series solution, respectively. The implications of this constraint effect to the CTOD fracture criterion and the energy release rate are then discussed.

2. THEORETICAL CONSIDERATION

Fracture of a flawed specimen or structure is a process in which the material separates at the crack tip. Thus, to interpret the fracture behavior it is natural to examine the stress and strain fields near the crack tip. The mechanics solution for this problem under linear elastic assumptions was obtained by Williams (1957). Using (r,θ) centered at the crack tip as the coordinate system, the stress and strain can be written as

$$\sigma_{ij} = \sum_{n=1}^{\infty} A_n r^{\frac{n}{2}-1} f_{ij}^{(n)}(\theta) \quad (1)$$

$$\varepsilon_{ij} = \sum_{n=1}^{\infty} A_n r^{\frac{n}{2}-1} g_{ij}^{(n)}(\theta) \quad (2)$$

where σ_{ij} is the stress tensor, ε_{ij} the strain tensor, $f_{ij}^{(n)}(\theta)$, and $g_{ij}^{(n)}(\theta)$ are the angular functions of the nth term for the stress and strain, respectively.

Equations (1) and (2) are asymptotic solutions. That is, the contribution of the lower order terms increases relative to the higher order terms as $r \to 0$. Furthermore, the first or the lowest order term is singular as $r \to 0$. Since fracture starts from the crack tip ($r = 0$), one typically assumes that the first term in (1) and (2) dominates as $r \to 0$. Therefore,

$$\sigma_{ij} \cong \frac{K_I}{\sqrt{2\pi r}} f_{ij}^{(1)}(\theta)$$

$$\varepsilon_{ij} \cong \frac{K_I}{\sqrt{2\pi r}} g_{ij}^{(1)}(\theta) \quad (3)$$

In (3), K_I is equal to $A_1\sqrt{2\pi}$. It is obvious from (3) that the fracture event can be quantified by a single parameter K_I, as long as the fracture process is controlled by stress, strain or a combination of stress and strain, e.g. strain energy. The coefficient K_I is proportional to the applied load and the critical K_I corresponding to the instant of fracture is defined as K_C.

Based on (3), material testing community has designed a testing procedure for the determination of K_C for engineering materials. The testing procedure is designed in a way such that certain restrictions such as specimen geometry and size requirements must be satisfied. The particular K_C determined following the standard testing procedures and satisfying the requirements is labeled as K_{IC} or the fracture toughness of the material (see ASTM E399).

Note that the above discussion and the concept of fracture toughness introduced by Irwin (1957) and defined by the ASTM testing standards are purely based on the validity of (3). In the general case, however, the critical stress or strain near a crack tip that dictates the fracture may not be sufficiently quantified by only the first term in the Williams' series as shown by (3). This can be rationalized by considering a fracture process zone at a loaded crack tip. Since the critical distance r_c, where the stress/strain dictates the fracture event, is equal or larger than the size of the process zone, r_c is *finite*. When r_c is *finite*, the stress/strain at r_c are better represented by more than just the first terms from (1) and (2). If one focuses on the Mode I loading case, one needs only looking into the stress/strain ahead of the crack tip, i.e. along $\theta = 0$ degrees. Assuming that two non-vanishing terms in (1) and (2) are sufficient to characterize the stress/strain at the critical distance, one arrives

$$\sigma_{yy}|_{\theta=0} = \frac{A_1}{\sqrt{r}} + 3A_3\sqrt{r} = \frac{K_I}{\sqrt{2\pi r}} + 3A_3\sqrt{r} \quad (4)$$

$$\varepsilon_{yy}|_{\theta=0} = \frac{1-v^*}{E^*}\frac{K_I}{\sqrt{2\pi r}} - \frac{v^*}{E^*}T \quad (5)$$

where $E^* = E$ and $v^* = v$ for plane stress, and

$E^* = \dfrac{E}{1-v^2}$ and $v^* = \dfrac{v}{1-v}$ for plane strain conditions.

In (5), $T = 4A_2$ and is the so-called T-stress. Along $\theta = 0$ degrees the second term in Eq.(1), or the T-stress, vanishes. Thus, three terms or two non-vanishing terms from Williams' expansion is retained in (4).

Equation (4) shows that the opening stress is controlled by two mechanics parameters K_I and A_3 --- K_I can be interpreted as the applied loading level and A_3 a function of specimen geometry and loading configuration. Accordingly, equation (5) can be interpreted as two parameters, K_I and T, control the opening strain. Note that both A_3 and T are proportional to K and the applied load for a given geometry due to the linearity of the problem.

Adopting the fracture criterion, "the fracture event occurs when the opening stress at r_c in front of the crack tip reaches a critical value σ_C" or the RKR model (Ritchie, et al, 1973), equation (4) becomes

$$\sigma_c = \frac{K_c}{\sqrt{2\pi r_c}} + 3A_c\sqrt{r_c} \quad (6)$$

where K_C and A_C are the critical K_I and A_3, respectively, corresponding to the critical load. Equation (6) shows that the critical stress at r_C is controlled by two parameters, K_C and A_C, as contrast to a single parameter presented by equation (3).

Since A_C can be either positive or negative depending upon specimen or structural geometry, the magnitude of K_C varies with specimen geometry in order to achieve the same fracturing stress σ_C for the material to fail. In the special case of $A_3 = 0$, the opening stress is controlled by K_I alone, i.e. K-dominance is ensured. Therefore, for this type of specimen the fracture event is completely controlled by a single parameter K. It follows that the specimen size and geometry restrictions set forth in the ASTM fracture testing standard (ASTM E399) are meant to ensure this condition.

Using similar argument for stress controlled fracture, one arrives at the following equation for "strain" controlled fracture :

$$\varepsilon_C |_{\theta=0} = \frac{1-v^*}{E^*} \frac{K_I}{\sqrt{2\pi r}} - \frac{v^*}{E^*} T_C \qquad (7)$$

where ε_C is the critical opening strain at the critical distance r_C in front of the crack tip and K_C and T_C are the critical stress intensity factor and the critical T-stress, respectively, at the instant of fracture.

3. EXPERIMENTAL ANALYSIS

To verify the above discussion, specimens of three geometries, i.e. CT, SENT, and DCB, with various crack depth were tested. The specimens were chosen in such a way that a wide range of T or A_3 could be achieved. A PMMA plate with thickness t = 5.84 mm was selected as the material and tested at room temperature to study the *brittle* fracture behavior. Figure 1 shows the dimensions of the specimen configuration. The crack length, i.e. a/W, and the specimen designation can be found in the second column of Table 1.

Simple tension test following the ASTM E 8-93 and ASTM E132-92 standard procedures was performed to determine the elastic material properties for the PMMA. Test results show that E (Young's modulus) = 2.95 GPa, σ_{ys}= 44.6 MPa, and v (Poisson's ratio) = 0.34.

All fracture tests were carefully conducted according to ASTM E399-92 guidelines. The critical stress intensity factors K_C for each test specimen is determined using a finite element analysis (FEA) and the fracturing load P_Q as specified by the ASTM testing standard. The ABAQUS code was used for the FEA. A typical FEA mesh is shown in Fig. 2 for specimen CT1B. The specimen in the numerical analysis was loaded up to the load P_Q recorded from the test. The stress intensity factor at this load from the numerical analysis is designated as the critical intensity factor or K_C for the particular specimen. Only the "valid" test data, i.e. satisfying the ASTM requirements in crack front roundness and symmetry are used for later discussion. The thickness t = 5.84 mm meets the thickness requirement specified by the ASTM E399-92, that is, t > 2.5 (K_{IC} / σ_{ys})^2 , for this material where K_{IC} = 1.121 MPa*m$^{1/2}$ is obtained from specimen CT3c which is the only CT specimen tested that satisfy all the E399 requirements including the a/W=0.45 ~ 0.55 rule.

4. DETERMINATION OF T-STRESS AND A_3 VALUES

There are several methods for the determination of T-stress for a given specimen geometry. In this work, we adopted finding the σ_{xx} along the crack face to determine the T-stress because of its simplicity. Using equation (1) and setting $\theta=180^0$, one has

$$\sigma_{xx} = 4A_2 = T \qquad (8)$$

Figure 3 shows the distribution of σ_{xx} along the crack face from the finite element analysis for CT specimen with a/W=0.295, i.e. specimen CT1b. The numerical data shown in Fig. 3 indicates that near the crack tip, there exists a nearly constant σ_{xx} region, i.e. x/a = -0.001 to –0.01, and this σ_{xx} value is chosen as the T-stress.

The A_3 term in the William's series expansion was determined by employing equation (4), along with the aid of FEA results. Equation (4) can be rewritten as

$$\sigma_{\theta\theta}|_{\theta=0} \cdot \sqrt{2\pi r} = \left(3\sqrt{2\pi} \cdot A_3\right) \cdot r + K_I \qquad (9)$$

Equation (9) is a simple linear expression,

$$f(x) = m \cdot x + n \qquad (10)$$

Since f(x) can be determined by FEA, A_3 can then be obtained from the slope m, i.e. $A_3 = \frac{m}{3\sqrt{2\pi}}$ through a linear curve fit. One example is shown in Fig. 4 for the CT1b specimen.

The apparent fracture toughness K_C, T-stress and A_3 at the critical load P_Q for each specimen tested are listed in Table 1. The T-stress is then converted to the dimensionless constraint parameter B using $B = \frac{T\sqrt{\pi a}}{K_I}$. The values of B for the test samples at the fracturing load are also included in Table 1.

5. INTERPRETATION OF THE TEST RESULTS

Assuming the fracture event is controlled by an opening stress, equation (6) predicts that the functional form in the K_C and A_3 plane is a straight line. Therefore, using the critical K_I and A_3 in Table 1 for both CT and SEN specimens, a straight line is fitted to Eq. (6) and presented in Fig. 5. This fitted line can be interpreted as the material failure curve for PMMA and σ_C = 7.57 MPa and r_C = 2.32 mm for this batch of PMMA. Notice the slope of the straight line, which indicates that specimens having a lower or more negative A_3 fail with higher apparent fracture toughness K_C.

Similarly, assuming the fracture event is controlled by an opening strain, equation (7) predicts that the functional form in the (K_C , T) space is a straight line. Therefore, using the critical K_I and T in Table 1 for both CT and SEN specimens, a straight line is fitted to equation (7) and presented in Fig. 6. This fitted line can be interpreted as the material failure curve for PMMA and ε_C = 2000 x 10^{-6} and r_C = 0.86 mm for this batch of PMMA. Notice the slope of the straight line, which indicates that specimens having lower T values fail at lower apparent fracture toughness K_C.

The implication from the above analysis is that the fracture toughness is specimen geometry dependent. Thus, the fracture toughness value determined following the ASTM E399-92 procedures is simply a particular toughness of the material, which satisfies all the ASTM test procedure requirements. When this particular ASTM fracture toughness value is used for specimens of other geometry or structures it may or may not predict the fracture event accurately or even conservatively.

It is noted that a specimen of deep crack is regarded as high constraint relative to a specimen of shallow crack, as commonly interpreted by the fracture mechanics community. In general, a deep cracked specimen has a higher T-stress than that from a shallow crack under the same applied K. Thus, higher T implies higher constraint. From the data in figures 5 and 6 it can be concluded that the relationship between the constraint level T or A_3 and the apparent fracture toughness K_C in this class of materials is *" high constraint specimens fail with higher fracture toughness values"*. This conclusion is opposite to the behavior of elastic-plastic metals when failure is governed by large-scale plasticity. Detailed discussions on this issue can be found in Chao and Zhang (1997).

Note that the test data for DCB specimens are not included in figures 5 and 6 because the fracture path is not along the symmetric plane in these specimens. Thus, the application of equation (4) to (7) to the problem is not valid. It can be shown, on the other hand, that when higher order terms are included in the determination of stress or strain near a crack tip the maximum circumferential stress (or strain) may not always occur ahead of the crack tip, i.e. θ=0 degrees, for

Mode I loading conditions. This is true when the contribution from the higher order terms is not negligible, i.e. when the contribution of the T-stress or A_3 are relatively large. It follows that if a critical circumferential stress (or strain) fracture criterion is appropriate then the fracture path should deviate from $\theta = 0$ degrees. This interesting point is indeed the case for the DCB specimens tested. The analysis and interpretation for this crack curving behavior will be reported in a follow-up paper.

6. CTOD AS A FRACTURE CRITERION

The fracture mechanics community has adopted the concept of a critical crack tip opening displacement (CTOD) as a fracture criterion, in addition to the fracture toughness or K_{IC} (see, e.g. ASTM E 1290-93). We will prove here that the critical CTOD also varies with specimen geometry or constraint.

Assuming the crack propagates in its original direction, the crack opening displacement, $V(r)$ for Mode I is obtained from the Williams series solution by setting $\theta = \pi$ and can be rewritten as

$$V = \frac{4K_I \sqrt{r}}{\sqrt{2\pi} E^*} - \frac{4A_3 r^{\frac{3}{2}}}{E^*} \qquad (11)$$

If there exists a constant critical CTOD for the initiation of fracture for a given material, then for different combination of K_I and A_3 or different specimens there exists a critical distance r_C behind the crack tip where the CTOD's from all specimens are equal to the critical CTOD. In other words, the crack face opening profiles from any two specimens under the fracturing loads should intersect each other or completely overlap if a critical CTOD exists.

Mathematically, let us assume there are two specimens with (K_{I-1}, A_{3-1}) and (K_{I-2}, A_{3-2}) at the fracturing loads, respectively. We shall set $V_1 = V_2$ to solve for the critical distance r_C. It follows

$$V_1 - V_2 = \left(\frac{4K_{I-1}\sqrt{r}}{\sqrt{2\pi}E^*} - \frac{4A_{3-1}r^{\frac{3}{2}}}{E^*} \right) - \left(\frac{4K_{I-2}\sqrt{r}}{\sqrt{2\pi}E^*} - \frac{4A_{3-2}r^{\frac{3}{2}}}{E^*} \right) = 0 \qquad (12)$$

or simplified as

$$\sqrt{r}\left[(K_{I-1} - K_{I-2}) - \sqrt{2\pi}(A_{3-1} - A_{3-2})r\right] = 0 \qquad (13)$$

The solutions of the above equation are either

$$r = 0$$

or

$$r = \frac{K_{I-1} - K_{I-2}}{\sqrt{2\pi}(A_{3-1} - A_{3-2})} \qquad (14)$$

As shown in Fig. 5, the slope of the K_C-A_C curve is negative, which makes the solution of (14) impossible and $r = 0$ is the only possible solution. Physically it implies that the crack opening profiles from two different specimens at the respective critical loads do not intersect each other and consequently a common critical CTOD for the two specimens dose not exist.

7. CRITICAL ENERGY RELEASE RATE G_C

In this section we assume that the energy release rate G_C for fracture initiation is a constant for a given material and investigate the relation between G_C and the critical stress intensity factor K_C. Recall that

$$G = \lim_{\Delta a \to 0} \left(\frac{1}{\Delta a} \int_0^{\Delta a} V \sigma_{\theta\theta} \big|_{\theta=0} dx \right) \qquad (15)$$

Using (4) and (11), equation (15) becomes

$$G = \frac{K_I K_I'}{E^*} + \frac{3\sqrt{2\pi} K_I' A_3}{E^*} \Delta a + \frac{3\sqrt{2\pi} K_I A_3'}{4E^*} \Delta a + \frac{3\pi A_3 A_3'}{4E^*} \Delta a^2 \qquad (16)$$

where K_I and A_3 are the critical values as the crack length equal to a and K_I' and A_3' are as crack length $a+\Delta a$. Note that when $\Delta a \to 0$, $K_I' \to K_I$ and $A_3' \to A_3$. So, we have,

$$G = \frac{K_I^2}{E^*} + \frac{15\sqrt{2\pi} K_I A_3}{4E^*} \Delta a + \frac{3\pi A_3^2}{4E^*} \Delta a^2 \qquad (17)$$

Ignoring the second higher order term Δa^2, one has

$$G = \frac{K_I^2}{E^*} + \frac{15\sqrt{2\pi} K_I A_3}{4E^*} \Delta a \qquad (18)$$

or it can be written in the form

$$K_I^2 = -\frac{15\sqrt{2\pi}\Delta a}{4} K_I A_3 + G \cdot E^* \qquad (19)$$

Thus, if there is a critical G, then K_I^2 has a linear relationship with $K_I A_3$. And also from this linear relationship, one can obtain the critical G or G_C. Note that both Eq.(18) and (19) state that the critical energy release rate G_C is a function of both K_I and A_3 or it varies with the constraint of the specimen. Figure 7 shows the test data and the linear fit of the data using (19). The critical energy release rate obtained from the fit to the data is $G_C = 2.486 \times 10^{-4}$ MPa*m. Note that G_{IC} obtained from the specimen CT3c, which satisfies all ASTM E399 requirements, is 3.767×10^{-4} MPa*m. This number is higher than 2.486×10^{-4} MPa*m since A_3 is not zero for standard ASTM specimens.

8. CONCLUSION

In this paper the variation of apparent fracture toughness with specimen configuration or constraint is investigated. The test data confirm that the critical opening stress or a critical opening strain is the valid fracture criterion. Furthermore, two non-vanishing terms in the Williams' series solution for characterizing the stress/strain ahead of the crack tip can be used to quantify the constraint behavior. Since the fracture toughness varies with specimen configuration, it follows directly that (1) a critical CTOD does not exist in general, and (2) the relation between the critical energy release rate and the apparent fracture toughness must be modified to include the constraint parameter.

9. ACKNOWLEDGEMENTS

This work is sponsored by the NSF/EPSCoR Cooperative Agreement No. EPS-9630167 to the University of South Carolina. The support from NSF is greatly appreciated.

10. REFERENCES

ASTM E8-93, 1993, "Standard Test Methods for Tension Testing of Metallic Materials," *Annual Book of ASTM Standards*, Vol.03.01, pp.130-146.

ASTM E132-92, 1993, "Standard Test Method for Poisson's Ratio at Room Temperatures," *Annual Book of ASTM Standards*, Vol.03.01, pp.323-325.

ASTM E399-92, 1993, "Standard Test Method for Plane-Strain Fracture Toughness of Metallic Materials," *Annual Book of ASTM Standards*, Vol.03.01, pp.509-539.

ASTM E1290-93, 1993, "Test Method for Crack-Yip Opening Displacement (CTOD) Fracture Toughness Measurement," *Annual Book of ASTM Standards*, Vol.03.01, pp.952-961.

Chao, Y.J. Yang, S., and Sutton, M. A., 1994, "On the Fracture of Solids Characterized by One or Two Parameters: Theory and Practice," Journal of Mechanics and Physics of Solids, Vol.42, pp629-647.

Chao, Y.J. and Zhang, X., 1997, "Constraint Effect in Brittle Fracture", 27th National Symposium on Fatigue and fracture, ASTM STP 1296, R.S. Piascik, J.C. Newman, Jr. and D.E. Dowling, Eds., American Society for Testing and Materials, Philadelphia, pp.41-60.

Hancock, J.W., 1992, "Advances in Characterization of Elastic-Plastic Crack-Tip Fields," Topics in Fracture and Fatigue, Editor: A.S. Argon, Springer-Verlag, pp.59-98.

Irwin, G.R., 1957," Analysis of Stresses and Strains near the End of a Crack Traversing a Plate." Journal of Applied Mechanics, Vol.24, pp.361-364.

O'Dowd, N.P. and Shih, C.F., 1992," Family of Crack-Tip Fields Characterized by a Triaxiality Parameter-II. Fracture Applications," Journal of the Mechanics and Physics of Solids, Vol.40, pp.939-963.

Rechardson, D. E. and Goree, J. G., 1993, "Experimental Verification of a New Two-parameter Fracture Model", Fracture Mechanics: 23rd Symposium, ASTM STP 1189, R. Chona, Ed., American Society for Testing and Materials, Philadelphia, pp.738-750.

Ritchie, R.O., Knott, J.F., and Rice, J. R., 1973, "On the Relationship between Critical Tensile Stress and Fracture Toughness in Mild Steel." Journal of the Mechanics and Physics of Solids, Vol.21, pp.395-410.

Williams, M.L.,1957," On the Stress Distribution at the Base of a Stationary Crack," Journal of Applied Mechanics, Vol.24, pp.109-114.

Yang, S., Chao, Y.J., and Sutton, M. A., 1993, "Higher Order Asymptotic Crack Tip Fields in a Power-Law Hardening Material," Engineering Fracture Mechanics, Vol. 45(1), pp.1-20.

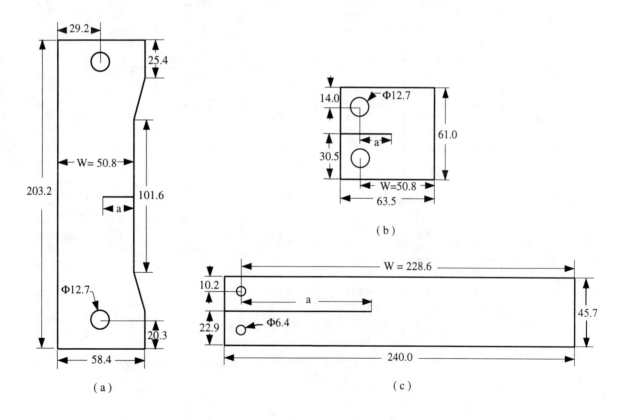

**Figure 1. Dimensions of test specimens (a) SEN (b) CT (c) DCB
(All dimensions are in 'mm', and specimen thickness t = 5.84mm)**

Table 1: K, T and A_3 values for SEN, CT, and DCB specimens

Specimens		a/W	K_I (MPa*m$^{1/2}$)	T (MPa)	A_3 (MPa*m$^{1/2}$)	B	Did Crack Curve?
SEN	1b*	0.296	0.944	-1.333	0.161	-0.307	No
	2a	0.405	1.016	-0.790	-3.168	-0.198	No
	2b	0.402	1.039	-0.817	-3.151	-0.199	No
	3a	0.504	0.959	-0.374	-5.038	-0.111	No
	3b	0.503	0.993	-0.393	-5.093	-0.112	No
	4a	0.598	0.849	0.058	-7.076	0.021	No
	4b	0.588	0.998	0.015	-8.672	0.005	No
CT	1a	0.298	1.267	2.051	-12.090	0.353	No
	1b	0.295	1.036	1.648	-9.549	0.345	No
	1c	0.296	1.355	2.169	-12.826	0.348	No
	2c	0.400	1.075	2.222	-10.883	0.523	No
	2d	0.396	1.085	2.219	-10.917	0.515	No
	3a**	0.496	1.006	2.076	-10.736	0.581	No
	3c	**0.485**	**1.121**	**2.327**	**-12.112**	**0.578**	**No**
	4a**	0.596	1.060	2.037	-13.583	0.593	No
	4b	0.602	1.266	2.425	-16.253	0.594	No
	5a	0.698	1.310	2.334	-23.131	0.595	No
	5b	0.699	1.139	2.042	-20.252	0.599	No
DCB	1a	0.216	1.005	5.463	-17.902	2.140	Yes
	1b	0.201	1.055	5.501	-18.837	1.979	Yes
	2a	0.304	1.071	6.959	-20.121	3.036	Yes
	3b	0.398	0.935	6.576	-17.664	3.762	Yes
	4a	0.497	1.053	8.074	-20.161	4.581	Yes
	4b	0.495	1.145	8.764	-23.875	4.567	Yes
	5a	0.594	1.029	7.897	-20.947	5.013	Yes
	5b	0.599	1.015	7.812	-20.595	5.048	Yes
	6a	0.697	1.012	8.155	-22.097	5.701	Yes
	6b	0.701	1.084	8.756	-21.118	5.730	Yes
	7b	0.07	1.335	4.676	-19.834	0.788	Yes
	7c	0.064	1.227	3.839	-17.789	0.670	Yes
	7f	0.0439	0.951	-3.324	-16.120	0.177	No
	8c-1	0.0738	1.153	3.851	-17.063	0.768	Yes
	8c-2	0.0738	1.163	3.884	-17.181	0.768	Yes
	8d-1	0.0836	1.127	4.530	-16.903	0.985	Yes
	8d-2	0.0836	1.142	4.590	-17.144	0.985	Yes
	8e-1	0.103	1.102	5.347	-19.180	1.321	Yes
	8e-2	0.1076	1.186	5.854	-17.934	1.372	Yes

Note : SEN1b*, CT3a** and CT4a** are invalid data per ASTM E399

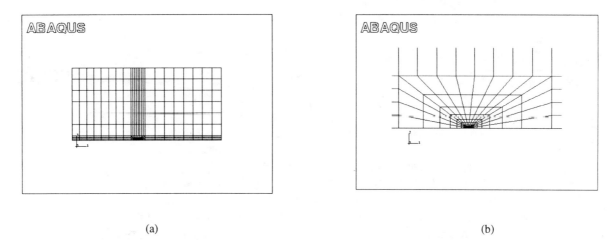

Figure 2. Typical FEA mesh (CT1b specimen) (a) global mesh (b) mesh near the crack tip

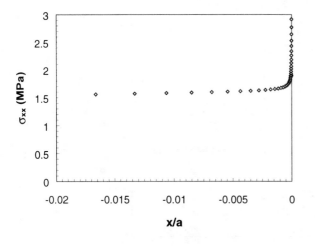

Figure 3. σ_{xx} distribution along the crack face from finite element analysis (CT1b)

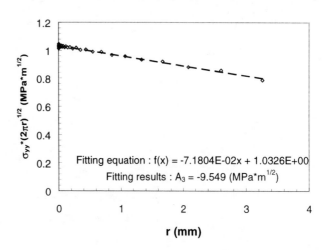

Figure 4. A_3 determination from finite element analysis (CT1b)

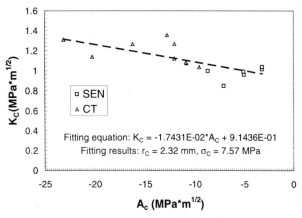

Figure 5. PMMA test data (SEN, CT) in the K_c and A_c plane

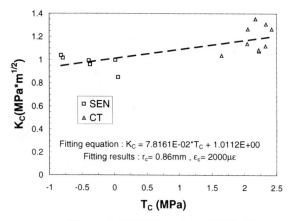

Figure 6. PMMA test data (SEN, CT) in the K_c and T_c plane

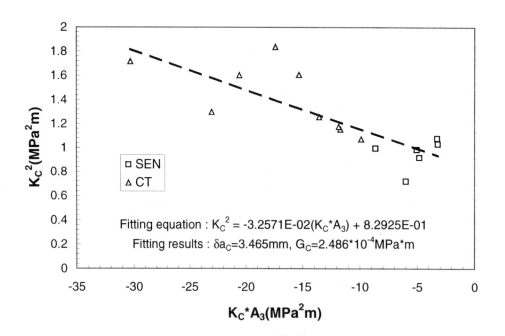

Figure 7. Curve fit of K_C^2 vs $K_C * A_C$ to get critical G_c and δa_c
(Based on Eq.(19))

FRACTURE MECHANICS EVALUATION FOR AND EVOLUTIONARY DEVELOPMENT OF THE BREAK PRECLUSION CONCEPT

E. Roos, K.-H. Herter
Stattliche Materialprüfungsanstalt (MPA)
University of Stuttgart
Stuttgart, Germany

ABSTRACT

In recent years piping of the primary circuit of nuclear power plants (NPP) were part of extensive research to demonstrate the leak-before-break (LBB) behavior and the break preclusion concept for safety related components. The integrity of these components has to be guaranteed even in the presence of defects of limited size both under operational and accident loading conditions.

Results from several research projects carried out at MPA Stuttgart are reported. The experimental, analytical and numerical investigations involve piping sections of ferritic and austenitic materials with dimensions comparable to the primary circuit. As regards to the austenitic components an overview of the experimental, analytical and numerical results is presented.

Furthermore the conclusions for the evolution of the break preclusion concept and the application to systems which have been requalified in the older nuclear power plants to comply with the break preclusion concept are shown.

INTRODUCTION

In the last 25 years within the German reactor safety research program tremendous efforts were made to develop and to verfiy methods to describe the load bearing capacity as well as the fracture behaviour of primary circuit piping of nuclear power plants (Kussmaul et. al. 1990, Bartholomé et. al. 1995). Together with similar R&D efforts performed in other countries, like the U.S. NRC "Degraded Piping Program", the IPIRG program, the NUPUC program in Japan and further programs a huge amount of experimental data and numerical analyses are available (Wilkowski et. al. 1998).

RECENT FRACTURE MECHANICS RESEARCH ACTIVITIES

Most recently within the scope of research projects at MPA Stuttgart component tests on austenitic straight pipes with nominal diameter down to a size of 50 mm diameter were carried out (Roos et. al. 1997, Stadtmüller 1997, Diem 1997). Circumferential flaws were artificially applied to the inner surface of the pipe by pre fatigue cracking of spark eroded notches. The aim of the investigations were to improve the basis of the safety assessments and proofs of integrity also for austenitic pipes with smaller diameters especially the load bearing behaviour as well as the failure and fracture behaviour (crack initiation, crack growth, leak-before-break)

The experimental results for the pipe tested under internal pressure and superimposed bending load showed that: (1) even the cracked pipes could be extremely deformed up to bending angels of $25°$, *Fig. 1*, (2) pipes with deep part through cracks and even through wall cracks considerable exceeded the ASME collapse load, *Fig 2 and 3*, (3) failure due to a large break (crack instability) did not occur in any of the tests conducted, (4) the experimental maximum load could be assessed by limit analysis within reasonable accuracy.

NUMERICAL ANALYSIS

Limit load analyses are usually the most widely used procedures for the assessment of degraded piping components and for the calculation of failure loads, which take into account only the strength characteristics of the materials used. A prediction with respect to crack initiation or crack growth is not possible. However fracture mechanics methods, like the R6-method, allows the toughness of the material to be taken into account in calculations on the basis of fracture mechanics data (Kussmaul et. al. 1990), *Fig. 4*. Using fracture mechanics methods more detailed predictions about the failure and fracture behaviour are possible.

Most of the common fracture mechanics evaluation procedures are based on one parametric fracture mechanics procedures. It is known that the crack resistance or fracture behaviour is

Fig. 1: Pipe with DN80 after test

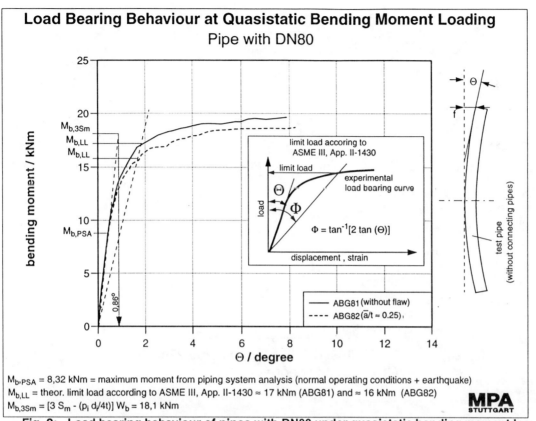

Fig. 2: Load bearing behaviour of pipes with DN80 under quasistatic bending moment loading

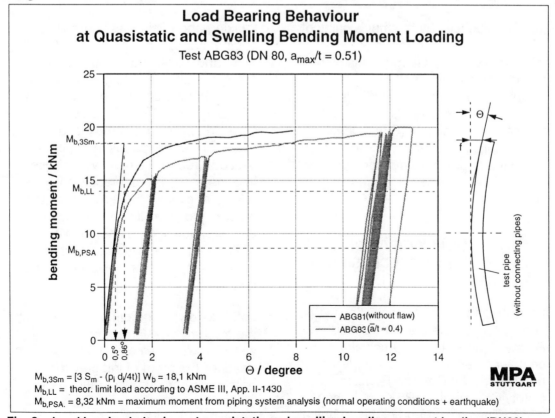

Fig. 3: Load bearing behaviour at quasistatic and swelling bending moment loading (DN80)

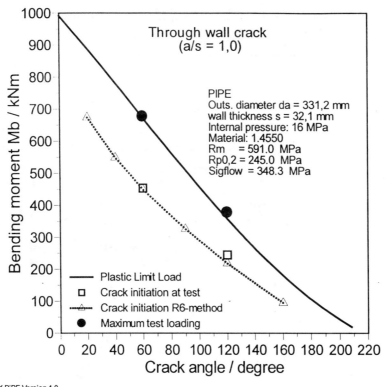

Fig. 4: Bending moment versus crack angle

mainly influenced by the multiaxiality of stress state and therefore has to be taken into consideration. Elastic plastic fracture mechanics (EPFM) and evaluation procedures including the multiaxialitiy of stress state across the ligament are available (Clausmeyer, Kussmaul, Roos 1991). According to *Fig. 5* the appropriate methodology for application to piping components is shown.

EVOLUTIONARY DEVELOPMENT OF THE BREAK PRECLUSION CONCEPT

Parallel to the research activities in the late 1970ies in Germany the Basis Safety Concept (break preclusion concept) was developed and adopted in principle by the German Reactor Safety Commission (RSK guidelines 1982). The background of the technical thinking and basis was published in original papers (Kussmaul et. al. 1979, 1983 and 1984; Bartholomé et. al. 1983; Schulz 1983) In Germany thus the break preclusion concept became a legal requirement. For practical reasons an upper limit of the leakage area of 0.1A (A corresponds to the cross section area of the pipe) was chosen. In general the break preclusion concept is applied to the large diameter primary piping and to the branch connections down to an internal diameter of 200 mm. The concept was in principle also applied to pipes down to a diameter of 50 mm.

In the following the different steps for the fracture mechanics procedures which have to be carried out and to be used for the proof of integrity (break preclusion) are shown in *Fig. 6*.

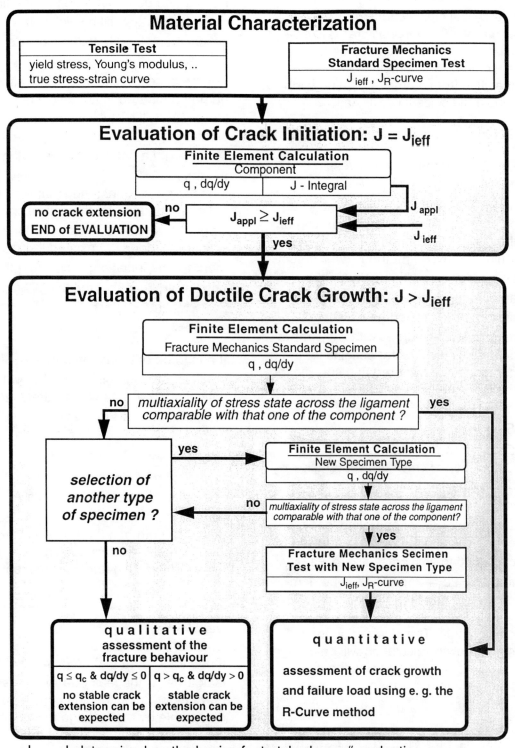

$J_{ieff} = J_i$ determined on the basis of „stretched-zone" evaluation

$q_c \approx 0.3$ for steels

Fig. 5: Flow chart for the evaluation of degraded components

Step 1

Calculation of the critical crack length:
The critical crack length $2c_{crit}$ (critical length of a through wall crack) is calculated with the maximum loading (maximum load combination to be specified from normal operating conditions and accident conditions.

Step 2

Definition of the initial crack size:
The initial crack size (crack depth a_a and crack length $2c_a$) is defined in a component specific way depending on the capabilities of non-destructive examination (NDE) methods (well detectable crack size with crack depth a_{NDE} and crack length $2c_{NDE}$). The initial crack size is defined as $a_a = a_{NDE} \cdot S_{NDE}$ and $2c_a = 2c_{NDE} \cdot S_{NDE}$ with S_{NDE} representing a safety factor specific for the component considered and specific for the NDE method used.

Step 3
Crack growth calculations:
Calculation of crack growth (crack depth Δa and crack length $2\Delta c$) of the initial crack size (depth a_a and length $2c_a$) under normal operating loading conditions and the corresponding load cycles up to final crack size $a_o = a_a + \Delta a$ and $2c_o = 2c_a + 2\Delta c$. It has to be demonstrated, that there will be only limited growth in crack depth as well as in crack length. The time period for the crack growth calculation will be assumed to be at least twice the specified life time of the plant.

Step 4
Safety factors S have to be fixed in a plant and component specific way. Especially the requirements for the NDE inspections have to be adapted.

Leak before break (LBB) behaviour:
- $\underline{a_o/s < 1}$ [1]
It has to be demonstrated that the final crack length $2c_o$ will be well below the critical crack length, that means $2c_o < 2c_{crit}/S$.
For the case of $2c_o \geq 2c_{crit}/S$ it has to be demonstrated that the final crack length $2c_o$ will be well below the critical crack length for a part through crack ($2c_{crit}^*$) with a crack depth of a_o, that means $2c_o < 2c_{crit}^*(a_o/s)/S$. Otherwise it has to be demonstrated that the critical moment used for the critical crack length $2c_{krit}$ (critical length of a through wall crack) does not exceed the moment for the initial crack size (depth a_a and length $2c_a$), that means $M_{crit}(a_a, 2c_a) / M_{crit}(a/s=1) > 1$.

- $\underline{a_o/s = 1}$
It has to be demonstrated that the final crack length $2c_o$ will be well below the critical crack length with a safety factor of S, that means $2c_o < 2c_{crit}/S$

Step 5
Calculation of leak area:
The calculation of the leak area A_o is based on the critical crack length $2c_{crit}$ (critical length of a through wall crack, step 1) and the crack opening displacement (COD) under normal operating loading conditions.

Step 6
Calculation of leak rate:
It has to be demonstrated, that for the leak area A_o the leak rate $\dot{m}_o$ under normal operating loading conditions is detectable by the leakage monitoring system (LMS) installed, that means $\dot{m}_o > \dot{m}_{LMS}$.

---

[1] s - wallthickness

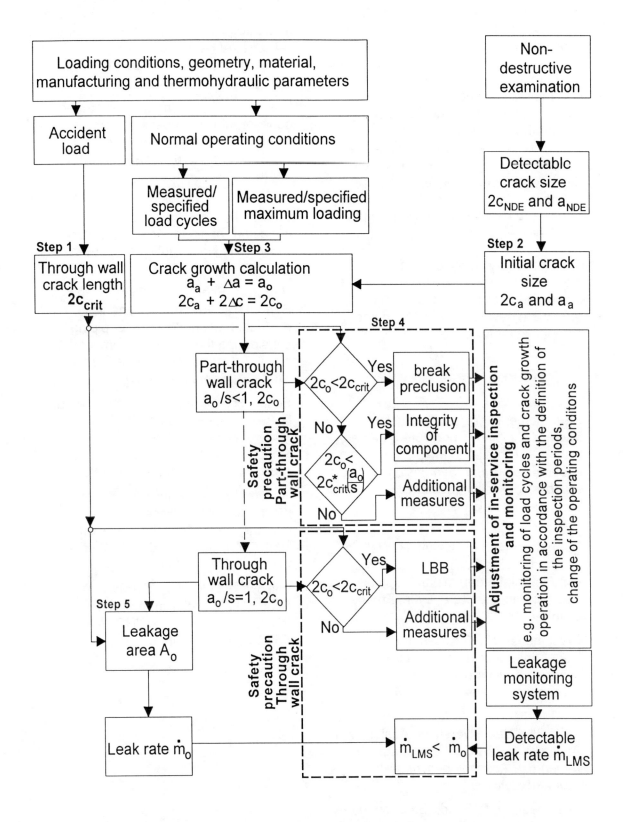

Fig. 6: Fracture mechanics procedure witin proof of integrity of piping components

SUMMARY

The experimental and numerical results from investigations carried out on austenitic piping with nominal diameter down to an internal diameter of 50 mm are introduced and discussed together with their assessment using fracture mechanics methods and the application for the proof of integrity (break preclusion) of piping components. The results obtained confirmed the procedures for the proof of integrity by updating information on load, deformation and failure behaviour of austenitic piping degraded by circumferential flaws.

The necessary steps for the fracture mechanics procedures which have to be carried out used for the proof of integrity are developed with respect to a complete safety concept.

REFERENCES

Bartholomé, G., et. al., "Exclusion of fracture in piping of pressure boundary, Part 2: Application to the primary coolant piping", *Int. Symposion on reliability of reactor pressure components,* IAEA-SM-269/7, Stuttgart, March 1983

Bartholomé, G., E. Bazant, R. Wellein, W. Stadtmüller, D. Sturm, "German experimental programs and results", *Spec. Meeting on LBB*, Lyon, France October 1995

Clausmeyer, H., K. Kussmaul, E. Roos, "Influence of stress state on the failure behaviour of cracked components made of steel", *Appl. Mech. Rev. Vol. 44*, No. 2, 77-92

Diem, H., "Bauteilversuche Austenit mit Analytik", *MPA/VGB-Forschungsvorhaben 6.1, Abschlußbericht*, MPA Stuttgart, Juli 1997

Kussmaul, K., D. Blind, "Basis saftey – a challenge to nuclear technology", *IAEA spec. Meeting*, Madrid, 1979

Kussmaul, K. et. al., "Exclusion of fracture in piping of pressure boundary, Part 1: Experimental investigations and their interpretation", *Int. Symposion on reliability of reactor pressure components,* IAEA-SM-269/7, Stuttgart, March 1983

Kussmaul, K., "German basis safety concept rules out possibility of catastrophic failure", *Nuclear Engineering International* 12 (1984), 41-46

Kussmaul, K., D. Blind, E. Roos, D. Sturm, "The leak before break behaviour of pipes", *VGB Kraftwerkstechnik 70*, No. 7, July 1990, 465-477

Roos., E., H. Kockelmann, K.-H. Herter, X. Schuler, W. Zaiss, J. Bartonicek, "Fracture mechanics evaluation of small diameter piping considering the latest experimental results", *German-Japanese Seminar*, Stuttgart, September 1997

Schulz,H., "Current position and actual licensing, decisions on LBB in the Federal Republic of Germany", *CSNI Report No. 82, NUREG/CP-0051,*Monterey, September 1983

Stadtmüller,W. "Beitrag zur bruchmechanischen Absicherung von Rohrleitungen aus hochzähem Werkstoff", *BMBF Vorhaben 1500 964, Abschlußbericht*, MPA Stuttgart, Juni 1997

Wilkowski, G.M., R.J. Olson, P.M. Scott, "State-of-the-art report on piping fracture mechanics", *NUREG/CR-6540*, January 1998

MECHANICAL PROPERTIES FOR FRACTURE ANALYSIS OF MILD STEEL STORAGE TANKS

R. L. Sindelar, P. S. Lam, G. R. Caskey, Jr.
Savannah River Technology Center
Westinghouse Savannah River Company
Aiken, South Carolina

L. Y. Woo
Georgia Institute of Technology
Atlanta, Georgia

ABSTRACT

Mechanical properties of 1950's vintage, A285 Grade B carbon steels have been compiled for elastic-plastic fracture mechanics analysis of storage tanks (Lam and Sindelar, 1999). The properties are from standard Charpy V-notch (CVN), 0.4T planform Compact Tension (C(T)), and Tensile (T) specimens machined from archival steel from large water piping. The piping and storage tanks were constructed in the 1950s from semi-killed, hot-rolled carbon steel plate specified as A285 Grade B. Evaluation of potential aging mechanisms at both service conditions shows no loss in fracture resistance of the steel in either case.

Site and literature data show that the A285, Grade B steel, at and above approximately 70°F, is in the upper transition to upper shelf region for absorbed energy and is not subject to cleavage cracking or a brittle fracture mode. Furthermore, the tank sidewalls are 1/2 or 5/8-inch thick, and therefore, the J-resistance (J_R) curve that characterizes material resistance to stable crack extension under elastic-plastic deformation best defines the material fracture toughness. The J_R curves for several heats of A285, Grade B steel tested at 40°F, a temperature near the average ductile-to-brittle (DBTT) transition temperature (CVN @ 15 ft-lb), are presented. This data is applicable to evaluate flaw stability of the storage tanks that are operated above 70°F since, even at 40°F, crack advance is observed to proceed by ductile tearing.

INTRODUCTION

Mild carbon steel with specification ASTM A285 is a common material of construction for vessels in the petroleum and nuclear industries. Storage tanks were constructed between 1951 and 1956 from hot-rolled carbon steel plate specified as ASTM A285 Grade B. Extensive analyses and experimental investigations have demonstrated tank integrity in full consideration of potential service-induced degradation mechanisms, including stress corrosion cracking (Marra, et. al., 1995).

The operating temperature of the storage tanks is 70°F and above, placing the carbon steel in the upper transition region where ductile tearing would be the failure mode. The Department of Energy (DOE) Tank Structural Integrity Panel has recommended J_R Analysis or Deformation Plasticity Failure Analysis Diagrams as elastic-plastic fracture mechanics (EPFM) analysis tools to evaluate integrity of storage tanks (Bandyopadhyay, et al., 1997). The approach allows determination of critical flaw size under conditions where stable crack extension would precede a ductile tearing instability. A J-integral fracture mechanics analysis has been performed to evaluate flaw stability using material J_R curves to characterize the fracture toughness and to set the criterion for a cut-off to the J_R curve (Lam and Sindelar, 1999).

The validity and limitations of the fracture mechanics analysis depend, in part, upon the available mechanical property data. The steel suppliers provided tensile properties for each heat of steel that was used in construction of the tanks. Impact properties were measured for one sample of this steel only. Fracture mechanics analyses require measurement of fracture toughness. Procedures had not been established in the early 1950's to measure fracture toughness; therefore, J_R curves are not available for the specific heats of steel in the storage tanks. However, J_R curves for fracture toughness have been measured on specimens of carbon steel pipe that were made from plates in the 1950's to the same specifications as the storage tanks. The application of this combined database to analysis of the storage tanks has been demonstrated through fundamental materials understanding described in this report. Limitations to the application of this data have been identified and additional fracture tests proposed to address the limitations.

EVALUATION OF 1950'S VINTAGE A285, GRADE B CARBON STEEL IN STORAGE TANKS AND PIPING

The storage tanks and the cooling water piping were constructed of ASTM A285, Grade B carbon steel during the 1950s. Based on evaluation of the composition, fabrication, and service conditions of the storage tanks and cooling water piping, the mechanical property data from the pipe sections are judged to be applicable to analysis of the storage tanks.

Composition

Storage Tanks

The carbon steel conformed to specification ASTM A285-50T, Grade B firebox quality (see Table 1) including the 7/8-inch plates that form the bottom knuckle region of one of the two tank types. The average, maximum, and minimum constituents reported for 21 heats of the steel are listed in Table 2. The material was melted in an open-hearth furnace, semi-killed, and then hot rolled into plate.

Piping

Large diameter piping was built to a 1950's edition of the ASME, Section VIII, Unfired Pressure Vessel Code. Portions of the piping system that were fabricated from A285 carbon steel were removed for mechanical testing. Table 3 shows the composition of four different heats of steel and two weldments.

The compositions of the steels in the piping are also within the specifications for ASTM A285-50T, Grade B. The carbon contents of the pipe steels correspond to the most frequently occurring carbon contents in the tank steels, 0.10 – 0.14 wt % carbon, but do not cover five of the 21 heats of steel with carbon contents in the range 0.15 to 0.20 wt %. On average the manganese contents in the piping steels are slightly higher than in the tank steels leading to higher manganese carbon ratios for the pipe steels. The manganese carbon ratios for the tank steels have a bimodal distribution with peaks at 3.8 and 4.6 Mn/C. The manganese carbon ratios for the pipe steel correspond to the higher peak at 4.6. Overall, the sections of pipe steel most closely resemble those tank steels with lower carbon and higher manganese to carbon ratios.

Fabrication

Storage Tanks

Tank construction conformed to the Rules for Construction of Unfired Pressure Vessels, Section VIII of the ASME Boiler Construction Code 1949 or 1952. Welding procedures and welding operator qualifications were in compliance with Section IX of the ASME Code.

The wall thickness of the steel plates in the tanks were: 1/2-inch for top and bottom plates; 1/2- or 5/8-inch for side plates; 1/2- or 7/8-inch for knuckle plates joining the bottom and sides.

Tank inspections included visual, radiographic, and leak testing. All welds were visually inspected upon completion of the weld and/or after each pass if requested. The welds had to be approved before radiographic inspection could begin. All welds affecting the ability of the tank to retain liquids or gasses were radiographed by methods that met the accuracy required by the Code. This included welds to and in manholes, nozzles, sleeves, or couplings attached to or penetrating the steel shell. All repaired welds were radiographed.

Piping

The large diameter piping had a wall thickness of 0.5 inch. It is assumed that the pipe was fabricated by roll forming and seam welding. Sections of piping were joined by butt welds made by shielded metal-arc welding with AWS E6010 electrodes welding from the outside of the pipe, back gouging, and then rewelding from the inside of the pipe. Radiographic inspections were not performed on the original pipe welds and were not required by any of the applicable piping standards at the time of construction.

Service Conditions and Effects of Service on Mechanical Properties

A potential difference between the storage tanks and cooling water piping is in the exposure of the materials to service conditions that could potentially affect the mechanical properties of the A285 Grade B carbon steel. Several degradation mechanisms that potentially could affect either the mechanical properties of the steel or the load bearing capacity of the tanks have been identified. These are: corrosion (general and pitting); thermal embrittlement; radiation embrittlement; and hydrogen embrittlement. They have been evaluated (Marra et. al., 1995 and Bandyopadhyay, et. al., 1997) and a summary of the results is below.

Corrosion

In-service inspection and laboratory testing have shown that general corrosion and pitting are insignificant. Ultrasonic inspections of the tank walls indicated that no detectable thinning had occurred in over 25 years of operation. In addition, corrosion coupons immersed in the tanks for approximately 15 years showed little evidence of general corrosion. Likewise, no significant pitting has occurred. Only broad, shallow pitting has been observed. Shallow pitting would have insignificant effects on the mechanical properties of the material.

Thermal Embrittlement

Both elevated and low temperature environments may result in the embrittlement of carbon steels. Embrittlement is characterized by an increase in the strength and hardness of the material with a corresponding loss in ductility and toughness.

Table 1 - ASTM Requirements for Chemical Composition for A285-50T, Grade B Firebox Quality

For plates ≤ 0.75" thickness	Composition, wt. %			
	C max	Mn max	P max	S max
	0.2	0.8	0.035	0.04

Table 2 - Chemical Composition of Storage Tank Plates

	Composition, wt. %			
	C	Mn	P	S
Average Composition of 21 Heats	0.12	0.48	0.01	0.028
Maximum Composition from 21 Heats	0.20	0.58	0.015	0.037
Minimum Composition from 21 Heats	0.08	0.37	0.007	0.020

Table 3 - Composition of Material from Cooling Water Piping

Material	Specimen ID	Composition, wt%					
		C	Mn	P	S	Si	Cu
A285 Pipe	P5	0.14, 0.14a, 0.148b	0.56 0.56a	0.006 0.007a	0.029 0.073a	0.09 0.063a	0.045 0.047a
A285 Pipe	P6	0.12	0.56	0.007	0.020	0.12	0.096
A285 Pipe	P7	0.12	0.54	0.007	0.027	0.11	0.170
A285 Pipe	P8	0.10	0.58	0.006	0.027	0.10	0.100
A285 Weld	CW11	0.09	0.56	0.008	0.015	0.14	0.095
A285 Weld	CW12	0.09	0.54	0.008	0.017	0.13	0.100

Notes:
1. The composition reported from wet chemistry analysis in 1983
2. [a]1998 Analysis Using Wavelength Dispersive X-Ray Flourescence Spectroscopy (WDS). The WDS analysis also included Ni (0.032 %), Cr (0.030%), and Mo (0.003%)
3. [b]1998 Analysis Using Carbon Analyzer

The temperatures experienced by the storage tanks (measured temperatures of 84 to 146°F) are well below those needed for elevated temperature embrittlement (200-500°F) and above those needed for low temperature embrittlement (DBTT). Therefore, thermal embrittlement is not considered to be a significant degradation mechanism.

Radiation Embrittlement

Radiation embrittlement of the ferritic steels such as the carbon steels of the storage tanks arises from displacement of atoms in the steel by neutron irradiation or exposure to high-energy gamma radiation. The embrittlement is characterized as a reduction in ductility and/or an increase in the ductile to brittle transition temperature with a loss in upper shelf absorbed energy, as measured by standard CVN testing.

The highest estimated damage level in a storage tank is less than 4.0E-7 dpa and is well below the level of 1.0E-5 dpa where changes in the mechanical properties of ferritic steels due to radiation damage have been observed.

Hydrogen Embrittlement

Hydrogen embrittlement of carbon steel may occur through formation of methane gas from radiolytically or cathodically produced hydrogen that diffuses into the steel and reacts with the carbon. The reaction results in severe loss of ductility and strength for the steel. However, data from the American Petroleum Institute demonstrates that carbon steel at temperatures less than 500°F (260°C) and pressures less than several hundred atmospheres can perform safely for an indefinite time (API, 1977). Since operating temperatures and pressures in the tanks are well within the parameters, the mechanism is insignificant.

None of the degradation mechanisms significantly affect the mechanical properties of the A285 steel in storage tank service.

This conclusion also applies to the pipe service since the piping carried cooling water at temperatures ranging from 40°F to 180°F (4.4°C to 82.2°C), and at pressures ranging up to 70 psig. Therefore, considering the composition, fabrication, and service conditions of the A285 steel, the properties of A285 steel from the piping are applicable to the tanks.

MECHANICAL PROPERTY DATA OF A285, GRADE B MATERIAL

The tanks operate above a minimum temperature of 70°F in order to avoid the potential for brittle fracture. This temperature corresponds to upper transition to upper shelf behavior for A285 steel and thus failure could only occur through ductile tearing. Elastic-plastic analysis must be used to characterize the deformation of the thin wall tanks. The J-resistance (J_R) curve that characterizes material resistance to stable crack extension under elastic-plastic deformation is used to define the material fracture toughness.

A database is being developed to quantify the fracture toughness and provide a statistical base for flaw stability analysis. The initial data available for the database is from testing of A285 carbon steel from the cooling water piping performed by Materials Engineering Associates, Inc. Specimens were machined from four pieces of pipe and two weld regions. The mechanical properties of the archival A285 carbon steel were characterized through the following tests:

1. Static and dynamic tensile
2. Static and dynamic compact tension fracture toughness 0.4 C(T)
3. Charpy V Notch

The specimens were oriented in either the L-C orientation or the C-L orientation. The L-C orientation positioned the crack plane perpendicular to the rolling direction, while C-L orientation positioned the plane parallel to the rolling direction. Both the tensile test and fracture toughness tests were conducted with all specimens in the L-C orientation. The Charpy V Notch test was conducted in both orientations. The specimen orientations are illustrated in Figure 1.

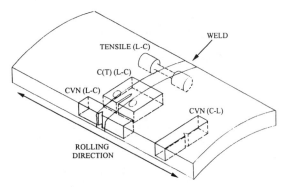

Figure 1 – Orientation of Test Specimens

Static and dynamic tests accounted for normal and seismic loads. The dynamic loading rates for the material were too high compared to the rates expected in the tanks under seismic conditions and therefore the results of the static testing only are provided in the following sections.

Tensile Properties

Storage Tanks

The ASTM specification A285-50T for the tensile properties of A285, Grade B steel is shown in Table 4 below. ASTM Standard A20-50 specified tensile specimens with an 8-inch gauge length machined with their axes parallel to the longitudinal direction of the plate. The range of values from triplicate tests of the 21 heats of steel from tank construction are within the ASTM specification for yield strength, tensile strength, and elongation.

Piping

The testing was in accordance with ASTM E8 specification for tension testing of metallic materials with the subsize tensile specimen, 0.25-inch diameter. The results of the static tensile tests are summarized in Table 5. All specimens were in the L-C orientation and were tested at 40°F (4.4°C), the minimum service temperature for the pipe.

The tensile specimens for the weld material were machined so that the gage section contained base, heat-affected-zone, and weld metal. All failures of these specimens occurred in the base metal. Therefore, neither the weld metal nor the heat-affected-zone is the weak link in the weld area.

Fracture Properties

Fracture of carbon steels may occur by ductile rupture or microvoid coalescence, by brittle or cleavage fracture, or by ductile tearing interrupted by brittle fracture. Decreasing temperature, increasing constraint, and rapid loading rate promote brittle fracture. The transition from ductile to brittle fracture is also a material property that depends on grain size and composition of the steel, as discussed later.

Storage Tanks

Fracture characteristics of the A285 carbon steel used in construction of the tanks were investigated after construction to evaluate the susceptibility of the steel to brittle fracture and establish temperature limits for operation of the tanks. The nil-ductility transition temperature (NDTT) was −20± 10°F as measured by the ASTM E-208 Drop Weight Test with 0.5-inch thick non-standard type P-2 specimens.

Charpy V-Notch (CVN) specimens were machined from an archival plate of A285 steel from the construction of the tanks and tested over the temperature range −30 to +111°F. The ductile-to-brittle transition temperature (DBTT), the temperature at 15 ft-lb absorbed energy, was +45°F and the upper shelf energy impact energy was estimated at 62 ft-lb. The orientation of the test specimens with respect to the plate rolling direction was not reported. Based on the relatively low upper shelf energy impact energy (USE), the CVN specimens were probably in the T-L orientation.

An extensive database of absorbed energy from Charpy V-notch testing is reported in the literature for A285, Grade C steel (Hamel, 1958). Data applicable to the storage tanks can be developed by using the 33 data in which the grain size is greater or equal to ASTM number 6 and carbon content is less than or equal to 0.22 wt.%. The range in DBTT is –35 to 60°F and the average is 27°F. This data is consistent with the site-specific tank plate steels and the piping steels.

Piping

Charpy V-notch testing was also performed using the four heats of plate material and two weld materials. The results shown in Table 6 are consistent with the literature results and the tank plate materials.

Fracture toughness specimens of the piping steel were machined to the ASTM E399 configuration for a specimen with thickness equal to 0.394 inch. The tests were in accordance with the applicable portions of ASTM E813 specification, the standard test method for J_{Ic}. All specimens were pre-cracked in accordance with ASTM E399 requirements.

Static fracture toughness tests were conducted at a stress intensity rate of 40 ksi $\sqrt{in}$ /min; the results are summarized in Table 7. Testing occurred at 40°F (4.4°C) with all specimens in the L-C orientation. All fractures were ductile over the entire crack extension range except for weld specimen CW11-2 where fracture began and continued in a ductile manner but changed to cleavage after significant crack extension. The tests did not meet the validity requirements of ASTM E-813 because of the ductility of the steel.

The results of the static 0.4 C(T) tests provided fracture properties in the form of J-resistance or J_R curves. J is the energy made available at the crack tip per unit crack extension (Δa). The calculated value of J was a modified J (J_m), where J_m was the deformation theory J (J_d) adjusted by a term that accounted for the elastic plastic failure. For small crack extensions on the order of 1mm, J_m was equal to J_d. However, at larger crack extensions the difference between the two J values was significant with J_m believed to produce values that were more geometry independent. Therefore, the modified J was used for all static tests.

J_{Ic} is the energy at the onset of crack initiation and was calculated from the power law equation at a fixed crack extension of 0.2 mm. The value of the fixed crack extension was chosen based on past experience and approximates the maximum blunting extension attainable with low strength structural steels. Using the value for J_{Ic}, the elastic initiation fracture toughness K_{Jc} may be calculated from the following equation:

$$K_{Jc} = \left[\frac{E \cdot J_{Ic}}{(1-\nu^2)}\right]^{0.5}$$

where $\nu = 0.3$ and $E = 3 \times 10^7$. The values for K_{Jc} are shown in Table 7. The average toughness for the pipe material was 205 ksi$\sqrt{in}$ with a standard deviation of 42 ksi$\sqrt{in}$. For the weld material, the average toughness was 148.5 ksi$\sqrt{in}$ with a standard deviation of 7.8 ksi$\sqrt{in}$.

The J_m vs. Δa curves under static loading conditions are shown in Figure 2. The data was fit to a power law equation of the form, $J = C (\Delta a)^n$ where C and n are constants. The values of C and n are listed in Table 8.

SEM and Optical Microscopy

Fracture surfaces of four compact tension specimens were examined: two static pipe specimens (P5-2 and P8-2), one static weld specimen (CW11-2), and one dynamic pipe specimen (P6-8). The surfaces were inspected for evidence of ductile or brittle fracture.

The ductile nature of the crack growth is evident in the optical photograph (upper left) and the electron micrographs in Figure 5 of specimen P5-2. The optical photograph shows plastic deformation or lateral contraction of the specimen in the region of crack advance (Δa). The high magnification electron micrographs show dimpled rupture or microvoid coalescence, a characteristic of ductile failure. The only indication of brittle fracture was seen in the dynamic fracture specimen, where a transition from ductile to cleavage fracture was beginning in the region of maximum crack extension.

Dependency of Mechanical Properties on Material and Test Conditions

Several material or test condition parameters are identified that can affect the mechanical properties of the A285 steel and the applicability of the pipe material test results to analysis of the storage tanks.

Effects of Composition on Tensile Properties

Carbon and manganese are the main compositional variables that influence the tensile properties, yield strength, tensile strength, and ductility of A285 carbon steel. Both carbon and manganese raise the strength and lower the ductility of hot rolled carbon steel. However, the reported elongations of the tank and pipe steels can not be compared because the measured elongation is sensitive to the shape of the tensile specimen and its gauge length. The tensile specimens for the pipe steel were sub-sized round bar specimens, whereas the tensile specimens for the tank steels were flat specimens with an 8-inch gauge length, which was the standard test specimen in the early and mid 1950s.

The carbon contents of the 21 heats of carbon steel in the storage tanks range from 0.08 to 0.20 wt.% carbon with 14 of the 21 analyses in the range of 0.09 to 0.14 wt.% carbon. The four sections of piping also have carbon analyses within this latter range. The manganese to carbon ratios for the tank steels range from less than 2.6 to 5.8 and have a bimodal distribution with

Table 4 - ASTM Requirements for Tensile Properties for A285-50T, Grade B Firebox Quality

For plates ≤ 0.75" in thickness	Mechanical Property Ranges		
	Tensile (ksi)	Yield (ksi)	% Elongation in 8-inches
	50-60	27	27

Table 5- Static Tensile Test Data Summary for A285 Carbon Steel from the Cooling Water Piping

Material	Specimen ID	Yield Stress			Ultimate Stress	Elongation	Reduction In Area
		Upper (ksi)	Lower (ksi)	0.2 % (ksi)	(ksi)	(%)	(%)
A285 Pipe	P5-2	43.0	36.6	37.1	58.0	39.2	70.0
A285 Pipe	P6-2	38.3	34.7	35.6	57.8	39.6	70.4
A285 Pipe	P7-2	42.8	35.6	35.9	59.6	37.8	63.1
A285 Pipe	P8-2	43.4	36.8	37.5	57.8	39.4	62.8
A285 Weld	CW11-2	47.9	43.8	45.0	62.8	35.5	66.8
A285 Weld	CW12-2	49.4	46.5	47.2	63.5	23.0	66.8

Note: Static loading rate approximately 1×10^3 psi/sec; 40 - 80 seconds to upper yield stress

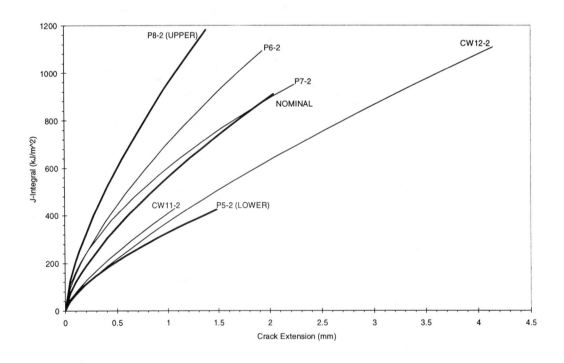

Figure 2 - J_R Curves for A285 Grade B Steel at 40°F

Table 6 - Charpy V Notch Impact Test Data Summary for A285 Carbon Steel from the Cooling Water Piping

Material	Specimen ID	41J (30 ft-lb) Temperature (°C)	41J (30 ft-lb) Temperature (°F)	Shelf Energy Level L-C orientation (J)	Shelf Energy Level L-C orientation (ft-lb)	Shelf Energy Level C-L orientation (J)	Shelf Energy Level C-L orientation (ft-lb)	Energy Level at 40°F (J)	Energy Level at 40°F (ft-lb)
A285 Pipe	P5-2	14	57	149	110	52	38	18	13
A285 Pipe	P6-2	15	59	285	210	76	56	14	10
A285 Pipe	P7-2	1	34	230	170	65	48	54	40
A285 Pipe	P8-2	-2	28	285	210	83	61	95	70
A285 Weld	CW11-2	-8	18	127	94	N/A	N/A	72	53
A285 Weld	CW12-2	-15	5	134	99	N/A	N/A	85	63

Table 7 - Static Fracture Toughness Data Summary for A285 Carbon Steel from the Cooling Water Piping

Material	Specimen ID	J_{Ic} in-lb/in²	J_{Ic} kJ/m²	K_{Jc} ksi$\sqrt{in}$	K_{Jc} MPa$\sqrt{m}$	Failure Type
A285 Pipe	P5-2	650	113.8	146	160.9	Ductile
A285 Pipe	P6-2	1370	240	213	233.6	Ductile
A285 Pipe	P7-2	1369	239.8	212	233.5	Ductile
A285 Pipe	P8-2	1844	322.9	247	271.0	Ductile
A285 Weld	CW11-2	714	125.1	154	168.7	Ductile/Cleavage
A285 Weld	CW12-2	620	108.5	143	157.1	Ductile

Note: For both ductile and ductile/cleavage failure types, J is the value at the initial maximum load point which, on a tensile test, would correspond to the upper yield point. For cleavage failure types, J is measured at the point of cleavage.

Table 8 – Values of C and n for the Power Law Fit to the J_R data

Specimen Number	C (Newton/(mm)$^{n+1}$)	n
P5-2	328.1	0.6578
P6-2	704.2	0.6688
P7-2	601.2	0.571
P8-2	951.5	0.6716
CW11-2	408	0.7346
CW12-2	372.8	0.7668

peaks at Mn/C ratios of 3.8 and 4.6. Manganese to carbon ratios for the piping are between 4.0 to 5.8, overlapping the higher peak in the bimodal distribution for the tank steel. As a consequence of the higher manganese contents (0.54-0.58 wt % Mn), the pipe steels have slightly higher strengths than the tank steels for the same carbon contents.

Effects of Composition on Impact and Fracture Toughness Properties

Carbon has been shown to raise the transition temperature and lower the upper shelf energy of carbon steels whereas manganese and silicon have the opposite effect. Consequently, the Mn/C ratio can be used as an indicator of the impact properties; fracture toughness is expected to vary with composition in the same manner.

The DBTT for the pipe was in the range of -25°F to 50°F (-32°C to 10°C), comparable to that of the tank steel (45°F).

The relation between Mn/C ratio and fracture toughness is evident in results from compact tension tests conducted at 40°F on steel from the pipe (Tables 3 & 7). The static fracture toughness increased from 650 to 1844 in-lb/in^2 as the Mn/C ratios increased from 4.0 to 5.8. The upper shelf energy measured in CVN tests (Table 6) shows a similar relation to the Mn/C ratio.

Effects of Microstructure on Fracture Toughness Properties

The ferrite grain size and volume fraction pearlite, influence the fracture of carbon steels and are controlled by the composition and finishing temperature during hot rolling. A three fold increase in ferrite grain size from 4.6×10^{-4} inch to 12.1×10^{-4} inch increased the brittle fracture transition temperature from –42°C to +17°C in a 0.11 wt% carbon steel (Burns and Pickering, 1964); a similar trend was seen in A285 Grade C steel (Hamel, 1958). Within the range of carbon content in the storage tank and cooling water pipe steels (0.08 to 0.20 wt% C), a small variation in volume fraction pearlite would be expected. The only available microstructures, from plate P7 from the pipe and a sample from in-situ metallography of a storage tank, have comparable grain sizes and volume fractions of pearlite.

Effects of Thickness on Fracture Toughness Properties

Constraint, which includes out-of-plane or thickness effects, has an effect on the fracture parameter value at failure and the J_R curve. For highly constrained conditions, fracture usually occurs at a lower value of the fracture parameter than for low constraint.

The C(T) specimens machined from the piping had a thickness of 0.394 inch. Since the thickness of these specimens is less than the thickness of the tank walls (0.5 or 0.625 inches), the constraint effects on the fracture parameters needs to be considered. Research is being conducted to formulate an advanced fracture mechanics methodology (J-A2) that allows for consideration of constraint effects in ductile materials (Chao and Lam, 1998 and Chao, et. al., 1999).

The J-A2 methodology is being developed to identify fracture parameters that are independent of specimen geometry, such as crack depth and sample thickness.

Orientation Effects on Fracture Toughness Properties

The texture formed during hot rolling of carbon steel has a pronounced effect on the fracture toughness and impact energy. This is demonstrated in the impact tests on the pipe steel where the orientation had a significant effect on the upper shelf energy but a negligible one on the lower shelf energy. L-C specimens had an upper shelf energy three to four times greater than that for the C-L specimens. Consequently, fracture toughness at temperatures in the upper shelf region would be lower in the T-L orientation than in the L-T orientation of rolled steel plates.

Effects of Temperature

The lower bound temperature of the storage tanks studied is 70°F (21.1°C), which is well above both the NDT and the DBTT. At this temperature, the tank steels will exhibit upper transition to upper shelf behavior. The initial growth of structural flaws would be stable extension by ductile tearing under sufficiently high mechanical loads. Therefore, fracture analyses based on ductile failure or on elastic-plastic tearing instability criteria are applicable to storage tanks and best represent material behavior.

The response of A285 Grade B steel to loading at 40°F (4.4°C) is conservative when compared to the response at actual operating temperatures, ≥70°F (21.1°C). An elastic-plastic fracture analysis using the results of the 40°F (4.4°C) tests would provide a conservative estimate of flaw stability.

Effects of Loading Rate

The loading rate in tensile and fracture toughness testing of carbon steel affects both yield strengths and fracture toughness (K_{IC}) values. In general, static yield strength are lower than dynamic yield strengths, and fracture toughness value of K_{IC} under dynamic loading are lower than the static values in the fracture transition temperature region. Crack resistance (J_R) curves under dynamic loading were not developed in these tests. For seismic conditions, it is important that dynamic testing reflect the loading rate appropriate to the seismic response of the tanks.

CONCLUSIONS

An initial database of elastic-plastic fracture toughness properties has been established for 1950s vintage A285,

Figure 3 – Fracture Surface of P5-2

Grade B steel. The role of material composition, tank temperature, constraint, orientation, and loading rate effects on the mechanical properties has been presented.

ACKNOWLEDGEMENT

Our colleague Dr. Bruce J. Wiersma is the lead investigator for service effects to the storage tanks and has provided much of the background work for this report. This research was supported in part by an appointment to the U. S. Department of Energy Scientists Emeritus Research Participation program at the Westinghouse Savannah River Company administered by the Oak Ridge Institute for Science and Education. This work was funded by the U. S. Department of Energy under contract No. DE-AC09-96SR18500.

REFERENCES

American Petroleum Institute Report "Steels for Hydrogen Service at Elevated Temperatures and Pressures," API-941, 2nd Edition, 1977.

Bandyopadhyay, K., S. Bush, M. Kassir, B. Mather, P. Shewmon, M. Streicher, B. Thompson, D. van Rooyen, and J. Weeks, "Guidelines for Development of Structural Integrity Programs for DOE High-Level Waste Storage Tanks," BNL-52527, UC-406, prepared by Brookhaven National Laboratory for the United States Department of Energy, January 1997.

Burns, K. W. and F. B. Pickering, "Deformation and Fracture of Ferrite-Pearlite Structures," Journal of the Iron and Steel Institute, Nov. 1964, pp. 899-906.

Chao, Y. J. and P. S. Lam, "On the Use of Constraint ParameterA2 Determined from Displacement in Predicting Fracture Event," Engineering Fracture Mechanics, Vol. 61, pp. 487-502, 1998.

Chao, Y. J., X. K. Zhu, P. S. Lam, M. R. Louthan, and N. C. Iyer, "Application of Two-Parameter J-A2 Description to Ductile Crack Growth," submitted to the 31st National Symposium on Fatigue and Fracture Mechanics, June 29-July 1, 1999.

Hamel, F. B., "An Investigation of the Impact Properties of Vessel Steels (A Progress Report)," Proc. Div. Of Refining of American Petroleum Institute, Vol. 38, No. 3, (1958) pp. 239-257.

Lam, P. S. and R. L. Sindelar, "J-Integral Based Flaw Stability Analysis of Mild Steel Storage Tanks," in proceeding of ASME Pressure Vessel and Piping Conference, July 1999.

Marra, J. E., H. A. Abodishish, D. M. Barnes, R. L. Sindelar, H. E. Flanders, T. W. Houston, B. J. Wiersma, F. G. NcNatt, Sr., C. D. Cowfer, "Savannah River Site (SRS) High Level Waste (HLW) Structural Integrity Program," in proceedings of ASME Pressure Vessel and Piping Conference, July 1995.

J-INTEGRAL BASED FLAW STABILITY ANALYSIS OF MILD STEEL STORAGE TANKS

P.-S. Lam and R. L. Sindelar
Savannah River Technology Center
Westinghouse Savannah River Company
Aiken, South Carolina

ABSTRACT

The J-integral fracture methodology was applied to evaluate the stability of postulated flaws in mild steel storage tanks. The material properties and the J-resistance (J_R) curve were obtained from the archival A285 Grade B carbon steel test data. The J-integral calculation is based on the center-cracked panel solution of Shih and Hutchinson (1976). A curvature correction was applied to account for the cylindrical shell configuration. A finite element analysis of an arbitrary flaw in the storage tank geometry demonstrated that the approximate solution is adequate.

INTRODUCTION

Mild steel storage tanks are widely used in the petrochemical and nuclear industry. The A285 Grade B carbon steel is one of the commonly used materials of construction in such tanks. A key element in assessing the fitness-for-service of aged mild steel storage tanks is the flaw stability analysis. In applications where the normal operating temperature of the tanks is 70°F and above, the material of construction (A285 Grade B carbon steel) falls in the upper transition region where ductile crack growth occurs. The most appropriate fracture mechanics methodology to evaluate such structures is the J-integral based elastic-plastic fracture analysis in which stable tearing is taken into consideration. This approach best represents the actual materials behavior. This paper presents the J-T flaw stability methodology and the results of the analyses, where T is the tearing modulus.

The analyses utilize the comprehensive materials property database for A285 carbon steel. The mechanical and fracture properties including the effect of age related degradation on the properties have been assimilated and discussed by Sindelar, Lam, Caskey, and Woo (1999). The flaw stability evaluation for the mild steel storage tank presented in this paper utilizes the lower bound J_R curve from the above referenced paper. This lower bound curve is considered to provide a conservative analysis when applied to tanks where the minimum operating temperature is higher than the test temperature of 40°F. Both temperatures are within the regime where crack extension is stable, crack advance proceeds by ductile tearing, and where toughness or resistance to ductile crack propagation increases as the temperature rises. Therefore, fracture resistance at the tank operating temperature would be greater than that obtained at 40°F.

The analytical solution for a center-cracked panel developed by Shin and Hutchinson (1976) was primarily used to evaluate the J-integral values under applied load. The J-integral calculation and curvature correction procedure developed by Lam, Sindelar, and Awadalla (1993) was used to account for the tank geometry. The stress intensity factor due to residual stress was also included. The J-integral value at which flaw instability occurs is taken from the lower bound specimen data for J_R curve when the crack extension reached 1.5 millimeters and was well within the stable growth regime. The instability crack length corresponding to this critical J value versus applied load can therefore be obtained.

A finite element analysis for an arbitrary flaw in tank geometry has been performed. The analysis demonstrated that the center-cracked panel solution is adequate for the flaw stability analysis of this type of mild steel storage tanks.

J-T FLAW STABILITY METHODOLOGY

J-T Analysis of Flaw Stability

The tearing stability of the material is characterized by the tearing modulus (T), which is defined by:

$$T = \frac{E}{\sigma_o} \frac{dJ}{da}$$

where J is the value of J-integral, σ_o can be the 0.2% yield stress, E is the Young's modulus, and da is an incremental crack extension. Instability flaw lengths are based on the loading conditions and calculated stresses and are determined by an elastic-plastic J-integral or J-T analysis. The crack growth (J ≥ J_{IC}) is stable if $T < T_R$, where T_R is the tearing modulus of the material. As schematically shown in Figure 1, the intersection point of the applied J-T curve and the material J-T curve will determine the crack growth stability limit.

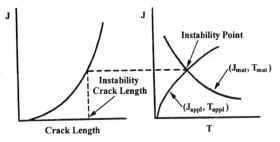

FIGURE 1 – J-T METHODOLOGY FOR INSTABILITY CRACK LENGTH

Development of Material J-T for Flaw Stability Analysis

The J_R curves were developed from the 0.4C(T) fracture toughness tests of the A285 (Sindelar et al., 1999). A lower bound J_R curve for A285 compact tension specimen tests is shown in Figure 2. The material J_R curve was obtained from a power law fit to the experimental data:

$$J = C(\Delta a)^m$$

where Δa is the crack extension. The material parameters C and n are determined through curve fitting. For this lower bound curve in Figure 2, C= 328.1 and m= 0.6578. It should be noted that the last data point represents the termination of the compact tension test, rather than the rupture failure of the specimen. The power law formulation of the $J(\Delta a)$ obtained from material testing can be plotted with its tearing modulus, $T(\Delta a)$, to produce the material J-T curve.

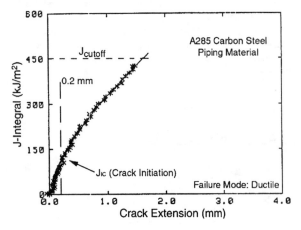

FIGURE 2 - LOWER BOUND J_R CURVE FOR A285 CARBON STEEL PIPING MATERIAL

Determination of J-Δa Cut-off for J-controlled Growth in C(T) Specimens

Stable crack growth occurs under conditions where additional deformation is needed to maintain the appropriate level of strain concentration at the crack tip (Hutchinson and Paris, 1979). The J-integral is an appropriate parameter for characterizing crack growth provided increments in the strain field that are proportional to applied load are greater than increments which are nonproportional to the applied load. These conditions may be expressed as,

$(dJ/da)(b/J) \equiv \omega \gg 1$, where b is the uncracked ligament size.

Crack extension in C(T) specimens generally limits (in ASTM standards E399, E813, and E1152) the region of the plastically blunted crack tip in relation to the in-plane dimension of the specimen (remaining ligament). Tough materials (large J value) such as austenitic stainless steels or carbon steels do not generally meet the criteria when tested above the nil-ductility transition temperature (NDTT) for small planforms. Data at high crack extension from these specimens can be applied in elastic-plastic fracture analyses if J-controlled growth can be established.

A program to measure the fracture toughness of austenitic stainless steel was completed in the early 1990's. Test results from the program can be applied to evaluate J-controlled growth in the 0.394T planform specimen from the carbon steel test program. The measurement of an austenitic stainless steel specimen in both the large planform (0.394 × 1T, the width of the specimen is 2 inches measured from the load line to the back face of the specimen) and small planform (0.394T, the width of the specimen is 0.788 inches measured from the load line to the back face of the specimen) allows a direct comparison of J (deformation theory) versus crack extension. Deviation of the small specimen J_R curve from the large specimen curve would indicate the point (Δa) at which the toughness (defined by the deformation theory) is changing due to size effects. It was found that the J values from the small planform deviate markedly from the large planform values at crack extensions greater than 3 millimeters. Above this point, the J values from the small planform specimen are lower (conservative) compared to the large planform results. The point of deviation between the large and small planform results is suggested to indicate the limit of validity of J deformation theory for the 0.4T planform specimen. For the lower bound carbon steel specimen (Fig. 2), the value of the ω factor is 1.5 at 3-millimeter crack extension for this 0.4T planform. Note that the value of ω is significantly less than the proposed value ($\omega \gg 1$).

In some applications including the present case, the material J-T curve does not intersect with the applied J-T curve, unless the material J-T curve is extrapolated extensively. Under these circumstances, a cut-off J value is conservatively used (rather than extrapolation) to determine the instability crack length. Therefore, in the current flaw stability analysis, a cut-off J-integral value as shown in Figure 2 is taken as 450 kJ/m² which corresponds to a crack extension at about 1.5 millimeters in a 0.4T planform (Sindelar et al., 1999).

Center-Cracked Panel Solution (Shih and Hutchinson, 1976)

The Ramberg-Osgood power law for stress and strain can be expressed as $\frac{\varepsilon}{\varepsilon_o} = \frac{\sigma}{\sigma_o} + \alpha\left(\frac{\sigma}{\sigma_o}\right)^n$, where σ_o is the 0.2% yield stress and ε_o is the corresponding elastic strain. Figure 3 shows a uniaxial tension test result for a specimen representing a material property lower bound based on its J_R curve and J_{IC} value (Fig. 2). The true stress-true strain curve is characterized with $\alpha=17.176$ and n= 3.585 (Sindelar, et al., 1999).

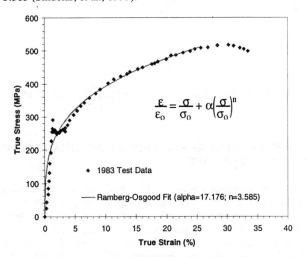

FIGURE 3 - RAMBERG-OSGOOD FIT FOR A LOWER BOUND A285 SPECIMEN

A solution of J-integral for a center-cracked panel (CCP) with a finite width (2b) under plane stress condition for Ramberg-Osgood materials was obtained by Shih and Hutchinson (1976):

$$\frac{J}{\sigma_o \varepsilon_o a(1-a/b)} = \psi \left(\frac{P}{P_o}\right)^2 g_1\left(\frac{a_{eff}}{b}, n=1\right) + \alpha \left(\frac{P}{P_o}\right)^{n+1} g_1\left(\frac{a_{eff}}{b}, n\right)$$

Note that the total J-integral can be composed of two parts, the elastic portion (J_{el}^{ccp}) and a plastic portion (J_{pl}^{ccp}). Therefore,

$$J_{el}^{ccp} = \psi \sigma_o \varepsilon_o a(1-a/b)\left(\frac{P}{P_o}\right)^2 g_1\left(\frac{a_{eff}}{b}, n=1\right)$$

$$J_{pl}^{ccp} = \alpha \sigma_o \varepsilon_o a(1-a/b)\left(\frac{P}{P_o}\right)^{n+1} g_1\left(\frac{a_{eff}}{b}, n\right),$$

where a is the half crack length, b is the half specimen width, $\varepsilon_o = \sigma_o / E$,

$a_{eff} = a + \varphi\, r_y$, $P \leq P_o$ (Kumar and Shih, 1980),

$a_{eff} = (a_{eff})_{P=P_o}$, $P > P_o$,

$\varphi = \frac{1}{1+(P/P_o)^2}$,

$P_o = 2(b-a)\sigma_o$ is the lower bound limit load,

$P = 2b\sigma^\infty$ is the applied load corresponding to a remote stress σ^∞,

$r_y = \frac{1}{2\pi}\left(\frac{n-1}{n+1}\right)\left(\frac{K_I}{\sigma_o}\right)^2 = \frac{a}{2\pi}\left(\frac{n-1}{n+1}\right)\left(1-\frac{a}{b}\right)\left(\frac{P}{P_o}\right)^2 g_1\left(\frac{a}{b}, 1\right)$ for plane stress,

$\psi = \frac{a_{eff}}{a}\left(\frac{b-a}{b-a_{eff}}\right)$,

and

$$g_1\left(\frac{a}{b}, 1\right) = \pi\left[1 - 0.5\frac{a}{b} - 0.37\left(\frac{a}{b}\right)^2 - 0.044\left(\frac{a}{b}\right)^3\right]^2$$

For this material, the values for $g_1(a/b, n=3.585)$ can be calculated according the procedure outlined in Shih and Hutchinson (1976):

a/b	g_1(a/b, n=3.585)
0	6.133
1/8	4.152
1/4	3.156
1/2	1.984
3/4	1.385
1	0.916

A simpler solution for an infinite plate can also be found in Shih and Hutchinson (1976). That solution is also adequate for this type of large tank geometry (the radius to thickness ratio is greater than 800).

Curvature Correction

This analytic solution (J^{ccp}) provides a basis for constructing an approximate solution (J^{cur}) for the sidewall of a storage tank by the application of a curvature correction factor (Lam, Sindelar, and Awadalla, 1993). The correction factors (Y) can be derived from a linear elastic stress intensity factor (K) solution of Tada, Paris, and Irwin (1973). Assumptions have been made: 1) The correction factor for J-integral is Y^2 since elastic J-integral is proportional to K^2; and 2) Same correction factor is applied to the elastic portion of J-integral as well as to its plastic portion ($J_{el}^{cur} = Y^2 J_{el}^{ccp}$ and $J_{pl}^{cur} = Y^2 J_{pl}^{ccp}$).

For an axial flaw opened by a hoop stress(σ_H), the stress intensity factor for a crack with length 2a in a cylinder with mean radius R and thickness t is (Tada, et al. 1973)
$K_I = \sigma_H\sqrt{\pi a}\, Y_1(\lambda)$, where $\lambda = \frac{a}{\sqrt{Rt}}$. Therefore, the correction factor for an axial crack is

$$Y_1(\lambda) = \sqrt{1 + 1.25\, \lambda^2} \text{ for } 0 < \lambda \leq 1, \text{ or}$$
$$Y_1(\lambda) = 0.6 + 0.9\, \lambda, \text{ for } 1 \leq \lambda \leq 5.$$

For a circumferential crack with length 2a or angle 2θ subjected to a longitudinal stress σ_L, the stress intensity factors are (Tada et al. 1973)

$K_I = \sigma_L\sqrt{\pi a}\, Y_2(\lambda \text{ or } \theta)$, where

$$Y_2(\lambda) = \sqrt{1 + 0.3225\, \lambda^2} \text{ for } 0 < \lambda \leq 1,$$

$$Y_2(\lambda) = 0.9 + 0.25\, \lambda, \text{ for } 1 \leq \lambda \leq 5,$$

and

$$Y_2(\theta) = \sqrt{\frac{2}{\eta}\frac{1}{\pi\theta}}\, f(\theta), \text{ for } \lambda > 5,$$

in which

$$\eta^2 = \frac{t/R}{\sqrt{12(1-\nu^2)}}$$

and

$$f(\theta) = \theta + \frac{1 - \theta \cot\theta}{2 \cot\theta + \sqrt{2} \cot\left(\frac{\pi - \theta}{\sqrt{2}}\right)}$$

Procedure of Combining J-integral Solutions

(1) For a given applied remote stress, calculate the CCP solution of Shih and Hutchinson for various crack lengths. The J-integral (J^{ccp}) is composed of an elastic portion (J_{el}^{ccp}) and a plastic portion (J_{pl}^{ccp}), that is, $J^{ccp} = J_{el}^{ccp} + J_{pl}^{ccp}$.

(2) The plastic zone size correction (or small scale yielding correction) is applied to J_{el}^{ccp} (Kumar and Shih, 1980).

(3) The CCP plate solution is corrected for the curvature of the shell or cylindrical structure. The approximated J-integral values for the tank shell (J_{el}^{cur} and J_{pl}^{cur}) are
$J_{el}^{cur} = Y^2 J_{el}^{ccp}$ and $J_{pl}^{cur} = Y^2 J_{pl}^{ccp}$, respectively.

(4) The contributions of fracture parameters from the other sources, such as thermal stress or residual stress, can be combined in the sense of linear elastic fracture mechanics. The elastic portion of J-integral (J_{el}^{cur}) in (4) above is first converted to K_I^{appl}, the Mode I stress intensity factor due to applied load:
$K_I^{appl} = \sqrt{E J_{el}^{cur}}$, for the plane stress condition.

(5) A residual stress contribution can be included in the J-integral solution using the formula advanced by Green and Knowles (1994). The residual stress distribution is a self-equilibrium, symmetric pattern with maximum tension ($+\sigma_r$) on the edges and maximum compression ($-\sigma_r$) in the mid-section of the plate. The through-thickness variation from tension-compression-tension is assumed to be a cosine shape. The σ_r value is taken to be the yield stress of the base metal. The maximum stress intensity factor is $K_{max}^{res} = 0.43 \sigma_r \sqrt{\pi t}$. Note that K_I^{res} is saturated to a maximum value when the crack is extended in length only a fraction of the plate thickness. Therefore, the residual stress of this type is not subject to curvature correction.

(6) The total elastic portion of J is calculated as
$J_{el} = \frac{1}{E}\left(K_I^{appl} + K_{max}^{res}\right)^2$.

(7) The plastic portion of J remains unchanged, that is, $J_{pl} = J_{pl}^{cur}$.

(8) Finally, the total J-integral of the crack is $J = J_{el} + J_{pl}$.

Based on the calculation procedure described above, the instability crack length as a function of applied stress is shown in Figure 4. The remote applied stress is perpendicular to the crack and is up to the yield stress of the material (256 MPa or 37.1 ksi). Both solutions for an axial crack (2b is the height of the storage tank) and for a circumferential crack (2b is the circumference of the storage tank) are presented.

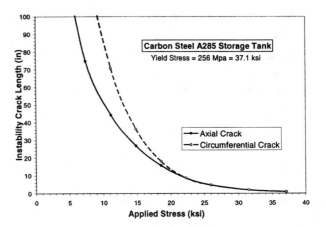

FIGURE 4 - INSTABILITY CRACK LENGTH FOR A MILD STEEL STORAGE TANK

FINITE ELEMENT ANALYSIS OF AN ARBITRARY FLAW IN A STORAGE TANK

As a demonstration of the J-Integral fracture methodology applied to a flaw in a tank and to validate the application of the CCP solution to the tank configuration, a finite element analysis was performed for an arbitrary flaw in storage tank geometry. This flaw has an arc length of about 16 inches (the projection length is about 13 inches perpendicular to the direction of the applied stress).

The finite element region was chosen such that the flaw is away from the edges of the model to minimize the boundary effects. The mesh shown in Figure 5 was generated with MSC/PATRAN (1996), a finite element analysis pre/post-processor. The near crack tip elements were refined for accurately evaluating the J-integral. In addition, it was designed for a potential crack extension analysis for the right-end crack tip.

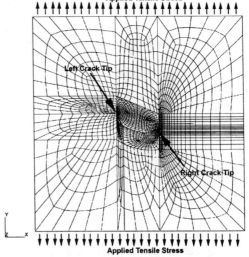

FIGURE 5 - FINITE ELEMENT MESH FOR AN ARBITRARY CRACK

This finite element model contains 2096 four-noded plane stress elements with 2192 nodes before the multi-point constraints are applied (for example, for zipping the nodes ahead of the crack tip in the direction of crack growth). The true stress-true strain curve is characterized by the Ramberg-Osgood power law. The mechanical properties for input to the finite element program are: The Young's modulus is 208 GPa, the Poisson's ratio is 0.333, the yield stress is 256 MPa, and the Ramberg-Osgood parameters α and n are, respectively, 17.176 and 3.585 (Fig. 3).

The ABAQUS finite element program (1998) was used for the analysis. Because the radius to thickness ratio is extremely large (> 800), the curvature effect of the tank is then insignificant and the plane stress elements were used for calculation. The J-integral values were obtained at each load level up to the yield stress (256 MPa or 37.1 ksi) as shown in Figure 6. It can be seen that the cut-off J (450 kJ/m^2) corresponds to an applied stress level equal to 58% of the yield stress, or 150 MPa (22 ksi). This stress level is equivalent to an instability flaw size about 10-inch long according to Figure 4 which was based on CCP solution. This demonstrates that the CCP solutions are adequate for approximating J-integral solutions for a complex flaw configuration in storage tanks. Furthermore, it provides a methodology for assessing the fitness-for-service of the existing storage tanks for known loading conditions and safety margins.

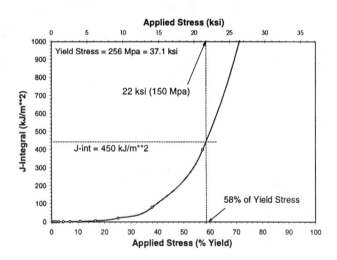

FIGURE 6 - J-INTEGRAL SOLUTION FOR AN ARBITRARY CRACK

CONCLUSION

Elastic-plastic fracture methodology based on J-integral was used to investigate the flaw stability in mild steel storage tanks. Finite element analysis of an arbitrary flaw in tank geometry showed that the analytical solution for a center-cracked panel is appropriate and adequate. Because the material and the applied J-T curves do not intersect unless by extrapolation, a conservative, cut-off J value corresponding to 1.5-millimeter stable crack growth was used to define the instability crack length. The resulting relationship between the instability flaw size and the applied stress can be used to provide guidelines for assessing the fitness-for-service of existing storage tanks for known loading conditions and safety margins.

ACKNOWLEDGEMENT

The authors acknowledge the many helpful discussions with our colleague Dr. B. J. Wiersma. This work was supported by the U. S. Department of Energy under contract No. DE-AC09-96SR18500.

REFERENCES

ABAQUS/STANDARD, Version 5.7, Hibbitt, Karlsson & Sorensen, Inc., Pawtucket, Rhode Island, 1998.

Green, D. and Knowles, J., "The Treatment of Residual Stress in Fracture Assessment of Pressure Vessels," Journal of Pressure Vessels Technology, Vol. 116, pp. 345-352, 1994.

Hutchinson, J. W. and Paris, P. C., "Stability Analysis of J-Controlled Crack Growth," Elastic-Plastic Fracture, ASTM Special Technical Publication 668, pp. 37-64, 1979.

Kumar, V. and Shih, C. F., "Fully Plastic Crack Solutions, Estimation Scheme, and Stability Analyses for the Compact Specimen," Fracture Mechanics: Twelfth Conference, ASTM STP 700, American Society for Testing and Materials, 1980, pp. 406-438.

Lam, P. S., Sindelar, R. L., and Awadalla, N.G., "Acceptance Criteria for In-service Inspection of Heat Exchanger Head and Shell Components," in Fatigue and Fracture of Aerospace Structural Materials, ed. A. Nagar and A.-Y. Kuo, American Society of Mechanical Engineers, AD-Vol. 36, pp. 43-57, 1993.

MSC/PATRAN, Version 7.5, MacNeal-Schwendler Corporation, Los Angeles, California, 1996.

Shih, C.F. and Hutchinson, J.W., "Fully Plastic Solutions and Large Scale Yielding Estimates for Plane Stress Crack Problems," Trans. American Society of Mechanical Engineers, Journal of Engineering Materials and Technology, Series H, Vol. 98, pp. 289-295, 1976.

Sindelar, R. L., Lam, P. S., Caskey, G. R. Jr., and Woo, L. Y. "Mechanical Properties for Fracture Analysis of Mild Steel Storage Tanks," Proceedings of the American Society of Mechanical Engineers Pressure Vessels and Piping Conference, Boston, Massachusetts, August 1999.

Tada, H., Paris, P.C. and Irwin, G.R., The Stress Analysis of Cracks Handbook, Second Edition, pages 33.3, 33.4, 33.6 and 34.1, Paris Productions Incorporated (and Del Research Corporation), Saint Louis, MO, 1985.

EFFICIENT AND RELIABLE ANALYSIS OF STRUCTURAL COLLAPSE BY USING A DIFFERENTIAL QUADRATURE FINITE ELEMENT METHOD ALONG WITH A GLOBAL SECANT RELAXATION-BASED ACCELERATED EQUILIBRIUM ITERATION PROCEDURE

Chang-New Chen
Department of Naval Architecture and Marine Engineering
National Cheng Kung University, Tainan, Taiwan

ABSTRACT

The differential quadrature finite element method which uses the differential quadrature techniques to the finite element discretization is used to analyze the static elastic-plastic collapse of structural problems with linear or nonlinear deformations. The incremental loading procedure in conjunction with the use of equilibrium iteration is used to update the response history. The equilibrium iteration can be carried out by using the global secant relaxation-based accelerated constant stiffness iteration procedure which is efficient and reliable even the load stage is up to the near collapse level. Sample problems are analyzed. Numerical results demonstrate the algorithm.

INTRODUCTION

An efficient and reliable incremental/iterative solution procedure, based on using a global secant relaxation technique to adjust the constant stiffness prediction and improve the convergency, for nonlinear finite element problems which have geometric and/or material nonlinearities has been proposed by the author (Chen, 1990).

Efficient and reliable solution of a static nonlinear finite element system with equilibrium iteration is an important topic in the area of scientific and engineering computation since advanced design is a challengeable trend in current and future technology. The global secant relaxation-based accelerated constant stiffness iteration technique can be incorporated into the static nonlinear finite element solution procedure to construct an efficient and reliable algorithm.

In this paper the numerical procedure for solving static elastic-plastic material failure finite element problems by using the incremental loading procedure and the global secant relaxation-based accelerated constant stiffness iteration technique are presented. It has been proved that the algorithm is efficient and reliable. It is also thought that the accuracy is high due to the adoption of equilibrium iteration. Consequently, the problems of structural collapse can be efficiently and reliably solved by this incremental/iterative numerical procedure.

In order to further reduce the computational cost, the differential quadrature finite element method (DQFEM) (Chen, 1998) proposed by the author is used to carry out the finite element discretization. This approach can reduce the arithmetic operations in calculating the static or dynamic incremental equilibrium equations.

The method of DQ defines a set of nodes in a problem domain. Then a derivative or partial derivative of a variable function at a node with respect to a coordinate is approximated as a weighted linear sum of all the function values at all nodes along that coordinate direction (Bellman and Casti, 1971). The DQ has been generalized which leads to the generic differential quadrature (GDQ) (Chen, 1998). The weighting coefficients for a grid model defined by a coordinate system having arbitrary dimension can also be generated. The configuration of a grid model can be arbitrary. In the GDQ, a certain order derivative or partial derivative of the variable function with respect to the coordinate variables at a node is expressed as the weighted linear sum of the values of function and/or its possible derivatives or partial derivatives at all nodes.

The DQ and GDQ have been extended which results in the extended differential quadrature (EDQ) (Chen, 1998). In the EDQ discretization, a discretization can be defined at a point which is not a node. The points for defining the discretizations are discrete points. In using the EDQ to FEM, the integration points are typical discrete points. A node can also be a discrete point. Then a certain order derivative or partial derivative of the variable function with respect to the coordinate variables at an arbitrary discrete point can be expressed as the weighted linear sum of the values of function and/or its possible derivatives or partial derivatives at all nodes. By using DQ, GDQ and EDQ, the DQFEM has been developed.

DQFEM DISCRETIZATION

FEM Discretization

For a deformable body under elastic-plastic deformation condition, let Δu_i denote the incremental displacement vector, the incremental form of Green-Saint Venant strain is defined as follows:

$$\Delta \varepsilon_{lm} = \frac{1}{2}(\Delta u_{l,m} + \Delta u_{m,l} + \Delta u_l \Delta u_m) \quad (1)$$

For elastic-plastic analysis, the incremental strain $\Delta \varepsilon_{lm}$ at some stage after initial yielding can be separated into elastic component $\Delta \varepsilon_{lm}^e$ and plastic component $\Delta \varepsilon_{lm}^p$

$$\Delta \varepsilon_{lm} = \Delta \varepsilon_{lm}^e + \Delta \varepsilon_{lm}^p \quad (2)$$

Introducing the elastic constitutive matrix D_{kjlm} into the generalized Hooke's law, the incremental stress $\Delta \sigma_{kj}$ can be obtained

$$\Delta \sigma_{kj} = D_{kjlm}(\Delta \varepsilon_{lm} - \Delta \varepsilon_{lm}^p) \quad (3)$$

From the theory of plasticity, it is known that the yield surface can be described by the following function relation:

$$F(\sigma_{kj}, \varepsilon_{kj}^p, \kappa) = 0. \quad (4)$$

in which κ is a strain hardening parameter. And by adopting F as the plastic potential and considering the normality rule, the increment of plastic strain during plastic deformation can be expressed as

$$\Delta\varepsilon_{kj}^p = \Delta\lambda A_{kj} \qquad (5)$$

where $\Delta\lambda$ is a scalar parameter, and

$$A_{kj} = \frac{\partial F}{\partial \sigma_{kj}}$$

Using (3) and (4), the following elastic-plastic constitutive relation can be obtained (Zienkiewicz, 1977)

$$\Delta\sigma_{kj} = D_{kjlm}^{ep} \Delta\varepsilon_{lm} \qquad (6)$$

where

$$D_{kjlm}^{ep} = (D_{kjlm} - \frac{D_{kjrs}A_{rs}A_{pq}D_{pqlm}}{A_{rs}A_{pq}D_{rspq} + b})$$

$$b = -\frac{1}{\Delta\lambda}(\frac{\partial F}{\partial \varepsilon_{kj}^p}\Delta\varepsilon_{kj}^p + \frac{\partial F}{\partial \kappa}\Delta\kappa), \quad \Delta\lambda = \frac{A_{kj}D_{kjlm}\Delta\varepsilon_{lm}}{A_{kj}D_{kjlm}A_{lm} + b}$$

For isotropic hardening materials described by work hardening hypothesis, b is equivalent to the strain hardening rate H'.

The total Lagrangian formulation is used for carrying out the finite element discretization. For load stage o, let $u_i^{o,n}(x_h)$, $\varepsilon_{lm}^{o,n}(x_h)$ and $\sigma_{kj}^{o,n}(x_h)$ denote the updated displacemen strain and stress of iteration step n, respectively, in a finite element domain Ω^e. Also let $\Delta u_i^{o,n+1}(x_h)$, $\Delta\varepsilon_{lm}^{o,n+1}(x_h)$ and $\Delta\sigma_{kj}^{o,n+1}(x_h)$ denote the incremental displacemen strain and stress of iteration step $n+1$, respectively. Then the updated displacement $u_i^{o,n+1}(x_h)$, strain $\varepsilon_{lm}^{o,n+1}(x_h)$ and stress $\sigma_{kj}^{o,n+1}(x_h)$ of iteration step $n+1$ can be expressed as the following form:

$$u_i^{o,n+1}(x_h) = u_i^{o,n}(x_h) + \Delta u_i^{o,n+1}(x_h)$$
$$\varepsilon_{lm}^{o,n+1}(x_h) = \varepsilon_{lm}^{o,n}(x_h) + \Delta\varepsilon_{lm}^{o,n+1}(x_h) \qquad (7)$$
$$\sigma_{kj}^{o,n+1}(x_h) = \sigma_{kj}^{o,n}(x_h) + \Delta\sigma_{kj}^{o,n+1}(x_h)$$

Let f_i^o denote the body force of load stage o. The equilibrium condition after updating the incremental information of iteration step $n+1$ can be expressed as the following equation:

$$[\sigma_{kj}^{o,n+1}(\delta_{ij} + u_{i,j}^{o,n+1})]_{,k} = -f_i^o \qquad (8)$$

In the above equation, the updated $u_{i,j}^{o,n}$ and $\sigma_{kj}^{o,n}$ are necessary for defining $u_{i,j}^{o,n+1}$ and $\sigma_{kj}^{o,n+1}$. Also let n_k and T_i^{o+1} denote the outward unit normal vector and traction force, respectively, on the element boundary. The traction condition is expressed by

$$T_i^{o,n+1} = n_k \sigma_{kj}^{o,n+1}(\delta_{ij} + u_{i,j}^{o,n+1})$$

Using the interpolation function, denoted as $\Psi_\alpha(x_h)$, the incremental and updated displacements in an element can be discretized as follows:

$$\Delta u_i^{o,n+1}(x_h) = \Psi_\beta(x_h)\Delta u_{i\beta}^{o,n+1}$$
$$u_i^{o,n+1}(x_h) = \Psi_\beta(x_h)u_{i\beta}^{o,n+1} \qquad (9)$$

A Galerkin procedure can be constructed by using the equilibrium condition of (8) and the interpolation function $\Psi_\alpha(x_h)$ which results in obtaining the following discretized equation:

$$([\sigma_{kj}^{o,n+1}(\delta_{ij} + u_{i,j}^{o,n+1})]_{,k}, \Psi_\alpha) = (-f_i^o, \Psi_\alpha)$$
$$- <(T_i^{o,n+1} - n_k\sigma_{kj}^{o,n+1}(\delta_{ij} + u_{i,j}^{o,n+1})), \Psi_\alpha> \qquad (10)$$

In the above equation $(.,.)$ and $<.,.>$ represent the integrations over the element and element boundary, respectively. Integrating by parts for (10) and considering the traction condition, then introducing (1), (6), (7) and (9) into the resulting equation the equilibrium equation for iteration step $n+1$ of load stage o for an element can be obtained which is expressed as the following form:

$$k_{i\alpha h\beta}^{o,n}\Delta u_{h\beta}^{o,n+1} = r_{i\alpha}^{o,n} \qquad (11)$$

where

$$k_{i\alpha h\beta}^{o,n} = (\sigma_{kj}^{o,n}\Psi_{\beta,j}\delta_{ih}, \Psi_{\alpha,k})$$
$$+ (D_{kjlm}^{ep}(\delta_{ij} + u_{i,j}^{o,n})(\delta_{lh} + u_{h,l}^{o,n})\Psi_{\beta,m}, \Psi_{\alpha,k}) \qquad (12)$$

is the tangent stiffness matrix updated at the end of iteration step n, and

$$r_{i\alpha}^{o,n} = (f_i^o, \Psi_\alpha) + <T_i^{n+1}, \Psi_\alpha>$$
$$+ <T_i^{n+1}, \Psi_\alpha> - (\sigma_{kj}^{o,n}(\delta_{ij} + u_{i,j}^{o,n}), \Psi_{\alpha,k}) \qquad (13)$$

a force vector resulted from subtracting the internal force vector updated at end of iteration step n from the force vector due to body force and traction force. It should be mentioned that the concept of weak formulation is considered when assemble the incremental element stiffness equations into the overall incremental stiffness equation. It should also be mentioned that the residual force vector can be expressed by the incremental response quantities and external forces.

For the elastic-plastic analysis a proper returning scheme must be used to set the final stress state on the yield surface if the resulting stress state defined by the elastic trial stress lies outside of the elastic region enclosed by the yield surface.

EDQ Discretization

The EDQ is used to discretize the derivatives or partial derivatives of the displacements with respect to the coordinate variables at integration points or certain other points interested in carrying out the FEM formulation. Since the displacements can be expressed by using the interpolation functions the derivatives or partial derivatives of the interpolation functions with respect to the coordinate variables can be expressed by the weighting coefficients. Thus the EDQ weighting coefficients for one element in an element group can also be used to discretize all other elements in that element group. It can reduce the CPU time required for calculating the discrete dynamic equilibrium equations.

Consider the two-dimensional elements with the displacements being approximated by using certain analytical functions of two coordinate variables. The dimensions for defining the discrete point and node can be different. By adopting a one-dimensional node identification method to express both the discrete point and node, the EDQ discretization for a $(m+n)$th order partial derivative of the displacements u_i at the discrete point L can be expressed by

$$\frac{\partial^{(m+n)}u_{iL}}{\partial x^m \partial y^n} = D_{Lr}^{x^m y^n}\tilde{u}_{ir}, \quad r = 1, 2, ..., N_D \qquad (14)$$

where N_D is the number of degrees of freedom and $\tilde{u}_{iL}$ the values of displacements and/or its there possible partial derivatives at the N_N nodes. Each of the displacements can be a set of appropriate analytical functions denoted by $\Upsilon_p(x,y)$. The substitution of $\Upsilon_p(x,y)$ in (14) for any displacement leads to a linear algebraic system for determining $D_{Lr}^{x^m y^n}$. The set of analytical functions can also be expressed by a tensor having an order other than one. The displacements can also be approximated by

$$u_i(x,y) = \Psi_p(x,y)\tilde{u}_{ip}, \quad p = 1, 2, ..., N_D \quad (15)$$

where $\tilde{u}_{ip}$ are the values of displacements and/or there possible partial derivatives at the N_N nodes, and $\Psi_p(x,y)$ their corresponding interpolation functions. Adopting the set of $\Psi_p(x,y)$ as each displacement $u_i(x,y)$, the same procedure can also be used to find $D_{Lr}^{x^m y^n}$. And the $(m+n)$th order partial differentiation of (15) at discrete point L also leads to the EDQ discretization equation (14) in which $D_{Lr}^{x^m y^n}$ is expressed by

$$D_{Lr}^{x^m y^n} = \frac{\partial^{(m+n)} \Psi_r}{\partial x^m \partial y^n}\Big|_L \quad (16)$$

The displacements can also be approximated by

$$u_i(x,y) = \Upsilon_p(x,y) c_{ip}, \quad p = 1, 2, ..., N_D \quad (17)$$

The constraint conditions at all discrete points can be expressed as

$$\tilde{u}_{iq} = \chi_{qp} c_{ip} \quad (18)$$

where χ_{qp} are formed by the values of $\Upsilon_p(x,y)$ and their possible partial derivatives at all nodes. Using (17) and (18), the weighting coefficients can also be obtained

$$D_{Lr}^{x^m y^n} = \frac{\partial^{(m+n)} \Upsilon_{\bar{q}}}{\partial x^m \partial y^n}\Big|_L \chi_{r\bar{q}}^{-1}$$

In (17), the unknown coefficients and appropriate analytical functions can also be expressed by certain other tensors having orders other than one.

It should be mentioned that the coordinate variables x and y can be either physical or natural. It should also be mentioned that the appropriate analytical functions can be formed by using certain basis functions defined by the coordinate variables independently. The basis functions can be the polynomials, sinc functions, Lagrange interpolated polynomials, Chebyshev polynomials, Bernoulli polynomials, Hermite interpolated polynomials, Euler polynomials, rational functions, ..., etc. To solve problems having singularity properties, certain singular functions can be used. The problems having infinite domains can also be treated.

EQUILIBRIUM ITERATION

A generic nonlinear equation system can be solved by the nonlinear iteration. The generalized nonlinear iteration techniques such as the nonlinear Jacobi, Gauss-Seidel, successive over-relaxation(SOR), Peaceman-Rachford iterations, ..., ect. are typical iteration methods.

The generalized-linearized methods such as the Newton, secant and Steffensen iterations are simplified nonlinear iterative methods. The nonlinear iteration can be carried out by combining a generalized-linearized method with a certain linear iterative method. By adopting a linear iterative procedure as the primary iteration and a generalized-linearized method as the secondary iteration, in the nonlinear iteration, it results in a linear-nonlinear iteration scheme. The Jacobi-, Gauss-Seidel-, SOR- and Peaceman-Rachford-, etc. Newton, secant and Steffensen iterations are typical linear-nonlinear iteration schemes.

By reversing the roles of the linear iterative method and the generalized-linearized method it leads to the composite nonlinear-linear iteration procedure with the generalized-linearized method as the primary iteration and the linear iterative method as the secondary iteration. The Newton-, secant- and Steffesen-, etc. Jacobi, Gauss-Seidel, SOR, and Peaceman-Rachford iterations are typical nonlinear-linear iteration schemes. The composite nonlinear-linear iteration can be generalized to solve the generalized-linearized system not only by the linear iterations but also by certain other solvers such as the direct methods. The quasi-Newton methods, modified Newton-Raphson methods and accelerated modified Newton-Raphson methods are also generalized-linearized iteration schemes.

Certain popular FEM nonlinear equilibrium iteration techniques are summarized (Chen, 1998). The mathematical procedures of numerical algorithms are described.

The Newton-Raphson Method

In using the Newton-Raphson method to the equilibrium iteration, the stiffness matrix K_{rs} and residual force vector have to be updated for each iteration step. The stiffness matrix can thus be constructed. Then the incremental stiffness equation for the iteration step $n+1$ of a specific load stage o can be expressed by

$$K_{rs}^{o,n} \Delta U_s^{o,n+1} = R_r^{o,n} \quad (19)$$

where $K_{rs}^{o,n}$ is the stiffness matrix, $\Delta U_s^{o,n+1}$ the incremental displacement vector and $R_r^{o,n}$ the residual force vector. By solving (19) to obtain $\Delta U_s^{o,n+1}$, the updated displacement vector of iteration step $n+1$ can be obtained

$$U_s^{o,n+1} = U_s^{o,n} + \Delta U_s^{o,n+1} \quad (20)$$

Quasi-Newton Methods

Quasi-Newton methods use finite difference approximations to the derivative operations to update a secant stiffness matrix at each iteration step. The information of two consecutive iteration steps is necessary in order to construct a secant stiffness matrix. It can overcome the possible difficulty caused by the exact evaluation of derivative operations in updating a tangent stiffness matrix to obtain a stiffness matrix for the Newton-Raphson iteration.

Broyden's method is a generalization technique of the secant method. Let $\bar{K}_{rs}^{o,n}$ denote the secant stiffness matrix of this method. $\bar{K}_{rs}^{o,n}$ can be updated by using the following equation

$$\bar{K}_{rs}^{o,n} = \bar{K}_{rs}^{o,n-1} + \frac{\left[(R_r^{o,n-1} - R_r^{o,n}) - \bar{K}_{rt}^{o,n-1}\Delta U_t^{o,n}\right]\Delta U_s^{o,n}}{\Delta U_t^{o,n} \Delta U_t^{o,n}} \quad (21)$$

Using $\bar{K}_{rs}^{o,n}$ to replace $K_{rs}^{o,n}$ in (19), the incremental displacement vector $\Delta U_s^{o,n+1}$ can be predicted by this quasi-Newton method.

Using Sherman and Morrison's matrix inversion formula, $\left(\bar{K}_{rs}^{o,n}\right)^{-1}$ can be directly updated without decomposing $\bar{K}_{rs}^{o,n}$.

This can improve the numerical efficiency significantly. Consider a nonsingular matrix A_{rs} and two vectors p_r and q_s, the inverse of $A_{rs} + p_r q_s$ is

$$(A_{rs} + p_r q_s)^{-1} = A_{rs}^{-1} - \frac{A_{rl}^{-1} p_l q_m A_{ms}^{-1}}{1 + q_l A_{lm}^{-1} p_m} \quad (22)$$

Letting $A_{rs} = \bar{K}_{rs}^{o,n-1}$, $p_r = (R_r^{o,n-1} - R_r^{o,n}) - \bar{K}_{rt}^{o,n-1} \Delta U_t^{o,n}$ and $q_s = \Delta U_s^{o,n}$, (21) together with (22) implies that

$$\begin{aligned}(\bar{K}_{rs}^{o,n})^{-1} &= \left(\bar{K}_{rs}^{o,n-1} + \frac{[(R_r^{o,n-1} - R_r^{o,n}) - \bar{K}_{rt}^{o,n-1} \Delta U_t^{o,n}] \Delta U_s^{o,n}}{\Delta U_l^{o,n} \Delta U_t^{o,n}}\right)^{-1} \\ &= (\bar{K}_{rs}^{o,n-1})^{-1} - \frac{(\bar{K}_{rl}^{o,n-1})^{-1} \{[(R_l^{o,n-1} - R_l^{o,n}) - \bar{K}_{lt}^{o,n-1} \Delta U_t^{o,n}] \Delta U_m^{o,n}/\Delta U_t^{o,n} \Delta U_t^{o,n}\} (\bar{K}_{ms}^{o,n-1})^{-1}}{1 + \Delta U_l^{o,n} (\bar{K}_{lm}^{o,n-1})^{-1} [(R_m^{o,n-1} - R_m^{o,n}) - \bar{K}_{mt}^{o,n-1} \Delta U_t^{o,n}]/\Delta U_t^{o,n} \Delta U_t^{o,n}} \\ &= (\bar{K}_{rs}^{o,n-1})^{-1} - \frac{[(\bar{K}_{rl}^{o,n-1})^{-1}(R_l^{o,n-1} - R_l^{o,n}) - \Delta U_r^{o,n}] \Delta U_m^{o,n} (\bar{K}_{ms}^{o,n-1})^{-1}}{\Delta U_l^{o,n} (\bar{K}_{lm}^{o,n-1})^{-1} (R_m^{o,n-1} - R_m^{o,n})} \\ &= (\bar{K}_{rs}^{o,n-1})^{-1} + \frac{[\Delta U_r^{o,n} - (\bar{K}_{rl}^{o,n-1})^{-1}(R_l^{o,n-1} - R_l^{o,n})] \Delta U_m^{o,n} (\bar{K}_{rs}^{o,n-1})^{-1}}{\Delta U_l^{o,n} (\bar{K}_{lm}^{o,n-1})^{-1} (R_m^{o,n-1} - R_m^{o,n})}\end{aligned} \quad (23)$$

In using this quasi-Newton method to the equilibrium iteration, the stiffness matrix of the first iteration step, for each time stage, has to be updated. The stiffness matrix can thus be formed and decomposed. Then (23) is used to update the inverse matrices of all subsequent iteration steps. This updating procedure involves only matrix multiplication at each iteration step. The condition for $(\bar{K}_{rs}^{o,n})^{-1}$ to be singular is that $\Delta U_l^{o,n}$ and $(R_m^{o,n-1} - R_m^{o,n})$ are orthogonal relative to $(\bar{K}_{lm}^{o,n-1})^{-1}$.

Modified Newton-Raphson Methods

In using the standard modified Newton-Raphson scheme to the equilibrium iteration, only the stiffness matrix of the first iteration step, for each time stage, is necessary to be updated. The stiffness matrix can thus be constructed. Letting K_{rs}^o denote the stiffness matrix of the first step of time stage o. Then the incremental stiffness equation can be expressed by

$$K_{rs}^o \Delta U_s^{o,n+1} = R_r^{o,n} \quad (24)$$

In applying (24) to the equilibrium iteration, K_{rs}^o can also be replaced by a stiffness matrix formed at a certain other iteration response stage in the incremental/iterative integration solution.

Accelerated Modified Newton-Raphson Methods

The convergence of modified Newton-Raphson methods can be improved by using certain procedures to adjust the incremental response vector obtained by the modified Newton-Raphson prediction at each iteration step. The accelerated modified Newton-Raphson schemes using a global secant relaxation (GSR) procedure, proposed by the author (Chen, 1990), are described.

Through the introduction of an implicit secant stiffness matrix, denoted as $\tilde{K}_{rs}^{o,n}$, the incremental displacement vector obtained by solving (24) can be used to construct a secant relation for hardening or softening response behaviours. This secant relation is shown to have the following form:

$$\tilde{R}_r^{o,n} = R_r^{o,n} \pm \tilde{K}_{rs}^{o,n} \Delta U_s^{o,n+1} \quad (25)$$

in which $\tilde{R}_r^{o,n}$ is a residual force vector after the modified Newton-Raphson prediction. The GSR method uses an accelerator defined by minimizing certainly defined system error to scale the incremental displacement vector. Let $\omega^{o,n+1}$ denote this accelerator, the updated displacement vector after acceleration can be expressed as

$$U_s^{o,n+1} = U_s^{o,n} + \omega^{o,n+1} \Delta U_s^{o,n+1} \quad (26)$$

and a residual force vector after acceleration can be expressed by the following form:

$$\bar{R}_r^{o,n+1} = R_r^{o,n} \pm \omega^{o,n+1} \tilde{K}_{rs}^{o,n} \Delta U_s^{o,n+1} \quad (27)$$

This residual force vector provides important information for defining the system error.

In defining the system error, consistency and reliability have to be considered, in the sense that the defined error must be able to better reflect the true error existing in the system. The Euclidean norm of $\bar{R}_r^{o,n+1}$ is a good quantity for representing the system error. This quantity can be expressed as

$$E_f^{o,n+1} = \bar{R}_r^{o,n+1} \bar{R}_r^{o,n+1} \quad (28)$$

By using (25) and (27) in (28), then minimizing $E_f^{o,n+1}$ with respect to $\omega^{o,n+1}$, an accelerator denoted as $\omega_f^{o,n+1}$ can be obtained which shows to have the following form:

$$\omega_f^{o,n+1} = \mp \frac{R_r^{o,n}(R_r^{o,n} - \tilde{R}_r^{o,n})}{(R_s^{o,n} - \tilde{R}_s^{o,n})(R_s^{o,n} - \tilde{R}_s^{o,n})} \quad (29)$$

The related acceleration scheme is GSR-MR. By using a diagonal matrix with all diagonal elements having the same value to replace the implicit secant stiffness matrix, (29) will result in the formula representing the Generalized Aitken accelerator proposed by Cahill (1992).

It is worth mentioning that no mathematical approximation is involved in deriving the accelerator $\omega_f^{o,n+1}$. Therefore, in considering the exact description of the secant relation of (25) and the direction of the incremental response vector, the resulting iterative procedure is believed to be a highly consistent secant improvement-based iteration scheme.

The energy norm defined as the inner product of residual

force vector $\bar{R}_r^{o,n+1}$ and the incremental displacement vector caused by $\bar{R}_r^{o,n+1}$, along the linear deformation surface is also used to evaluate the system error. This error can be expressed as

$$E_e^{o,n+1} = \bar{R}_r^{o,n+1}(\tilde{K}_{rs}^{o,n})^{-1}\bar{R}_s^{o,n+1} \quad (30)$$

By using the same procedures as those used in defining $\omega_f^{o,n+1}$, another accelerator denoted as $\omega_e^{o,n+1}$ can be obtained:

$$\omega_e^{o,n+1} = \mp \frac{\Delta U_r^{o,n+1} R_r^{o,n}}{\Delta U_s^{o,n+1}(R_s^{o,n} - \tilde{R}_s^{o,n})} \quad (31)$$

The improvement scheme using $\omega_e^{o,n+1}$ as the accelerator is GSR-MW.

It should be noted that the accelerator $\omega_e^{o,n+1}$ is shown to have the same form as the single-parameter accelerator proposed by Crisfield (1984) though the fundamental concepts of acceleration, the mathematical formulations and the resulting iterative algorithms are actually different. In investigating the formulation procedures of these two accelerators, it is believed that this similarity is caused by the fact that both of these two approaches use the secant relation to approximately describe the deformation behaviours of the two states used to construct the secant relation.

It is also valuable to investigate the difference of convergency performances between GSR-MR and GSR-MW theoretically. As already mentioned previously, the mathematical formulation of defining $\omega_f^{o,n+1}$ is absolutely consistent. Therefore, the resulting iterative scheme is believed to be highly reliable and the convergency performance should be good. On the other hand, inconsistency does exist in the formulation procedure in defining $\omega_e^{o,n+1}$ since a mathematical approximation of using $\tilde{K}_{rs}^{o,n}$ to predict the incremental response vector is used to construct the energy norm-based system error. This approximation will result in obtaining a less reliable evaluation of the system error, which will lead to a less reliable acceleration scheme with less promising convergency performance. And the inconsistency will be even severe for solving non-linear finite element problems if large time increments are used.

The accelerated constant stiffness iteration is one of the accelerated modified Newton-Raphson methods. This scheme uses the linear elastic stiffness matrix to predict the initial incremental displacement vector for all incremental/iterative steps.

The linear equation systems existing in the generalized-linearized DQFEM nonlinear iterations can be solved by using a certain direct or iterative solver. The most commonly used direct solvers are Gauss elimination, Cholesky decomposition and frontal method. Various techniques including the sparse implementation strategies, the domain decomposition and the parallel implementation was cosidered in implementing an efficient direct solution procedure into a DQFEM computer program. There are also many iterative solvers that can be used to solve a linear equation system. Among the indirect solvers the preconditioned conjugate gradient (PCG) methods have been attracting lots of the finite element programmers. In solving large linear equation systems, the PCG methods can offer promising performances due to the substantial reductions in computer memory requirements and the function of taking the advantage of vector and parallel processing strategies on computers that support these features. The iterative solvers possess a relatively high degree of natural concurrency, with the predominant operations in PCG algorithms being saxpy operations, inner products and matrix-vector multiplications. Among the PCG algorithms, the stabilized and accelerated version of the biconjugate gradient method, which is an extension of the conjugate gradient method to nonsymmetric systems, is one of the most commonly used iterative solvers. The element-by-element solution procedure is also a useful algorithm which has considerable operation count and I/O advantages since no overall stiffness matrix is needed to be formed.

Explicit Predictor-Corrector Iteration

The author has also proposed a diagonal stiffness-based predictor-corrector procedure for iteratively solving linear or nonlinear finite element equation systems (Chen, 1995). It is an explicit iteration procedure in the nonlinear iteration. Instead of using an assembled overall stiffness matrix, this method only uses the diagonal elements of the overall stiffness matrix to predict the incremental displacement vector in carrying out the iterative solution. Consequently, only the diagonal elements of the element stiffness matrices are needed to be calculated. Thus the computer memory requirement can be minimized. Let $K_s^{o,n}$ denote the vector representing the set of the diagonal elements in the stiffness matrix $K_{rs}^{o,n+1}$. Then, by using $K_s^{o,n}$ and referring (19) the following equation can be constructed.

$$K_{(s)}^{o,n}\Delta U_s^{o,n+1} = R_s^{o,n} \quad (32)$$

Equations (37) and (31) represent the explicit predictor-corrector equilibrium iteration procedure. The discrete equation system can also be solved without updating the vector representing the set of the diagonal stiffness elements for each iteration step. Then, by modifying (29) the following equation can be constructed

$$K_{(s)}^o\Delta U_s^{o,n+1} = R_s^{o,n} \quad (33)$$

Equations (33) and (26) represent a different version of the explicit predictor-corrector equilibrium iteration procedure.

For solving a linear equation system such as the incremental equation of the step-by-step procedure, which might adopt the unbalance load correction instead of adopting the equilibrium iteration, the overall stiffness can be formed and used to calculate the residual force vectors required for the predictor-corrector iteration. However, it needs to save the overall stiffness matrix in the computer memory unit. Let ΔF_r^o denote the incremental load vector. The incremental equation of the step-by-step procedure for load stage o is expressed by

$$K_{rs}^o \Delta U_s^o = \Delta F_r^o \quad (34)$$

Then the residual force vectors used to define the scaling factor can be calculated by

$$R_r^{o,n} = \Delta F_r^o - K_{rs}^o \sum_{k=1}^{n} \Delta U_s^{o,k} \quad (35)$$

$$\tilde{R}_r^{o,n} = R_r^{o,n} - K_{rs}^o \Delta U_s^{o,n+1} \quad (36)$$

The predictor-corrector iterative procedure needs less computer storage space. It is also suitable for vector and parallel implementation. In applying this iterative procedure to solve a generic equation system, the amplification of longer period errors can be prevented by the introduction of a GSR correction which contributes to greater numerical stability. Thus all longer and shorter period errors can be effectively eliminated by this predictor-corrector solution. Numerical results have proved that this iterative procedure has good numerical stability. It is also an efficient algorithm. This iterative procedure can also be used to the multi-grid solution in which the longer period errors, which may not be efficiently eliminated by the fine grid iteration, can be substantially reduced by the use of a coarse grid correction. The coarse grid can adopt either iterative or direct procedure.

In implementing the DQFEM analysis program, various phases including preprocessing, calculation of elemental discrete equations, incorporation of boundary conditions, solution of system equations and postprocessing can be parallelized. However, the assembly of elemental discrete equations can not take the advantage of parallel operation efficiently.

NUMERICAL RESULTS

The first sample problem been solved involves the static elastic-plastic analysis of a square plate with a square cutout, subjected to uniformly distributed axial load. The deformation is assumed to be linear. The description of the model problem can be seen in Fig. 1 in which the thickness of the plate is $1mm$. Elastic-perfectly plastic material with Young's modulus E being equal to $21000kg/mm^2$, yield stress $\sigma_Y = 36.3kg/mm^2$ and Poisson's ratio $\nu = 0.3$, was considered. The modelling considers the symmmetry property. Eight-node quadratic element adopting 3×3 quadrature rule was used to discretize one quarter of the plate domain. The mesh is shown in Fig. 2. The accelerated constant stiffness iteration was used to efficiently and reliably carry out the iterative computation in which Tresca's yield criterion was used to detect the plastification of integration points. The constant stiffness matrix was formed at the beginning of the incremental/iterative solution. The convergence indicator is defined as the ratio of residual norm to the norm of load vector, which was selected to be 10^{-6}. The response history was updated up to a near collapse load stage under which the structure showed globally unstable behavior due to the entire loss of elastic restoring capability caused by the formation of an unstable failure mechanism. The load-displacement curve with the displacement at point A for the plate model with $a/W = 0.5$ is shown in Fig. 3 in which the predicted value of the collapse load is $0.38\sigma_Y$. Convergency tests were carried out by using a single larger load increment of $0.3\sigma_Y$ to perform the equilibrium iteration. The convergence indicator is defined as the ratio of residual norm to the norm of load vector. The convergency curves of three different iteration algorithms were plotted. They can be seen in Fig. 4. It shows that the GSR acceleration has good numerical stability and convergency performance though the load increment is large. It also shows that GSR-MR performs the best. The collapse load analysis was also carried out for the model problems with $a/W = 0.15, 0.3$ and 0.8, respectively. The predicted collapse loads for the four model analyses were plotted in Fig. 5. They were compared with the results of Hodge's bound solutions (Hodge, 1959). It shows that the results of DQFEM solutions tend to be close to the upper bound of the results obtained by classical limit strength analysis, which reflects the engineering reality inherent in the DQFEM discretization of using displacement model.

The second problem been solved involves the elastic-plastic postbuckling analysis of a rectangular girder framework which can be seen in Fig. 6. Different structural models, which possess different plastic buckling behaviors, can be obtained by changing the material constants for horizontal or vertical members. Two dimensional plane stress theory was used to describe the mechanics behavior of the structure. Due to symmetry, one quarter of the structural domain was adopted for the finite element analysis. Eight-node quadratic finite element model adopting the standard 3×3 quadrature rule was used to discretize the analyzing domain. The finite element mesh is shown in Fig. 7. A uniformly distributed forced displacement is applied downward on line A-B. In detecting the plastification of integration points, Tresca's yield criterion was used. And the incremental/iterative procedure using the GSR-based accelerated constant stiffness iteration scheme was used to update the response history. Well converged results were used to study the plastic buckling behavior of the structure. Three model problems were analyzed. The results of response axial forces of the vertical members were plotted. They can be seen in Fig. 8. It shows that the model 1 problem, in which both horizontal and vertical members are elastic-perfectly plastic materials with significant difference of material stiffnesses, possess a rather bifurcation type buckling behavior with which the buckling stress is very close to the yield stress. By using the same material properties for all structural members, a model 2 problem can be obtained. This model 2 problem shows limit point buckling behavior with which the buckling stress just below the yield level. The results reflect the effects of two dimensional mechanics behavior to the two different buckling models. The two-dimensional effects are even significant for model 2 problem. By considering that the structural material has a constant strain hardening, a model 3 problem can be obtained. The results of plastic buckling analysis for this model problem are also shown in Fig. 8.

CONCLUSIONS

Efficient and reliable static elastic-plastic finite element analysis is an important topic in the field of scientific and engineering computation. This paper integrates the DQFEM and accelerated equilibrium iteration techniques proposed by the author which results in obtaining excellent algorithms for analyzing the static nonlinear structural or continuum mechanics problems. Sample collapse analyses of an elastic-plastic collapse problem with linear deformation and an elastic-plastic postbuckling problem by using the approach adopting the accelerated constant stiffness iteration technique proved that the integrated numerical procedure is efficient and reliable.

REFERENCES

Bellman, R.E. and Casti, J., 1971, "Differential Quadrature and Long-term Integration," J. Math. Anal. Appl., Vol. 34, pp. 235-238.

Cahill, E., 1992, "Acceleration Techniques for Functional Iteration of Non-linear Equations", IMACS Conference on Mathematical Modelling and Scientific Computing, Bangalove, INDIA.

Chen, C.N., 1990, "Improved Constant Stiffness Algorithms for the Finite Element Analysis", Proc. NUMETA 90, pp. 623-628, Swansea, UK.

Chen, C.N., 1995, "A Global Secant Relaxation (GSR) Method-Based Predictor-Corrector Procedure for the Iterative Solution of Finite Element Systems", Comput. Struct., Vol. 54, pp. 199-205.

Chen, C.N., 1998, The Development of Numerical Solution Techniques Based on the Differential Quadrature, A. P. Publications, 1998.

Chen, C.N., 1998, "Newton-Raphson Techniques in Finite Element Methods for Non-linear Structural Problems", a chapter in GORDON and BREACH International Series in Engineering, Technology and Applied Science, Volumes on Structural Dynamic Systems Computational Techniques and Optimization (ed. C. T. Leondes), Gordon and Breach.

Crisfield, M.A., 1984, "Accelerated and Damping the Modified Newton-Raphson Method", Comput. Struct., Vol. 18, pp. 267-278.

Hodge, P.G., 1959, "Plastic Analysis of Structures", McGraw-Hill, New York.

Zienkiewicz, O.C., 1977, The Finite Element Method, McGraw-Hill, New York.

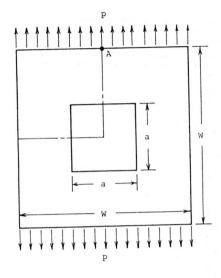

Fig. 1. A square plate with a square cutout.

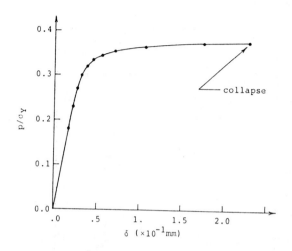

Fig. 3. Load-displacement curve of point A for $a/W = 0.5$.

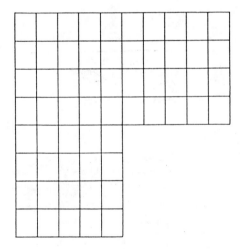

Fig. 2. The mesh.

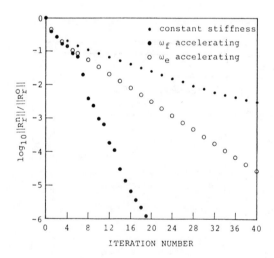

Fig. 4. The convergence of equilibrium iteration. (load increment $\Delta p^1 = 0.3\sigma_Y$)

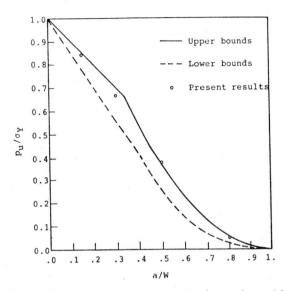

Fig. 5. Ultimate collapse loads of the square plate with a square cutout.

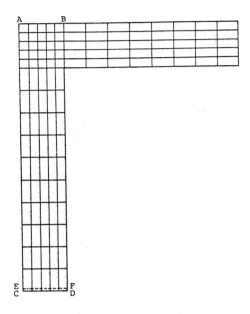

Fig. 7. The mesh.

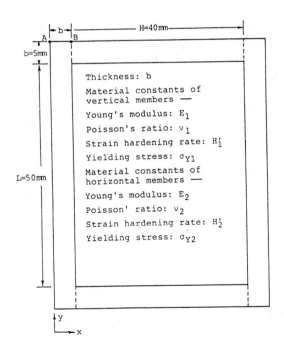

Fig. 6. A rectangular girder framework.

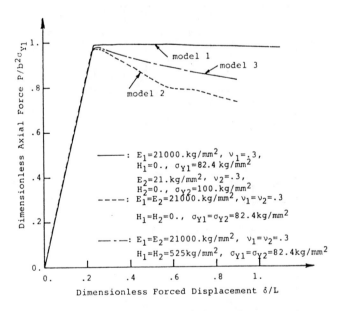

Fig. 8. Response history curves of the vertical members.

152

PIPING FRACTURE MECHANICS

Introduction

G. M. Wilkowski
Engineering Mechanics Corporation of Columbus
4181 Maystar Way
Hilliard, OH 43026

Piping fracture mechanics typically involves two basic types of analyses. One is when there is an actual flaw to evaluate, called flaw evaluation procedures for nuclear piping and engineering critical assessments (ECA) in the oil and gas industry. The second type of flaw analysis is for a hypothetical flaw for design purposes. Leak-before-break analysis in the nuclear and other industries typically is such an assessment. Of the papers in this session, the paper by Han et al. on "Limit Analysis for Pressurized Pipelines with Local Wall-Thinning" deals with the practical problem of flaw evaluation for either nuclear, oil, gas, or petrochemical plant piping. The papers by Smith on "The Opening Area Associated with a Crack That Is Subjected to a General Tensile Stress Distribution" and Miura on "Consideration of Internal Pressure on Pipe Flaw Evaluation Criteria Based in a New J-Estimation Scheme for Combined Loading" deal with leak-before-break aspects. These papers have more direct application to the nuclear industry.

Consideration of Internal Pressure on Pipe Flaw Evaluation Criteria Based on a New *J*-Estimation Scheme for Combined Loading

Naoki Miura
Materials Science Department
Central Research Institute of Electric Power Industry
Tokyo, Japan

Yeon-Ki Chung
Engineering Research Department, Regulatory Research Division
Korea Institute of Nuclear Safety
Taejon, Korea

ABSTRACT

Considering a rational maintenance rule of Light Water Reactor piping, reliable flaw evaluation criteria are essential to determine how a detected flaw is detrimental to continuous plant operation. Ductile fracture is one of the dominant failure modes to be considered for carbon steel piping, and can be analyzed by elastic-plastic fracture mechanics. Currently analytical results are provided as the flaw evaluation criteria using the load-correction factors such like the Z-factor in the ASME Code Section XI. The present correction factors were conventionally determined taken conservatism and simplicity into account, however, the effect of internal pressure, which would be an important factor under actual plant condition, was not adequately considered. Recently, a *J*-estimation scheme, "LBB.ENGC" method for ductile fracture analysis of circumferentially through-wall-cracked pipes subjected to combined loading was newly developed to have a better prediction with more realistic manner. This method is explicitly incorporated the contribution of both bending and tension due to internal pressure by means of the scheme compatible with an arbitrary combined loading history. In this paper, the effect of internal pressure on the flaw evaluation criteria for carbon steel pipes was investigated using the new *J*-estimation scheme. A correction factor based on the new *J*-estimation scheme was compared with the present correction factors, and the predictability of the current flaw evaluation criteria was quantitatively evaluated in consideration of internal pressure.

INTRODUCTION

To substantiate reliability of aged nuclear power plants, the importance of a rational maintenance rule has been widely recognized. The ASME Boiler and Pressure Code Section XI has an initiative in the maintenance code for Light Water Reactor (LWR) components, and some efforts to establish maintenance standards have been made in Japan in consideration with specific condition, database, and methodology (Saikawa et al., 1995, Imamura et al., 1995, Iida et al., 1996, and Iida, 1996). One of the key issues in the maintenance codes is a flaw evaluation method, which is used to determine the acceptability of flawed components for continued service. Since ductile fracture is one of the dominant failure modes to be considered for LWR piping, the fracture behavior of a flawed pipe is analyzed by elastic-plastic fracture mechanics. Pipe flaw evaluation criteria are provided using load-correction factors such like the Z-factor in the ASME Code Section XI to predict fracture loads.

When actual plant piping conditions are considered, the effect of the internal pressure acting as a tensile load occurs. The combination of bending load typical of a seismic event and tension loading due to an internal pressure is then considered to be a basic mode of pipe fracture analysis. The present load-correction factors were conventionally determined using *J*-estimation schemes taken conservatism and simplicity into account by some means. However, the effect of internal pressure could not be adequately considered because the theoretical investigation on the methodology to evaluate ductile fracture behavior under combined loading was insufficient. Recently, a *J*-estimation scheme, "LBB.ENGC" method to evaluate ductile fracture behavior of circumferentially through-wall-cracked pipes subjected to combined loading was newly developed (Miura, 1998 and Miura, 1999). This method is explicitly incorporated the contribution of both bending and tension. The method was applied to full-scale pipe fracture tests, and the ability to investigate the effect of internal pressure on pipe fracture was validated.

In this study, the load-correction factor for carbon steel pipes taken combined loading into account is newly derived using the LBB.ENGC method. The effect of internal pressure on the load-correction factor is quantitatively investigated. The derived load-correction factor is then compared with the present load-correction factors, and the predictability of the current flaw evaluation criteria is evaluated in consideration of internal pressure. It is noted that circumferential through-wall cracks are particularly focused to investigate the effect of internal pressure under the same situation as the development of the present load-correction factors.

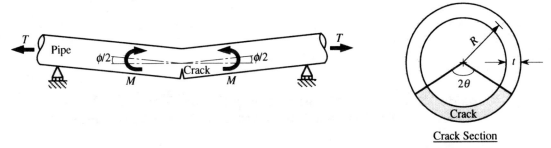

Figure 1 Circumferentially Through-Wall-Cracked Pipe Subjected to Combined Bending and Tension

DERIVATION OF PIPE FLAW EVALUATION CRITERIA

Significance of Load-Correction Factor

Consider a circumferentially through-wall-cracked pipe subjected to combined bending and tension due to internal pressure shown in Fig. 1. Two different fracture modes are to be considered for the cracked pipes made of actual structural materials. When pipe material shows relatively higher J-Resistance, the mode of pipe fracture is the plastic collapse provided by the Net-Section Collapse criterion. Here dominant fracture parameter is the net section stress, and fracture load can be evaluated by the limit load analysis. On the other hand, if material J-Resistance may become lower than driving force for crack extension during loading, pipe fracture occurs after a certain amount of stable crack extension. This type of fracture is more probable for carbon steel pipes. Then the mode of pipe fracture is the ductile fracture controlled by the J/T criterion. Dominant fracture parameter is the J-integral, and fracture load can be predicted by elastic-plastic fracture analysis. Table 1 shows the relation between the modes of fracture and the related fracture criteria.

The load-correction factor, Z, is defined as the ratio of the Net-Section Collapse load to the maximum load evaluated based on the elastic-plastic fracture analysis. When Z is larger than unity, ductile fracture is supposed to occur, and Z indicates the degree of the reduction of the load-carrying capacity from the Net-Section Collapse load. Once Z is identified, the fracture load corresponding to the ductile fracture can be easily evaluated as the Net-Section Collapse load divided by Z. Alternatively, an allowable crack size for a specific applied load under plastic collapse condition can be directly converted into the same allowable crack size for the applied load multiplied by Z under ductile fracture condition.

The concept of Z was first introduced in the ASME Code Section XI, Articles IWB-3640, IWB-3650, Appendices C and H. Z was analyzed using the modified GE/EPRI estimation scheme (Zahoor, 1987) for circumferentially through-wall-cracked pipes subjected to pure bending, and was formulated as a function of the nominal diameter, OD (in inches), and the radius-to-thickness ratio, R/t, for ferritic/austenitic base/weld metal (Section XI Task Group, 1986 and Novetech Corporation, 1988). In case of carbon steel base metal, Z in the ASME Code, Z_{ASME}, is,

$$Z_{ASME} = 1.20\left[1 + 0.021\,A\,(OD - 4)\right] \quad (1)$$

$$A = \begin{array}{l}[0.125\,(R/t) - 0.25]^{0.25},\ \text{for}\ 5 \le R/t \le 10 \\ [0.4\,(R/t) - 3.0]^{0.25},\ \text{for}\ 10 < R/t \le 20\end{array} \quad (2)$$

Another load-correction factor called G-factor, G, was independently developed (NUPEC, 1989 and Asada et al., 1990). G was prepared using the R6 procedure option 2 (Milne, 1986), and could be applied to Japanese carbon steels.

$$G = \begin{array}{l}1,\ \text{for}\ 2 \le OD < 6 \\ [0.692 - 0.0115\,OD] + [0.188 + 0.0104\,OD]\log_{10}(\theta_{max}) \\ \qquad,\ \text{for}\ 6 \le OD \le 30\end{array} \quad (3)$$

where θ_{max} is the half crack angle at the maximum load (in degrees) and is tabulated as a function of OD and the initial half crack angle, θ_0. The Net-Section Collapse load is to be calculated using θ_{max}.

The Japan Power Engineering and Inspection Corporation (JAPEIC) improved the above G-factor to obtain the simple expression of the load-correction factor, Z_{JAPEIC}, as a part of the draft of the Japanese pipe flaw evaluation criteria (Iida et al., 1996). The Z_{JAPEIC} for carbon steel is tentatively,

$$Z_{JAPEIC} = 0.2885\,\log_{10} OD + 0.9573 \quad (4)$$

Figure 2 shows the comparison of present load-correction factors. Z_{ASME} gives generally higher value than the others. Z_{JAPEIC} is more conservative than G because of the exclusion of the θ_{max} dependence as well as the different material properties used for analyses. (Material properties used to derive Z_{JAPEIC} are not published at this time.)

Table 1 Relation between Fracture Modes and Fracture Criteria

Fracture Mode	Plastic Collapse	Ductile Fracture
Fracture Criterion	Net-Section Collapse Criterion	J/T Criterion
Dominant Parameter	Net-Section Stress	J-Integral
Analysis Method	Limit Load Analysis	Elastic-Plastic Fracture Analysis

Load-Correction Factor Taken Account of Combined Load

In the derivation of the load-correction factors, the effect of combined loading due to internal pressure was not adequately considered. On the development of Z_{ASME}, the effect of internal pressure was taken no account. G was developed under a fixed internal pressure condition, however, internal pressure was not regarded as a factor affecting the load-correction factor. The reason of these was that there were no appropriate methodologies to evaluate ductile fracture behavior under combined loading.

Recently, a J-estimation scheme, LBB.ENGC method to evaluate the ductile fracture behavior of circumferentially through-wall-cracked pipe subjected to combined loading was newly developed (Miura, 1998 and Miura 1999). It was ascertained that the method could well predict ductile fracture behavior under combined loading through the comparison with full-scale pipe fracture tests. Consequently, it is expected that using the LBB.ENGC method can derive a reliable load-correction factor taken account of combined load.

The load-correction factor taken account of combined load, Z_C, is defined as,

$$Z_C = \text{Max}\left[1, M_{NSC} / M_{max}\right] \tag{5}$$

M_{NSC} in the above equation is the Net-Section Collapse moment given by,

$$M_{NSC} = 4 \sigma_f R^2 t \, h_p(\theta, s) \tag{6}$$

$$h_p(\theta, s) = \cos\left(\frac{\theta}{2} + \frac{\pi}{2} s\right) - \frac{1}{2} \sin \theta \tag{7}$$

$$s = \sigma_t / \sigma_f \tag{8}$$

where σ_f is the flow stress, R is the mean radius, t is the thickness, θ is the half crack angle, s is the nondimensional tensile stress, and σ_t is the tensile stress due to internal pressure. While M_{max} is the maximum moment predicted by the LBB.ENGC method.

LBB.ENGC Method

In the LBB.ENGC method, a circumferentially through-wall-cracked pipe subjected to bending moment, M, and tensile load, T, shown in Fig. 1 is considered. The beam theory is enforced to the pipe of nonlinear elastic material under the deformation theory of plasticity. Stress distribution can be determined so as to be in equilibrium with the applied M and T. The existence of circumferential through-wall crack is replaced by a pipe section with reduced thickness at the cracked section. The reduced thickness section is expected to simulate the reduction of the compliance due to the existence of the crack. Concerning J-integral, the Combined-Load η-Factor Solution (Miura et al., 1998) is applied to take an arbitrary combined loading condition into account.

The summary of the evaluation procedure by the LBB.ENGC method is as follows. The detailed technical basis is presented in the references (Miura, 1998 and Miura, 1999).

(Total rotation due to crack, ϕ)

$$\phi = \phi_{EL} + \phi_{PL} \tag{9}$$

where ϕ_{EL} and ϕ_{PL} are the elastic and plastic components of ϕ.

(Elastic rotation due to crack, ϕ_{EL})

$$\phi_{EL} = \sigma_b I_b(\theta) / E + \sigma_t I_t(\theta) / E \tag{10}$$

$$\sigma_b \approx \frac{M}{\pi R^2 t}, \quad \sigma_t = \frac{T}{2 \pi R t} \tag{11}$$

$$I_b(\theta) = 4 \int_0^\theta \theta F_b^2(\theta) \, d\theta \tag{12}$$

$$I_t(\theta) = 4 \int_0^\theta \theta F_b(\theta) F_t(\theta) \, d\theta \tag{13}$$

where σ_b and σ_t are the bending and tensile stresses, respectively, E is the Young's modulus, F_b and F_t are the compliance functions of stress intensity factors for bending and tension provided by Klecker (1986), respectively.

(Plastic rotation due to crack, ϕ_{PL})

for $M \neq 0$,

$$\phi_{PL} = \left(\frac{t}{t_e}\right)^{n-1} \left(\frac{\pi}{4 \widehat{K}_M}\right)^n \alpha \frac{\kappa^n}{\kappa_1} \left(\frac{\sigma_b}{\sigma_0}\right)^{n-1} \phi_{EL} \tag{14}$$

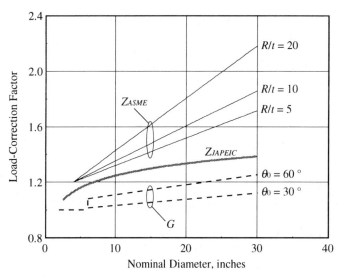

Figure 2 Present Load-Correction Factors

$$\kappa = \frac{1}{2}\left[1 + \sqrt{1 + \frac{16 \widehat{K}_M (\widehat{K}_T g_T - \widehat{K}_M g_M)(\frac{\sigma_t}{\sigma_b})^2}{\widehat{K}_T^2 g_T^2}}\right] \quad (15)$$

for $M = 0$,

$$\phi_{PL} = (\frac{t}{t_e})^{n-1} (\frac{\pi}{2\widehat{K}_M})^n \alpha \frac{\kappa^n}{2\kappa_1} (\frac{\sigma_t}{\sigma_0})^{n-1} \phi_{EL} \quad (16)$$

$$\kappa = \frac{\sqrt{\widehat{K}_M (\widehat{K}_T g_T - \widehat{K}_M g_M)}}{\widehat{K}_T g_T} \quad (17)$$

$$\kappa_1 = \kappa (n=1) \quad (18)$$

$$\widehat{K}_M = \frac{\sqrt{\pi}}{2} \Gamma(\frac{1}{2n}+1) / \Gamma(\frac{1}{2n}+\frac{3}{2}) \quad (19)$$

$$\widehat{K}_T = \frac{\sqrt{\pi}}{2} \Gamma(\frac{1}{2n}+\frac{1}{2}) / \Gamma(\frac{1}{2n}+1) \quad (20)$$

$$g_M = \frac{n+1}{n} \frac{2}{\pi} \quad (21)$$

$$g_T = \frac{1}{n} + \frac{2}{\pi} \quad (22)$$

$$\frac{t}{t_e} = \begin{cases} \left[1 - (1-\frac{\pi}{4})\sin 2\theta\right] \cos \frac{\pi}{2} \cdot \frac{s}{h_p(s=0)}, & \text{for } \theta < \frac{\pi}{4} \\ \frac{\pi}{4} \cos \frac{\pi}{2} \cdot \frac{s}{h_p(s=0)}, & \text{for } \theta \geq \frac{\pi}{4} \end{cases} \quad (23)$$

where Γ is the Gamma function, t/t_e is the ratio of the thickness to the reduced thickness, and σ_0, α, and n are the reference stress, material constant, and hardening exponent, respectively, which are related to the following Ramberg-Osgood relation,

$$\frac{\varepsilon}{\varepsilon_0} = \frac{\sigma}{\sigma_0} + \alpha (\frac{\sigma}{\sigma_0})^n, \quad \varepsilon_0 = \frac{\sigma_0}{E} \quad (24)$$

(Total J-integral, J)

$$J = J_{EL} + J_{PL} \quad (25)$$

where J_{EL} and J_{PL} are the elastic and plastic components of J.

(Elastic J-integral, J_{EL})

$$J_{EL} = K_I^2 / E \quad (26)$$

where K_I is the stress intensity factor by Klecker (1986).

(Plastic J-integral, J_{PL})

$$J_{PL} = \int_0^{\phi_{PL}} \eta_p M \, d\phi_{PL} + 2\int_{\theta_0}^{\theta} \gamma_p J_{PL} \, d\theta + 2\int_{s_0}^{s} \mu_p J_{PL} \, ds \quad (27)$$

$$\eta_p = -\frac{\partial h_p / \partial \theta}{2 R t h_p} \quad (28)$$

$$\gamma_p = \frac{\partial^2 h_p / \partial \theta^2}{2 \partial h_p / \partial \theta} \quad (29)$$

$$\mu_p = \frac{\partial^2 h_p / \partial s \partial \theta}{2 \partial h_p / \partial \theta} \quad (30)$$

where s_0 is the initial s and h_p is given in Equation (7).

By integrating the above equations together with the J-R curve as a criterion of crack extension, ductile fracture behavior can be analyzed in consideration of both combined loading and growing crack. The LBB.ENGC method has ability to consider an arbitrary combined loading history by nature. Pressure induced bending is also taken into account through the equilibrium of the stress distribution at the crack section. In this study, constant tensile load due to internal pressure followed by external monotonic bending load is supposed.

ANALYSIS OF LOAD-CORRECTION FACTOR

Condition of Analysis

The LBB.ENGC method was applied to analyze the load-correction factor taken account of combined loading, Z_C.

Material properties used for the analysis were determined based on the database of Japanese carbon steel pipes. They were nearly the same as those employed to develop the G-factor (NUPEC, 1989 and Asada et al., 1990). The stress-strain relation came from STS480 carbon steel at 300°C, which gave the most conservative failure assessment curve in

Table 2 Material Properties Used for Analysis

Young's Modulus	196 GPa
Flow Stress	389 MPa
Stress-Strain Relation	$\frac{\varepsilon}{\varepsilon_0} = \frac{\sigma}{\sigma_0} + \alpha(\frac{\sigma}{\sigma_0})^n$, $\varepsilon_0 = \frac{\sigma_0}{E}$
σ_0	244 MPa
α	4.412
n	3.389
J-R Curve	$J = J_{in} + C(\Delta a)^m$, Δa in [mm]
J_{in}	0.1187 MN/m
C	136.3
m	0.9234
Design Stress Intensity	139 MPa

Table 3 Matrix of Analysis

Nominal OD	Outer Diameter	Nominal Thickness	Thickness	Initial Half Crack Angle	Membrane Stress Design Stress Intensity	Item to Be Investigated
2.5 B	76.3 mm	Sch. 80	7.0 mm	5, 10, 15, ···, 90 °	0.00, 0.34, 0.50	Diameter Dependence & Crack Angle Dependence
4 B	114.3 mm		8.6 mm			
6 B	165.2 mm		11.0 mm			
10 B	267.4 mm		15.1 mm			
16 B	406.4 mm		21.4 mm			
20 B	508.0 mm		26.2 mm			
26 B	660.4 mm		34.0 mm			
30 B	762.0 mm	-	33.0 mm			
26 B	660.4 mm	Sch. 80	34.0 mm	60 °	0.00, 0.34, 0.50	Thickness Dependence
		Sch. 100	41.6 mm			
		Sch. 120	49.1 mm			
		Sch. 140	56.6 mm			
		Sch. 160	64.2 mm			
6 B	165.2 mm	Sch. 80	11.0 mm	60 °	0.00, 0.05, 0.10, ···, 0.50	Pressure Dependence
16 B	406.4 mm		21.4 mm			
26 B	660.4 mm		34.0 mm			

the R6 procedure option 2. While the J-R curve was obtained from that of STS410 carbon steel at 300°C, which was the lowest J-R curve. It is noted that the yield stress and the flow stress used for the development of the G-factor were design values. However, the confusing use of the experimental stress-strain relation and the design tensile properties may prevent the essential expression of deformation behavior. Therefore, the experimental yield and flow stresses were used in this analysis. Table 2 shows the material properties used for the analysis. The flow stress was defined as the average of the yield stress and the ultimate strength.

Table 3 shows the matrix of the analysis. θ_0, OD, the nominal thickness (or the radius-to-thickness ratio, R/t), and the normalized membrane stress P_m/S_m (where P_m is the membrane stress due to internal pressure and S_m is the design stress intensity) were considered as the parameters to investigate the effect on Z_C. The range of θ_0 was 5 to 90°, and the range of OD was 2.5 to 30 inches. The nominal thickness was almost Schedule 80 and partially up to Schedule 160. The thickness of 30-inch diameter pipe was determined based on an actual configuration since it was not specified. P_m/S_m was set to 0 (no internal pressure), 0.34 (maximum value obtained in actual plant survey), and 0.5 (maximum allowable value under normal operating condition). More detailed division was separately applied to examine the effect of P_m/S_m. The number of the total analysis was 472.

Results of Analysis

Figures 3(a) and (b) show the typical predicted relations between bending moment and rotation for 6 and 26-inch diameter pipes, respectively. The bending moment is normalized by the Net-Section Collapse moment for uncracked pipe under pure bending, $M_{UC} = 4\sigma_f R^2 t$. These relations have similar configuration, and the maximum moment decreases with increasing P_m/S_m. For the case of $P_m/S_m = 0.50$, the maximum moment is about 13% lower at $\theta_0 = 30°$ and about 31% lower

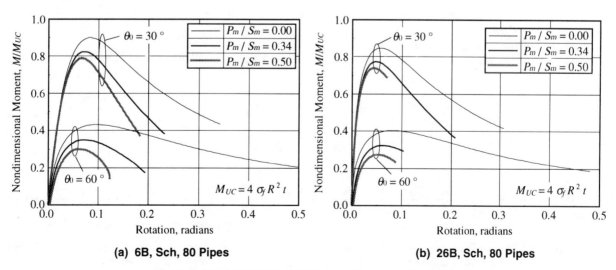

(a) 6B, Sch, 80 Pipes   (b) 26B, Sch, 80 Pipes

Figure 3 Typical Relations between Moment and Rotation

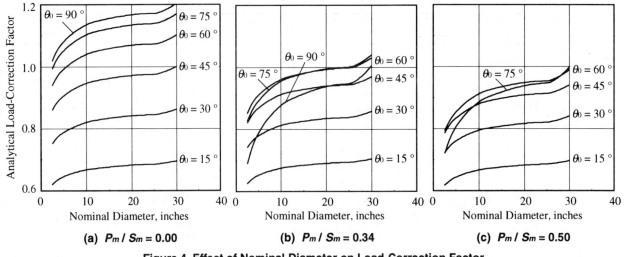

(a) $P_m / S_m = 0.00$ (b) $P_m / S_m = 0.34$ (c) $P_m / S_m = 0.50$

Figure 4 Effect of Nominal Diameter on Load-Correction Factor

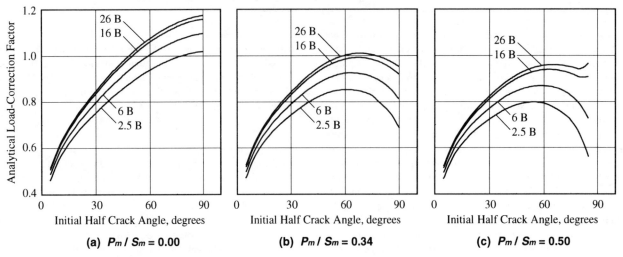

(a) $P_m / S_m = 0.00$ (b) $P_m / S_m = 0.34$ (c) $P_m / S_m = 0.50$

Figure 5 Effect of Initial Half Crack Angle on Load-Correction Factor

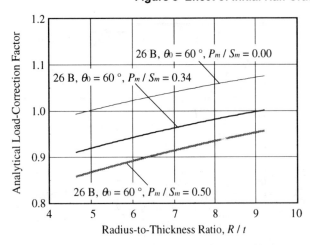

Figure 6 Effect of Radius-to-Thickness Ratio on Load-Correction Factor

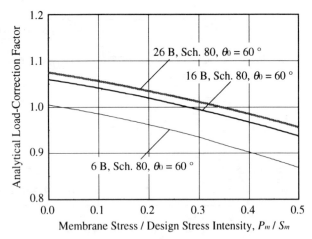

Figure 7 Effect of Normalized Membrane Stress on Load-Correction Factor

at $\theta_0 = 60°$ than that without membrane stress.

Figures 4(a) to (c) show the effect of OD on Z_C for three different P_m/S_m values. As defined in Equation (5), Z_C is effective only in case of $Z_C > 1$, however, here the ratio of the Net-Section Collapse moment to the predicted maximum moment by the LBB.ENGC is conventionally regarded as the analytical load-correction factor. θ_0 was fixed to 15, 30, ···, 90° and OD was varied from 2.5 to 30-inch. The case of $P_m/S_m = 0.50$ and $\theta_0 = 90°$ was excluded because the case fell into the plastic collapse without any bending. The nominal thickness was Schedule 80. These figures show that larger OD gives generally larger Z_C regardless of the value of P_m/S_m or θ_0. The dependence of OD on Z_C seems to be significant when θ_0 is larger. OD, P_m/S_m and θ_0 influence whether Z_C is larger than unity or not. The change of slope at 30-inch diameter is caused by the larger R/t at that point.

Figures 5(a) to (c) show the effect of θ_0 on Z_C for three different P_m/S_m values. OD was fixed to 2.5, 6, 16, and 26-inch and θ_0 was varied from 5 to 90°. The case of $P_m/S_m = 0.50$ and $\theta_0 = 90°$ was also excluded. The nominal thickness was Schedule 80. These figures show that Z_C increases monotonously with increasing θ_0 in case of $P_m/S_m = 0.00$, however, Z_C attains to the maximum value at a certain θ_0 for positive P_m/S_m values. The value of θ_0 at the maximum Z_C ranges between 55 to 70° and depends on both P_m/S_m and OD.

Figure 6 shows the effect of R/t on Z_C for three different P_m/S_m values. OD and θ_0 was fixed to 26-inch and 60°, respectively. The nominal thickness was varied from Schedule 80 to 160. R/t was 9.2 for Schedule 80, and was 4.6 for Schedule 160. It can be seen that Z_C increases almost linearly with increasing R/t from the figure.

Figure 7 shows the effect of P_m/S_m on Z_C. OD was fixed to 6, 16, and 26-inch, and θ_0 was fixed to 60°. P_m/S_m was changed from 0.00 to 0.50. The figure also shows that Z_C has a linear correlation with P_m/S_m. It can be ascertained that the similar correlation is also satisfied for different OD, θ_0, or R/t from Figs. 7 to 9. In the past study (NUPEC, 1989 and Asada et al., 1990), an increase of P_m/S_m due to internal pressure slightly reduced the load-correction factor. This is because the R6 option 2 was applied to identify the load-correction factor under combined loading, though the accuracy of the evaluation using this method was not adequately confirmed. Therefore, it could be seemed that the method gave too conservative prediction under combined loading condition.

The above analysis on the effective factors is summarized as Table 4. It can be found that Z_C is generally larger for larger diameter, larger crack angle, smaller thickness, and smaller pressure.

Table 4 Effect of Factors on Load-Correction Factor

Smaller ⇐	Load-Correction Factor ⇒	Larger
Smaller ⇐	Diameter ⇒	Larger
Smaller ⇐	Crack Angle ⇒	Larger*
Larger ⇐	Thickness ⇒	Smaller
Larger ⇐	Pressure ⇒	Smaller

* : $\theta_0 = 55$ to 70° for positive pressure

DEVELOPMENT OF NEW LOAD-CORRECTION FACTOR

Formulation of Load-Correction Factor

Based on the results of the analysis, the formulation of Z_C was examined. Z_C can be generally expressed as a function of OD, θ_0, R/t, and P_m/S_m,

$$Z_C = Z_C(OD, \theta_0, R/t, P_m/S_m) \qquad (31)$$

However, it would not be appropriate to consider all variables in Equation (31) from the point of codification. The present formula to evaluate Z_C needs not only accuracy, but also simplicity. The present load-correction factors were attempted to determine the individual evaluation formulae based on the specific concept. The parameters to be considered for the load-correction factors were listed in Table 5 with their applicable range.

In this study, the effect of the parameters is treated as follows. The main parameter affecting Z_C is OD. The effect of θ_0 is difficult to be included in the formula in an uniform manner. If an upper applicable

Table 5 Considered Parameters and Applicable Range for Load-Correction Factors

Load-Correction Factor	Parameters to Be Considered	Applicable Range			
		OD	R/t (or Schedule)	θ_0	P_m/S_m
Z_{ASME}	Nominal Diameter, OD Radius-to-Thickness Ratio, R/t	4 to 37B	5 to 20	0 to 90°	Not Provided
G	Nominal Diameter, OD Half Crack Angle at M_{max}, θ_{max} (Function of OD and θ_0)	2 to 30B	Sch. 80 to 160	0 to 60°	Not Provided
Z_{JAPEIC}	Nominal Diameter, OD	2.5 to 30B	Sch. 80 to 160	0 to 60°	Not Provided
Z_C	Nominal Diameter, OD Radius-to-Thickness Ratio, R/t Normalized Membrane Stress, P_m/S_m	2.5 to 30B	Sch. 80 to 160	0 to 60°	0 to 0.5

range of θ_0 is limited to 60°, then Z_C is bounded by the maximum Z_C for an arbitrary θ_0, $Z_{C,max}$, which is nearly equal to Z_C at $\theta_0 = 60°$, see Fig. 5.

$$Z_{C,max} \approx Z_C (\theta_0 = 60°) \qquad (32)$$

The effects of both R/t and P_m/S_m seem to be independent from Figs. 6 and 7, respectively. Consequently, Equation (31) can be rewritten as,

$$Z_C \leq Z_{C,max} = f_1 (P_m / S_m) \cdot f_2 (R / t) \cdot f_3 (OD) \qquad (33)$$

As for the effect of P_m/S_m, a function f_1 is to be defined so that it can be equal to 1 at $P_m/S_m = 0$, and it can present the tendency of the reduction of Z_C with increasing P_m/S_m. Using the least square method, the analytical result shown in Fig. 6 leads to,

$$f_1 (P_m / S_m) = 1.00 - 0.226 \, P_m / S_m \qquad (34)$$

Regarding to the effect of R/t, a function f_2 is to be defined so that it can be equal to 1 at $R/t = 9.2$ (which is equivalent to 26-inch Schedule 80 pipe), and it can present the tendency of the increase of Z_C with increasing R/t. The analytical result shown in Fig. 7 is similarly reduced, and,

$$f_2 (R / t) = 0.826 + 0.0189 \, R / t \qquad (35)$$

When we consider the effect of OD, the tentative load-correction factor, ζ, is introduced. ζ is the analytical load-correction factor excluding the effect of both P_m/S_m and R/t,

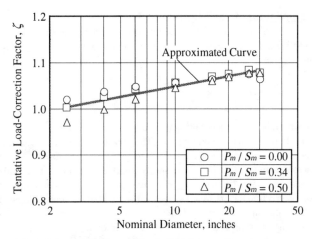

Figure 8 Load-Correction Factor Excluding Effects of Internal Pressure and Radius-to-Thickness Ratio

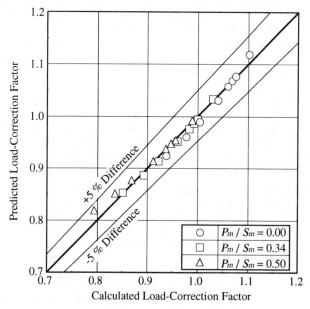

Figure 9 Comparision between Calculated and Predicted Load-Correction Factors

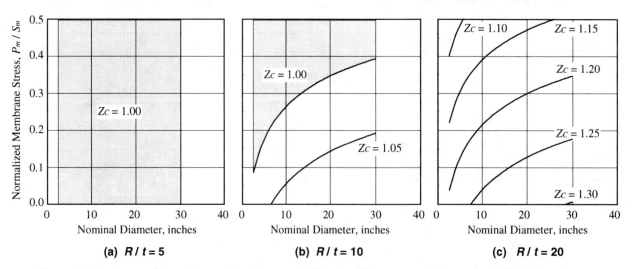

Figure 10 Contours of Load-Correction Factor with Nominal Diameter and Normalized Membrane Stress

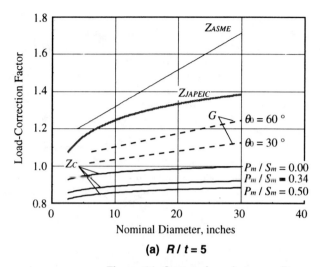
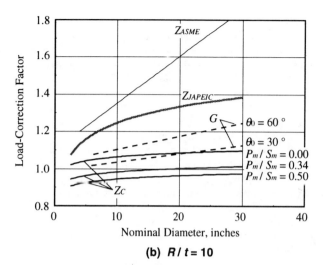

Figure 11 Comparison between Proposed and Present Load-Correction Factors

$$\zeta = \frac{Z_{C,max}}{f_1(P_m/S_m) \cdot f_2(R/t)} \quad (36)$$

Figure 8 shows the relation between ζ and OD. It can be seen that ζ is determined by OD only, and is independent of other factors. This relation can be simply approximated by the following curve,

$$f_3(OD) = \zeta = 0.977 + 0.0707 \log_{10} OD \quad (37)$$

By integrating Equations (34), (35), and (37), then the formula to evaluate Z_C under combined loading is defined as,

$$Z_C = [1.00 - 0.226 \, P_m/S_m] \times [0.826 + 0.0189 \, R/t] \\ \times [0.977 + 0.0707 \log_{10} OD] \quad (38)$$

Again it is noted that Z_C should be unity in case of $Z_C < 1$. The parameters to be considered and their applicable range for Z_C are also listed in Table 5. Figure 9 shows the comparison of the predicted Z_C by Equation (38) and the calculated Z_C at $\theta_0 = 60°$, which gives less conservatism. (Points under $Z_C = 1$ are also plotted for comparison.) Predicted Z_C is in good agreement with calculated Z_C for a wide range of OD and P_m/S_m. The difference is less than 3%.

Figures 10(a) to (c) show the contours of Z_C obtained by Equation (38) as a function of OD and P_m/S_m for three different R/t values. For $R/t = 5$, Z_C is always 1 for the entire applicable range ($OD = 2.5$ to 30-inch and $P_m/S_m = 0$ to 0.5). The area where $Z_C = 1$ decreases as R/t increases. Higher Z_C appears at larger OD and smaller P_m/S_m.

Comparison with Present Factors

The load-correction factor under combined loading, Z_C, was compared with the present load-correction factors, Z_{ASME}, G, and Z_{JAPEIC}, shown in Figs. 11(a) and (b). In case of $R/t = 5$, Z_C is always less than the other factors independently of OD as well as P_m/S_m. In case of $R/t = 10$, Z_C at $P_m/S_m = 0.00$ is slightly larger than G at $\theta_0 = 30°$ for smaller OD, however, Z_C still locates relatively lower region than the others. Since the material properties used to develop G, Z_{JAPEIC}, and Z_C are considered to be substantially same, the difference is wholly due to the accuracy and conservatism of the evaluation methods. The difference between Z_{ASME} and Z_C may be resulted by the different J-estimation schemes and the different material properties used for the analysis.

Figure 12 shows the ratio of Z_C to Z_{JAPEIC}. Z_{JAPEIC} is just taken as a representative present load-correction factor. The ratio corresponds to the degree of conservatism included in the present load-correction factor. In other words, if Z_C is substituted for Z_{JAPEIC}, an assumed fracture load can be reduced by the ratio. The figure shows that the ratio is somewhat affected by R/t and P_m/S_m. In the range shown in the figure, it is 5 to 7% at OD is 2.5-inch, and 21 to 28% at OD is 30-inch.

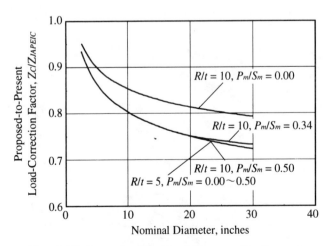

Figure 12 Example of Ratio of Proposed to Present Load-Correction Factor

CONCLUSIONS

In this study, the load-correction factor for carbon steel pipes taken combined loading into account was investigated. The J-estimation scheme, LBB.ENGC method newly developed to evaluate ductile fracture behavior of circumferentially through-wall-cracked pipes subjected to combined loading was used to analyze the load-correction factor. The effect of internal pressure as well as other factors on the load-correction factor was quantitatively investigated. It was found that the load-correction factor was generally larger for larger diameter, larger crack angle, smaller thickness, and smaller pressure. The formula to evaluate the load-correction factor under combined loading was developed based on the results of the analysis. The new load-correction factor was then compared with the present load-correction factors. As a result, it could be ascertained that the present load-correction factors in less consideration of combined loading were still conservative under combined loading condition. The new load-correction factor seemed to be promising to make more reliable flaw evaluation possible.

REFERENCES

Asada, Y., et al., 1990, "Leak-Before-Break Verification Test and Evaluations of Crack Growth and Fracture Criterion for Carbon Steel Piping," International Journal of Pressure Vessel and Piping, 43, pp. 379-397.

Iida, K., 1996, "Draft of New Maintenance Standards for LWR in Japan," ASME PVP, Vol. 339, pp. 13-24.

Iida, K., et al., 1996, "New Maintenance Code for Operating Nuclear Power Plants in Japan (Draft)," Proceeding of the International Workshop on the Integrity of Nuclear Components, Taejon, Korea, pp. 9/1-9/22.

Imamura, Y., et al., 1995, "Development of Nuclear Plant Operation and Maintenance Code in Japan," ASME PVP, Vol. 313-2, pp. 395-397.

Klecker, R., et al., 1986, "NRC Leak-Before-Break (LBB.NRC) Analysis Method for Circumferentially Through-Wall Cracked Pipes Under Axial Plus Bending Loads," NUREG/CR-4572.

Milne, I., et al., 1986, "Assessment of Integrity of Structures Containing Defects," CEGB R/H/R6-Revision 3.

Miura, N., 1998, "Approximate Evaluation Method for Ductile Fracture Analysis of Circumferentially Through-Wall-Cracked Pipe Subjected to Combined Bending and Tension," Proceeding of the 2nd International Workshop on the Integrity of Nuclear Components, Tokyo, Japan, pp. 209-228.

Miura, N., et al., 1998, "J-R Curves from Circumferentially Through-Wall-Cracked Pipe Tests Subjected to Combined Bending and Tension - Part I: Theory and Numerical Simulation," Transactions of the ASME, Journal of Pressure Vessel Technology, 120, pp. 406-411.

Miura, N., 1999, "Approximate Evaluation Method for Ductile Fracture Analysis of Circumferentially Through-Wall-Cracked Pipe Subjected to Combined Bending and Tension," Submitted to Nuclear Engineering and Design.

Novetech Corporation, 1988, "Evaluation of Flaws in Ferritic Piping," EPRI NP-6045, Final Report.

Nuclear Power Engineering Test Center (NUPEC), 1989, "Proving Test on the Integrity of Carbon Steel Piping in LWRs," Summary Report of Proving Tests on the Reliability for Nuclear Power Plant.

Saikawa, K. et al., 1995, "Status of Efforts for Establishing Maintenance Standards for Nuclear Power Plant Facilities in Japan," ASME PVP, Vol. 313-2, pp. 117-124.

Section XI Task Group for Piping Flaw Evaluation, 1986, "Evaluation of Flaws in Austenitic Steel Piping," EPRI NP-4690-SR, Special Report.

Zahoor, A., 1987, "Evaluation of J-Integral Estimation Scheme for Flawed Throughwall Pipes," Nuclear Engineering and Design, 100, pp. 1-9.

THE OPENING AREA ASSOCIATED WITH A CRACK THAT IS SUBJECTED TO A GENERAL TENSILE STRESS DISTRIBUTION

E. Smith

Manchester University-UMIST Materials Science Centre,
Grosvenor Street, Manchester, M1 7HS, United Kingdom.

ABSTRACT

A key element in the development of a leak-before-break case for a component in a pressurized system, is an estimation of the size of through-thickness crack that will give a measurable leakage under normal operating conditions. This requires a knowledge of the crack opening area. The paper focuses on this issue and provides an expression for the crack opening area associated with a two-dimensional crack that is subjected to a general tensile stress field, which simulates the stress distribution arising from a combination of applied and residual stresses. The results for particular distributions are used to test the viability of a simplified procedure that has been used to estimate the crack opening area. This is based on the determination of the crack opening displacement at the crack centre coupled with the assumption that the crack opening profile conforms to an elliptical shape. The results from this paper's general analysis shows that the procedure gives reasonably accurate crack opening predictions for a wide range of crack sizes and loading conditions.

INTRODUCTION

There are two key elements when developing a leak-before-break argument for a pressurized system as part of an overall structural integrity case for the system. These elements are: (a) The size of through-thickness crack that is stable under accident loading conditions, e.g. a seismic event, and (b) The size of through-thickness crack that gives a measurable leakage under normal operating conditions. The viability of a leak-before-break case depends on the ability to demonstrate that there is an acceptable margin between these two critical crack sizes.

Consequently, estimation of the crack opening (leakage) area for normal operating conditions is particularly important, recognising that the crack opening might be due to a combination of pressure induced and residual stresses. For a wide range of situations, an analysis based on elastic behaviour of the material is probably adequate and the present paper proceeds from this basis. Thus the paper provides an expression for the crack opening area associated with a two-dimensional crack that is subjected to a general tensile stress field, this simulating the stress distribution arising from a combination of applied and residual stresses. The results are used to test the viability of a simplified procedure that has been used[1] to estimate the crack opening area. This is based on the determination of the crack opening displacement at the crack centre coupled with the assumption that the crack opening profile conforms to an elliptical shape, an assumption that is exact for the case of a uniform tensile stress. Thus we show that this procedure gives reasonably accurate crack opening area predictions for a wide range of crack sizes and loading conditions.

THEORETICAL ANALYSIS

The very simple model that will be analysed in this paper is shown in Figure 1. We have a two-dimensional crack of length 2c in an infinite solid that is subjected to the applied tensile stress (in the crack's absence)

$$\sigma(x) = \sigma \sum B_n \left(\frac{|x|}{h}\right)^n = \sigma \sum_{n=0}^{m} B_n \left(\frac{c}{h}\right)^n \left(\frac{|x|}{c}\right)^n \quad (1)$$

with n and m being integers, and the B_n are constants. $\sigma(x)$ simulates the stress distribution due to a combination of applied (e.g. pressure induced) and residual stresses and is symmetric about the crack centre. The stress intensity factor at a crack tip due to the stress $\sigma(|x|/h)^n$ is given by the expression[2]

$$K(c) = \sigma \left(\frac{c}{h}\right)^n c^{1/2} \frac{\Gamma\left(\frac{n}{2}+\frac{1}{2}\right)}{\Gamma\left(\frac{n}{2}+1\right)} \quad (2)$$

where Γ is the Gamma function.

Following the arguments developed by Streitenberger and Knott[3], it then follows that the crack opening area A_n due the stress $\sigma(|x|/h)^n$ is given by the expression

$$A_n = \frac{4}{E_o} \int_0^c (\pi\lambda)^{1/2} K(\lambda) d\lambda \quad (3)$$

where $E_o = E/(1-v^2)$, with E = Young's modulus, v is Poisson's ratio and with $K(\lambda)$ being given by relation (2) with c replaced by λ.

It thus follows, from relations (2) and (3), that

$$A_n = \frac{4\sigma c^2}{E_o} \left(\frac{c}{h}\right)^n \frac{\pi^{1/2} \Gamma\left(\frac{n}{2}+\frac{1}{2}\right)}{(n+2)\Gamma\left(\frac{n}{2}+1\right)} \quad (4)$$

whereupon the crack opening area A due to the general stress distribution given by relation (1) is

$$A = \frac{4\sigma c^2}{E_o} \sum_{n=0}^{m} B_n \left(\frac{c}{h}\right)^n \frac{\pi^{1/2} \Gamma\left(\frac{n}{2}+\frac{1}{2}\right)}{(n+2)\Gamma\left(\frac{n}{2}+1\right)} \quad (5)$$

This expression allows the crack opening area to be obtained for a general stress distribution, assuming that it can be approximated by the power law distribution given by relation (1).

Now the relative displacement ϕ_o of the crack faces at the crack centre due to the stress distribution $\sigma(x)$ is[2]

$$\Phi_o = \frac{4}{\pi E_o} \int_0^c \sigma(x) \ln\left[\frac{c+(c^2-x^2)^{1/2}}{c-(c^2-x^2)^{1/2}}\right] dx \quad (6)$$

Thus if $\sigma(x) = \sigma(x/h)^n$ where n is an integer,

$$\Phi_o(n) = \frac{4\sigma}{\pi E_o} \int_0^c \left(\frac{x}{h}\right)^n \ln\left[\frac{c+(c^2-x^2)^{1/2}}{c-(c^2-x^2)^{1/2}}\right] dx \quad (7)$$

whereupon evaluation of the integral, after substituting $x = c\sin\theta$, gives

$$\Phi_o(n) = \frac{4\sigma c}{\pi E_o} \left(\frac{c}{h}\right)^n \frac{\pi^{1/2} \Gamma\left(\frac{n}{2}+\frac{1}{2}\right)}{(n+1)\Gamma\left(\frac{n}{2}+1\right)} \quad (8)$$

Consequently for the stress distribution given by relation (1), the relative displacement ϕ_o is given by the expression

$$\Phi_o = \frac{4\sigma c}{\pi E_o} \sum_{n=0}^{m} B_n \left(\frac{c}{h}\right)^n \frac{\pi^{1/2} \Gamma\left(\frac{n}{2}+\frac{1}{2}\right)}{(n+1)\Gamma\left(\frac{n}{2}+1\right)} \quad (9)$$

With the elliptical profile assumption, the crack opening area $A_{APP} = \pi\phi_0 c/2$, whereupon relation (9) gives

$$A_{APP} = \frac{2\sigma c^2}{E_o} \sum_{n=0}^{m} B_n \left(\frac{c}{h}\right)^n \frac{\pi^{1/2}\Gamma\left(\frac{n}{2}+\frac{1}{2}\right)}{(n+1)\Gamma\left(\frac{n}{2}+1\right)} \quad (10)$$

It then follows from relations (5) and (10) that the ratio of the approximate crack opening area A_{APP}, determined via the elliptical profile assumption, to the actual area A is

$$\frac{A_{APP}}{A} = \frac{1}{2} \frac{\sum_{n=0}^{m} B_n \left(\frac{c}{h}\right)^n \frac{\Gamma\left(\frac{n}{2}+\frac{1}{2}\right)}{(n+1)\Gamma\left(\frac{n}{2}+1\right)}}{\sum_{n=0}^{m} B_n \left(\frac{c}{h}\right)^n \frac{\Gamma\left(\frac{n}{2}+\frac{1}{2}\right)}{(n+2)\Gamma\left(\frac{n}{2}+1\right)}} \quad (11)$$

Now let us consider some special cases. For the case where

$$\sigma(x) = \sigma\left[1 - \frac{|x|}{h}\right] \quad (12)$$

i.e. where $B_o = 1$ and $B_1 = -1$ with all the other B_n being equal to zero, relation (11) shows that

$$\frac{A_{APP}}{A} = \frac{1 - \frac{c}{\pi h}}{1 - \frac{4c}{3\pi h}} \quad (13)$$

and we see that $A_{APP} \sim A$ provided that the crack size c is not unduly large compared with the scale h of the stress distribution, e.g. $A_{APP}/A = 1.18$ for $c = h$. For the case where

$$\sigma(x) = \sigma\left[1 - \left(\frac{x}{h}\right)^2\right] \quad (14)$$

i.e. where $B_o = 1$, $B_1 = 0$ and $B_2 = -1$ with all the other B_n being equal to zero, relation (11) shows that

$$\frac{A_{APP}}{A} = \frac{1 - \frac{1}{6}\left(\frac{c}{h}\right)^2}{1 - \frac{1}{4}\left(\frac{c}{h}\right)^2} \quad (15)$$

and we again see that $A_{APP} \sim A$ provided that the crack size c is not unduly large compared with the scale h of the stress distribution, e.g. $A_{APP}/A = 1.11$ for $c = h$.

DISCUSSION

The paper has provided a general expression for the crack opening area A associated with a two-dimensional crack that is subjected to a general tensile stress field, which simulates the stress distribution arising from a combination of applied and residual stresses; the general tensile stress field has been represented by a power law variation. Furthermore the relative displacement at the crack centre has been determined, and coupling this with the assumption of an elliptical opening profile enables an approximate value A_{APP} for the crack opening area to be obtained; this simple procedure has been used by some workers[1]. A comparison of A and A_{APP} for two particular stress variations has shown that the simple procedure gives reasonable crack opening area estimates provided that the crack size is not unduly large compared with the scale of the stress distribution. However, it should be pointed out, by way of a cautionary comment, that although the simple procedure gives reasonably accurate estimates of the crack opening area, these estimates in fact slightly overestimate the area. This means that for a prescribed crack opening (leakage) area, the critical crack size associated with a critical leakage is slightly underestimated which means that there is a degree of non-conservatism in the context of a leak-before-break argument. However, this comment has been made on the basis of the exact area A having been calculated on the basis of an elastic analysis. If A is calculated on the basis of an elastic-plastic analysis, the value of A is increased[4] and this will bring the A_{APP} (still calculated via an elastic analysis) value and A value even closer and it is then likely that the non-conservatism will be removed. Whether or not this indeed the case is an issue that the author will be addressing in the near future.

REFERENCES

1. S. Rahman, N. Ghadiali, G. Wilkowski and N. Bonora, PVP-Vol. 304 ASME(1995) 149.

2. H. Tada, P.C. Paris and G.R. Irwin, "The Stress Analysis of Cracks Handbook", Del Research Corporation, Hellertown, Pa, USA (1973).

3. P. Streitenberger and J.F. Knott, Int. Jnl. of Fracture, 76 (1995/96) R.49.

4. E. Smith, PVP-Vol. 373 ASME (1998) 485.

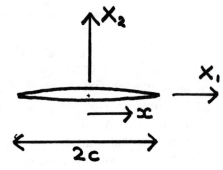

FIGURE 1 The model analysed in the paper

Limit Analysis for Pressurized Pipelines with Local Wall-Thinning

Lianghao Han[*]
Institute of Nuclear Energy Technology, Tsinghua University, Beijing 100084

Yinpei Wang Cengdian Liu
East-China University of Sci.&Tech., Shanghai 200237

Abstract: Local wall-thinning is a very common volume flaw on the surface of oil and gas transmission pipeline. Predicting the remaining strength of pipelines with local wall-thinning is essential for the determination of design tolerance, integrity assessment and effective maintenance action. In this paper, 3D elastic-plastic finite element analysis was used to establish limit load for pipelines with local wall-thinning, and some geometric parameters which were not considered in existing widely-accepted criterion ANSI/ASME B31.G used in the assessment of corrosion damage were also investigated. The results were presented for a wide range of dimension of pipeline and local wall-thinning using appropriate non-dimension parameters, and compared with ANSI/ASME B31.G. It was shown that the weakening effect of circumferential size of local wall-thinning on strength was appreciable, particularly for simultaneously short and deep local wall-thinning. The effect of circumferential size was a function of local wall-thinning dimensions and the loading conditions. Based on numerical results, a critical circumferential size of local wall-thinning was given, and the scope of application for ANSI/ASME B31.G was proposed.

KEY WORDS: Limit Load, Corrosion, Pipeline

Notation
P Numerical limit pressure for pipe with groove
P^* Numerical limit pressure for pipe without groove
$2L$ Length of local wall-thinning
2θ Width of local wall-thinning
t Thickness of pipe
d Remaining thickness of pipe
R_i Internal radius of pipe
R_o External radius of pipe

1. Introduction

Local wall-thinning is a very common defect of oil and gas transmission, which can result from erosion, corrosion, mechanical damage, or blend grinding to remove crack-like flaws. The local wall-thinning not only produces stress concentration, but also reduces the integrity of the structure, threatens the safe operation of pipeline and even brings out damage accident. Therefore, it is significant to do more detail and systematical research on local wall-thinning in order to predict the remaining strength of pipeline with local wall-thinning.

The existing widely-accepted criterion used in the assessment of corrosion damage, ANSI/ASME B31.G[1], was developed over 20 years ago. It was based on a semi-empirical fracture mechanical relationship conceived by Maxey[2,3]. It was very conservatively based on experimental data limited to the thin-wall pipes with narrow machined slots in the external pipe wall. It did not consider the effect of circumferential size of local wall-thinning.

The present work was concerned with determination of limit load of pipe with rectangular local wall-thinning using 3D finite element. The effect of circumferential dimension of local wall-thinning was also investigated. In order to investigate the applicability of B31.G, the numerical analysis results were compared with the results from B31.G code.

2. Limit analysis

Limit analysis calculates the maximum load that a given structure of perfectly-plastic or ideally rigid-plastic material can sustain. The loading is assumed to vary proportionally with a single factor. The maximum sustainable load is called the limit load, and when this load is reached the deformations become unbounded and structure becomes a mechanism. The limit analysis can give a more realistic evaluation of the limit loads and provides valuable guidance in the design and

---

[*] Corresponding author: Tel: +86-10-6278-4809, E-mail: lhaohan@263.net

safety assessment of pipeline.

The limit load is usually defined as the load giving the final convergent solution in the non-linear procedure or, if a non-zero plastic modulus is used, the load given by a specified limit load criterion, such as the tangent intersection method illustrated in figure 1. The load at the intersection of two tangent lines drawn from elastic and plastic part of load-deformation curve is taken as limit load.

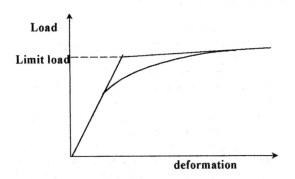

Figure 1 Tangent Intersection Method for Defining Limit Load

3. Finite Element Modeling and Geometry

In order to get limit load of pressurized pipeline with local wall-thinning, there had some assumptions. (1)the material behaved in an elastic-perfectly plastic manner(2) the material obeyed the von Mises yield criterion with isotropic hardening rule (3) the material was initially homogenous and isotropic and remained so during plastic deformation. (4) the loading is a uniform internal pressure with the pipe ends subjected a uniformly distributed pressure.

Part-through rectangular groove was studied in the analysis, Figure 2 illustrated configuration of part-through rectangular groove.The finite element analysis used the EMRC/NISA95[1] software. Three-dimensional models(Figure3) were developed to simulate the behaviour of pipe with rectangular grooves. Considering the symmetry, only quadrant of pipe was taken as finite element model. The corresponding displacement constraints was imposed on the symmetry boundary. The mesh was selected as aresult of a series of convergence studies, the details of which were omitted for brevity. Internal pressure load increased to a level at which the required load increment was less than a specified minimum value of 10^{-5} of the total applied load.

To define the limit load of pipe with part-through rectangular grooves, the curve of load-deformation must be obtained firstly, then limit load(pressure) could be determined according to tangent intersection method. Figure 4 showed two typical load-deformation curves of points as marked in figure 2 . As shown in figure 4, when using elastic-perfectly plastic material model, typically there was not a significant increase in collapse pressure when higher levels of deformation(such as equivalent plastic strain, radial displacement). In the analysis, equivalent plastic strain was taken as a failure parameter.

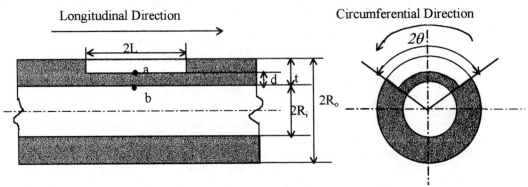

Figure 2 Geometry of a Pipe with a 3D Rectangular Groove

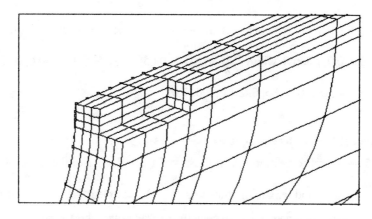

Figure 3 Finite Element Mesh Used in Parameter Study

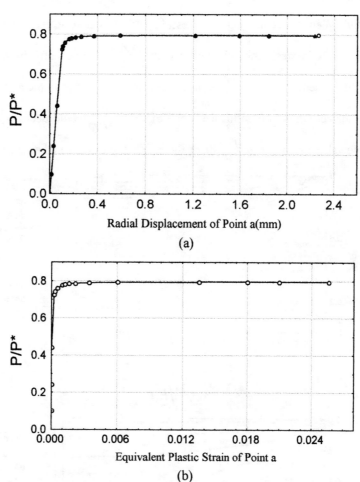

(a)

(b)

Figure 4 Typical Load-Deformation Curves

4. Finite Element Analysis Results

Considering convergence problem, a bilinear hardening material model with a low value of plastic modulus $E_p = 10 MPa$ in practical analysis, and elastic modulus was $E = 2.1 \times 10^5 MPa$. The radius ratio $R_o/R_i = 1.2$. The yield stress σ_s was $200\,MPa$. Poisson's ratio was 0.3. To represent the results in general form, non-dimensional parameters were employed: $p = P/P^*$, $L/\sqrt{R_o t}$, θ/π, d/t. A comprehensive survey was made for the following combinations of geometric parameters: relative remaining wall thickness d/t =0, 0.1, 0.2, 0.3, 0.4, 0.5, 0.6, 0.7, 0.8, 0.9; relative width θ/π =0, 0.02, 0.05, 0.2, 0.3, 0.4, 0.5, 0.6, 0.7, 0.8, 0.9; relative length $L/\sqrt{R_o t}$ =0, 0.2, 0.4, 0.5, 0.8, 1.0, 2.0, 3.0.

4.1 The Effect of Circumferential Size of Grooves

Figure 5 presented some plots which illustrated the influence of the defect width θ/π on limit pressure at some specified remaining thickness d/t. It was clear that the defect width weakening effect on limit load was appreciable, especially for deep and short defect. When the remaining thickness (d/t >0.8) was relatively large, the defect width θ/π did not affect the limit pressure of pipe apparently; when the remaining thickness (d/t <0.5) was relatively small and the defect width θ/π was relatively small, the influence of width θ/π was apparently. From figure 5, it could be found that the influence of width θ/π on limit pressure was relative to the length of defect. With the increase of defect length, the influence scope of width θ/π would decrease. When θ/π was more than some value (for example θ/π >0.7), the defect could be treated as axisymmetrical.

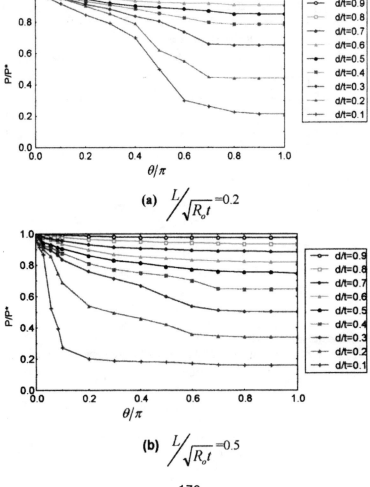

(a) $L/\sqrt{R_o t}$ =0.2

(b) $L/\sqrt{R_o t}$ =0.5

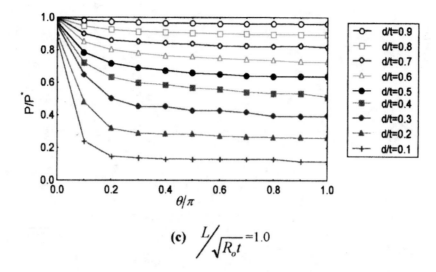

(c) $L/\sqrt{R_o t} = 1.0$

Figure 5 The Computational Curves of the Limit Loads P/P^* vs. θ/π

4.2 The comparision of ANSI/ASME B31.G and Numerical Analysis

Failure equations incorporated into B31.G code were derived based on a failure mechanism calibrated by extensive testing of vessels with narrow machined slots, and were justed by fitting burst-test results for pipe with narrow corrosion. In addition, experimental data were limited to thin-wall(0.25in.-0.50in.)pipes containing small corrosion areas. In order to investigate failure equations of B31.G,numerical analysis results were compared with results from B31.G in figure 6. Failure equations for rectangular shape in B31.G were equations (1) and (2) as following, where SMYS represents specified minimum yield stress. Equation(1) could be substituted with equation (3) in non-dimensional form.

$$P = 1.1 SMYS \frac{d/t}{1 - M^{-1}(1 - d/t)} \qquad (1)$$

$$M = \sqrt{1 + 1.6 L^2/R_o t} \qquad (2)$$

$$p = \frac{P}{P^*} = \frac{d/t}{1 - M^{-1}(1 - d/t)} \qquad (3)$$

As shown in figure 6, the curves presented the results for equation (3) at specified remaining wall thickness d/t, points presented the results from finite element analysis. From figure 6, it could be found : when defect width $\theta/\pi = 0.1$, the results from numerical analysis were consistent with B31.G; when remaining wall thickness ($d/t > 0.7$) was relatively large, the numerical analysis results were also consistent with B31.G; when defect width was relatively large and remaining wall thickness was relatively small ($d/t < 0.7$), there would exist apparent difference between numerical analysis results and B31.G, the reason may lie in B31.G did not consider the circumferential size of corrosion.In fact,with the increase of width and depth of defect , the longitudinal stress from internal pressure will increase in thinning area, and consequently the remaining strength of pipe will decrease.

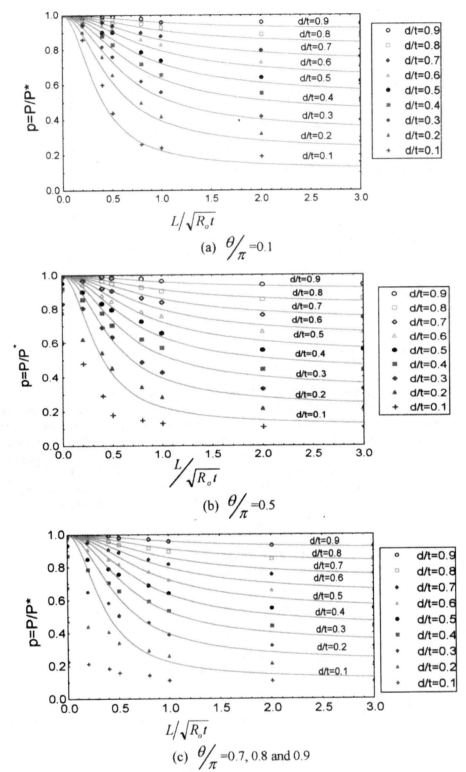

Figure 6 The Comparision between B31.G and Numerical Analysis

5. Conclusion and Discussion

In this paper, a numerical limit analysis was carried out for pipeline with groove on the outside surface. Limit analysis results were compared with widely-accepted ANSI/ASME B31.G code, and weakening effect of width of local wall-thinning was studied.

The weakening effect of circumferential size of thinning area varies depending on two factors, longitudinal stress and remaining material restrain,: On the one hand, with the increase of width and depth of defect, the cross-section of thinning area will decrease, and the longitudinal stress in thinning area will increase, consequently the limit load will decrease; On the other hand,the center of thinning area is also subjected to the restrain from remaining material.When circumferential size of thinning area is relative small,the most dangerous point(the center of thinning area) will obtain larger restrain from remaining material surrounding thinning area,the increase of circumferential size will result in larger weakening effect;When circumferential size of thinning is relative large,the remaining material restrain to center of thinning area will be close to constant and the change of longitudinal stress is relative small, so the weakening effect of circumferential size will be close to constant.

When circumferential size of local wall-thinning was small,the increase of longitudinal stress in thinning area is relative small, the numerical analysis gave same results as ASME B31.G, so it was conservative to assess pipe with local wall-thinning with ASME B31.G; When circumferential size of local wall-thinning was relative large, the increase of longitudianl stress from internal pressure was relative large in thinning area,and the limit load would decrease largely, the effect of longitudinal stress could not be neglected, So B31.G may give non-conservative evaluation in some case ,especially when there exsits larger longitudinal stress .

When defect is relative short, the center of thinning area(the most dangerous point) will subject to larger longitudinal strain from remaining material surrounding thinning area,the weakening effect of width of thinning area will be apparent; When defect is relative long,the longitudinal strain to center of thinning area is relative small,

References

1. ANSI/ASME B31.G 1991. *Manual for Determining the Remaining Strength of Corroded Pipelines*, ASME. New York.
2. Maxey W. A., Kiefner J. F. et al., "Ductile Fracture Initiation, Propagation and Arrest in Cylindrical Vessels," ASTM STP 514, p.p. 70-82,1972.
3. Kiefner J.F. Maxey, W. A. et al., "Failure Stress Levels of Flaw in Pressurized Cylinders," ASTM STP 536, p.p. 461-481,1973.
4. NISA, Engineering Mechanics Research Corporation, Michigan,American,1995.
5. J.F.Kiefner, and A.R.Duffy, "Summary of Research to Determine the Strength of Corroded Areas in Line Pipes," NG-18 Paper No.39, Battelle Columbus Laboratories (July,1971).

WELD RESIDUAL STRESSES AND EFFECTS ON FRACTURE/FATIGUE

Introduction

Pingsha Dong
Battelle
Columbus, OH 43201-2693

Over the last decade, welding-induced residual stresses have received an increasing attention in pressure vessel and piping research community. The primary driving force can be attributed to the fact that recent advances in structural integrity assessment of welded components demand more accurate information on weld residual stress state as a priori. The needs become more evident as welding repairs are increasingly used in aging power plants. As such, both repair procedure development and the subsequent safety assessment require a better understanding of the welding effects on structural components. As some of the advanced computational modeling techniques as well as new and improved experimental methods become available over the recent years, more accurate residual stress information can now be obtained for various structural integrity assessment applications. In the following four sessions, a diverse collection of eighteen papers from all over the world cover provide the state of the art summary from advanced modeling and experimental techniques for residuals stress characterization and the effects on fracture and fatigue in pressure vessel and piping components.

In the first session, a collection of five papers addresses various fundamental issues related to finite element based residual stress modeling techniques and their applications in predicting residual stresses and distortions in typical applications. The papers by Junek et al and Juliien et al discussed the effects of weld and HAZ microstructure in residual stress predictions. A more complex joint configuration (moment frame connection) was addressed by Zhang et al in the third paper and the effects of residual stress triaxiality on propensity for brittle fracture were also discussed. The fourth paper by Chao et al discussed residual stress and distortion prediction issues associated with aluminum alloys by considering a butt joint. Residual stresses in pipe girth welds were discussed by Janosch et al in the last paper.

The second session contains five papers. The first one discussed a proposed treatment of weld residual stresses for fitness-for-service assessments with reference to API RP-579 Appendix E. Although not related directly to welding, the second paper by Martin discussed finite element procedures for modeling residual stresses induced by cold rolling process. Residual stresses induced by hot iso-static pressing (HIP) between two dissimilar materials were discussed by Parker. A high strength aluminum alloy butt joint was analyzed by Preston et al. The fifth paper by Leung discussed residual stress analysis for a TIG joint in a high vacuum tube.

The papers in Session III, the first three papers discussed the effects of local PWHT and proof loading effects on weld residual stresses. In many pressure vessel and piping weld applications, post weld heat treatment (PWHT) are required for either improving weld property or relieving weld residual stresses. For example, effects of local PWHT on weld residual stresses were discussed by Murakawa et al. The last paper by Brust provided a state of the art summary on classical and emerging fracture parameters for welded structures. Some of the issues on incorporating weld residual stresses were also discussed. Additional structural integrity issues were covered in Session IV on crack opening area (COA) for LBB assessment, HAZ fracture behavior, etc.

RESIDUAL STRESS SIMULATION
INCOPORATING WELD HAZ MICROSTRUCTURE

LUBOMÍR JUNEK
Vítkovice Institute of Applied
Mechanics Brno Ltd., Czech Republic

MAREK SLOVÁČEK
Vítkovice Institute of Applied
Mechanics Brno Ltd., Czech Republic

VLADISLAV MAGULA
Welding Research Institute Bratislava,
Slovak Republic

VLADISLAV OCHODEK
VŠB Technical University Ostrava,
Dept.of Mechanical Technology, Czech
Republic

ABSTRACT

This paper deals with the numerical simulation of welding. The introduction discusses the challenging problems under the solution and describes the numerical simulation welding procedure used. At the beginning the paper analyses is detail individual HAZ zones from the point of view of grain distribution. Starting with the description of the HAZ zones of a simple overlay the paper continues with the HAZ zone description of a multilayer welding. It reveals the critical places in welding where relevant defects resulting from the welding can be expected. It stresses the necessity to apply an "in situ" welding diagram with the metallurgical calculation including the effect of the grain sizes to the phase transformations. The part dealing with the stress analyses discusses the changes in mechanical properties during welding. It outlines the possible development approach for the assessment of the residual stresses. At the end, the paper briefly describes and gives the applications for the industrial use having been used for the above mentioned analysis of the welding. The results of the applications used with a brief commentary are given.

INTRODUCTION

Numerical simulation of welding is one of the most complex tasks within the subject of mechanics. When working out the problem of welding the person in charge is required to have a wide range of theoretical background in material engineering and continuum mechanics. The person responsible for calculation has to have long-term practical experience in using the finite element method (FEM) for working out nonlinear tasks.

It is also necessary for the person to possess good laboratory techniques for determining complex material characteristics of individual phases up to high temperatures, a suitable laboratory for carrying out experimental measurements during the process of welding, effective computer equipment for processing a large quantity of data with proper analysis procedures to mathematically calculate the theoretical welding problem in the corresponding way. Without satisfying above requirements, the person responsible for weld residual stress simulation may obtain results that are highly distorted from reality. This is because choosing the wrong mesh (FEM) or not using a suitable linear element or boundary condition, or in case the time step is badly chosen at iterative calculations, the person responsible for calculation can obtain false results even if all theoretical knowledge about welding is all right. Welding tasks cannot be left at stating the residual stresses only because this does not say anything to the customer about how the construction will be affected as far as the limit states are concerned. Therefore, it is inevitable to include in the welding problem report an expertise on residual stresses with more than comparing the stresses to allowable limits. All this determines the character of complex mathematical calculation of the welding problem.

The program SYSWELD was used for welding simulation by the French company, Framasoft. The process of welding simulation using the SYSWELD program consists of three overlapping stages. During the first one, using a specific modulus, the complete CCT diagram is defined. This diagram is the basis of the metallurgical calculation. The result of the first stage is the coefficient of transformational equations that serve as the direct access into the second stage in which – after the thermo-physical qualities are defined – the thermal and metallurgical calculations of the welding process are carried out. The usual equation of heat conduction in the SYSWELD program is modified by the influence considering both transformation latent heat and latent heat at the change of state. The result of the second stage is the non-stationary temperature fields, the percentual distribution of individual phases calculated on the basis of the warming-up speed and the cooling speed and the size of austenite grains. In the second stage there is one specific part – hardness calculation in a defined area, in which the hardness is calculated on the basis of chemical composition, percentual distribution of phases and the speed of cooling through characteristic temperatures. The results of the second stage are stored in their temporary files. In these they enter the third part, in which – after the mechanical qualities of individual phases (depending on temperature) are defined – the general, elastic, plastic and thermal deformations are calculated. On the basis of these the calculation of resulting mechanical qualities of the structure and residual stresses is carried out. The last – the fourth – indirect stage must be the very important assessment of the residual stresses.

METALLURGICAL ANALYSIS

The goal of the metallurgical calculation is to define the percentual proportions of individual phases in the heat affected zone (HAZ). The elementary phases for the steel welding in the iron-carbon composition are ferrite, bainite, pearlite, martensite and austenite. Except these elementary phases it is possible to identify – on the metallographic cut – also other phases, for instance tempered martensite – sorbite, upper bainite, lower bainite etc. Each of the phases mentioned above has different mechanical qualities and its own influence on the resulting general utility qualities of steels. For the stage of heating-up it is necessary (for the requirements of the metallurgical calculation) to exactly define the dependence of transformational temperatures AC_1 and AC_3 on the speed of heating-up. This dependence defines the HAZ during the numerical simulation. The stage of cooling is based on knowing the welding CCT diagrams. This is very important because the diagram specifies the influence of fast cooling on the phase transformation in the HAZ. Knowing the dependence of the austenite grain size on the thermal input, which is possible to calculate when evaluating the CCT diagram.

It is possible to divide the heat affected zone of welds (applicable for carbon steels with carbon contents less than 0.30%) into three main regions [1] : the partial grain-refining, grain-refining and grain-coarsening regions. Figure 1 shows the major regions of HAZ. The partial grain-refining region is subjected to a peak temperature between the effective lower and upper transformation temperatures AC_1 and AC_3. The prior pearlite (P) colonies in this region transformed to austenite (γ) and expanded slightly into the prior ferrite (α) colonies upon heating to above the AC_1 temperature and then decomposed into very fine grains of pearlite (P) and ferrite (F) during cooling. Figure 2 shows the sketch of this process. The grain-refining region is subjected to a peak temperature just above the effective upper transformation temperature AC_3, thus allowing austenite grains to nucleate. Grain refining much improves plastic qualities of the material in the area of the weld. Finally, the grain-coarsening region is subjected to a peak temperature well above the AC_3 temperature, thus promoting the coarsening of austenite grains. This region of HAZ can call the most degradation part too. Usually the cracks that sometimes accompany welding initiate in this region. Although the HAZ contains both fine and coarse grains, its average grain size is much smaller than the coarse columnar grains of the fusion zone. Therefore, if the fusion zone of a weld pass is replaced by the HAZs of its subsequent passes, the fusion zone of this weld pass is „grain-refined". Such grain refining is often desired in the multi-pass welding of carbon and alloy steels.

The "higher-carbon steels" are carbon steels with more than 0.30% carbon. The welding of medium- and high-carbon steels is more difficult than that of mild steels. This is because with higher carbon contents the tendency to form hard, brittle martensite in the weld HAZ is greater, and hence under-bead cracking is more likely to occur.

The importance of influence of gradual grain refining on mechanical qualities increases with multi-pass welding. In the weld HAZ a very complicated and complex structure can be found, often with very different mechanical qualities. These actually depend on phase distribution and size of the grain (for the structure scheme of multi-pass welding see Figure 3). Numbers 1,2,3 and 4 represent individual beads laid down. A more detailed analysis of the coarse-grained region (for multi-pass welding) can describe characteristic parts of this region that it is - in the HAZ - composed of. Place A shows the coarse-grained region of the basic material that went through two degradation temperature cycles caused by laying down beads 1 and 3. It is not affected by heat anymore. Place A is constantly in the HAZ. Part B is the coarse-grained region of the basic material that went through a temperature cycle similar to annealing to remove the tension under the AC_1 temperature caused by laying down the third bead. Part C of the coarse-grained area of the additional material is the place that went through a heat treatment similar to annealing to remove inner tension when laying down bead no. 4. Part D shows the coarse-grained area of the basic material that did not go through any temperature influence of the upper beads and is permanent in the material. Part E shows the original coarse-grained area of the additional material, where - when laying down bead no. 4 – the temperature impact was similar to the normalizing annealing above the AC_1 temperature was carried out accompanied by grain refinement. Place F shows the coarse-grained part of the additional material that went through two degradation temperature cycles. The last characteristic area is place G, where the basic material went through grain coarsening when laying down the first bead and during laying down bead no. 2. This is where a process similar to normalizing annealing resulting in grain refinement took place. This description of the coarse-grained region makes it much easier to define the places that require a special attention when analyzing residual stresses – the most critical part of welds proves to be the coarse-grained part of the HAZ. The metallographic analyses indicate that the cracks under weld deposits are mostly initiated in the coarse-grained area of the basic material because the chemical composition of additional materials is much more homogeneous and steady. This fact reduces the number of places to analyze. These are the main reasons for using the CCT diagrams – part of which is the influence of the austenite grain dimension on the phase transformations – for the numerical simulation of welding.

WELDING CCT DIAGRAMS

The method of measuring the points of phase transformation for plotting the ARA diagrams was introduced at Welding Research Institute Bratislava in 1983 and it has been used since. An example of the CCT diagram is shown in Figure 5. The most suitable method for real welding conditions has proved to be the Granjon's in-situ method based on the principle of thermal analysis[2,3]. The measurement principle consists of bead deposition and measuring the thermal cycles in the heat affected zone using thermocouples. The local extremes of thermal cycles are emphasized by derivation. A hole is drilled in the area of supposed heat affected zone, where a thermocouple type chromel-alumel ϕ 0.5 mm will be located. Electrode OK 67.13 containing 26% Cr and 21% Ni, assuring pure austenitic weld overlay is used for surfacing. Diverse heat inputs are used so that the characteristic interval of thermal cycle $\Delta t_{8/5}$ would vary within the interval of 1 to 40 s. It is altered by the change of current and welding speed. The maximum temperature of thermal cycle is always above 1400°K.

TEMPERATURE AND STRESS ANALYSIS

Simulation of fusion welding has only one load – temperature. This is caused by the influence of the heat source on the construction. The temperature flow equals arc welding here. The problems of temperature specifications for calculation models were discussed at the PVP 98 conference in San Diego last year[4].

The impact of temperature history on the calculation of taut includes the change of the Young Modulus, yield stress and thermal expansion dependent on temperature. The influence of metallurgical history is included in the following four factors[5]:

1) The impact of metallurgical structure and the size of austenite grains on the mechanical qualities. Resulting mechanical qualities of the calculated structure in HAZ are derived from the mechanical qualities of individual phases. Each phase has entirely different material qualities and a big influence on the resulting mechanical qualities. In the corresponding literature [7,8] it is specified that the yield stress of austenite and ferrite is between 200 and 300 MPa, pearlite between 250 and 400 MPa, bainite between 400 and 800 MPa and martensite between 800 and 1200 MPa at 300°K. The results of our measurements correspond to those found in the corresponding literature. Values of yield stress are dependent on temperature and they are necessary for calculation.

2) The impact of different expansion and contraction of the α-phases and the γ-phase formed as a result of different temperature qualities during transformations. From thermal expansion the thermal strain ε^{th} is calculated from the well known formula $\varepsilon^{th} = \alpha \cdot (\Delta T)$, in which α is the thermal expansion. The use of εth in dependence on the temperature and phases is very practical because it makes it possible to detect the phenomenon of metal having a zero thermal strain at the moment of solidification. Thermal strain evolution during welding is shown on Figure 4. At the moment of solidification, austenite is specified by $\varepsilon^{th} = 0$ and then its shrinkage is considered and its influence on the weld surrounding (a). A specific area – from the point of view of thermal strain – is the place where the basic material melts with the metal added. The temperature of the basic material very rapidly rises to the temperature of melting – its extension affects the originating coarse-grained area of HAZ. During smelting all deformations (including stress) are removed and at the moment of solidification the area has a zero thermal strain (b). We have to follow a similar pattern also during such transformations when the phase that is being transformed also has a zero thermal strain. Since the phase transformations take place at different temperatures, the resulting thermal strain HAZ has qualities typical for the dilatometrical test (c). Thermal strain of the basic material, where there are no phase transformations, has linear qualities. The analysis of the process of thermal strain in individual phases of welding has an essential impact on the correct simulation of welding conditions.

3) The impact of different deformation strain hardness during metallurgical transformations. The plastic qualities of material result in the movement of dislocations. If a dislocation hits a barrier – for example a boundary of grains, alloying elements, segregation and so on, its movement stops and only after adding another – bigger – amount of energy other slip mechanisms can be started out. During transformations the level of plastic deformations and deformation strain hardness of material decreases because of the movement of dislocations. Each phase in the structure has a different strain hardness character. To specify the material strain hardness as a whole, two deformation parameters have an initial impact - α-phases and γ-phase. In case of martensite transformation the deformations are very small and the deformation strain hardness of martensite is modeled using the memory coefficient of the material specifying the anisotropic qualities of martensite.

4) The impact of transformation plasticity. During metallurgical transformations, that take place in the field of stress, a plastic deformation originates and causes the stress reduction to zero. This is caused by the change of volume during austenite disintegration into α-phases. Although, in steel, austenite forms a cubical square-centered crystal grid, the α-phases have a cubical space-centered crystal grid. Each grid occupies a different volume in the space. The factor describing this important phenomenon must be included into the formula for total proportional deformation. Our experience is that if transformational plasticity is not included into the calculation the difference in results of residual stresses could be as high as 120 MPa.

Total proportional deformation can be calculated as follows:

$$\varepsilon_t = \varepsilon_e + \varepsilon_{th} + \varepsilon_{pc} + \varepsilon_{tp}$$

where
- ε_t - total proportional deformation
- ε_e - elastic part of deformation
- ε_{th} - temperature part of deformation
- ε_{pc} - plastic part of deformation
- ε_{tp} - transformational plasticity

ANALYSIS OF RESIDUAL STRESSES

Analysis of residual stresses is an essential part of welding simulation. It is the most important part for the customer because he/she needs to know to what extent the construction has been impacted by welding, if a crack can occur either during welding or later during the device's operation, if it is necessary to use annealing to get rid of stress or if the construction can be used without taking any risks. For all these reasons this part is very important. It is clearly evident that in this part it is not enough to know the distribution of maximum stresses in the construction because it is not possible to use the limit criteria as with classical static analysis and explicitly say that 650 MPa of residual stress level is either too little or too much.

For constructions that are not designed for cyclical operation there seem to be three possibilities of residual stress analysis – calculation of the stress intensity factor K_I, local approach and calculation of micro-volume defect when modern damage models are used during welding. The first possibility describes the qualities of an already existing crack in the stress field. The method is very often presented in literature.[6] To specify the K_I factor we chose the process of creating a crack in the fields of residual stress in the FEM mesh – gradually in several places, and carried out the K_I calculation here using the J integral. This has one disadvantage – it is that the J integral is defined only for plain so it is possible to use it only for plain calculations. Some material engineers would also often object that the K_I factor is not a material constant and the crack has different behavior in the material – they doubt this way of analysis. We think that the following two ways are more prospective for future analysis, although working on their development has not been finished yet. Using the local approach makes it possible for us to utilize 100% of the result that the SYSWELD program provides for welding simulation – the levels of total plastic strain and grain size in individual areas of HAZ. We have verified the grain size calculation before but we have not used it for assessment yet. We are in the stage of discussing on what basis of experimental work should we specify the criteria of analysis of the simulation results so that both plastic strain and grain size are taken into account. Numerically, it is very easy to obtain the result, but the criteria for analysis will take a lot of effort. The last prospective possibility is the one that Welding Research Institute Bratislava has been working on – it is the development of a new non-linear model of welding that takes into account not only the combination of kinematic and isotropic strain hardness of the material but also the degree of damage during welding. Theory of damage is concerned with all materials at low as well as high temperatures under any kind of load. Knowing the stress and strain history for a given volume element of a structure, the damage laws provide, by integration with respect to time, the damage evolution in the element up to point of macroscopic crack initiation[10].

If the construction is engaged in a cyclical operation it is necessary to carry out the fatigue analysis. It is well known that residual stresses have a significant effect on the asymmetry of cycles, which has a very negative influence only with those devices that work in the regime of high-cycle fatigue. In the construction – external loading – a re-distribution of stresses takes place so the peak stress concentrations move into those areas with a lower level of stress. Here they have a negative effect on durability. For the analysis it is necessary to know the distribution of stress for the operation conditions. The influence of welding should be shown in the fatigue curve. We have used and described this process in our publications [9].

MENDING THE BODY OF THE PRESSURE VESSEL

The pressure vessel was manufactured from the low carbon steel. (0.20% C, 0.85%Mn, 0.30%Si, max.0.30%Ni, max.0.40%Cr, max. 0.30%Cu, max.0.025 S and P). The goal of this was to specify the levels of residual stress after mending the crack close to the weld of neighboring splints of the cylindrical casing of the pressure vessel, to analyze the levels of residual stress and to specify the impact on the pressure vessel's lifetime. The vessel was annealed during the manufacturing process so the manufacturing or operational residual stress was not taken into account. Figure 6 shows the FEM mesh. Figure 8 the detail of the repair area with elements. The sketch of the repair is shown on Figure 7. The suggested way of mending the defect was simulated on a level task with a detailed metallurgical, temperature and thermoplastic calculations taking into account also the grain size for phase transformations. In the first part of the task a verification program was prepared – all material and input data having an impact on the calculation were verified in it. The basis for the verification was an experimental measuring of temperature cycles to verify the temperature load, the level of HAZ, grain size in individual parts of the HAZ, hardness and geometry of the laid down beads on simple models[4]. These were simulated numerically and the results were compared to those of experiments. Material qualities of individual phases depending on the temperature for the welding needs were measured in the VÍTKOVICE, company. The condition for beginning the second part of the process was a satisfactory correspondence of the simple models' calculation results with experiments. In the second part of the process a simulation of the whole mending process was carried out including laying down individual beads. The impact of individual beads on the phase transformations (Figure 9) and levels of residual stress was monitored (Figure 10). Fig. 9 shows the results of the metallurgical calculation. It reveals the percentage development of the distribution of individual phases during welding. The Figure also gives the percentage of the distribution of martensite, bainite, and the dominanting tempering phase. The sum of all the phases, at any point, must give 100 %, which is represented by 1 in Fig. 9. The Figure of the final structure after welding demonstrates the effect of individual beads to the individual phase distribution in HAZ. The Figure shows the percentage phase distribution after the second, fifth, twentieth and the last, 37th, bead. The phase distribution was verified indirectly by means of the hardness measurement. Fig. 10 shows the development of individual components of the residual stresses in MPa when laying down individual beads. The final technology proposal was offered to the authorities in charge to evaluate.

MENDING THE PARTING PLANE OF THE PRIMARY AND SECONDARY CIRCUITS OF STEAM GENERATOR

For working out this task the SYSWELD program was used as a means of supporting a brand new technology for mending the parting plane of the cylindrical collector of steam generator for Slovak power plants. The parting plane of the steam generator is the integrity border between the primary and secondary spaces of a power plant. This part of steam generator is very important. The new mending technology designer was Welding Research Institute Bratislava (Slovak Republic). During the whole project the designer of the mending technology and the manufacturer of the welding simulation had to cooperate very closely because many different possible ways of mending proved to be possible during the process. Individual variants of the mending technology were being changed during the process according to the numerical simulation results. The results were not compared to the measurements but to individual variants. Neither the phase transformations nor the grain size was included into the calculation because the subject dealt with the simulation of austenitic steel qualities. Only material characteristics were specified according to measurements. For the simulation of the variants plain-strain models were used because they made it possible to get the results in real time. The plain-strain model of one variant simulation is shown on Figure 11. The technology under the development proposes to deposit thin layers of high-percentage nickel-material (80 % of Ni) on the parting plane of the header. The filler material used shows remarkably better mechanical properties than the basic material does. That is also a reason why the maximum residual stresses, after the groove repair, are in the filler material and the maximum equivalent plastic deformations in the basic material (Fig. 12). Within the procedure of the simulation the effect of individual beads on the development of equivalent plastic deformation at the critical points of the sealing groove were carefully monitored. The resulting mending technology of the parting plane was simulated on a spatial model (Figure 13 and 14). Final result of mending after restoration of selling grove from 3D model is shown on Figure 15.

SUGGESTION FOR INTERNAL FLAP DIMENSION FOR AUSTENITIC WELDS - DISTORTION PREDICTION

The closing pipeline flap is a cylindrical body made of carbon steel (0.20% C, 0.85%Mn, 0.30%Si, max.0.30%Ni, max.0.40%Cr, max. 0.30%Cu, max.0.025 S and P) onto which three layers of austenitic anti-corrosion material were welded. The customer demanded very strict production tolerances of minimum and maximum thickness of the austenitic weld so it was necessary to suggest the geometry of internal diameter numerically before welding – on the basis of simulating welding all three layers using the SYSWELD program. The manufacturer of the flaps required the result within three weeks. For the simulation a transversal cross section of the flap was chosen because it made it possible to simulate all the passing process of the electrode – on the internal perimeter of the flap. An elliptical non-stationary heat source was used – specially treated for a level task. Its dimensions were specified according to corresponding literature and practical experience (welding such an electrode in the Vitkovice company). The paths of the heat source were circles of different radius. Molten area and temperature field distribution during welding shows Figure 16. The beginning and end of each welding was recorded and taken into account on the model and it was – every time – moved 30º. In this calculation phase transformations were taken into account only in the basic material. The result of this calculation was movements of the internal diameter of the carbon part of the flap on its whole perimeter – after welding all three layers. On the basis of this the manufacturers designed the internal dimension of the flap before welding in such a way that the tolerance required by the customer was fulfilled.

CONCLUSION

For the welding simulation usually the plain tasks are used but with the rapid development of computer technology also spatial models are used more and more often. According to our experience we can conclude as follows:

Plain models
- it is difficult to specify the temperature load to correspond the cooling proportions in the HAZ,
- they provide us with much more precise and detailed results in the area of the HAZ because it is possible to choose a very detailed distribution of the MKP mesh, most significantly in the HAZ,
- they do not include the initial and final position of the heat source during welding,
- the condition of level deformation appears to be very conservative,
- they are practical for optimum welding, for verification of some results and for studying,

Spatial models
- require longer time machines to work out the results and efficient computer equipment,
- it is possible to specify temperature load more precisely and in an easier way,
- they are limited as far as the number of welded beads with regard to obtaining results in a realistic time is concerned,
- they can take into account the impact of the initial and final position of the heat source on the resulting residual stress,
- they are more suitable for specifying total deformations of the construction
- to obtain metallurgical results they require a very careful distribution of the FEM mesh in the HAZ,
- it is not possible to use them for variant tasks.

To successfully handle the problem of welding simulation it is necessary to have a high level of theoretical knowledge in several scientific subjects. Before all it is material engineering – to understand material characteristics during the process of welding, then it is the mechanics of rigid bodies for the internal verification of the stress results, then it is the welding technology and finally the experimental measuring to define the examinations on measuring non-standard material characteristics of individual phases up to high temperatures. Orientation in several scientific subjects requires the manufacturer of the welding simulation to have more theoretical knowledge.

REFERENCES
1. S. Kou, „Welding Metallurgy", students book, Madison, University of Wisconsin, August 1987.
2. H. Grajon, S. Debiez, R. Gaillard.: Soudage et Techniques Conn., 1968,3/4.
3. F. Kiraly.: Welding CCT diagrams of steels and weld joins, Welding Research Institute Bratislava, Slovak Republic, 1980.
4. L.Junek et al., The Effect of Repair Welding Residual Stress on Steam Generator Lifetime, In PVP-Vol.373, Fatigue, Fracture, and Residual Stress 1998, page 377, San Diego
5. L.Junek : Numerical Simulation of Welding and Heat Treatment, doctor thesis, Brno, June 1998, Czech Republic
6. P.Dong et al., Residual Stresses in Strength-Mismatched Welds and Implications on Fracture Behaviour, In PVP-Vol.373, Fatigue, Fracture, and Residual Stress 1998, page 351, San Diego
7. E.I.Kazancev, "Promyslenoje peci" (Moscow Metallurgia 1975)
8. Goldswith, Waterman and Hirschhorn, Handbook of Thermophysical Properties of Solid Materials, vol. 2, (The McMillan company, NY, 1961).
9. L. Junek at al., Steam Generator Weld Repair and Their Influence on Its Lifetime, 50th Annual Assembly International Institute of Welding, San Francisco, July 1997
10. J. Lemaitre, J.L. Chaboche : Mechanics of Solid Materials, Cambridge University press, 1990

11. J Becka : Analysis of grain-coarsening region degradation in HAZ of low alloy steels (in Czech), Zvaranie 5/1994, Bratislava

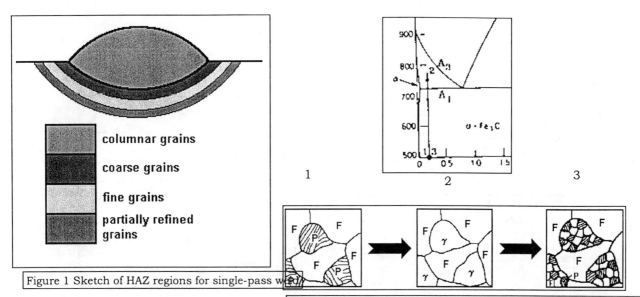

Figure 1 Sketch of HAZ regions for single-pass weld

Figure 2 Schematic illustration for partial grain refining region in the

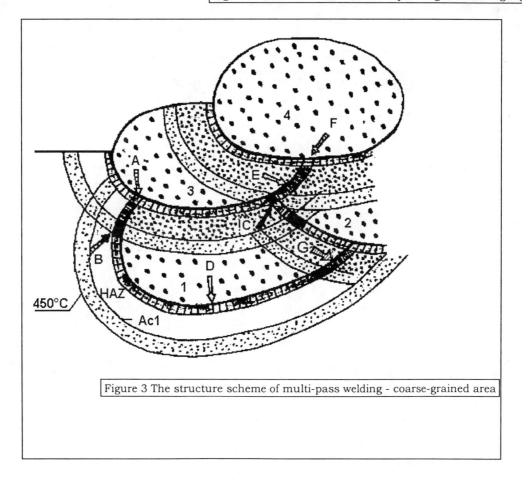

Figure 3 The structure scheme of multi-pass welding - coarse-grained area

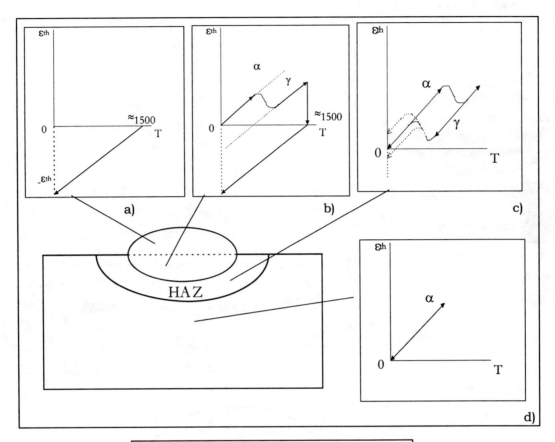

Figure 4 Thermal strain evolution during welding

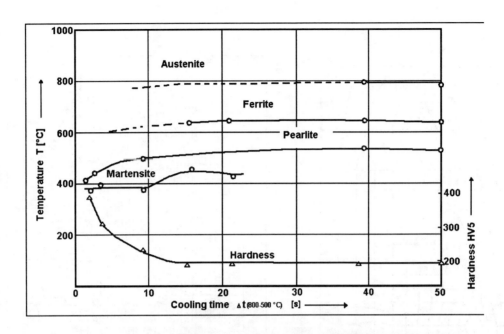

Figure 5 „In situ" CCT welding diagram example
(0.18% C, 0.89%Mn, 0.30%Si, 0.01%Mo, 0.05%Ni, 0.08%Cr, 0.03%Al)

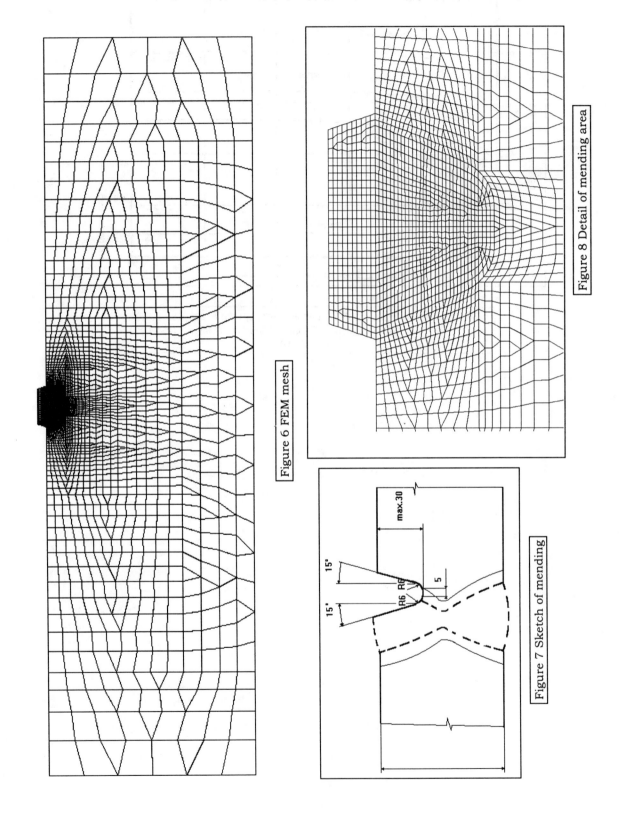

Figure 6 FEM mesh

Figure 8 Detail of mending area

Figure 7 Sketch of mending

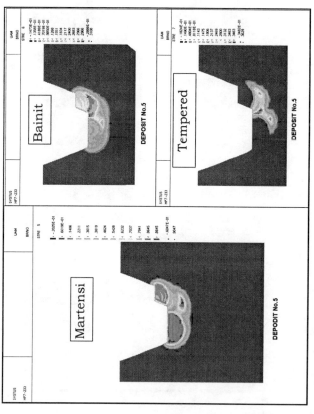
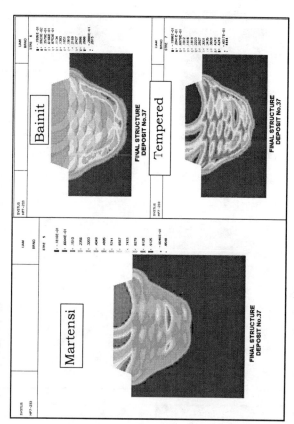
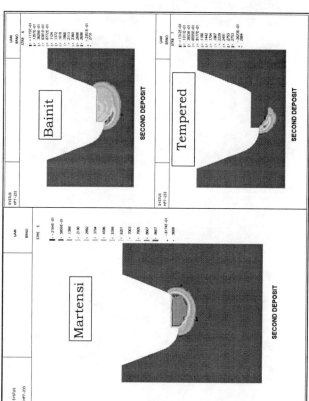
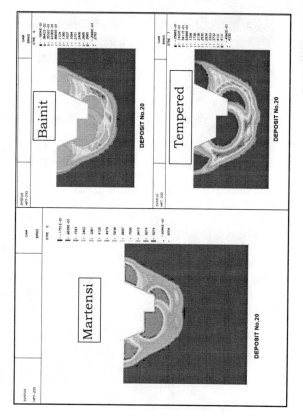

Figure 9 Phase proportion during mending

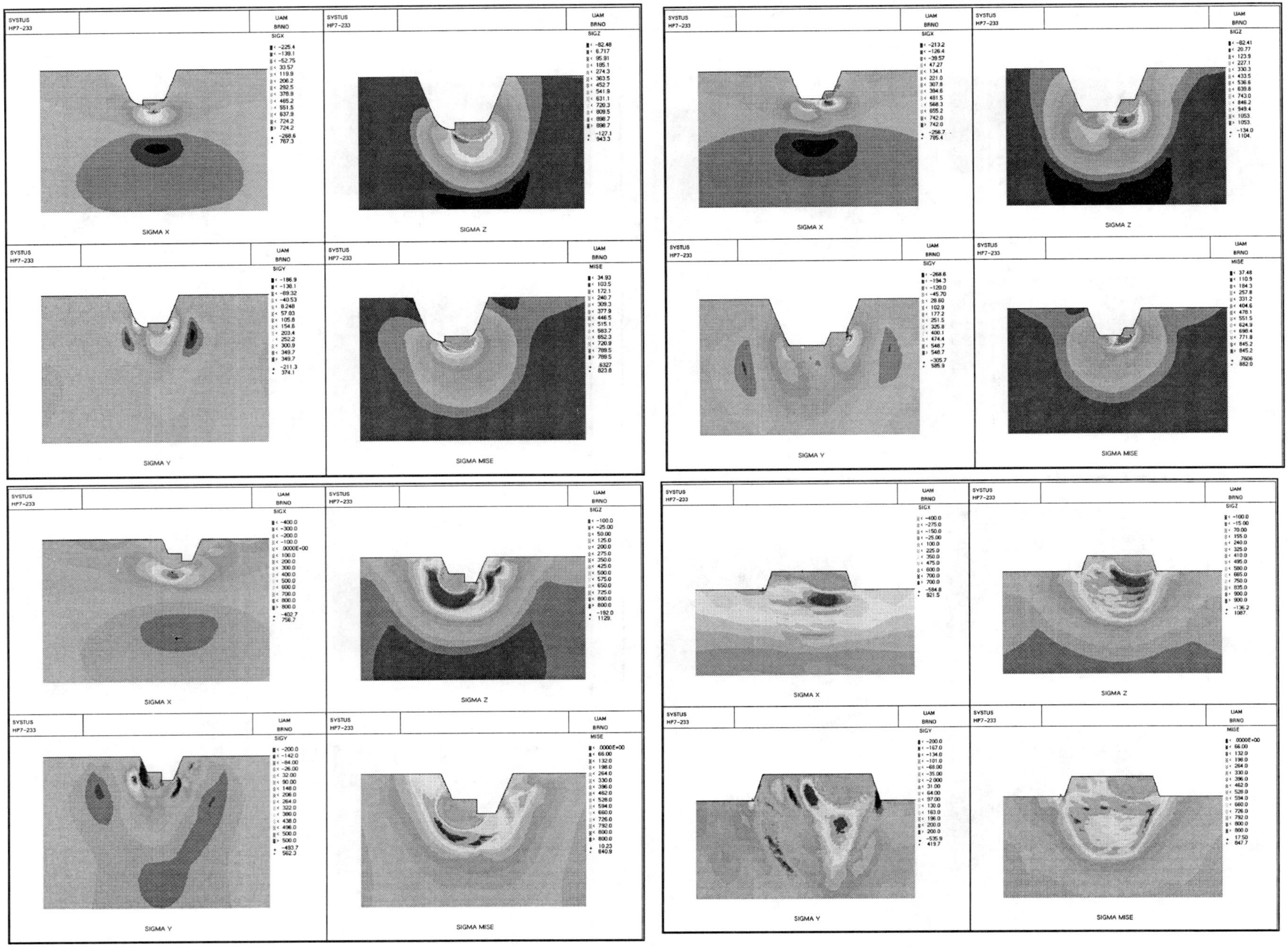

Figure 10 Residual stress distribution during mending

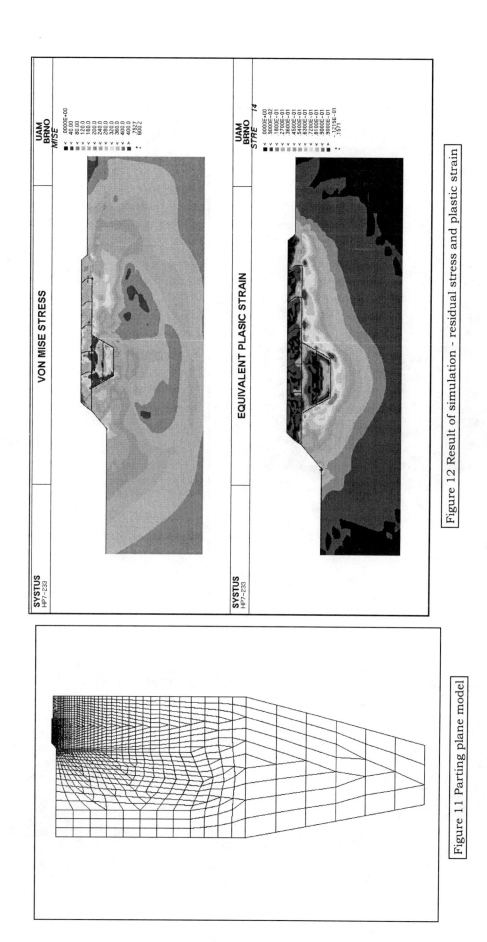

Figure 12 Result of simulation - residual stress and plastic strain

Figure 11 Parting plane model

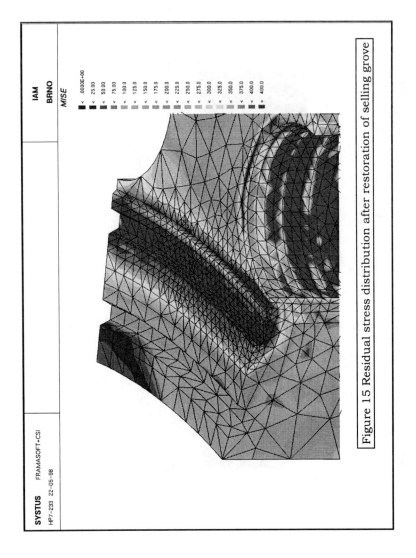

Figure 15 Residual stress distribution after restoration of selling grove

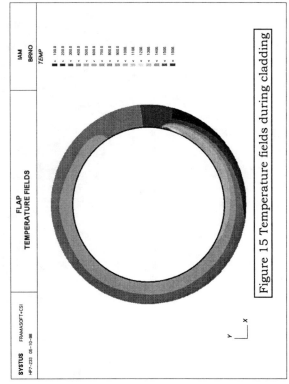

Figure 15 Temperature fields during cladding

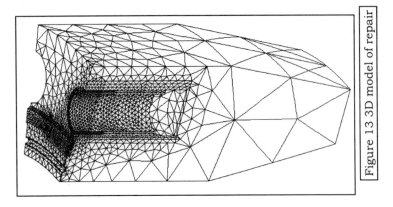

Figure 13 3D model of repair

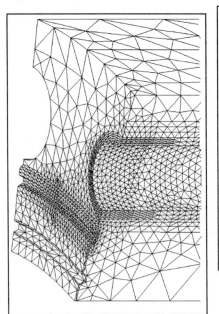

Figure 14 Detail of repair area with elements

ON THE VALIDATION OF THE MODELS RELATED TO THE PREVISION OF THE HAZ BEHAVIOUR

Y. VINCENT, JF. JULLIEN, N. CAVALLO, L. TALEB
URGC Structures, INSA de Lyon, 20, Avenue Albert Einstein, 69621 Villeurbanne cedex, France

V. CANO, S. TAHERI
EDF-EDF, 1, avenue du Général de Gaulle 92141 CLAMART cedex, France

Ph. GILLES
Tour FRAMATOME, 92084 Paris La défense Cedex, France

Abstract: Numerical simulations of a representative test of welding process are presented in this paper. A French vessel steel, which involves metallurgical phase transformations, is considered in this work. The aim is to validate the thermo-metallo-mechanical models taking into account the metallurgical transformations in the finite element codes Sysweld (Framatome) and *Code_Aster* (EDF).

The test is performed on a thin disc submitted to a thermal loading by means of a CO_2 laser beam, which leads to metallurgical transformations. The thermal, metallurgical and mechanical numerical results have been compared to the experimental results (temperatures, sizes of transformed zones, displacements and residual stresses). The simulation results of the two codes well reproduce the experimental results.

I INTRODUCTION

Electricité de France, Framatome and bureau de contrôle des chaudières nucléaires, had led a research program on welding finite-element simulation for the determination of nuclear components residual stresses and in particular for some steel with metallurgical transformations. One of the main stage of this work is to validate the mechanical models implanted in finite-element codes within which the metallurgical transformations effects are incorporated. This is the purpose of the present paper. The two codes used here are Sysweld (Framatome) and *Code_Aster* (EDF).

In particular this study is relative to the behaviour of materials situated in Heat Affected Zone (H.A.Z). In this aim INSA realised a test described in [1] in order to obtain a HAZ simular to those produced during welding process. This test is realised on a thin disc made of a carbon manganese steel (16MND5 in AFNOR norm) of 5mm thickness and 160mm of diameter. It consists of applying an axisymetric thermal load and in acquisition maximum measurements in order to validate the numerical simulations. The main advantage of this experimental welding simulation is the elimination of the two difficulties associated in welding process : the 3-dimensional aspect due to the mobility of heat source and the creation of a melting zone. By keeping only the creation of HAZ phenomena, the model's validation taking into account phase transformations becomes easier.

In the first section, one describes the experimental test. In the second section, one presents the two thermo-metallo-mechanical modellings implanted in the two codes. The last section is dedicated to the presentation of the results. One compares the simulation results with the experimental data.

II. / EXPERIMENTAL DEVICE, TEST AND RESULTS [1]

II.1 / Disc size definition

The thermal load is chosen to produce a martensitic transformation and a bainitic transformation during the cooling process in the centre of the disc and through its thickness. Therefore, the maximum temperature must be higher than 850°C which is the temperature of the end of austenitization for the considered steel. Every point whose the maximum temperature is higher than 750°C must have a minimum cooling rate of 10°C/s. Thermal load is applied by a $CO2$

laser. The heat flow transmitted to the disc by the laser is assumed to be axisymetric and have a defined form. Numerical preliminary simulations allowed to estimate the thickness (5mm) and the diameter of disk (160mm) in order to produce measurable stresses and displacements.

II.2 / Disc elaboration

The disc is submitted to a preliminary thermal cycle to relax internal stresses. The disk is manufactured, rectified and polished to reduce the residual stresses of the disc sides as much as possible.

The initial structure of the disc is essentially bainitic with a weak proportion of ferrite.

II.3 / Experimental procedure

The disc support is made of three alumina shafts whose the extremities are pointed to reduce the contact surface (see figure 1). The disc is heated by the CO2 laser during 70s and is cooled by natural convection and radiation.

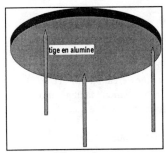

Figure 1

The experimental device is presented in the following figure.

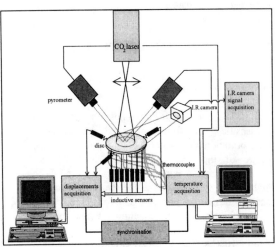

Figure 2

II.4 / Experimental measurements and results [1]

On the lower face of the disc (see figure 3), the temperature is obtained by thermocouples placed each 2mm from the centre, along the radial direction. The number and the position of the thermocouples have been chosen so that to allow by the use of a non-linear inverse method [2] the estimation of the thermal load in the entire specimen.

Seven inductive transducers allow the measurements of vertical displacements on the lower face for different radius as a function of time. Three displacement sensors are equally distributed around the circumference to measure the specimen's diameter variation and to verify the axisymetry of the thermal load.

Final shapes of the two sides are recorded along four diameters.

The residual stresses on upper and lower faces are measured by X-ray diffraction.

Finally, a micrographical analysis has been done that reveals several zones (see fig. 3) :
- Zone 1 : the steel is totally austenitized in heating. At the end in cooling, the steel contains bainite and martensite.
- Zone 2 : the steel is partially transformed in heating. At the end in cooling it contains ferrite, bainite and martensite.
- Zone 3 : this zone is affected micro-structurally by the heating but without any transformation. There is a tempering of the bainite. In this part, the phenomenon is characterised by a decrease of 15 % of the Vickers' hardness.
- Zone 4 : there is no transformation.

III. / MODELLING

III.1 / Boundary conditions

The problem is axi-symmetric. For the thermal part, the flow estimated by the inverse method is imposed on the upper face on a radius of 37mm (point A fig. 3) from the centre of the disc. In heating the boundary conditions, on the upper face from the radius 37 to 80 mm and on the lower and lateral faces are conditions of type natural convection and radiation. In cooling, all the faces are cooled by natural convection and radiation.

The initial metallurgical state of material is bainitic and ferritic.

For the mechanical part, the vertical displacement of the node B (see fig. 3) on the lower face situated at r=72mm is supposed equal to zero to simulate experimental conditions of simple support.

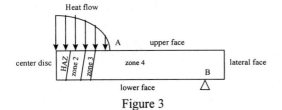
Figure 3

III.2 / Thermal and metallurgical modelling

Thermal and metallurgical calculations:

Heat diffusion is treated by the two codes with an enthalpic formulation. The thermo-physical properties (conductivity and enthalpy) depend on the temperature and on the metallurgical proportions of each phase.

In Sysweld, metallurgical and thermal calculations are coupled : at each temperature, the phase proportions are calculated. By a linear mixture law, the thermal characteristics are determined from the data of the austenitic and ferritic phases. The enthalpy of each phase includes the inertia effects and the latent heat of transformation.

In *Code_Aster*, there is no coupling : the phases proportions are determined in post-treatment of the thermal simulation. In order to take the metallurgical transformation effects into account on the enthalpy and the conductivity, these properties are not considered the same in heating and in cooling for the HAZ. The two enthalpic curves (in heating and in cooling) include too the latent heat of transformation.

Note that the values of the latent heat of transformation between ferritic and austenitic phases are not the same in *Code_Aster* and in Sysweld.

Transformation kinetic :

The constitutive equations giving the transformation kinetics used in Sysweld and *Code_Aster* are presented respectively in the papers [3] and [4].

The different transformations, which must be taken into account, are :
- the transformation of bainite and ferrite into austenite in heating,
- the transformation of austenite into bainite and martensite in cooling,
- the tempering of bainite into the zone 3.

The kinetics of diffusional transformations (austenitic, bainitic) are given by following relations :

In Sysweld :
$$\dot{Z}_i = f_i(\dot{T}) \frac{Z_{eq}(T, Ac1, Ac3) - Z_i}{\tau(T)}$$

In *Code_Aster* :

- For austenitic transformation (in heating) :
$$\dot{Z}_\gamma = \frac{Z_{eq}(T, Ac1, Ac3) - Z_\gamma}{\tau(T, Ac1, Ac3)}$$

- For bainitic transformation (in cooling) :
$$\dot{Z}_b = g_b(T, \dot{T}, Z_b) \frac{\langle T - Ms \rangle}{T - Ms}$$

Where :
T is the temperature, Z_i the proportion of phase i (i=b for bainite and i=γ for austenite), Z_{eq} the proportion of phase i at equilibrium, $f_i(\dot{T})$ a function allowing the transformation evolution control, $\langle X \rangle$ is the positive part of X, τ a temperature function.
Ms is the temperature at the beginning of the martensitic transformation. Ac1 and Ac3 are the quasi-static temperatures corresponding respectively at the beginning and the end of the austenitic transformation.
Note that the bainitic transformation is impossible when the martensitic transformation begins.

In *Code_Aster*, the kinetic of the bainitic transformation is not given by an analytic function. An interpolation procedure is used from TRC diagram.

However if the two codes use different kinetics, the parameters identification (Ac1, Ac3, τ) was done from the same experimental database.
For example, the temperatures Ac1 and Ac3 are :

Code_Aster: Ac1=716°C, Ac3=802°C

Sysweld: Ac1=705°C, Ac3=795°C

For the martensitic transformation, the two codes use the Koistinen-Marburger kinetic [5] :

$$Z_m = (1 - Z_b - Z_f)[1 - \exp(\beta \langle Ms - T \rangle)]$$

Where β is a material parameter.
The values of Ms and β are taken from [11] and are equal to : Ms=365°C and $\beta = -2.47 \cdot 10^{-2} \, °C^{-1}$.

Remark : A mechanical simulation showed that the tempering phenomenon is important to well describe the final state of stresses at the limit of the zones 2 and 3 (see figure 3). In Sysweld the tempering is described by a pseudo-transformation of metal parent. In *Code_Aster*, this phenomenon is not taken into account but the initial metallurgical composition is chosen such as at the end of cooling, the material located in tempering zone has a yield stress equal to the measured one. We assumed that a decrease of 15% of the Vickers' hardness (see section II.2) is associated with the same rate of decrease of the yield stress of the initial material.

III.2 / Mechanical modelling

The mechanical models included in *Code_Aster* and Sysweld are described in [6], [7] and [8].

They are unified constitutive models, written within the classical framework of irreversible thermodynamics processes using material state variables [9]. They include elastic plastic constitutive laws with isotropic hardening R and von Mises criterion of plasticity involving metallurgical effects. The two calculations were performed at finite strains. In Sysweld finite strains are treated by a reactualised Lagrange. In *Code_Aster*, the model at finite strains has been proposed by Simo and Miehe [10] and has been extended to take the metallurgical effects into account. The main aspect of the proposed model in [10] is that it allows to obtain exactly the numerical solution at finite rotations.

These two models describe several mechanical effects associated to metallurgical changes. These effects are :

- Mechanical characteristics modifications (especially plastic and thermal expansion). In particular, yield stress is modified. A linear mixture law is used for a ferritic phases mixture and a non-linear one for a austenito-ferritic mixture. The non-linearity intervenes through a function G(Z) depending on the proportions of ferritic phases Z.

$$\sigma_y = (1 - G(Z))\sigma_{y\gamma} + G(Z)\sigma_{y\alpha}$$

Where $\sigma_{y\gamma}$ and $\sigma_{y\alpha}$ are the yield stresses respectively of austenite and ferritic mixture.

- Thermal expansion and volume modification associated to structural transformation. The thermo-metallurgical strain is written as:

$$\varepsilon^{Th} = [Z.\alpha_f + (1 - Z).\alpha_\gamma](T - T_{ref}) - (1 - Z).\Delta\varepsilon_{f\gamma}$$

Where α_γ is the thermal expansion coefficient for austenite, α_f the thermal expansion coefficient for ferritic phases, $\Delta\varepsilon_{f\gamma}^{Tref} = \varepsilon_f^{Tref} - \varepsilon_\gamma^{Tref}$ the relative difference of volume between the two structures at the reference temperature Tref and Z the proportion of ferritic phases.

The thermal expansion coefficient and the relative difference of volume are taken from [11].

- Transformation plasticity phenomenon. It is the plastic flow arising from the variation of the phase proportions when a stress, even lower than the yield stress, is applied. The constitutive equations used in Sysweld and *Code_Aster* are :

Sysweld: $\dot{\varepsilon}^{pt} = \dfrac{3}{2}\tilde{\sigma}KF'(\sum_i Z_i)\sum_i \langle \dot{Z}_i \rangle$

Code_Aster: $\dot{\varepsilon}^{pt} = \dfrac{3}{2}\tilde{\sigma}\sum_i K_i F_i'(\sum_i Z_i)\langle \dot{Z}_i \rangle$

Where $\tilde{\sigma}$ is the deviator of the Cauchy stress tensor, Z_i the proportion of ferritic phases formed (here bainite and martensite), K and Ki constant material parameters, F and Fi given functions of Zi correspond to the ferritic phase i.
In the simulations, the function F is equal to $F(Z) = Z(1 - \ln Z)$ in Sysweld and $F(Z) = Z(2 - Z)$ in *Code_Aster* for the bainite and the martensite.
K is the same for all phases and equal to $0.7 \cdot 10^{-4}$ MPa for the two simulations. The K value is taken from [11].

- Hardening recovery associated to metallurgical phase change. Strain hardening due to dislocation structure in the material may be affected by atomic motion during structural change. Thus, the newly formed phase can have only a partial memory or even no memory at all, of the previous hardening.
To involve and describe the hardening recovery due to structural change, a parameter hardening r (associated with isotropic hardening R) is defined for each phase. Evolution laws include the hardening due to plastic strain and hardening recovery due to metallurgical transformations. These laws depend on the material parameter θ, which characterises hardening rate transmitted during the transformation. For transformations with diffusion (austenitic, bainitic and ferritic), θ is taken equal to zero (no memory). For transformations without diffusion (martensitic), θ is taken equal to one (perfect memory).

Remarks: The isotropic-hardening variable R (as a function of plastic strain) is linear in *Code_Aster* and given in Sysweld by the monotonic tension curve (point by point).
All the mechanical properties depend on the temperature.
The mechanical characteristics of different phases are taken from [12].

IV. / NUMERICAL SIMULATIONS AND COMPARISON WITH EXPERIMENTAL RESULTS

IV.1 / Thermal Result

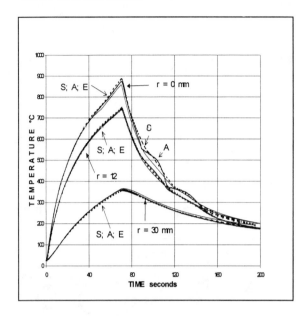

Figure 4

Experience : E --------
Code_Aster : A ———
Sysweld   : S --------

Figure 4 gives temperatures, simulated by the two codes and those measured experimentally, for different radius of disc (r=0, 12, 30 mm) on the lower face.

The comparison gives very satisfactory results. The evolution of temperatures calculated at heating by the two codes are completely superposed on experimental measurements until the time t=45s. At the end of heating t=70s, the maximum difference between the experiment and the simulation is respectively 16°C for *Code_Aster* and 29°C for Sysweld.

A variation of cooling rate can be remarked for the point situated in the HAZ (r=0mm) and is more pronounced for *Code_Aster* that for Sysweld and for the experiment. This variation is produced at the beginning of bainitic and martensitic transformations. It is related to the latent heat transformation at cooling, which has not the same value in *Code_Aster* and in Sysweld.

IV.2 / Metallurgical result

The size of totally and partially austenitized zones, obtained on the lower and upper faces, are given in the following table.

	Zone totally austenitized		
	Experiment	*Code_Aster*	Sysweld
Upper face	12mm	11.7mm	13mm
Lower face	9mm	7.9mm	9.8mm

	Zone partially austenitized		
	Experiment	*Code_Aster*	Sysweld
Upper face	14mm	15.4mm	15mm
Lower face	12mm	12.4mm	12.7mm

Figure 5

The results obtained by the two simulations are satisfactory and are practically the same. It is important to well find the size of the partially austenitized zone because the maximum residual stresses take place in the zone near to the border of the partially austenitized part (zone 2).

IV.3 / DISPLACEMENT

Figure 6 compares vertical displacements as a function of time obtained by the experiment, *Code_Aster* and Sysweld at the lower face (for r=1 mm and for r=30 mm).

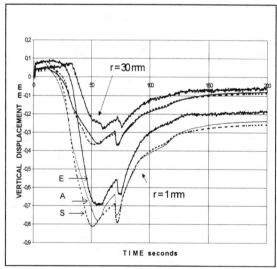

Figure 6

The evolution of calculated and measured displacements is nearly the same, despite some differences. Indeed one can notice that the two numerical simulations overestimate the maximal displacement range and that the fall of the central zone of the disc estimated by the simulations is produced earlier compared to the experiment.

In the beginning, the central zone of the disc rises because an important bending effect is induced by the existence of thermal strain gradient in the thickness.

At the instant t=10s, before the beginning of austenitic transformation, the displacement of the centre of the disc decreases and becomes negative. It is supposed that the displacement becomes negative when a sufficient part of upper surface of the disc is plastified. One can see this on the following figure which shows the plastic equivalent strain at t= 12.5s.

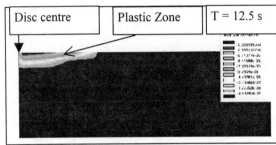

Figure 7

The difference between the experimental measurements and the numerical simulations can therefore be explained by the fact that the plasticity is produced earlier in the simulations than in the experiment. The temperature gradient in the thickness and the mechanical properties could explain so the observed difference.

Finally, between time t=50s and time t=75s, the evolution of the curve is due to austenitic transformation. The austenitic transformation begins at the centre of the disc on the upper face at t=42s and propagates in the disc with the increasing of the temperature. When the austenitized zone is sufficiently high, at t=55s, the volume diminution, due to the crystalline state change, generates an inversion in the slope of the displacement curve until t=70s (end of heating).

At t=71s the displacement of the centre of the disc decreases again because the temperature on the lower face continues to increase by conduction.

At cooling for t=75s, the central part of the disc retracts and the range of displacement reduces.

IV.4 / RESIDUAL DISPLACEMENT

Next figures present the residual displacement obtained on upper and lower faces by the two simulations and by the experiments.

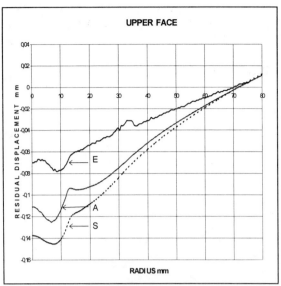

Figure 8

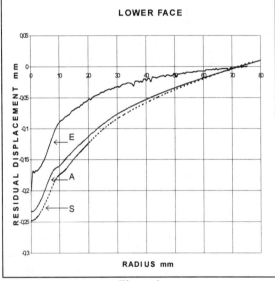

Figure 9

The results are satisfactory; the general shapes of simulated and experimental curves are the same.

The experimental curve shows a perturbation between 32 and 37 mm on upper face corresponding to the frontier of laser impact.

The numerical simulations slightly overestimate the residual displacement on the disc centre. The responsible causes of this gap are considered to be the same as those stated previously in IV.3.

IV.5 / RESIDUAL STRESSES

The next figures compare the residual stresses obtained by the two simulations and measured by X-ray diffraction.

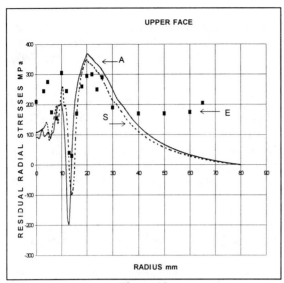

Figure 10

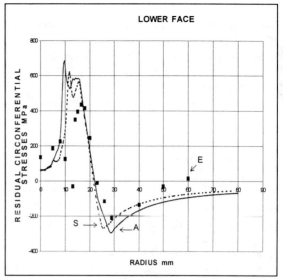

Figure 13

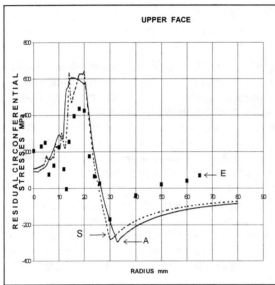

Figure 11

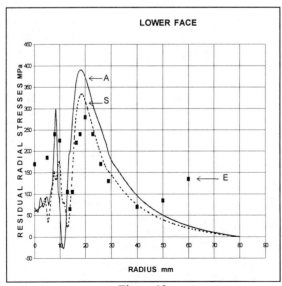

Figure 12

The general shapes of simulated curves are the same as experimental curves.

The disc before testing, has undergone a tempering in order to minimise initial stresses. Nevertheless, the radial stresses measured near to the border of the disc are not null. The analysis of this observation is going on.

The maximum simulated and measured stresses are located on the zone 2 and 3 (tempering zone, figure 3). The maximums calculated are however slightly upper with these measured.

It can be remarked that the circumferential stresses are higher than the radial stresses both for the numerical results and the experimental results.

V / CONCLUSION

The experimental device performed allowed the application of an axisymetric thermal load on a disc and the measurement during the test of temperatures and vertical displacements along a diameter of the lower face of the disc.

Furthermore, many usual measurements are done at the end of the test to obtain the size of totally and partielly austenitized zones and the residual stresses.

All these measurements allow the validation of each simulation stage and particularly the ability of the codes Sysweld and *Code_Aster* to estimate residual stresses in the case of structural transformations.

The results obtained by the two codes are nearly the same, despite the differences of modelling, and are in a good agreement with the experiment.

The three essential differents between *Code_Aster* and Syswel are :
- Thermal point of view analyses: The thermal calculated and the metallurgical calculated are

coupling in Sysweld and no coupling in *Code_Aster*,
- Metallurgical point of view : the phenomenon of tempering is taken in Sysweld into account and no in *Code_Aster,*
- Mechanical point of view : the finite strains are treated differently.

VI / REFERENCE

[1] Cavallo N., Contribution à la validation expérimentale de modèles décrivant la Z.A.T lors d'une opération de soudage. Thèse de L'INSA de Lyon, 1998, 211p.

[2] Blanc G., Raynaud M., Schau T.H., A guide for the use of the function specification method for 2D inverse heat conduction problems. International Journal of heat and mass transfers, 1997, Vol.13, pp. 703-716.

[3] Leblond J.B & Devaux J.C., A new kinetic model for anisothermal metallurgical transformations in steel including effect of austenite grain size, Acta Metallurgica, Vol.32, n°1, pp 137-146, 1986

[4] Waeckel F., Andrieux S, Bironneau L., Thermo-metallurgical modelling of steel cooling behaviour during quenching or welding, proceedings, fouth European Conference on residual stresses, Juin 4-6 1996

[5] Koïstinen D.O & Marburger R.E., Acta Metallurgica, 7, pp 50-60, 1959

[6] Razakanaivo A., Waeckel F., A viscoplastic model for numerical simulation of welding and post-weld heat treatment, euromech-mecamat, 3rd European mechanics of materials conference on mechanics and multi-physics processes in solids: Experiments, Modelling, Applications, Nov. 1998, Oxford.

[7] Leblond J.B., Mottet G. & Devaux J.C., A theoretical and numerical approach to the plastic behavior of steels during phase transformation, I : Derivation of general relations, II : Study of classical plasticity for ideal-plastic phases, Jour. Of the Mech. And Phys. of solids, Vol. 34, n°4, pp 395-432, 1986

[8] Leblond J.B., Devaux J. & Devaux J.C., Mathematical modelling of transformation plasticity in steels, I : Case of ideal-plastic phases, II : Coupling with strain-hardening phenomena, Int. Jour. Of Plasticity, Vol.5, pp 551-591, 1989

[9] Lemaitre J., Chaboche J.L., 1985, Mécanique des milieux continus, Editions Dunod

[10] Simo J.C, Miehe C., Associative coupled thermoplasticity at finite strains: Formulation, numerical analysis and implementation, Comp. Meth. Appl. Mech., 98, pp. 41-104

[11] Taleb L, Synthèse des résultats des essais de base de données pour les transformations uniques. Numéro : INSA-1/972/005a,09/06/98.

[12] Dupas P., Waeckel F., Recueil bibliographique de caractéristiques thermo-mécaniques pour l'acier de cuve, les revêtements inoxydables et les alliages 182 et 600 Clamart / Les Renardières: EDF/DER/IMA, 17/01/1994, note d'étude n)HI-74/93/097,46p.

RESIDUAL STRESS ANALYSIS AND FRACTURE ASSESSMENT OF WELD JOINTS IN MOMENT FRAMES

Jinmiao Zhang, Pingsha Dong, Frederick W. Brust

Battelle
505 King Avenue
Columbus, Ohio 43201-2693 USA
(614) 424-4690; Fax (614) 424-7618;
email - zhang@battelle.org

ABSTRACT

This paper summarizes the results of a recent study on the effects of residual stresses on the fracture behavior of weld joints in moment resistant frames. In particular, the effects of residual stresses on the fracture driving forces were investigated in detail. In the study, a typical beam-flange-to-column-flange weld configuration was analyzed using advanced weld modeling techniques. A series of fracture mechanics analyses were carried out to calculate the stress intensity factor and energy release rate. The results indicate that welding-induced residual stresses can significantly increase the fracture driving force at the weld root. Because of the presence of residual stresses, the weld joints tend to have less capability of plastic deformations and consequently are more vulnerable to brittle fracture under remote loading conditions.

INTRODUCTION

Following the 1994 Northridge earthquake, a wide spread of damage was discovered in the pre-qualified welded steel moment frames. Detailed inspections indicated that most of the structural damage occurred at the weld connections between the beam and column flanges (Refs. 1-4). In particular, the weld joints between the bottom beam flange and the column face suffered the most severe damage. Cracks mostly initiated at the weld root and propagated with very little indication of plastic deformation. The desired plastic hinges by design were not formed in the weaker beam away from the weld joint. Instead, brittle weld fracture was identified as the dominant failure mechanism. Mock-up full-scale and specimen tests in the laboratories also confirmed the occurrence of brittle fracture in the pre-Northridge weld connections (Refs. 5-10).

Welding-induced residual stresses are believed to be one of the factors contributing to the brittle fracture. Indeed, there exist ample evidence that residual stresses can play dominant role in the fracture process of highly-restrained welded joints (Refs. 11, 12). The design of welded moment resistant frame connections presents perhaps the most severe mechanical restraint conditions both during welding and in service. Consequently, the presence of high weld residual stress is expected. In addition, the triaxiality of the residual stress state in the joints can be significant such that the anticipated plastic deformation cannot develop before the fracture driving force reaches its critical value, resulting in brittle fracture.

This paper briefly summarizes the results of a recent study on the effects of residual stresses on the fracture behavior in moment resistant frame. A more comprehensive report on this topical investigation can be found in Ref. (13). The results summarized here include the residual stress distributions in a typical beam-flange-to-column-flange weld connection; the effects of residual stresses on the fracture driving forces (e.g., energy release rate and stress intensity factor); and the effects of beam material yield strength, weld defect size, and backing plate on the fracture driving forces.

Residual Stresses in Beam-to-Column Weld Connections

In this section, residual stress distributions in a typical beam-to-column weld connection are presented. The residual stress distributions were obtained by using an advanced weld modeling technique developed recently (see, e.g., Refs. 14-18). The obtained residual stress information was then fed into a subsequent fracture analysis for weld fracture assessment.

Residual Stress Model

A typical beam-flange-to-column-flange weld connection was modeled as shown in Figure 1. The model mimics an actual weld cross section as shown in the weld macrograph. Details of the weld joint configurations, such as the weld bead profiles (9 passes) and backing plate, were explicitly modeled. Note that the finite element mesh was extremely refined in the weld root area so that weld defects can be appropriately introduced in the subsequent fracture analysis.

The materials considered are as follows: A36 for the beam flange, A572-Gr.50 for the column, and E70-T4 for the weld. The mechanical properties of these materials are listed in Table 1. Note that there are yield strength mismatches between the beam (A36) and weld (E70T-4) materials, as well as between the weld and column (A572-Gr.50). Two different sets of yield and ultimate strengths were considered for A36 material, representing the upper (A36-High) and lower (A36-Low) limits of the strength variation commonly observed in this material. The welding parameters used can be found in Refs. (10,13).

Table 1. Mechanical Properties of Materials

Material		E (MPa)	<	Yield Stress (MPa)	Ultimate Strength (MPa)
A36	Low	220E3	0.3	260	460
	High	220E3	0.3	380	475
A572-Gr.50		220E3	0.3	400	540
E70T-4		220E3	0.3	415	495

Results and Discussions

The transverse residual stress distributions are shown in Figure 2. Also shown is a line plot of the residual stresses at the interface between the weld and column flange. Two tensile zones of stresses are clearly indicated, one in the top portion of the beam flange and another at the weld root. The line plot indicates that the tensile stress at the weld root is very high. This is due to the fact that a sharp notch exists at the weld root between the weld backing plate and column face. The size of tensile zone at weld root, however, is relatively small. It extends only about 2.5mm into the weld region. Beyond that region, a compressive stress zone is present, as shown in Figure 2.

FRACTURE ANALYSIS

The finite element model used for the fracture analysis of weld joint is identical to the one used for the residual stress characterization (Figure 1), except that a crack-like weld defect was introduced at the weld root. The procedure for the fracture analysis can be described as follows:

- After welding simulation, introduce a crack-like defect at the weld root
- Apply tensile load at the end of beam up to the yielding of the beam flange
- Calculate the energy release rate at the crack tip (G_{tip}) by using the crack closure integral method (virtual crack extension method, Ref. 19)
- Convert G_{tip} to stress intensity factor (K).

The weld defect was introduced after the completion of welding simulation. The defect can be viewed as a crack-like flaw, e.g., caused by lack of fusion. Field inspections indicate that crack-like weld flaws are common in the beam-to-column weld connections. The size of weld flaws observed in the inspection ranges from 0.5mm to 10mm (Ref. 8). In the model, the weld flaw was represented by defining duplicated nodes along the crack plane. These duplicated nodes were tied (bonded) together in the welding simulation and debonded when the crack was introduced. As a result, the as-welded residual stresses are re-distributed to achieve the stress-free condition at the crack surface. Two different sizes of weld defects were considered, 0.5mm and 2.5mm. External loads were then applied at the end of the beam flange by using a displacement control approach. The fracture driving force was determined by the energy release rate at the crack tip. In calculating the energy release rate, the virtual crack extension method (Ref. 19) was used by releasing the pair of bonded nodes at the crack tip (virtual crack extension) and calculating the work required to close them back. An extremely refined mesh was used in this area to ensure an adequate resolution.

Residual Stress Effects

The solutions of the energy release rate (G_{tip}) are shown in Figure 3a for the cases with and without residual stress effects. These solutions were obtained from a weld defect of 2.5mm and with A36-Low beam material. The external load (σ_p) was applied up to a level slightly beyond the initial yield stress of the beam material. The applied load plotted in the figure was normalized by the yield stress of the weld material (σ_{yw}). It is clear that the energy release rates are much greater for the case with residual stress effects than those without residual stresses. The presence of residual stresses leads to a non-zero initial value of the energy release rate and its more rapid increase as the load increases. However, the energy release rate ceases increasing (or increasing very slowly) after the load reaches the initial yield stress (indicated by the dashed line) of the beam material. This is different from the phenomenon commonly observed in the elasto-plastic fracture behavior for homogeneous materials in which the crack driving force (e.g., K_{IC}, J-integral) exhibits a sudden increase as the material is fully yielded. This can be explained by considering the yield stress mismatch effects. The weld joint analyzed represents a severe over-matched weld metal strength (60% higher than that of the beam material, see Table 1). The yield strength of the column material is relatively close to the weld strength. As the beam becomes fully yielded under the external tension, the further change in the stress state at the crack tip becomes negligible.

The corresponding stress intensity factor solutions are plotted in Figure 3b. Again, the effects of weld residual stresses are clearly seen. The fracture toughness (K_{IC}) of the weld material (E70T-4) ranges from 44 MPa*m$^{1/2}$ to 66 MPa*m$^{1/2}$ (Ref. 8), as shown in a shaded region. Thus, the present results considering the residual stress effects predict that brittle fracture can occur before the external load reaches the yielding stress of the beam material, while the results without considering residual stresses indicate that the crack remains stable even after the beam is fully yielded.

Beam Strength Effects

The effects of beam yield strength on the energy release rate are shown in Figure 4a. The corresponding stress intensity factor solutions are shown in Figure 4b. These results were obtained from a crack size of 2.5mm and with A36-Low and A36-High beam materials respectively. Weld residual stresses were considered in both cases. For A36-Low, the results are the same as those shown in Figure 3. These results indicate that the lower-strength beam material (A36-Low) causes even a slightly higher crack driving force before the beam is yielded. The initial crack driving force is also higher for A36-Low due to higher residual stresses at crack tip.

Defect Size Effects

The effects of weld defect size on the energy release rate and stress intensity factor are shown in Figures 5a and 5b respectively. The results were obtained from two different sizes of weld defects, 0.5mm and 2.5mm. The lower beam strength (A36-Low) was used in both cases along with the residual stress effects. For the cases without considering residual stresses, the larger defect size (2.5mm) results in a greater crack driving force than the smaller defect (0.5mm), as expected. However, when the residual stress effects are included, the small defect size yields even a larger fracture driving force despite the fact that the increase rate is smaller than that of the large defect. This is mainly because the initial value of the fracture driving force is much larger for the small defect situating in high weld residual stress region. This implies that small defects (as small as 0.5mm) can become critical if the surrounding residual stress state is not favorable.

Effects of Backing Plate

Additional analyses were also performed for the cases with and without backing plates. The analysis results are shown in Figures 6a and 6b for the energy release rate and stress intensity factor solutions. The results were obtained for both 0.5mm and 2.5mm defects. Residual stresses were included in both cases. The results indicate that, for the small defect (0.5mm), removing backing plate reduces the fracture driving force by about 5 to 8 percent. However, for the large defect (2.5mm), the backing plate removal shows almost no effects on the fracture driving forces.

CONCLUDING REMARKS

In this study, a series of finite element analyses were carried out to characterize the residual stresses in moment frame weld connections. The effects of residual stresses on the fracture behavior of the weld joints were investigated in detail. Advanced welding simulation techniques were used to quantify the residual stress development. Based on the results obtained, the following observations can be made:

- The transverse residual stresses in a beam-to-column weld connection are highly tensile in the top portion of the beam flange near the weld and at the weld root. The residual stress triaxiality is high in these regions, mainly caused by the severe structural constraints imposed on the weld. The tensile transverse residual stress is peaked at the notch tip between the backing plate and column face.

- The fracture driving force is very sensitive to the residual stress state at the weld root. The presence of high tensile residual stresses significantly elevates the fracture driving force. The results indicate that weld residual stresses can play a dominant role in the fracture process in the moment frame joints. Based on the cases analyzed, brittle fracture can be expected well before the beam reaches yielding. However, if the residual stresses are not considered, the crack may remain stable even after the beam is fully yielded.

- The yield strength of the beam material has some effects on the fracture driving force. As the beam is loaded up to the yielding point, the lower-strength beam material (A36-Low) produces less driving force than the higher one (A36-high) since the load transmitted to the weld region is lower at the yielding. However, at a given load less than the yield stress (e.g., $\sigma_p/\sigma_{yw}=0.5$), the fracture driving force produced by the lower strength beam is slightly larger. The initial fracture driving force is also slightly larger for the lower strength beam due to a higher residual stress at the weld root.

- The weld defect size effects should be evaluated with the consideration of both residual stress distributions and defect locations. For the two defect sizes studied, the small defect has even a higher fracture driving force if residual stress effects are considered.

- The removal of backing plate reduces the crack driving force for small defects at the weld root. However, this reduction is diminished as the weld defect size increases.

REFERENCES

1. Gates, W. E.; and Morden M. 1995. Lessons from Inspection, Evaluation, Repair and Construction of Welded Steel Moment Frames Following the Northridge Earthquake. *Technical Report: Surveys and Assessment of Damage to Buildings Affected by the Northridge Earthquake of January 17, 1994.* Report No. SAC-95-06: 3-1 to 3-79.

2. Naeim F.; DiJulio, R. Jr.; Benuska K.; Reinhorm, A. M.; and Li C. 1995. Evaluation of Seismic Performance of an 11-Story Steel Moment Frame Building during the 1994 Northridge Earthquake. *Technical Report: Analytical and Field Investigations of Buildings Affected by the Northridge Earthquake of January 17, 1994.* Report No. SAC-95-04, Part 2: 6-1 to 6-109.

3. Green, M. 1995. Santa Clarita City Hall: Northridge Earthquake Damage. *Technical Report: Case Studies of Steel Moment Frame Building Performance in the Northridge Earthquake of January 17, 1994.* Report No. SAC-95-07: 1-1 to 1-13.

4. Hajjar, J. F.; O'Sullivan, D. P.; Leon, R. T.; and Gourley, B. C. 1995. Evaluation of the Damage to the Borax Corporate Headquarters Building as a Result of the Northridge Earthquake. *Technical Report: Case Studies of Steel Moment Frame Building Performance in the Northridge Earthquake of January 17, 1994.* Report No. SAC-95-07: 2-1 to 2-76.

5. Whittaker, A.; Bertero V.; and Gilani A. 1996. Seimic Testing of Full-Scale Steel Beam-Column Assemblies. *Technical Report: Experimental Investigations of Beam-Column Subassemblages.* Report No. SAC-96-01, Part 1: 2-1 to 2-221.

6. Popov E. P.; Blonder M.; Stepanov L.; and Stozidar S. 1996. Full-Scale Steel Beam-Column Connection Tests. *Technical Report: Experimental Investigations of Beam-Column Subassemblages.* Report No. SAC-96-01, Part 2: 4-1 to 4-151.

7. Shuey, B. D.; Engelhardt, M. D.; and Sabol T. A. 1996. Testing of Repair Concepts for Damaged Steel Moment Connections. *Technical Report: Experimental Investigations of Beam-Column Subassemblages.* Report No. SAC-96-01, Part 2: 5-1 to 5-332.

8. Kaufmann, E. J.; and Fisher, J. W. 1995. Fracture Analysis of Failed Moment Frame Weld Joints Produced in Full-Scale Laboratory Tests and Buildings Damaged in the Northridge Earthquake. *Technical Report: Experimental Investigations of*

Materials, Weldments and Nondestructive Examination Techniques, Report No. SAC-95-08: 1-1 to 1-21.

9. Kaufmann, E. J.; and Fisher, J. W. 1995. A Study of the Effects of Material and Welding Factors on Moment Frame Weld Joint Performance using a Small-Scale Tension Specimen. *Technical Report: Experimental Investigations of Materials, Weldments and Nondestructive Examination Techniques*, Report No. SAC-95-08: 2-1 to 2-29.

10. Kaufmann, E. J. 1997. Dynamic Tension Tests of Simulated Moment Resisting Frame Weld Joints. *Steel Tips*: 1-24. Structural Steel Educational Council.

11. Brust, F. W.; Zhang, J.; Dong, P. 1997. Pipe and Pressure Vessel Cracking: The Role of Weld Induced Residual Stresses and Creep Damage during Repair. *Transactions of the 14th International Conference on Structural Mechanics in Reactor Technology (SMiRT 14)*, Lyon, France, Vol. 1, pp. 297-306.

12. Brust, F. W.; Dong, P.; Zhang, J. 1997. Influence of Residual Stresses and Weld Repairs on Pipe Fracture. *Approximate Methods in the Design and Analysis of Pressure Vessels and Piping Components*, W. J. Bees, Ed., PVP-Vol. 347, pp. 173-191.

13. Zhang, J.; Dong, P., "Residual Stresses in Welded Moment Frames and Implications on Structural Performance." *Proceedings of International Conference on Welded Construction in Seismic Areas*, Maui, Hawaii, October 6-8, 1998, pp. 57-75.

14. Brust, F. W.; Dong, P.; Zhang, J. 1997. A Constitutive Model for Welding Process Simulation using Finite Element Methods. *Advances in Computational Engineering Science*, S. N. Atluri and G. Yagawa, Eds., pp. 51-56.

15. P. Dong, J. K. Hong, J. Zhang, P. Roger, J. Bynum, and S. Shah, *Effects of Repair Weld Residual Stresses on Wide-Panel Specimens Loaded in Tension. Journal of Pressure Vessel Technology*, Vol. 120, No. 2, pp. 122-128 (1998).

16. Dong, P.; Zhang J.; and Brust, F. W. 1997. Residual Stresses in Strength-Mismatched Welds and Implications on Fracture Behavior. *IIW Doc. X-F 057-97*, IIW 50th Annual Assembly, San Francisco.

17. Zhang, J.; Dong, P.; Brust, F. W. 1997. Analysis of Residual Stresses in a Girth Weld of a BWR Core Shroud. *Approximate Methods in the Design and Analysis of Pressure Vessels and Piping Components*, W. J. Bees, Ed., PVP-Vol. 347, pp. 141-156.

18. Brust, F. W.; Dong, P.; Zhang, J. 1997. Crack Growth Behavior in Residual Stress Fields of a Core Shroud Girth Weld. *Fracture and Fatigue*, H. S. Mehta, Ed., PVP-Vol. 350, pp. 391-406.

19. Kanninen, M. F.; and Popelar, C. H. 1985. *Advanced Fracture Mechanics*. Oxford University Press, New York.

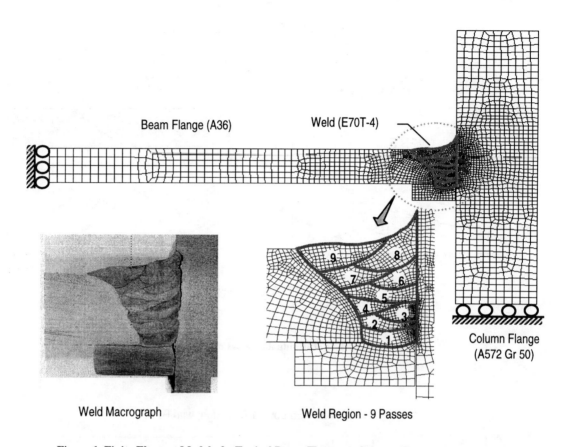

Figure 1. Finite Element Model of a Typical Beam-Flange-to-Column-Flange Weld Connection

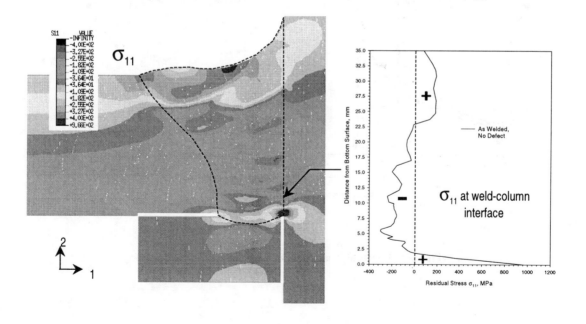

Figure 2. Transverse Residual Stress Distributions at Weld Joint

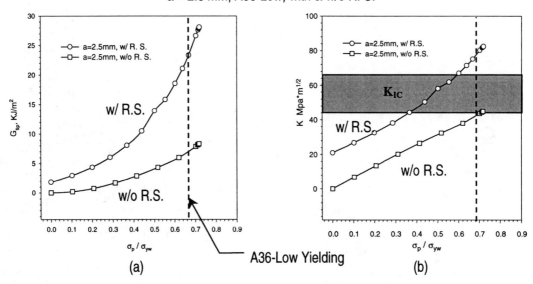

Figure 3. Effects of Residual Stresses on Crack Driving Force

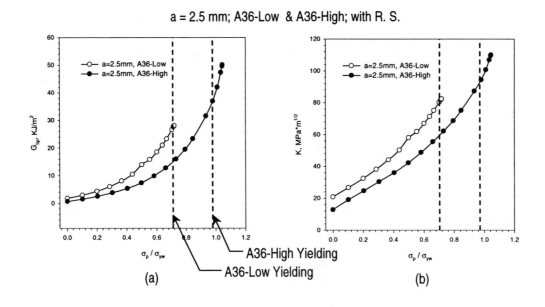

Figure 4. Effects of Beam Yield Strength on Crack Driving Force

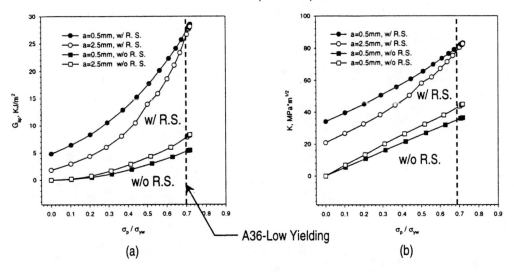

Figure 5. Effects of Weld Defect Size on Crack Driving Force

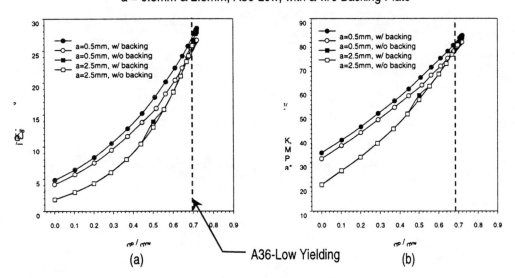

Figure 6. Effects of Backing Plate on Crack Driving Force

THERMO-MECHANICAL MODELING OF RESIDUAL STRESS AND DISTORTION DURING WELDING PROCESS

Yuh-Jin Chao and Xinhai Qi
Department of Mechanical Engineering, University of South Carolina
Columbia, SC 29208, USA
Tel: (803)777-5869, Fax : (803)777-0106
e-mail : chao@sc.edu

ABASTRACT

A three-dimensional finite element simulation of welding process is presented. The analysis adopts a de-coupled heat transfer and thermo-mechanical analysis for the determination of residual stress and distortion. Special features for the welding process are included in the modeling to reduce the computational time of the analysis. A butt-welded aluminum plate was analyzed and the results are compared with experimental data for the residual stress distribution. It is shown that the full three-dimensional model can provide some results that cannot be obtained by simplified two-dimensional simulation.

INTODUCTION

Welding is a popular joining method used in many industries, e.g. from offshore drilling platform to electronic packaging. By going through heating, melting and solidification processes, two workpieces are joined together using a torch, laser, electron beam or others as heat source. The highly localized, transient heat input and extremely non-uniform temperature fields in both heating and cooling processes cause non-uniform thermal expansion and contraction, and thus plastic deformation in the weld and its neighborhood. This results in residual stress and strain and permanent distortion of the welded structure. The residual stress in the weld region affects the service life of the structure, e.g. stress corrosion cracking of the weld, and the distortion directly influence the fabrication of the structure. Thus, control of residual stress and distortion from the welding process is extremely important in the manufacturing and construction industry. Besides the determination of the residual stress and distortion of the welded structure, modeling of the welding process can also be used to incorporate the welding procedures/pattern into design evaluation, optimization and manufacturing tools.

Due to its complexity, analytical solutions for the welding can rarely address the practical manufacturing processes. Numerical solutions, on the other hand, can provide more detailed results and address more realistic problems. Hibbitt and Marcal (1973) marked the first step in applying a two-dimensional finite element analysis (FEA) to predict residual stress in a weldment. Modeling efforts in the 70's and 80's are essentiall two-dimensional or very simple three-dimensional geometries due to the limitations on the computational time and cost of a full three dimensional model (see Masubuchi, 1980 and Karlsson, 1986). With the rapid advancement in computing speed from modern computers, numerical modeling of practical welding processes now becomes feasible. Recently, three-dimensional modeling using commercial finite element codes are reported by Tekriwal and Mazumder (1991), Michaleris and DeBiccari (1997), Dong, Y. et al (1997) and Dong, P. et al (1997). In using the commercial code, many subroutines must be added to handle the special features of the welding process, e.g. melting and annealing, which are not usually included in general-purpose finite element codes.

In this paper, a FEA program for welding process simulation is demonstrated. The FEA is three-dimensional and runs on a PC platform. A flow chart of a complete modeling of the welding process involving a heat transfer analysis and a subsequent thermo-mechanical analysis is shown in Fig.1. A moving heat source is modeled first to generate the temperature fields at various time steps during the welding process and then the temperature history is used for the calculation of the thermal stresses and displacement fields during and after the welding process. Unique features of the welding process are included in the code and appropriate convergence is controlled by the user with flexibility. The computer code is developed in order to study the welding procedures and for practical engineering applications.

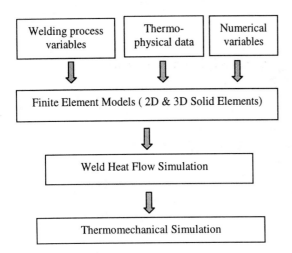

Figure 1 Flow chart of the welding process modeling

In the following sections we first outline the basic principles of the finite element analysis, then discuss special features of the code, present the results from the modeling of a gas metal arc welding of a rectangular plate, and finally compare the analytical results to experimental data. A discussion on why a three-dimensional modeling is needed is given.

FINITE ELEMENT FORMULATION

It has been shown by Argyris, et al (1982) and Papazoglou and Masubuchi (1982) that the difference between a coupled and de-coupled analysis for residual stress and distortion in a welding process is very small. Therefore, we have adopted a de-coupled finite element analysis. In the heat transfer analysis, the differential equation governing the unsteady heat conduction problems can be obtained by using the first and second law of thermodynamics. By assuming that the thermal conductivity λ is independent of spatial coordinate, one obtains:

$$\rho \, c \, \dot{T} = \dot{Q}_{int} + \lambda T_{,ii} \quad (1)$$

Where ρ = density (kg/m^3), c = thermal capacity (J/kg K), λ = thermal conductivity (W/m K), T = temperature (K), Q_{int} = internal heat generation per unit volume (J/m^3), dot ($\cdot$) denotes differentiation with respect to time t, comma (,) denotes partial differentiation with respect to a spatial coordinate. The heat input to the workpiece can go through the surface as q_i or as the internal heat source Q_{int}. The finite element formulations were then followed to solve the heat conduction problem. Several types of moving heat source applied to the surface are modeled to simulate the actual heat generated by the arc from the nozzle. For instance, a moving heat source with Gaussian distribution is included in the program for modeling the gas metal arc welding.

In the heat transfer analysis, the temperature history at each finite element node is stored and used for the subsequent thermo-mechanical analysis. The basic mechanics formulation starts from the constitutive equation, in incremental (differential) form for the thermal elastic-plastic material (Karlsson, 1986), i.e.

$$d\varepsilon^{total} = d\varepsilon^{elastic} + d\varepsilon^{plastic} + d\varepsilon^{thermal} + d\varepsilon^{creep} \quad (2)$$

The creep strain is for modeling the post-weld heat treatment used to reduce or change the distortion and reduce the residual stress of the welded structure. The finite element formulation uses a rate independent plasticity model with kinematic hardening to model the reverse plasticity and Bauschinger effect. Power law for the uniaxial stress-strain curve beyond the yielding is used for the non-linear material behavior. Von Mises yield criterion and an associated flow rule is used to determine the onset of yielding and the amount of incremental plastic strain.

SPECIAL FEATURES TO IMPROVE THE COMPUTATION SPEED AND ACCURACY

One of the drawbacks of using the commercial finite element codes for modeling welding process is the slow speed and non-existence of unique features in the welding process in the codes. In order to accelerate the computation, several methods are introduced in our welding process simulation.

(1) Flexibility in convergence tolerance: Using the virtual work theorem, the incremental form of the equilibrium equation at the (n+1) step can be written as:

$$[K]\{\Delta u\} = \{\Delta F\} + (F_n - f_n) \quad (3)$$

where $[K]$ is the tangent stiffness matrix; $\{\Delta F\}$ is the increment of external load; while F_n, f_n are the external and internal loads at the n-th step, respectively. In equation (3), F_{error} (=$F_n - f_n$) is the non-equilibrium load caused by various nonlinear factors, and it should be less than a given tolerance after several iterations. If F_{error} is too larger, the convergence would not be reached, and it is impossible to obtain accurate results. The convergence is set up in our FEM program if in any step

$$E_{rr} = \sqrt{\frac{\sum F_i^2}{\sum F_j^2}} < \varepsilon \quad (4)$$

where F_i is the non-equilibrium force at the free node i, F_j is the reaction force at the fixed node j on the boundary, and ε is a small tolerance value (Wang, et al, 1996).

Convergence of iteration is checked and compared with the tolerance value and the values of the last steps. If a persistent oscillation occurs, convergence of the iteration may become very slow. In the program, analysis will go on to the next step, with a warning signal recorded in the final output for the user to check, if tolerance is only slightly higher than the specified tolerance value. And, the analysis will be stopped only if tolerance value is much larger than the allowable tolerance. That is, convergence criterion of a few steps in the analysis for the whole problem may be slightly higher than the specified value, but a majority of iteration steps are still controlled under a small convergence criterion. It is found that with this flexibility the computation time for the complete analysis can be reduced and without sacrificing the accuracy of the results.

(2) During welding metal temperature can reach as high as melting point, but at these high temperatures the metal carries no strength (the elastic modulus is assumed negligible above the melting point) and hence does not contribute to residual stress calculation. For instance, the elastic modulus of mild steel drops considerably above 600 °C and therefore contribution to residual stress due to temperature above 600 °C is small. So, the temperatures above a cut-off point from the thermal analysis are reduced to the cut-off temperature during the mechanical analysis. In doing this, the iteration for higher temperature is avoided. Tekriwal and Mazumder (1991) have reported that similar results for residual stress are obtained when the temperature history is used with or without this modification, while computational time can be reduced considerably.

(3) *Weight factor for elastic to plastic transition*: A specific weight factor for elastic-plastic transition at loading and unloading stages is used in order to improve the accuracy and stability of the analysis, and thus greatly reduces iteration times especially at high temperature stages. When an element is in the transition between elastic stage and plastic stage, the step can be divided into two parts: the elastic deformation and the plastic deformation part, with a weight factor $\omega/(1-\omega)$ defining the relative portion of the two stages and $0 \leq \omega \leq 1$. Accurate determination of the weighting factor ω is important for improving the accuracy and convergence of the solution and for reducing the number of iterations. Generally, in the elastic-plastic analysis one has

$$\omega = (\sigma_y - \sigma_1)/(\sigma_2 - \sigma_1) \quad (5)$$

Where σ_y is the yield stress, σ_1 and σ_2 are the equivalent stress at the previous step and the current step, respectively. However, in thermal elastic-plastic analysis, the yield stress is a function of temperature, the factor ω could be less than zero or higher than one. Hence, the iteration of solution may become unstable.

To overcome this problem, the weighting factor for the transition between elastic and plastic stages is defined as follows (Wang, et al, 1996):

$$\omega = (\sigma_{y1} - \sigma_1)/(\sigma_{y1} - \sigma_1 + \sigma_2 - \sigma_{y2}) \text{ for loading, and}$$
$$\omega = 2\sigma_1/(2\sigma_1 + \sigma_2 - \sigma_{y2}) \text{ for unloading with re-yielding} \quad (6)$$

Where σ_{y1}, σ_1 are the yield strength and equivalent stress at the previous step; σ_{y2}, σ_2 are the yield strength and equivalent stress at the current step. The element reaches yielding at temperature T_y and the yield stress at that moment is given by $\sigma_y = \sigma_{y1}(1-\omega) + \sigma_{y2}\omega$. $T_1 \sim T_y$ is in the elastic stage and $T_y \sim T_2$ is in the plastic stage.

(4) *De-coupled heat transfer and thermo-mechanical analysis:* To speed up the computation, the welding simulation combines heat transfer analysis and thermal-mechanical analysis in a de-coupled manner. This de-coupling is valid when the relative displacements within the welding portions of the structure are small, assuring that displacements do not shift the welding electrode position and consequently the location of the heat flux applied to the model (Brown and Song, 1992).

In addition to the features discussed, dummy elements, latent heat, complex heat source models, local coordinate transformation, reduced integration, large deformation (for buckling analysis), and fixture relief after welding can be considered in this welding modeling as well, in order to simulation complex welding procedures and obtain more accurate results.

MODELING OF ALUMINUM PLATES JOINED BY GMAW

This section reports results from modeling a butt joint by gas metal arc welding (GMAW). The numerical results from our analysis are compared with the experimental data by Canas et. al. (1995). In addition, results which can only be obtained from a three-dimensional analysis and cannot be obtained by simplified two-dimensional analysis are reported. Figure 2 shows the geometry and welding configuration of the problem under consideration. A half model was used in the simulation due to symmetry. The FEM model consists of 4040 nodes and 2700 (8 nodes) solid elements in the heat transfer analysis, and 2040 nodes and 1350 solid elements in the thermo-mechanical analysis. Figure 3 shows the FEA Mesh for the thermo-mechanical analysis. The material is the aluminum alloy *Al*5083-O. A Gaussian distribution with an effective arc radius of 7 mm and efficiency of 64% is used to simulate the heat flux of GMAW. The heat is applied in the Z=0 plane and moves along the +X-direction. Material properties can be found in Canas et. al. (1995).

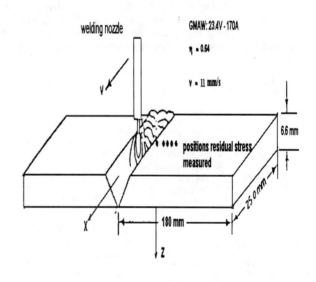

Figure 2 Configuration of the model

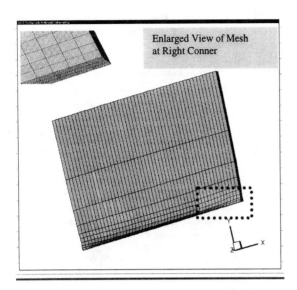

Figure 3 FEA mesh in the thermo-mechnical model

RESULTS FROM THE THERMO-MECHANICAL ANALYSIS AND COMPARISION TO THE EXPERIMENTAL DATA

Residual stresses of σ_{xx}, σ_{yy}, σ_{xy} at the centroids of the elements along a section in the middle of the plate (X, Y, Z) = (122.5, Y, 1) mm are shown in Fig.4 along with the experimental data. It is shown in figure 4 that the analytical results compare well with the test data.

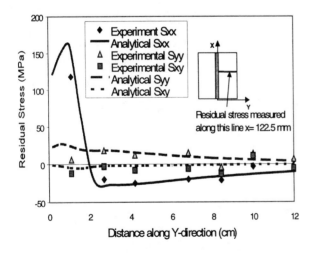

Fig. 4 Residual stresses: comparison of numerical and experimental data

Residual stresses from the centroids of the first elements along the weld line, i.e. (X, Y, Z)=(X, 2, 1) mm, which is near the center of the weld are shown in figure 5. It is seen that in the area of x =40 ~200 mm uniform values of the three components of residual stresses are obtained. However, near the edges of the plate, i.e. X = 0~40 mm and X = 200~250 mm, the residual stresses are not the same as those in the central area. In fact, the maximum residual stress occurs near the free edge, i.e. see σ_{yy} near X =0 and 250 mm in figure 5. It should be mentioned that a two-dimensional modeling can at most obtain the results shown in figure 4, which falls in the central portion of the welded plate. Results in figure 5 indicate the complete information on the residual stress can only be determined from the three-dimensional modeling.

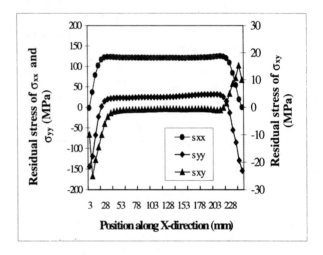

Figure 5 Residual stresses along weld line (x, 2, 1) mm

Figure 6 shows the residual stress distributions along the Y-direction in three sections; one near the starting edge X=7.5 mm, one in the central part X= 122.5 mm and one near the finishing edge X= 242.5 mm. Again, the difference in the residual stress distribution in the three sections can be seen.

DISTORTION

The computed shrinkage in the transverse direction and the longitudinal direction as well as the angular distortion β are listed in Table 1 along with the empirical predictions. The transverse shrinkage shown in Table 1 is the shrinkage of the plate transverse to the weld-line, i.e. along the Y-direction, averaged for the total length of the plate. The longitudinal shrinkage is the shrinkage at the center of the weld-line, i.e. along the X-axis. And the angular distortion β is defined in Fig. 7.

The empirical predictions are based on formulae from Masubuchi (1980), and Wang (1994), i.e.

Transverse shrinkage: $\Delta L = \dfrac{20.4 \cdot W \cdot 10^{-3}}{t \cdot v}$

Table 1 Comparison of distortion : FEA results vs. the empirical formulae

	Transverse shrinkage (mm)	Longitudinal Shrinkage (mm)	Angular Distortion (rad)
Empirical results	0.19	0.20	0.01
FEA results	0.24	0.17	0.007

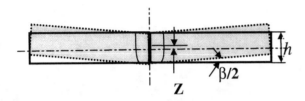

Figure 7 Angular distortion β of a butt-welded plate

Longitudinal shrinkage : $\Delta l = \dfrac{0.12 \cdot I \cdot L}{10^5 \cdot h}$

where ΔL = transverse shrinkage (mm); t = weld layer thickness for the particular pass (mm); v = welding travel speed (cm/min); W= I*U, I is welding current (Amp); U = welding voltage (Volt); Δl = longitudinal shrinkage (inch); h = plate thickness (inch); and L = length of the weld (inch).

Angular distortion: $\beta = \dfrac{12 \cdot \Delta L \cdot Z}{\delta^2}$

where, Z = the distance from the centroid of the filler material of the weld to the mid-plane of the plate, as shown in Fig. 7, and δ = the effective (accumulative) weld height. Detailed weld pool information is not provided by Canas, et al (1996) for the weld analyzed. Generally, Z is very small for a butt-weld joint of thin plates, because the temperature distributions at the top and bottom surfaces are very similar, and the transverse shrinkage at the top of the plate and the bottom of the plate are very close. A value of 0.2 mm (i.e. 0.2/6.6= 3% of the plate thickness) was used in the calculation. In addition, both t and δ equal to h was used because the weld is single pass.

The comparison in Table 1 indicates that the longitudinal shrinkage, transverse shrinkage and angular distortion from the FEA are close to the results calculated by the empirical equations. Note that the empirical formulae were derived from experimental data; thus partially verified our FEA modeling results.

CONCLUSIONS

Three-dimensional modeling of welding process is time consuming, however, essential for practical problems as indicated by the results in this paper. The butt-welded plate analyzed in the paper took less than 24 hours to complete both the heat transfer and thermo-mechanical analyses on a Pentium 233 MHz PC with 128 MB RAM. More results for the distortion and critical comparison of the computation efficiency will be presented later.

ACKNOLEGEMENT

This work is part of an active research program in *structural joints*, sponsored by NSF/EPSCoR Cooperative Agreement No. EPS-9630167 at the University of South Carolina. The support from NSF is greatly appreciate.

REFERENCES

Arrgyris, J.H., Szimmat, J. and William, K.J., 1982, "Computational aspects of welding stress analysis," Computer Methods in Applied Mechanical Engineering, No.33, 635-666.

Brown, S. and Song, H., 1992, "Finite Element Simulation of Welding Large Structures," ASME Journal of Engineering for Industry, Vol.114, pp441-451.

Canas, J., Picon, R., Paris, F., 1996, Blazquez, A. and Marin, J.C., "A Simplified Numerical Analysis of Residual Stresses in Aluminum Welded Plates," Computers & Structures, Vol.58, No.1, pp59-69.

Dong, Y., Hong, J.K., Tsai, C.L., and Dong, P., October 1997, "Finite element Modeling of Residual Stresses in Austenitic Stainless Steel Pipe Girth Welds," Welding Journal, 442s-449s.

Dong, P., Hong, J. K., Byum, J. and Rogers, P., 1997, Analysis of Residual Stresses in Al-Li Alloy Repair Welds, PVP-Vol.347, Approximate Methods in the Design and Analysis of Pressure Vessels and Piping Components, p.61-75, ASME

Hibbitt, H.D., and Marcal, P.V., 1973, "A numerical Thermo-Mechanical Model of the Welding and Subsequent Loading of a Fabricated Structure", Computers and Structures, Vol.3, No.5, pp.1145-1174.

Karlsson, L, 1986, "Thermal Stresses in Welding," Chapter 5, Thermal Stresses I, Volume I of Thermal Stress, editor R.B. Hetnarski, Elsevier Science Publishers.

Masubuchi, K., *Analysis of Welded Structures*, Pergamon Press, Oxford, U.K., 1980.

Michaleris, P. and DeBiccari, A., April 1997, "Prediction of Welding Distortion," Welding Journal, , 172s-181s.

Papazoglou, V.J. and Masubuchi, K., 1982, "Numerical Analysis of Thermal Stresses during Welding Including Phase Transformation Effects," ASME J of Pressure Vessel Technology, 104, 198-203.

Tekriwal, P. and Mazumder, J., July, 1991, "Transient and Residual Thermal Strain-Stress Analysis of GMAW," ASME Journal of Engineering Materials and Technology, Vol.113, pp336-343

Wang, J., Ueda, Y., Murakawa, H., Yuan, M.G., and Yang, H.Q., April 1996, "Improvement in Numerical Accuracy and Stability of 3-D FEM Analysis in Welding," Welding Journal, pp.129s-134s.

Wang, J. and Qi, X., 1994, "Calculation and Control of Angular Distortion in Heavy Plate Welding," *Welding*, No.8, pp.12-15.

STUDY OF THE MATERIAL PROPERTIES OF THIN PIPE BUTT WELDS (IN C-Mn AND STAINLESS STEEL) ON THE WELDING RESIDUAL STRESS DISTRIBUTION BY USING NUMERICAL SIMULATION

J. J. JANOSCH, D. LAWRJANIEC

Institut de Soudure, 57365 Ennery - France

ABSTRACT

Since decades, engineers have been interested in the residual stresses in welded structures. In fact, we do know their role in the failure behaviour or fatigue strength in that they will accelerate or delay cracking, depending upon whether they are tensile or compressive. In the field of welded structures, this knowledge is mainly "qualitative" and the objective of some present studies is to find a means for better "quantifying" the stresses. Ideally, one should be able to predict the behaviour of the structures as a function of the materials, the welding processes, the service conditions, etc.

In order to reach this objective, a possible approach is to separate the material-related effects (base material, HAZ, etc...) from those resulting from the external loads. The residual stresses, insofar as they add themselves to the stresses due to the loads, may be included in the latter.

The assessment of residual stresses, which is needed for the quantitative prediction of the behaviour of the welded structure, remains even now a delicate topic. As a matter of fact, although high-performance measurement methods are available for evaluating the residual stresses in the vicinity of a point on the surface, the knowledge of deeper fields requires removals of material which perturb the stress values we intend to measure. It is therefore necessary to correct the obtained results and, strictly speaking, this can be done only wherever the measured field is invariant with respect to the surface point under consideration. More precisely, the residual stress field must not display any important gradient parallel to the free surface which is supposed to be plane. If this is not the case, the corrections to be made with the removal of layers which are used by now are not applicable.

In a recent work, P. Ballard and A. Constantinescu [1] have shown on the basis of analytical examples that without adapted correction methods, quite wrong results can be obtained. And the weld does represent such a case, because it includes important gradients which impose a great care in the interpretation of the results of the measurements of the in-depth fields.

Hence, the advantage of having at our disposal new calculation methods which permit the welding operation and the weld residual stresses to be modelled as a function of the welding process, the restraint condition, the nature of the base and filler materials, etc.) and the operating conditions (test load, service load) in order to integrate more realistically the effect of mismatch and/or of the welding residual stresses in the methods of mechanical analysis, ref[2][3][4].

The present study proposes some examples of application of numerical simulation of (homogeneous or heterogeneous) butt welding of a thin-walled pipe (C-Mn and stainless steel). A particular analysis was conducted for evidencing the effect of the thermal, metallurgical and mechanical properties of the fabricated materials on the distribution of welding residual stresses in the weld and in the pipe assembly.

Additional investigations are conducted concurrently for studying the phenomena of residual stress relief and the development of the mechanisms of plasticity (in welds containing initial planar defects, or not) after the execution of the mechanical proof test. In that case, the maximum proof test pressure was specified by standards to ensure the service safety of the pipelines used for conveying hydrocarbides.

PRESENTATION

This study proposes some numerical simulations of automatic MAG butt welding of two 6 mm thick pipes in a diameter of 500 mm with a Vee groove preparation.

Four types of assemblies were studied:

Case 1 - Homogeneous welding of an assembly in C-Mn steel of A52 type (abbreviation CCC).
The yield strengths of the materials were respectively:
$\sigma y(PM)_{CMn}$ = 355 MPa and
$\sigma y(WM)_{CMn}$ = 510 MPa (overmatching).

Case 2 - Heterogeneous welding of an assembly in C-Mn steel (A52 steel) welded with an austenitic 308L filler metal (abbreviation CSC). The yield strengths of the materials were respectively: $\sigma y(PM)_{CMn}$ = 355 MPa and $\sigma y(WM)_{ss}$ = 350 MPa (evenmatching).

Case 3 - Heterogeneous welding of a mixed assembly composed of a pipe in A52 C-Mn steel and a pipe in Type 304L austenitic steel, welded with an austenitic 308L filler metal (abbreviation SSC). The yield strengths of the materials were respectively: $\sigma y(PM)_{CMn}$=355 MPa, $\sigma y(WM)_{ss}$= 350 MPa and $\sigma y(PM)_{ss}$ = 250 MPa (double mismatching).

Case 4 - Homogeneous welding of a pipe in Type 304L austenitic steel with a 308L filler metal (abbreviation SSS). The yield strengths of the materials were respectively: $\sigma y(WM)_{ss}$= 350 MPa and $\sigma y(PM)_{ss}$ = 250 MPa (overmatching).

The first phase of this study proposes, for each of these configurations, to determine the distribution of welding residual stresses while considering, or not, the presence of an initial defect at the weld root

In the second stage, this study proposes, according to these numerical results, to apply a representative oil-industry standard, in which the level of the proof testing loads applied to the pipes are established as a function of the weld qualities and the percentage of on-site NDT of welds. The aims of these investigations were:

- to evaluate, for circumferential welds, the evolution of the redistribution of welding residual stresses through the weld thickness after the proof test,
- to analyse in the specific case of a 3 mm long pre-existing crack located at the weld root, the redistribution of welding residual stresses through the weld thickness after the proof test.

MODELLING OF THE WELDING RESIDUAL STRESSES

SYSWELD software which is dedicated to the numerical simulation of welding, surface treatments and heat treatments while taking into account the metallurgical changes.

The code uses the finite element method and integrates the effects linked to the metallurgical changes in the thermal, mechanical and hydrogen diffusion analyses ref:[5]. The simulation is thus accomplished in successive stages:

- thermal and metallurgical calculation (determination of thermal cycles and proportions of metallurgical phases as a function of the position and time)
- mechanical calculation (calculation of the residual stresses and strains)
- calculation of hydrogen diffusion, integrating the effect of temperature, stresses and hydrogen entrapment (reversible or irreversible).

Thermometallurgical coupling

The strong coupling between the thermal and metallurgical analyses takes place at several points [6]; the metallurgical changes are dependent on the temperatures and on their rates and generate latent heats. These latent heats are considered as additional volumic heat sources. Moreover, the thermal properties vary as a function of the metallurgical phases. The thermal analysis is based on the resolution of the heat equation by the finite element method. When two or more phases are present, the heat content, the density and the thermal conductivity are approached by mixture rules.

Several models permit the description of the kinetics of metallurgical changes during heating and cooling. The first model is based on a purely phenomenological approach [6]; the second one corresponds to the Johnson-Mehl Avrami law and, lastly, the martensite transformation can be modeled with the Koïstinen-Marburger Law [7].

Mechanical analysis

The mechanical analysis uses the results of the thermometallurgical analysis. The effect of the mechanical parameters on the temperature distributions and the phase proportions is neglected. Except for the effect of the stresses and strains on the transformation kinetics discussed earlier, this hypothesis amounts to neglecting the strain energy in the thermal anlysis. Taking this term into account would produce temperature variations on the order of 2°C only for the processes under consideration. Therefore, this hypothesis is perfectly adequate.

The metallurgy has mainly two effects on the mechanical analysis:

- the metallurgical changes are accompanied by volume changes
- the material has a specific behaviour which is linked to its multiplane aspect

The volume changes resulting from the modifications of the crystal structure during the changes have an essential effect on the production of stresses [8].

Moreover, the behaviour law takes two phenomena into account:

- the metallurgical structures present have different mechanical properties

- the transformation plasticity which stems from the Greenwood-Johnson mechanism [10] and from the Magee mechanism [11].

The variation in plastic strain (ε^p) can be expressed as the sum of three terms which are respectively proportional to the variation in stress (σ), to the variation in temperature (θ) and to the variations in phase proportions (p) [9] : $\varepsilon^p = \alpha (...)\cdot \sigma + \beta (...)\cdot \theta + \gamma (...)\cdot p$

The first two terms of the right hand side of the equation represent the variation in conventional plastic strain (ε^{cp}) and the last one represents the variation in transformation plastic strain (ε^{tp}).

Theoretical research works [12, 13] using both analytical approaches and numerical approaches have permitted expressions of (ε^{cp}) and (ε^{tp}) to be obtained. This model, which was generalized in order to take into account the work hardening of the phases and the possible refining effects of work hardening during the metallurgical transformations [14] requires the knowledge of the basic mechanical behaviour of each phase.

APPLICATION

The two-dimensional axisymmetric calculations are carried out while considering that the weld beads are deposited simultaneously over the whole circumference of the pipe, figure (1). This permits the problems to be limited to a plane perpendicular to the welding direction.

The modelling of welding by means of the SYSWELD software devised for finite element analysis included two phases:
- The first phase consisted in a closely linked thermal/metallurgical analysis for the C-Mn steel, ref: [15] (in the specific case of the stainless steel modelling, we have considered a thermomechanical analysis only). The thermal properties of the weld (namely of the parent and filler metals) at any temperature between 20 C and 1500 C must be known. These properties are: thermal conductivity, specific heat and density of the materials. For the metallurgical calculation each C-Mn material (PM, WM) undergoes two metallurgical transformations, namely an austenitic transformation during heating and a transformation during cooling which leads to the presence of bainite at room temperature.
- The second phase of the simulation process includes a thermomechanical calculation based on the previous results. In this case, the thermal-elastic-plastic calculation accounts for the various physical phenomena associated with the metallurgical transformations (for C-Mn steel) ref:[15],.For this phase of modelling, it is necessary to know the mechanical properties of the materials in each phase, at any temperature (from 20 to 1500 C).

Validation of the modeling

The first validation of the calculations of (CCC) modelling was performed by comparing the numerical and experimental thermal cycles (heating and cooling) recorded in the HAZ between 800 and 500 C, table(1). The second validation was made by comparing the numerically determined phase distribution with the macrographs of the welds (showing the HAZ), figure (1).

The heat input used for the numerical simulation of the welding of a heterogeneous assembly (CSC) or mixed assembly (SSC) was so determined as to obtain, on the C-Mn pipe, the same HAZ dimension as the one which was measured on the C-Mn steel pipe (case 1)

The heat input used for the numerical simulation of the welding of a homogeneous assembly (SSS) was determined by comparing the measured and calculated thermal cycles in the HAZ of the weld, table (1).

ANALYSIS OF NUMERICAL RESULTS

Figures 3 to 6 present the isovalues of transverse and longitudinal residual stresses obtained by numerical simulation of these different cases of pipe assemblies :
- As welded,
- As welded after the proof test,
- As welded, with the presence of a 3 mm long pre-existing crack located in the fusion line at the root of the pipe butt weld,
- After the proof test, with the presence of a 3 mm pre-existing crack located in the fusion line at the root of the pipe butt weld.

Discussion

The numerical results revealed the following :

1. The overall distribution of the transverse welding residual stresses in the (CCC) assembly revealed that they were highly tensile on the internal surface of the pipe ($\sigma res = \sigma y_{C Mn}$) (up to weld mid-thickness) and then decreased towards the external surface where they became compressive. The longitudinal welding residual stresses (hoop stress) were higher in the filler metal than in the parent metal. This can be explained by the overmatching effect of the weld : M= $\sigma y_{WM}/\sigma y_{PM}$= 510/355=1.44).

2. The heterogeneous (CSC) butt welding (filler metal in austenitic steel) of two C-Mn pipes induced globally a strong decrease in the longitudinal and transverse residual stresses in the assembly as compared with the homogeneous welding (see -1-). The lowest longitudinal residual stresses were in the weld metal. This can be explained by the combined effect of the different metallurgical, thermal and mechanical properties of the welded materials (PM, HAZ, WM). Moreover, it was noted in this case that the local level of weld mismatch was complex and different as compared to the first study case (case CCC). In fact, the real M coefficient was equal to the ratio M=$\sigma y_{WM}/\sigma y_{HAZ}$=350/450=0.77 which

correspond to a local undermatching (σy_{HAZ}= 450 MPa in the C-Mn pipe).

3. The specific welding of the mixed butt assembly (SSC) induced an asymmetric distribution of the transverse and longitudinal welding residual stresses in the assembly. As a whole, the longitudinal and transverse residual stresses were markedly lower as compared to the (CCC) homogeneous welding (case 1). The lowest longitudinal and transverse residual stresses were in the (PM_{ss}) pipe and the (WM_{ss}) pipe. This was again justified by the heterogeneity of the properties of the welded materials and by the presence of a double mismatching in the weld:

- pipe side HAZ PM_{C-Mn} and WM_{SS}: $M = \sigma y_{WM}/\sigma y_{HAZ-PM}$ = 350/450 = 0.77, undermatching,
- pipe side PM_{SS} and WM_{SS} pipe side : $M = \sigma y_{WM}/\sigma y_{PM}$ = 350/250 = 1.4, overmatching.

The highest residual stresses were located in the HAZ (C-Mn) of the pipe butt weld.

4. Contrary to the butt welding of the C-Mn steel pipe, the overall distribution of the transverse welding residual stresses in the (SSS) assembly was tensile on the internal and external surfaces of the pipe. As in the case of the C-Mn steel, the longitudinal welding residual stresses (in (SSS) assembly) were higher in the filler metal than in the parent metal. This can be justified by the overmatching effect $M = \sigma y_{WM}/\sigma y_{PM}$ = 350/250 = 1.4.

5. In all the cases under study (case 1 to 4), the presence of a 3 mm long pre-existing crack at the WM-HAZ interface after welding produced a relief of the transverse residual stresses on the internal surface of the pipe. However, the crack tip was strongly subjected to tensile loads and the level of the local loading was proportional to the yield strength of the materials. Furthermore, an asymmetry in the distribution of the local transverse residual stresses was observed at the crack tip ; the asymmetry was more or less marked depending on the level of weld mismatch.

MODELLING OF A PROOF TEST PRESSURE

This study proposes a modelling of the redistribution of the residual stresses through the thickness of the circumferential weld on pipe, after the execution of the mechanical proof test. In that case, the maximum proof test pressure was specified by standards to ensure the service safety of the pipelines used for conveying hydrocarbides ref.[16].

This study proposes an example of interpretation of the numerical results as a function of the industrial application, which the design pressure and the proof test pressure for the pipe were respectively 36 bar and 54 bar (the joint efficiency was 0.85).

In a practical point of view, the numerical simulations were conducted by progressively raising the mechanical pressure inside these tubular assemblies up to the design pressure and then to the proof test pressure. Figure (2) shows the principle of pressure raising inside the pipe.

Figures (3) to (6) show, for the different cases under study (case 1 to 4), the evolution of the transverse and longitudinal residual stresses calculated in the circumferential weld of the pipe (while considering, or not, an initial crack in the weld) after application of the proof test pressure.

Discussion

The numerical results obtained after the application of a test pressure revealed the following:

- For all the cases under study, a significant relief of the longitudinal and transverse welding residual stresses was observed in the assembly. The average level of stress relief was about 50%. A strong decrease in the stresses was noted on the internal surface of the pipe, regardless of the case being studied.

- In the heterogeneous assemblies (CSC, SSC), a residual stress relief preferably located in the C-Mn steel was recorded. This was justified by the combined effect of the different mechanical properties of the involved materials, in particular a higher stiffness of the C-Mn pipe as compared to the Stainless steel pipe and filler metal.

- An asymmetrical relief of the transverse residual stresses was observed and localised at the tip of the initial crack (located at the WM-HAZ interface), which was more or less pronounced depending probably on the different cases of mismatch under study.

- An asymmetry of the local plastic zones at the tip of the initial crack (located at the WM-HAZ interface) was noted ; the asymmetry was more or less pronounced as a function of the different cases of mismatch being studied.

CONCLUSION

Thanks to the numerical simulation of welding, this study has permitted the prediction of the residual stress distribution and of the mismatch effect in a butt weld on a thin-walled pipe in a diameter of 500 mm, in the cases of homogeneous or mixed assemblies in C-Mn (A52) steel and/or austenitic (304L/308L) steels.

The numerical results showed the following:

- The overall distribution of the transverse welding residual stresses in the (CCC) assembly revealed that the stresses were highly tensile on the internal surface of the pipe ($\sigma res = \sigma y_{C\,Mn}$) (up to mid-thickness of the weld) and then decreased towards the external surface where they became compressive.

- The heterogeneous (CSC) butt welding (filler metal in austenitic steel) of two C-Mn pipes induced globally a strong decrease in the longitudinal and transverse residual stresses in the assembly as compared with the homogeneous welding.

- The specific welding of the mixed butt assembly (SSC) induced an asymmetry in the distribution of the (transverse and longitudinal) welding residual stresses in the assembly. As a whole, the longitudinal and transverse residual stresses were clearly lower as compared to the (CCC) homogeneous welding. The lowest longitudinal and transverse residual stresses were in the (PM_{ss}) pipe and the (WM_{ss}) pipe.

- Contrary to the butt welding of the C-Mn steel pipe, the overall distribution of the welding transverse residual stresses in the (SSS) assembly was highly tensile on the internal and external surfaces of the pipe. The longitudinal residual stresses were higher in the filler metal than in the parent metal, as in the case of the C-Mn steel.

- In all the studied cases, the presence of a crack at the WM-HAZ interface after welding produced a relief of the transverse residual stresses in the internal surface of the pipe. However, the crack tip was strongly subjected to tensile loads.

The numerical results obtained after the application of a test pressure revealed the following :

- In all the cases under study, a significant relief of the longitudinal and transverse welding residual stresses was observed in the assembly. The average amount of stress relief was about 50%. A strong decrease in the stresses was noted on the internal surface of the pipe, regardless of the case being studied.

- An asymmetrical relief of the local transverse residual stresses at the tip of the initial crack (located at the WM-HAZ interface) was noted ; the asymmetry was more or less pronounced as a function of the different cases of mismatch.

The results yielded by this study demonstrated the potential of these new weld modelling tools which permit not only to better understand the physical phenomena occurring during welding and its industrial applications, but also to favour the evolution of the existing recommendations, codes or standards in order to further increase the reliability of the welded constructions while optimising the manufacturing methods and the testing procedures applied during the service life of the pipelines.

REFERENCES

[1] "Inversion of subsurface residual stresses from surface stress measurements", P. Ballard, A. Constantinescu, J. of the Mech. and Phys. of Solids, Vol 42, n° 11, pp.1767-1787, 1988

[2] " Welding Residual Stress Modeling", doc: IIW-X-1349-96/XV-911-96, by J.J.Janosch Institut de Soudure (FMR), France.

[3] 1st annual report of " Residual Stress and Distortion Prediction in welded structures (RSDP)" Working Group, doc IIW-X/XV-RSDP-02-97, by J. J. Janosch, Institut de Soudure, France; R Gordon, Edison Welding Institute, USA.

[4] 2 nd annual report of " Residual Stress and Distortion Prediction in welded structures (RSDP)" Working Group, doc IIW-X/XV-RSDP-29-98 by J. J. Janosch, Institut de Soudure, France; R. Gordon, Edison Welding Institute, USA.

[5] " A general mathematical description of hydrogen diffusion in steels", J.B. Leblond, D. Dubois, I - Derivation of diffusion equations form BOLTZMANN-TYPE transport equations. Acta metall, vol 31, n° 10, pp 1471-1478, 1983, II - Numerical study of permeation and determination of trapping parmeters. Acta metall. vol. 31, n° 10, pp 137-146, 1983.

[6] " A new kinetic model for anisothermal metallurgical transformations in steel including effect of austenit grain size",J.B. Leblond & J. Devaux, 1984, Acta Metallurgica, vol 32, n° 1, pp 137-146.

[7] D.P. Koistinen & R.E. Marburger, 1959, Acta Metallurgica 7, pp 50-60

[8] " Transient thermal stresses of weld heat-affected zone by both-ends fixed bar analogy", K. Satoh, 1972, Trans. of Japan Welding Society, n° 1, pp 125-134

[9] " A theoritical and numerical approach to the plastic behaviour of steels during phase transformation, I : derivation of general relations", J.B. Leblond, G. Mottet & J.C. Devaux, 1986 Journal of the Mechanics and Physics of Solids, vol. 34, n° 4, pp 395-409

[10] " The deformation of metals under small stresses during phase transformation", G.W. Greenwood & R.H. Johnson, 1965, Proc. Roy. Soc. A283, pp 403-422

[11] " Transformation kinetics, microplasticity and aging of martensite in Fe-31 Ni", C.L. Magee, 1966, Ph. D. Thesis, Carnegie Institut of technology, Pittsburgh (USA)

[12] " A theoritical and numerical approach to the plastic behaviour of steels during phase transformation. II : study of classical plasticity for ideal -plastic phases", J.B. Leblond, G. Mottet & J.C. Devaux, 1986, Journal of the Mechanics and Physics of solids, vol. 34, n° 4, pp 411-432

[13] "Mathematical modelling of transformation plasticity in steels, I : case of ideal-plastic phases", J.B. Leblond, J.C. Devaux, 1989, International Journal of Plasticity, vol. 5, pp 551-572

[14] "Mathematical modelling of transformation plasticity in steels, II : coupling with strain hardening

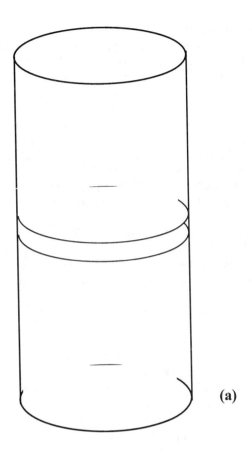 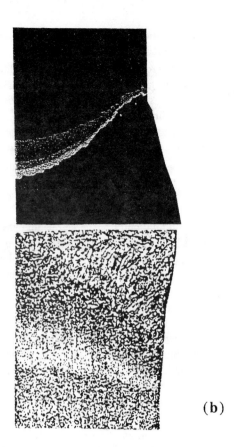

figure (1): (a) Numerical welding simulation of two 6 mm thick pipes in a diameter of 500 mm.
(b) Comparison of the numerical determined phase distribution with the macrographs of the butt welds

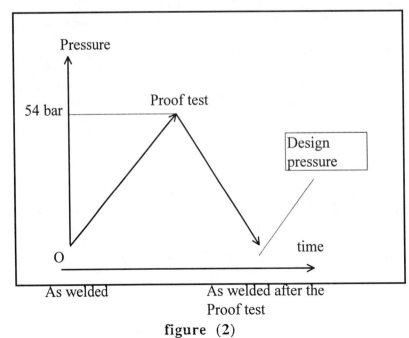

figure (2)

Isovalues of transverse residual stresses σyy (MPa)   *Isovalues of longitudinal residual stresses σzz (MPa)*

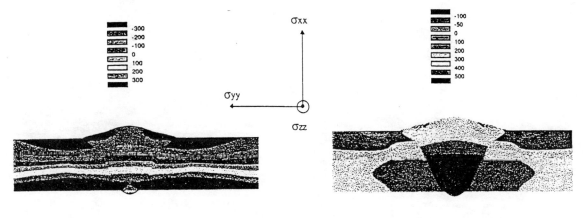

- As welded,

 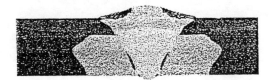

- As welded, with the presence of a 3 mm long pre-existing crack located in the fusion line at the root of the pipe butt weld,

- As welded after the proof test,

- After the proof test, with the presence of a 3 mm pre-existing crack located in the fusion line at the root of the pipe butt weld.

figure (3)

Homogeneous welding of an assembly in C-Mn steel of A52 type (abbreviation CCC).

Isovalues of transverse residual stresses σyy (MPa)   *Isovalues of longitudinal residual stresses σzz (MPa)*

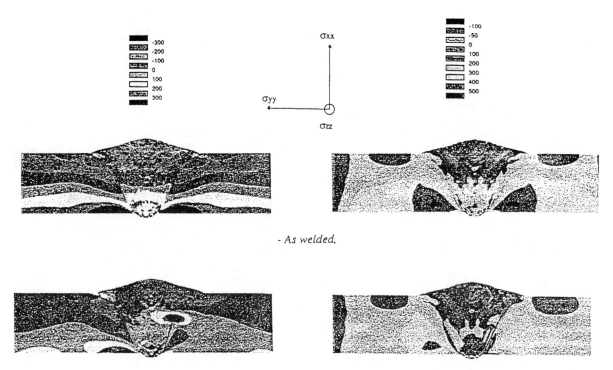

- As welded,

- As welded, with the presence of a 3 mm long pre-existing crack
located in the fusion line at the root of the pipe butt weld,

- As welded after the proof test,

- After the proof test, with the presence of a 3 mm pre-existing crack
located in the fusion line at the root of the pipe butt weld.

figure (4)

Heterogeneous welding of an assembly in C-Mn steel (A52 steel) welded with an austenitic 308L filler metal (abbreviation CSC).

Isovalues of transverse residual stresses σyy (MPa) *Isovalues of longitudinal residual stresses σzz (MPa)*

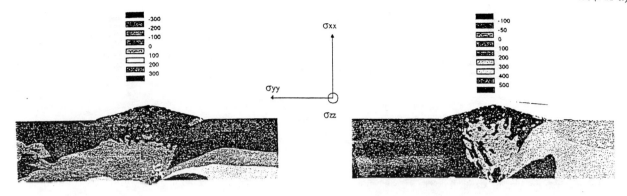

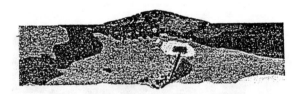 

- As welded.

- As welded, with the presence of a 3 mm long pre-existing crack located in the fusion line at the root of the pipe butt weld.

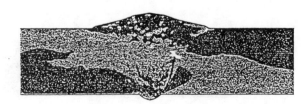 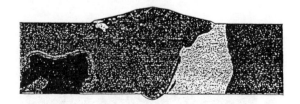

- As welded after the proof test.

- After the proof test, with the presence of a 3 mm pre-existing crack located in the fusion line at the root of the pipe butt weld.

figure (5)

Heterogeneous welding of a mixed assembly composed of a pipe in A52 C-Mn steel and a pipe in Type 304L austenitic steel, welded with an austenitic 308L filler metal (abbreviation SSC).

Isovalues of transverse residual stresses σyy (MPa)   *Isovalues of longitudinal residual stresses σzz (MPa)*

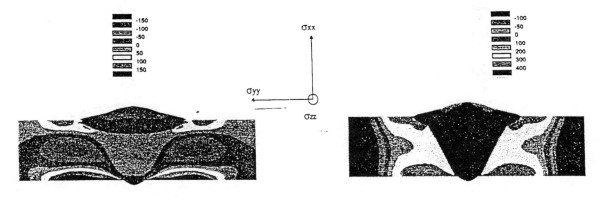

- As welded,

- As welded, with the presence of a 3 mm long pre-existing crack located in the fusion line at the root of the pipe butt weld,

- As welded after the proof test,

- After the proof test, with the presence of a 3 mm pre-existing crack located in the fusion line at the root of the pipe butt weld.

figure (6)

Homogeneous welding of a pipe in Type 304L austenitic steel with a 308L filler metal (abbreviation SSS).

PVP-Vol. 393, Fracture, Fatigue and Weld Residual Stress
ASME 1999

Welding Residual Stresses with Magnitudes Lower than the Material Yield Strength

Kyle Koppenhoefer, William Mohr, and
J. Robin Gordon
Edison Welding Institute
1250 Arthur E. Adams Drive
Columbus, Ohio 43221-3585
Tel: 614-688-5155
Kyle_Koppenhoefer@ewi.org

ABSTRACT

Appendix E of API RP-579 provides estimates of welding residual stresses for fitness-for-service analyses of planar flaws. These estimation techniques have been designed to provide simple approximations of the welding residual stress distributions in pressure vessel and piping geometries, based upon residual stress measurements and finite element analysis. The recommended distributions are either linear or bi-linear across the wall thickness and have been designed to provide conservative estimates, while reducing the magnitude of the estimate from the material yield strength whenever possible. Recent updates to the recommendations include reduction of residual stresses in girth welds for cylinders with large radius-to-thickness ratio and new estimates of the residual stresses around nozzle welds. These recommendations have been based on finite element studies, with an assessment of weld-to-weld variability derived from available test data.

INTRODUCTION

Typical fitness-for-service (FFS) assessments of welded structures assume residual stresses of yield level across the entire weld. Since most designs of these structures do not permit primary stresses to obtain this level, the welding induced residual stress becomes a dominate loading in FFS.

Assuming yield level residual stress across the entire weld eliminates the difficulties in developing a generalized, upper bound treatment of these stresses. Such universal guidelines are difficult to produce due to the significant weld-to-weld variation of residual stresses. For a generalized treatment of residual stress to be useful in FFS assessments, it must be sufficiently conservative to produce an upper bound estimate over a range of welds. However, excessive conservatism may result in unwarranted retirement of welded structures, or the repair of insignificant indications. A rational reduction of the excessive conservatism associated with assuming yield level residual stresses would produce significant financial benefits.

Previous work (Scaramangas, 1984; Mohr, 1996, Mohr, et al., 1997; Mohr, et al. 1998) indicates the residual stresses transverse to the welding direction are significantly less than yield for many cases. In contrast, this same work suggests residual stresses parallel to the welding direction obtain the material yield strength for most welds. Thus, we focus primarily on providing estimates of residual stress transverse to the welding direction.

This paper builds on previous work of the co-authors to propose residual stress distributions for welds that reduce the conservatism associated with a yield-level magnitude approximation. These distributions attempt to remain upper-bound estimates of the welding induced residual stresses for use with fitness-for-service assessments. The functional forms recommended here do not attempt to produce a best fit of residual stress through the thickness. As such many of these distributions are not self-equilibrating. Instead they are simple linear (or bilinear) equations.

WELDING SIMULATIONS

Koppenhoefer, et al. (1998) recently conducted welding simulations using a decoupled thermo-mechanical finite element procedure. The commercial finite element code ABAQUS® formed the basis for these simulations. The analysis matrix included single-V girth welds for two thicknesses (0.25 and 0.75 in.) at three different R/t values (see Table 1 for the entire analysis matrix).

Figure 1 shows the typical mesh for a three-pass weld. Eight-node axisymmetric elements provide the meshing capacity for all analyses conducted here. Simulation of multi-pass welds utilizes the element addition/deletion capability available in ABAQUS® to introduce elements during the analysis. This method activates elements in a zero stress state at the beginning of each weld pass.

As indicated in Table 1, the effects of hydrotesting on residual stress were examined. Following the welding simulation, an internal pressure, and the associated axial stress, is applied incrementally to produce the hydrotest loading. This internal pressure is equal to 1.5 times the maximum allowable working pressure (MAWP), where

$$MAWP = \frac{SEt}{R + 0.6t}$$

and S is the allowable stress for the material, E is the joint efficiency, R is the radius of the weld and t is the pipe thickness.

GIRTH WELDS

In this work, a simple linear distribution approximates the residual stress distribution near a girth weld. This distribution is formulated by specifying the residual stress values at the ID and OD of the pipe in the following equation:

$$\sigma^r(x) = \sigma_i^r + (\sigma_0^r - \sigma_i^r)\frac{x}{t}$$

where x is the local coordinate measured from the ID of the pipe, and $\sigma_0^r = 0.2\sigma_{ys}^r$.

This simple linear approximation does not capture the complexity of the residual stress distribution, as observed in thick-walled pipes (Scaramangas, 1984). Instead, the simple distribution proposed here provides an upper bound to the distribution. Instead of using a cosine function to fit the residual stress distribution for thick-walled pipes, a linear distribution is used to simplify loading calculations in FFS assessments (e.g., the stress intensity factor due to residual stress).

To determine σ_i^r, previous work by the co-authors (Mohr, 1996; Mohr, et al., 1997; Mohr, et al. 1998) classifies girth welds as either low or high heat input per pass, with 2 kJ/mm (50 kJ/in) providing the demarcation. In this previous work, the authors acknowledged the effect of net heat input per pass. However, this parameter is not typically recorded for welds. A more typically recorded parameter (i.e., out-of-position welding) forms the basis for classifying welds as low or high heat input. Any process used to weld steel pipe in these positions must use a heat input below 2 kJ/mm. Otherwise, the weld pool would fall off the pipe. Thus, during a fitness-for-service assessment it may be possible to determine the weld procedure used and classify the weld as either a low or high heat input.

Within these two categories of welds, previous work recommends a tri-linear function to estimate the residual stress at the ID of the pipe. For low heat input welds, this function is given by

(1) For $t \leq 20\ mm\,(0.75\ in)$

$$\sigma_i^r = \sigma_{ys}^r$$

(2) For $20\ mm\,(0.75\ in) < t < 70\ mm\,(2.75\ in)$

$$\sigma_i^r = \left(1.32 - 0.016\frac{t}{C_{ul}}\right)\sigma_{ys}^r$$

(3) For $t \geq 70\ mm\,(2.75\ in)$

$$\sigma_i^r = 0.2\sigma_{ys}^r$$

Where, σ_i^r equals the residual stress on the ID of the pipe, t is the thickness of the pipe, C_{ul} is a unit conversion (1.0 when thickness is in mm, 0.0394 when thickness is in inches), and σ_{ys}^r correspond to the yield strength used to estimate residual stress. This function approximates the upper bound of experimental data published by Mohr, et al.(1997) (see Figure 2).

For high heat input welds, the tri-linear function becomes,

(1) For $t \leq 50\ mm\,(2\ in)$

$$\sigma_i^r = \sigma_{ys}^r$$

(2) For $50\ mm\,(0.75\ in) < t < 100\ mm\,(4\ in)$

$$\sigma_i^r = \left(1.8 - 0.016\frac{t}{C_{ul}}\right)\sigma_{ys}^r$$

(3) For $t \geq 100\ mm\,(4\ in)$

$$\sigma_i^r = 0.2\sigma_{ys}^r$$

In this work, all residual stress distributions are functions of σ_{ys}^r instead of σ_{smys} (specified minimum yield strength - ASME B&PV Code Section II, Part D) to remove possible non-conservatism in these distributions. If specific yield strength data for the assessment material is available, then approximating σ_{ys}^r from σ_{smys} is not necessary. In cases where specific yield strength data is not available, σ_{ys}^r may be estimated from the following

$$\sigma_{ys}^r = C_{ys}\sigma_{smys}$$

where C_{ys} equals 1.2 for carbon steels and 1.5 for alloy steels.

Effects of *R/t*

The equations given previously for residual stress do not include the effects of R/t. To quantify these effects, a series of analyses were conducted for a range of R/t values (see Table 1). Figure 3 presents the transverse-to-weld residual stress distribution along the weld centerline for three R/t values. Figure 4 presents transverse-to-weld residual stress distributions along the weld toe for these same R/t values. These residual stress distributions show a moderate reduction of residual stress with increasing R/t due to a reduction in stiffness in the pipe. As shown in Figure 5, the hoop shrinkage produces a bending moment that gives rise to the transverse-to-weld residual stress. As the stiffness associated with this deformation mode decreases, the weld shrinkage about the circumference produces a lower bending moment through the pipe wall.

The residual stress distributions presented in Figures 3 and 4 indicate a 25% reduction between $R/t = 10$ and $R/t = 50$. From these residual stress distributions, the following reduction in residual stress may be recommended

(1) For $R/t \leq 10$

$$RF_{Rt} = 1.0$$

(2) For $10 < R/t < 50$

$$RF_{Rt} = 1.0625 - 0.00625 \frac{R}{t}$$

(3) For $R/t \geq 50$

$$RF_{Rt} = 0.75 .$$

Where this reduction factor may be applied directly to the internal residual stress calculated from the equations given previous such that

$$\sigma^r_{i-new} = RF_{Rt}\, \sigma^r_{i-old} .$$

This procedure reduces the residual stress independent of actual pipe thickness. However, experimental data shown in Figure 2 suggests internal residual stress values near yield for thin pipes. To maintain consistency with these experimental data, a modified thickness ($\hat{t}$) is proposed such that

(1) For $R/t \leq 10$

$$\hat{t} = t / C_{ul}$$

(2) For $10 < R/t < 50$

$$\hat{t} = \frac{t}{C_{ul}} + 3.9\left(\frac{R}{10t} - 1\right)$$

(3) For $R/t \geq 50$

$$\hat{t} = \frac{t}{C_{ul}} + 15.6 .$$

The internal residual stress for low heat input welds expressed in terms of $\hat{t}$ is

(1) For $\hat{t} \leq 20\ mm$

$$\sigma^r_i = \sigma^r_{ys}$$

(2) For $20\ mm < \hat{t} < 70\ mm$

$$\sigma^r_i = (1.32 - 0.016\hat{t})\sigma^r_{ys}$$

(3) For $\hat{t} \geq 70\ mm$

$$\sigma^r_i = 0.2\sigma^r_{ys}$$

Similarly, the internal residual stress for high heat input welds becomes

(1) For $\hat{t} \leq 50\ mm$

$$\sigma^r_i = \sigma^r_{ys}$$

(2) For $50\ mm < \hat{t} < 100\ mm$

$$\sigma^r_i = (1.8 - 0.016\hat{t})\sigma^r_{ys}$$

(3) For $\hat{t} \geq 100\ mm$

$$\sigma^r_i = 0.2\sigma^r_{ys}$$

The modification to include $\hat{t}$ shifts the transition region for the reduction of residual stresses to lower thickness values, as shown in Figure 6. Experimental data presented by Mohr, et al.(1997) may now be plotted in terms of $\hat{t}$ to show this modified approach still produces a conservative estimate of the experimental data (see Figures 7 and 8).

Alternative Method

Mohr, et al.(1997) proposed an alternative method utilizing the number of passes to predict the residual stress. This procedure may be impractical for most fitness-for-service assessments due to the inability to determine the number of weld passes.

Although it may be impossible to determine the number of weld passes for a FFS assessment, additional reduction in the estimated residual stresses may be possible. Finite element analyses conducted by Koppenhoefer, et al.(1998) indicate transverse-to-weld residual stresses at the ID decrease significantly for welds where each pass does not completely bridge the V-groove, as shown in Figure 9. Off-center weld passes that do not bridge the V-groove produce a bending moment in the opposite direction of passes that fill the V-groove. This reverse bending moment appears to reduce the transverse-to-weld residual stresses at the pipe ID. For all welding simulations that included these off-center weld passes, the transverse-to-weld residual stress did not exceed 60% of the yield strength, which occurred along the line starting at the weld toe. At the weld centerline, transverse-to-weld residual stresses were extremely small, as shown Figure 10.

The stress distributions shown in Figures 9 and 10 differ significantly from the distributions presented by Scaramangas (1984) for thick-walled pipes. However, the welds in thick-walled pipe (t=0.75 in) studied by Scaramangas contained 18 passes. In comparison, the welds examined here contained a maximum of 8 passes. The difference between the stress distribution presented here and the work of Scaramangas may be due to the significant difference in the number of passes.

The development of an alternative method based on existence of off-center weld passes may be possible. This method would require significantly less information regarding the weld produced. If the use of off-center weld passes could be determined for a specific weld, this alternative method could provide a reduction in ID residual stresses transverse to the weld.

Effects of Hydrotesting

Hydrostatic testing to 1.5 times MAWP reduces the parallel-to-weld residual stresses by approximately 30% at the weld centerline, as shown in Figure 11, while residual stresses at the weld toe show a 20% reduction. Since the parallel-to-weld residual stresses generate the transverse-to-weld residual stresses through the development of a bending moment in the pipe, the transverse-to-weld residual stresses also decrease (see Figure 12).

In this work, a hydrotest of 1.5 times MAWP produces a primary stress at the ID of $\sigma_{ys}/2$. This primary loading reduces the transverse-to-weld residual stress at the ID to approximately $0.8\sigma_{ys}$. Thus,

$$\sigma_i^{rh} = 1.3\sigma_{ys}^r - \sigma_i^p$$

where σ_i^{rh} is the residual stress at the ID after hydrotesting, and σ_i^p is the primary stress at the ID due to hydrotesting. To incorporate the possibility that σ_i^r calculated from previous equations may be less than σ_i^{rh}, the equation takes the form

$$\sigma_i^{rh} = \min\left[\sigma_i^r, \left(1.3\sigma_{ys}^r - \sigma_i^p\right)\right].$$

Similarly, the residual stress at the OD after hydrotesting may be determined from the following equation

$$\sigma_o^{rh} = \min\left[\sigma_o^r, \left(1.3\sigma_{ys}^r - \sigma_o^p\right)\right].$$

NOZZLE WELDS

To examine the residual stress that develops during welding of attachments between nozzles and spheres, axisymmetric welding simulations were conducted using the model shown in Figure 13. This model represents a pipe-to-spherical head weld without a pad. As shown in Figure 14, the transverse-to-weld residual stress along the path between points A and B show a strong bending component. However, the stresses at location C (the pipe end) equal approximately $\sigma_{ys}/2$. For this reason, a bilinear stress distribution is recommended such that

(1) For $x \leq \dfrac{t}{2}$

$$\sigma^r(x) = \sigma_i^r - 2\sigma_i^r \frac{x}{t}$$

(2) For $\dfrac{t}{2} < x \leq t$

$$\sigma^r(x) = 2\sigma_0^r \frac{x}{t} - \sigma_0^r$$

where, $\sigma_0^r = \sigma_{ys}^r$ and $\sigma_i^r = \sigma_{ys}^r$ (see Figure 15 for though-thickness distribution).

These recommendations for nozzle welds are based on limited finite element analyses. Thus, the distributions do not show any effects of the pipe or head geometry. Future work should refine these recommendations through a detailed parametric study of these weld joints

SUMMARY AND CONCLUSIONS

The residual stress distributions presented here represent a significant reduction from previous estimates. As such these proposed distributions for girth welds and attachments welds reduce the excessive conservatism associated with the assumption of residual stresses equal to yield across the entire weld joint.

For girth welds, a simple linear distribution is proposed where the residual stress at the inner surface of the pipe is a function of thickness, radius-to-thickness ratio, and the hydrotest pressure. The residual stress at the outer wall typically equals $0.2\sigma_{ys}^r$ (unless reduced by hydrotesting). Thus, the weld experiences a combined membrane and bending loading.

The proposed residuals stress distribution for nozzle welds is represented by a bi-linear function that equals the yield stress at the inner and outer surface of the pressure vessel, and equals zero at the midwall. The additional restraint associated with this geometry increases the residual stress in the pipe end (location C in Figure 13). These increased stresses limit the reduction in residual stress in comparsion with girth welds.

The inclusion of these residual stress distributions in API RP-579 should provide more accurate fitness-for-service assessments of pressure vessels and piping. By reducing excessive conservatism in fitness-for-service assesments, a significant financial saving may be obtained through life extension and less frequent weld repair.

ACKNOWLEDGEMENTS

This work was sponsored by the MPC under Agreement FS-II-5. David Osage, BP America, deserves sincere thanks for his guidance and input over the course of this endeavor. Wentao Cheng, EWI, provided considerable assistance by constructing the finite element models and processing many of the analyses in this work.

REFERENCES

Scaramangas, A., "Residual Stress in Girth Butt-Welded Pipes," Ph.D. Dissertation, Cambridge University, May 1984.

Mohr, William C., "Internal Surface Residual Stresses in Girth Butt-Welded Steel Pipes", PVP-Vol. 327, *Residual Stresses in Design, Fabrication, Assessment and Repair*, ASME 1996

Mohr, W.C., Michaleris, P., and Kirk, M.T., "An Improved Treatment of Residual Stresses in Flaw Assessment of Pipes and Pressure Vessels Fabricated from Ferritic Steels", PVP Vol. 359, *Fitness for Adverse Environments in Petroleum and Power Equipment*, ASME 1997.

Mohr, M.C., Koppenhoefer, K.C., Gordon, J.R., "Residual Stresses for Flaw Assessment of Ferritic Steel Pipes and Pressure Vessels", PVP Vol. 380, Fitness-for-Service Evaluations in Petroleum and Fossil Power Plants, ASME 1998.

Koppenhoefer, K.C., Feng, Z. and Cheng, W., "Incorporation of Residual Stresses Caused by Welding into Fracture Assessment Procedures," MPC FS-II-5, Materials Properties Council, September 1998.

Table 1. Analysis Matrix for Two-Dimensional Compendium of Residual Stresses.

Case No.	Weld Type	t (in.)	R/t	Condition	No. of Passes	Pipe Length	No. of Nodes	No. of Elements
1	Single-V Girth	0.25	10	as welded	3	20"	8860	2833
2	Single-V Girth	0.25	25	as welded	3	20"	8860	2833
3	Single-V Girth	0.25	50	as welded	3	20"	8860	2833
4	Single-V Girth	0.25	10	hydrotested to 19 MPa	3	20"	8860	2833
5	Single-V Girth	0.25	25	hydrotested to 8 MPa	3	20"	8860	2833
6	Single-V Girth	0.25	50	hydrotested at 4 MPa	3	20"	8860	2833
7	Single-V Girth	0.75	10		3	40"	9981	3224
8	Single-V Girth	0.75	25	as welded	3	40"	9981	3224
9	Single-V Girth	0.75	50		3	40"	9981	3224
10	Single-V Girth	0.75	10		5	40"	8063	2582
11	Single-V Girth	0.75	25	as welded	5	40"	8063	2582
12	Single-V Girth	0.75	50		5	40"	8063	2582
13	Single-V Girth	0.75	10		8	40"	8172	2621
14	Single-V Girth	0.75	25	as welded	8	40"	8172	2621
15	Single-V Girth	0.75	50		8	40"	8172	2621
16	Single-V Girth	0.75	10	hydrotested to 19 MPa	3	40"	9981	3224
17	Single-V Girth	0.75	25	hydrotested to 8 MPa	3	40"	9981	3224
18	Single-V Girth	0.75	50	hydrotested at 4 MPa	3	40"	9981	3224
19	Single-V Girth	0.75	10	hydrotested to 19 MPa	5	40"	8063	2582
20	Single-V Girth	0.75	25	hydrotested to 8 MPa	5	40"	8063	2582
21	Single-V Girth	0.75	50	hydrotested at 4 MPa	5	40"	8063	2582
22	Single-V Girth	0.75	10	hydrotested to 19 MPa	8	40"	8172	2621
23	Single-V Girth	0.75	25	hydrotested to 8 MPa	8	40"	8172	2621
24	Single-V Girth	0.75	50	hydrotested at 4 MPa	8	40"	8172	2621
25	Single-V Meridional	1	50	as welded	10	R = 50"	11454	3619
26	Single-V Meridional	1	200	as welded	10	R = 200"	10734	3357
27	Groove-Fillet Attachment	1	50	as welded	9	R = 50"	11001	3506

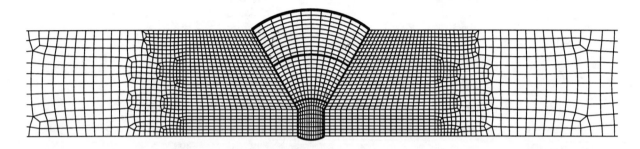

Figure 1. Finite Element Model of a Girth Weld in a $t = 0.25$ in. Pipe.

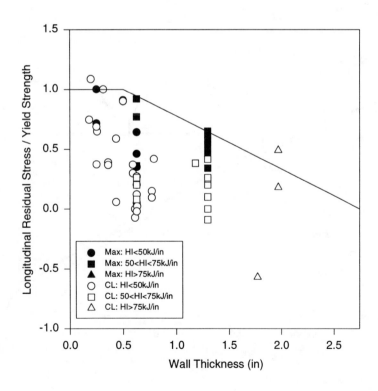

Figure 2. Inner Diameter Residual Stresses Transverse-to-Welding Direction Compared with Estimation Method [3].

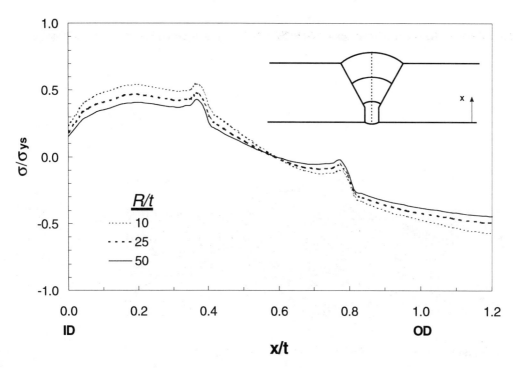

Figure 3. Through Thickness Distributions of Transverse-to-Weld Residual Stress for Three *R/t* Ratios (10, 25 and 50). The Dashed Line on the Sketch of the Weld Indicates the Path over which the Stresses were Calculated.

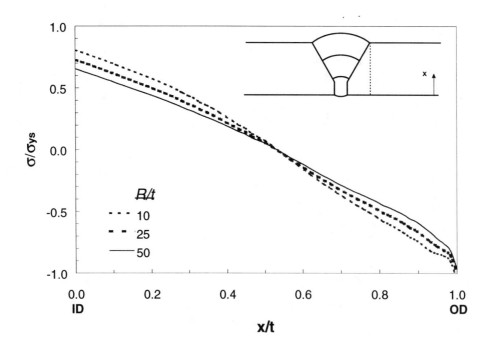

Figure 4. Through Thickness Distributions of Transverse-to-Weld Residual Stress for Three *R/t* Ratios (10, 25 and 50). The Dashed Line on the Sketch of the Weld Indicates the Path over which the Stresses were Calculated.

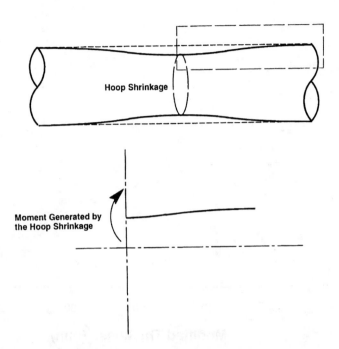

Figure 5. Basis for Development of Transverse-to-Weld Residual Stresses in a Girth Weld.

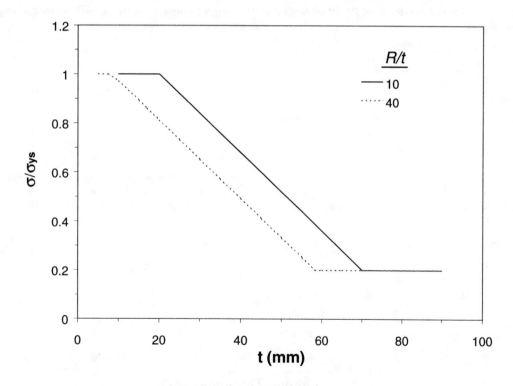

Figure 6. Effect of using the *R/t* modified thickness value, $\hat{t}$, on the residual stress predictions.

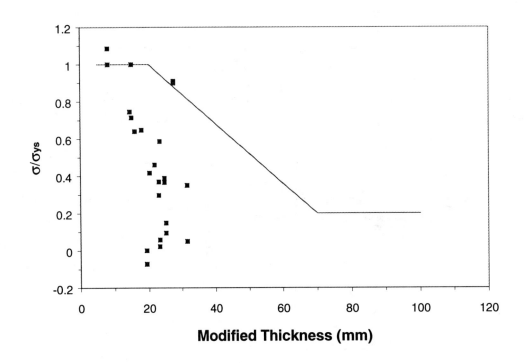

Figure 7. Transverse-to-Weld Residual Stress at ID of Pipe Compared with Estimation Method for Low Heat Input Welds

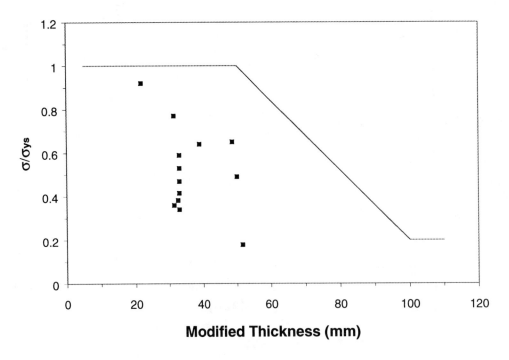

Figure 8. Transverse-to-Weld Residual Stress at ID of Pipe Compared with Estimation Method for High Heat Input Welds

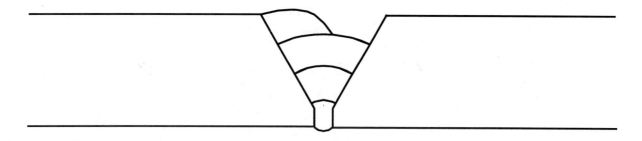

Figure 9.   Schematic of Weld Pass that does not Bridge the V-groove.

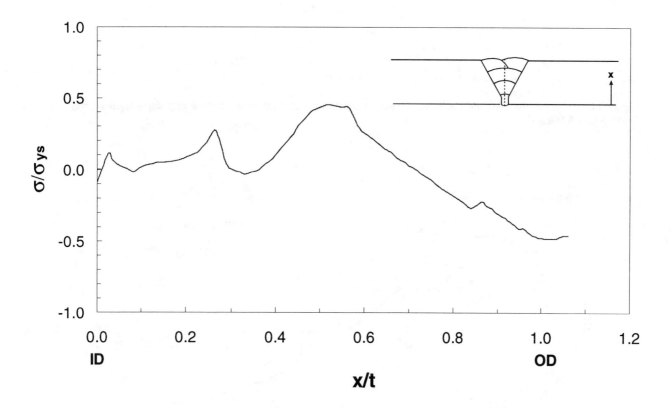

Figure 10.   Transverse-to-Weld Residual Stress in Girth Weld with Non-Centered Passes (t=0.75 in, R/t=10).

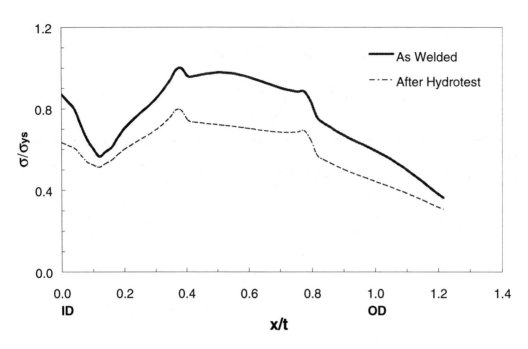

Figure 11 Parallel-to-Weld Residual Stress at Centerline of Girth Weld Before and After Hydrostatic Testing to 1.5 MAWP

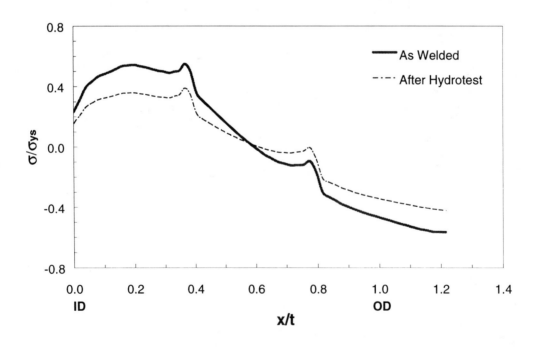

Figure 12 Transverse-to-Weld Residual Stress at Centerline of Girth Weld Before and After Hydrostatic Testing to 1.5 MAWP

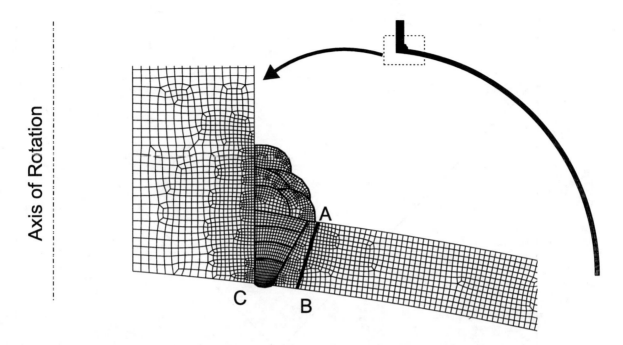

Figure 13.   Finite Element Model of an Attachment Weld between a Nozzle and Sphere

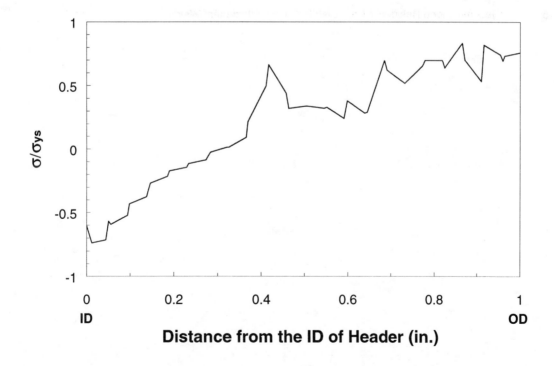

Figure 14.   Transverse-to-Weld Residual Stresses in Attachment Weld. In head, R/t=50, t=1.0 in. In the Pipe, ID=8 in., t=1.8 in.

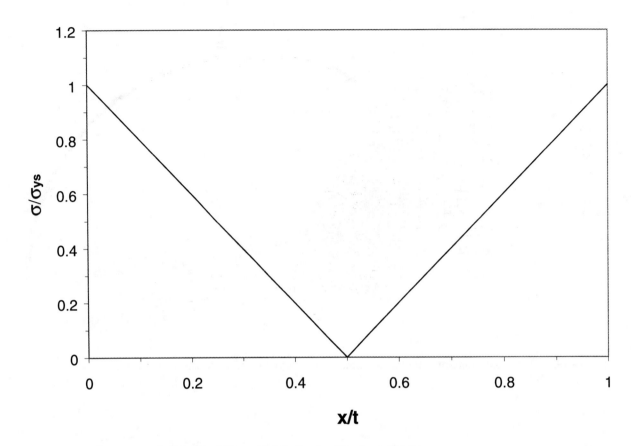

Figure 15.   Recommended Residual Stress Distribution for Attachment Welds

A Finite Element Evaluation of Residual Stress In a Thread Form Generated by a Cold-Rolling Process

by John A. Martin, Principal Engineer
Lockheed Martin Corporation, Schenectady, NY

ABSTRACT

This paper presents a finite element evaluation of residual stress in a thread form generated by a cold rolling process. Included in this evaluation are a mesh development study, methodology sensitivity studies, and the effects of applied loads on the stress in a rolled thread root. A finite element analysis of the thread forming process using implicit modelling methodology, incremental large deformation, elastic-plastic material properties, and adaptive meshing techniques was performed. Results of the study indicate the axial component of the residual stress in the thread root of the fastener is highly compressive. Results also indicate that a rolled threaded fastener loaded to an average tensile stress equal to yield through the cross-section will retain compressive stresses in the thread root. This compressive stress state will be advantageous when evaluating fasteners for fatigue and environmental concerns.

BACKGROUND

Many applications in the industrial world rely on threaded fasteners as a means for joining components. Depending on the application, fatigue behavior and/or the component's environment can dictate equipment life expectancy. Cold rolling manufacturing processes have been developed as a means of introducing beneficial residual stresses into thread roots, thereby increasing resistance to fatigue and environmental effects. However, the benefits are predominately determined by empirical means with very little numerical assessment of the residual stress magnitudes. Empirical testing efforts [1] to understand the fundamentals principles acting to generate the final stress state in a thread root fabricated by a rolling process are unable to quantitatively correlate changes in process parameters with changes in stress states.

Hence, potential design benefits that might be realized from a basic understanding of the residual stresses generated by a cold rolled thread forming process have, for the most part, not yet been established. This lack of predictive capability prompted efforts to develop an analytical program to determine thread root stresses due to the rolling process through the use of computer modelling. This paper is a continuation of the work of Reference 2 and reports the stress states in a thread root developed during and after cold rolling under simulated applied loadings.

DISCUSSION

This work supports an effort to gain more assurance of improved fastener fatigue resistance through the use of rolled fasteners. During the cold-forming thread rolling process, threads are formed on a round workpiece as it passes between rotating dies. This process generates a thread form without metal loss and with a higher yield strength due to the cold-working process. Testing has indicated that this process improves fatigue life [1,3,4], which is attributed to the creation of compressive residual stresses on the thread root surface. In an effort to verify this empirical result, an analytical evaluation is desired to obtain a fundamental analytical understanding of the thread rolling process to support design applications.

To obtain a better fundamental understanding of the principles behind the apparent increased performance brought about by thread rolling, finite element analysis is being performed. Reference 2 examined the residual stress state of a modified, oversized ACME thread root as part of a software qualification program supporting this analytical methodology. These analytical methodologies are extended to the evaluations of a thread form and include:

- Establishment of an adequate mesh for use in a 2-D plane strain finite element model representation of the thread form.
- Application of simulated end loading subsequent to residual stress state determination.
- Sensitivity studies to verify accuracy of the results.

The studies performed in this report primarily focus on the X-direction stress, or axial stress with respect to a loaded stud. Note the graphical output refers to the X-direction stress as the first component of stress (S_{11}). This stress state is of critical concern when addressing fatigue effects on studs.

Finite Element Model

For this evaluation the MARC software analysis package [5] with Lagrangian update capabilities, large strain formulation, plasticity/work hardening, adaptive meshing [6], and contact/friction options was used. In addition, an implicit, static, rate and time-independent analysis using rigid body motion as a means of providing loading to simulate a metal forming process was assumed.

The finite element modelling incorporated 2-D, four-noded, first-order (linear) isoparametric plane strain elements (MARC Element 11). This element provides for the finite strain plasticity and updated Lagrangian procedures necessary to address a highly non-linear contact problem.

Although the thread rolling process is a 3-D process, this initial investigation assumes a 2-D plane strain model is acceptable based on the localized stress effect created by the rolling process with respect to the overall thread and stud dimensions.

Material Properties

This study addresses commercially available ASTM Alloy 625 material. The elastic-plastic material properties used in the finite element analysis are shown in Table 1 and are the results of prior tensile testing [2].

Mesh and Model Development

Figure 1 provides the dimensions of the thread form used in this evaluation. These dimensions formed the basis for both the rigid body die model and the desired final deformed thread configuration.

The initial modelling effort consisted of a five thread 2-D plane strain model employing a standard rectangular mesh (Fig. 2), with symmetry boundary conditions invoked on the three sides not subject to die indentation.

The model loading consisted of a total indentation of 0.040" over 100 load increments with subsequent unloading over the same number of increments. This mesh configuration resulted in ill-conditioned elements causing convergence problems and lead to premature termination of the runs. An ill-conditioned element is one that distorts to such a degree that it turns itself inside out, provides an unacceptable aspect ratio, or generates a skew angle approaching or greater than 180 degrees [6]. Any of these conditions force the code to abort the solution and flag the user. Observations from this study included:

- The smaller the mesh, the more difficult convergence became due to the relatively large displacement to element size ratio. A series of models were run varying the number of iterations, the number of elements and element density, and the convergence tolerances without any success in approaching the desired deformation. The best results obtained were 50% completion of the desired deformation.

- A study using coarser elements resulted in program termination due to ill-conditioning elements. Again, a trade-off between number of elements and element size was assessed. With fewer elements, there were some cases that ran to completion, but the stresses reported across the element differed by such a large amount that the results were questionable. The results indicate that an acceptable aspect ratio for a rectangular element would undergo a transformation to a final element configuration whose shape is too deformed to prevent ill-conditioning of the elements prior to completion of the loading.

With an automatic mesh reconfiguration program unavailable, another avenue of mesh generation was applied. This alternate method entailed the use of material flow lines as the basis for mesh generation. When a material undergoes large plastic deformation the stress state in the surrounding region takes on a stress pattern that will reflect the deformed condition. This final stress field flow pattern was used to develop a workable mesh configuration.

An adequate mesh was developed by considering the flow field in the direction normal to the indentation. In this approach a finite element mesh is generated, so elements in the direction of the deformation initially have high aspect ratios (4:1 as modelled mesh). However, upon completion of the loading the final deformed element shapes will ideally approach a one-to-one element aspect ratio (0.5:1 final mesh aspect ratio). This modified mesh resulted in the model running to completion. The modified mesh concept is demonstrated in Fig. 3 for the entire plane strain model, and Fig. 4a through 4f for a single thread form development. Fig. 4a through 4f also demonstrate the adaptive meshing option.

Indentation Results

The y displacement (δ_y) results in this evaluation only reached a crest-to-root distance of 0.0552". Attempts to further deform the model to reach the desired crest-to-root distance (0.0718") resulted in ill-conditioned elements and/or convergence problems. To progress further would require either extensive mesh modifications or automatic remeshing capabilities. Although the finite element (FE) model final crest-to-root distance is less than desired, the trends observed and conclusions drawn are judged to be appropriate for thread rolling process evaluations.

Figure 5 reflects the symmetry of the model results; and, therefore, throughout this report only one location (center thread root surface node, node 72 see Fig. 4f) was selected to assess the relative effects of each variation on the particular parameter. Fig. 6 and 7 illustrate the first component of stress (S_{11}); that is axial stress, with respect to a thread orientation for the final loaded increment (inc. 99) and the fully unloaded increment (inc. 196), respectively. Comparison of the S_{11} values show a significant increase in compressive stress. For the fully loaded condition (Fig. 6), the compressive stress is actually beginning to decrease in the thread root region due to the flow of the material up into the flank portions of the thread. As the material flows up into the die and forms the crest, compressive stress in the flank region begins to increase. This increase in flank compressive stress and decrease in root stress is a result of the material flowing to reach an equilibrium state under the uniformly applied loading.

Figure 7, representing the fully unloaded condition, shows an increase in the compressive stress in the thread root subsequent to unloading. This increase in compressive stress reflects the amount of springback that occurs upon unloading. This springback is also partially responsible for the creation of the tensile stress along the upper top surface of the thread flank.

One can also observe that the depth of the compressive stress state during loading is relatively deep, as opposed to the close proximity to the surface of the high compressive region subsequent to the unloading. Although the reported stress is significantly greater than yield, this is attributed to a shift in the yield stress curve associated with the hydrostatic stress state that results from the high plastic deformation.

It is judged that although an increase in the depth of penetration would relax the compressive stress in the root existing during loading, the final unloaded residual stress state will be highly compressive. This judgement is supported by the results of the sensitivity studies.

Simulated Applied Stud Loading

In addition to understanding the residual stress state created by the thread rolling process, the goal of this study is to develop a process by which fatigue benefits derived from the rolling process can be quantitatively understood. One means of accomplishing this is to generate a 3-D rolled thread model, and gradually apply end loads while tracking the stress state at the thread root. Determination of the thread stress to load ratio can eventually be used to correlate with stud fatigue test results.

The first step toward that goal uses the results from the 2-D plane strain model and applies an incremental displacement driven loading. Two different loading conditions were examined. First, the final thread form geometry, with the residual stress state from an incrementally loaded/unloaded condition, was restrained at one end while the other end was incrementally displacement loaded. The criteria by which loading was considered complete was based on the average cross-section of the model reaching von Mises yield stress.

Figures 6 through 8 track the stress state of the center thread root throughout the entire loading/unloading and applied end loading to a yielded cross-section process. For all three cases the same legend is displayed to allow for easier tracing of the stress field changes.

Nodal values are extrapolated from Gauss point stresses and are used for the graphical display. Results indicate an additional benefit could be realized from further mesh refinement to more accurately calculate the residual stress states. These results demonstrate the formation of a compressive residual stress state in the rolled thread root that will remain slightly compressive, even when the model is axially loaded to yield through the cross-section. It is these results that provide promise for development of design correlations for fatigue in rolled threads.

The second case examined simulated a non-rolled (cut) thread with the rolled thread geometry without the residual stress state, and applied a similar incremental displacement end loading. Fig. 9 reflects the tensile nature of the non-rolled thread root under an end loading.

Comparison of these two cases emphasizes the distinct difference between the thread roots that have a compressive thread root stress state (rolled threads) and those that do not (cut threads).

Model Sensitivity Studies

Performance of this evaluation required assumptions to be made in an effort to obtain model convergence and reduce computer usage (CPU and memory requirements). Sensitivity studies were performed to assess the impact of these assumptions on the results and conclusions. These sensitivity studies varied the mesh density, the incremental loading/unloading, the coefficient of friction, and two convergence tolerances. In addition, a study assessing the reliability of the final element configuration was performed.

Mesh Density Study

Difficulties in developing a full thread form were due to the lack of an adequately refined mesh. To confirm the adequacy of the results presented, a mesh coarsening approach was taken. Similar to a mesh refinement, should a change in mesh result in less than a 10% change (arbitrarily selected) in reported stress values, then the initial mesh can be considered to be adequate to predict stress results. However, should the comparison reflect a change greater than 10%, then an additional mesh refinement should be applied to ensure accuracy of results.

Figures 10a and 10b plot S_{11} stress with respect to increments over the course of loading and unloading of the workpiece. As indicated by Fig. 10a, the coarser mesh model provides a residual stress approximately 40% less than the refined mesh model. From this result, it could be concluded that further mesh refinements would be required prior to acceptance of a quantitative representation of the thread root residual stress. However, the trend toward higher compressive residual stress with a finer mesh is assumed given these results. Although the finer mesh ,provided greater compressive stress, the runtime was three to four times greater than that of the coarse mesh. Given the conservative nature of the coarse mesh results, the subsequent sensitivity studies were performed with the coarse model primarily to conserve computer resources. The trends observed will not be affected by the use of the coarse mesh.

The highly irregular patterns observed in Fig. 10a are attributed to both the coarseness of the mesh when experiencing large plastic deformations, and the relaxed tolerances needed to ensure convergence. Node 72 is a surface node associated with elements that are highly distorted. As these elements are deformed with additional loading, the absolute displacement convergence tolerances used are compared to the incremental calculated values. In some cases the allowable may be just met, while in other increments it might be just missed. For the latter case additional iterations are required, thereby changing the difference between the allowable and reported values. It is this variability of the results caused by the selected convergence tolerance that contributes to the swing in results observed between increments. A confirmation of this observed condition is seen in Fig. 10b, a plot of the S_{11} stress for the adjacent node (1201) below Node 72 in the coarse model. For the refined model it is the third node below the surface, but in the same geometric location as Node 1201 in the coarse model. Notice that for this case there is less variability in the nodal results and approximately the same final residual stress state. This indicates the mesh below the highly distorted surface elements provides reasonably stable results.

Incremental Loading Study

The finite element mesh modelling was initially performed in a single loading/unloading method to facilitate mesh development. This precluded the occurrence of ill-conditioned elements after a series of loading/unloadings had occurred. In an actual rolling process the indentation/rolling occurs over a number of loading/unloading increments. This study assesses the effects of increasing the increments over which the total thread form is generated. The initial base case performed the 40 mil indentation in a single loading. This study compares those results with a 40 mil total indentation obtained in two (20 mils/set), four (10 mils/set) and five sets (8 mils/set) of loading/unloading intervals.

The results of this study (Fig. 11) show approximately the same trends for each series. Each subsequent loading pattern shows an increasing compressive residual stress value until the final loading increments. A slight decrease in compressive residual stress in the thread root is observed over the last loading increment. This change in compressive residual stress is attributed to the tendency for the coarser mesh to generate ill-conditioned elements. The ill-conditioning of the elements results in a large stress differential across the Gauss points making up the element, thereby skewing the reported surface results. It should be noted that with the exception of the single load step case, the multiple load cases appear to have approximately the same residual stress value. Further investigation indicates that the single load step case tends toward an ill-conditioned element; therefore, under predicting the final result. Results indicate that as the number of incremental loading/unloading increases, the greater the compressive residual stress value in the thread root.

Frictional Effects Study

Although Reference 2 results indicated minimal effects of coefficient of friction assumptions on the residual stress state, it was observed that the rigid body sliding along the elements making up the surface was one of the governing factors in generating ill-conditioned elements Although the impact on residual stress was small, delaying the onset of ill-conditioning, thereby maintaining better element configuration, would allow for a better prediction of residual stresses. Examination of the model results indicated that the incorporation of friction allowed for more uniform deformation in the surface elements, thereby maintaining a better aspect ratio. Figure 12 shows less fluctuation due to element distortion over the final few increments. This smaller fluctuation results in slightly deeper penetration and a higher compressive residual stress state. Again these results support the conservative nature of the results used in the simulated end loading study.

Convergence Tolerance Study

The relatively small element size coupled with the relatively large displacements created various convergence problems in the initial attempt to obtain a solution. Attempts involving both mesh changes/refinements and tolerance alterations were made to obtain a solution. Due to the difficulty in obtaining solutions, once a solution had been obtained it was necessary to perform a sensitivity on the corresponding tolerance value. The inability to obtain a converged solution with a tighter tolerance restricts the accuracy of this sensitivity study. Rather than tightening the tolerance (as required in a true tolerance study) a relaxation of the tolerance was examined similar to that in the mesh density. For the tolerance used (0.001) two additional cases were examined. One case relaxed the tolerance by five times the initial value (0.005), while the second case made an order of magnitude change (0.01). The results of this study are shown in Fig. 13.

The results of Fig. 13 indicate that convergence tolerance selection makes a difference in stress results only in the final loading step. The final step demonstrates that the more relaxed tolerances provide a greater compressive residual stress, than the tighter tolerance used in the base case. At first glance it would appear that the tightening of the tolerance should result in a greater compressive residual stress state. However, the tightening of the tolerance in the final steps of the coarse mesh initiates the process by which the elements begin to become ill-conditioned, and the solution begins to break down. This was concluded based on the large differential in stress values across a single element. The results of this study are not as conclusive as the other studies presented herein. However, the fact that there was no change in the results for the lower two tolerances and no trend toward ill-conditioned elements lends support for the reasonableness of the solution used to address the residual stress levels in a rolled thread root.

Iterative Solver Tolerance Study

Large deformation and the need for updating of the geometry with each iteration promoted the use of an iterative solver. A full Newton Raphson method was used for all the cases, with iterative solver tolerances modified where required. Maintaining the base case displacement convergence tolerance, the model was capable of running a study on the conjugate gradient solver tolerance by varying the base case value of 0.001 by an order of magnitude in either direction. For this study, the tighter the assumed tolerance the better behaved the model; and, hence, the greater the compressive residual stress value in the thread root. These results also support the slightly conservative nature of the quantitative stress values reported for the end loading results. The results of this study are found in Fig. 14.

Non-Positive Definite Study

Establishment of a mesh which demonstrated convergence was facilitated by an option that forced convergence once the predetermined number of iterations had been reached to allow the analysis to continue. This option is associated with a non-positive definite flag. This approach is only valid if the elements do not become so ill-conditioned that they turn inside-out. To confirm that there were no inappropriately formed elements, a run was made with the non-positive definite flag turned off. Fig. 15 demonstrates no difference between the two runs, thereby indicating the elements deforming over the duration of the run are not sufficiently ill-conditioned to invoke this option. This further supports the adequacy of the results presented.

Uncertainties

Based on the limited geometric representation of a rolled thread using 2-D plane strain modelling assumptions; additional confirmatory and theoretical work must be completed, including expansion of the results into 3-D, before final determination of the quantitative thread root residual stress state generated by cold-rolling. The extension of the 2-D work into 3-D addresses the following uncertainties:

- Differences created between indentation only and rolling/indentation modelling along with the influence of roll draft on the final stress state.

- Determination of the effects of increased thread form depth and thread configuration on final stress states.

- Determination of the contribution to and/or relaxation of residual stress that occurs as a result of a 3-D representation of the cold-forming process.

- The effects of geometric differences in thread configurations (helical versus cylindrical thread forms) and processes (through-feed versus in-feed processing).

- Residual stress state impact of evaluations made regarding the effects, if any, of load transfer through thread flank-to-flank contact.

The use of a metal forming software package (e.g., MARC/Autoforge) will eliminate the element ill-conditioning that comes with die penetration. It will also permit automatic mesh reconfiguration to provide improved solution accuracy.

None of these items are judged to significantly alter the nature of the residual stress state.

Summary

The results of this study provide an initial understanding of favorable fatigue test results demonstrated to date for rolled threads [1 and 4]. It also identifies the potential for increased fatigue resistance for design applications that incorporate rolled threads. Although additional analytical and possibly experimental qualification need to be addressed, there is a high degree of confidence that the end goal of developing design considerations will be attainable.

References:

[1] Kephart, AR, "Fatigue Acceptance Test Limit Criterion for Large Diameter Rolled Thread Fasteners", ASTM E-8 Sub-Committee Task Group 04-07 - Workshop on Fatigue and Fracture of Fasteners, May 6, 1997, St. Louis, MO.

[2] Martin, JA "Fundamental Finite Element Evaluation of a Three Dimensional Rolled Thread Form: Modelling and Experimental Results" - PVP-373 pg 457-467 presented at ASME Pressure Vessel and Piping Conference at San Diego July 26-30 1998

[3] Lin CS, Lourilliard JJ, and Hood AC, "Stress Corrosion Cracking of High Strength Bolting", SPS Technologies, Inc. ASTM STP 425, 1967, pgs 84 to 98

[4] Roach TS, Aerospace High Performance Fasteners Resist Stress Corrosion Cracking, SPS Technologies, Inc., *Material Performance*, Vol 23, No. 9 September 1984, pg 42

[5] MARC Software Version 6.2, MARC Analysis Research Corp. Palo Alto, Ca

[6] Martin, JA "A Mesh Density Study for Application to Large Deformation Rolling Process Evaluations" - PVP-373 pg 177-184 presented at ASME Pressure Vessel and Piping Conference at San Diego July 26-30 1998

Table 1: Material Properties at Room Temperature

Tensile Properties	ASTM Alloy 625
Ultimate Tensile Strength (ksi)	137.9
0.2% Yield Strength (ksi)	86.7
Elongation (%)	41.7
Reduction in Area (%)	52.0
Elastic Modulus, E (10e3Ksi)	29.8

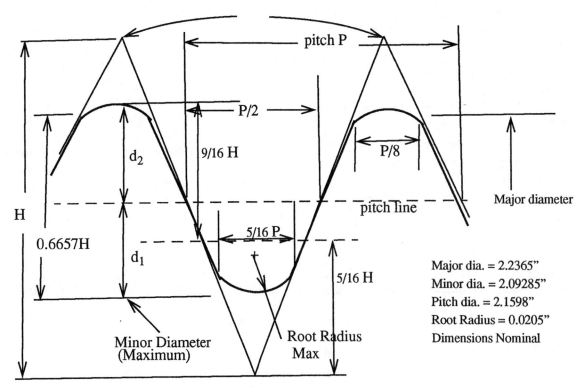

Figure 1 Geometric Configuration of Thread Form

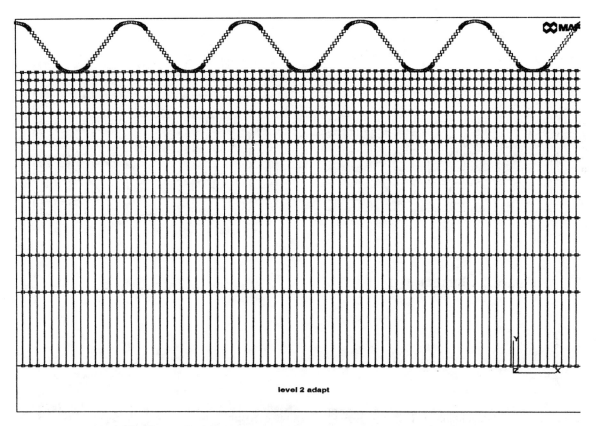

Figure 2: Initial Rectangular Finite Element Mesh

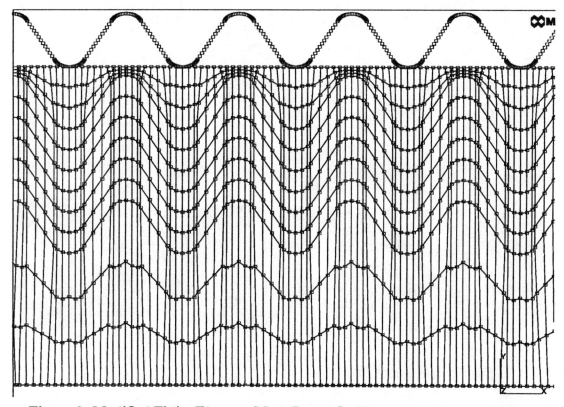

Figure 3: Modified Finite Element Mesh Based On Expected Deformed Shape

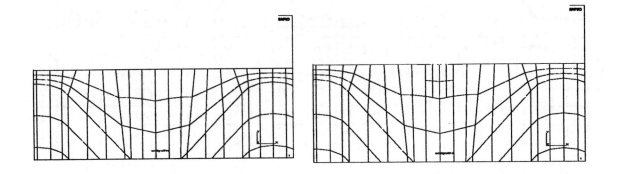

Figure 4a: Increment 0   **Figure 4b: Increment 2**

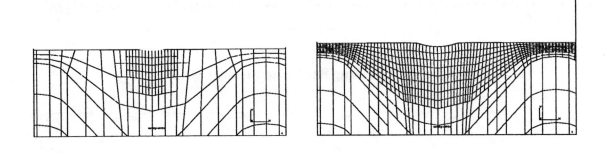

Figure 4c: Increment 4   **Figure 4d: Increment 11**

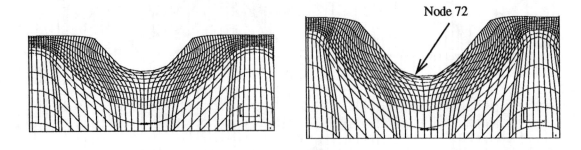

Figure 4e: Increment 75   **Figure 4f: Increment 196**

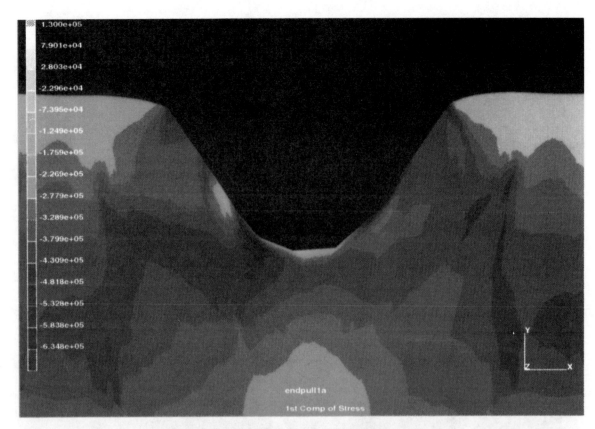

Figure 6: S_{11} Stress State At Full Indentation

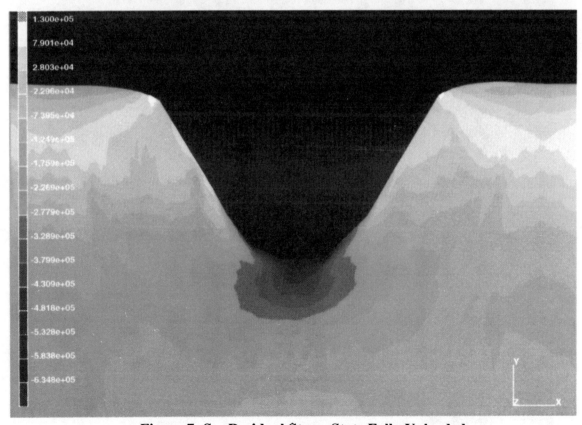

Figure 7: S_{11} Residual Stress State Fully Unloaded

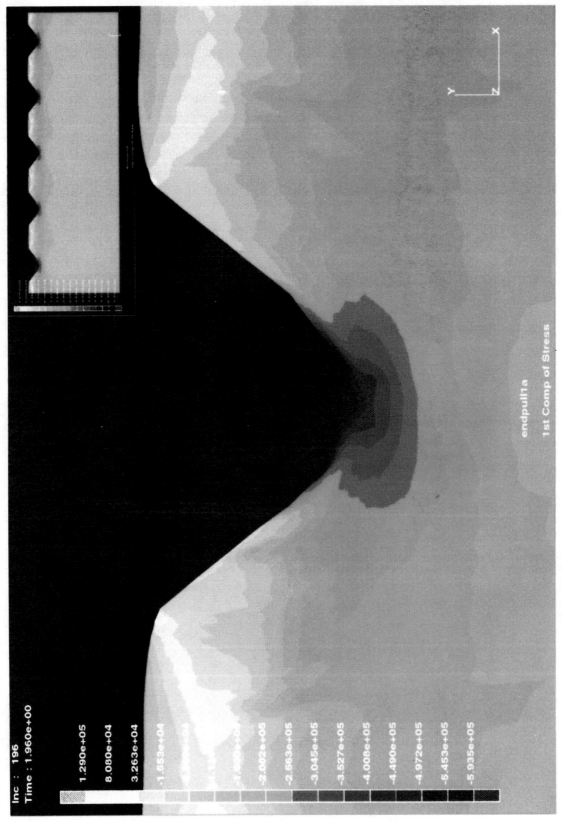

Figure 5: Residual Axial Stress State of Unloaded Rolled Thread Form

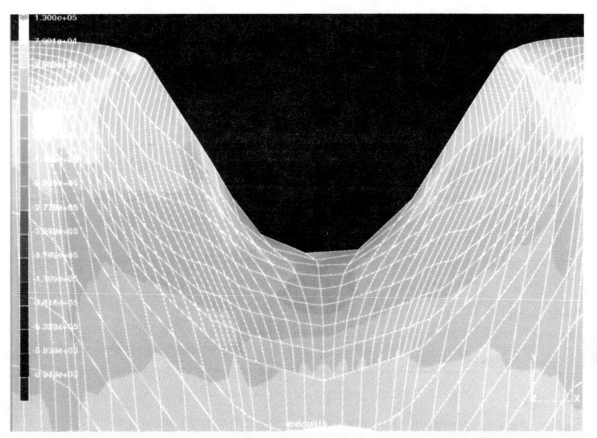

Figure 8: S_{11} Stress State w/ Applied End Load to Von Mises Sy Averaged Thru Cross-Section

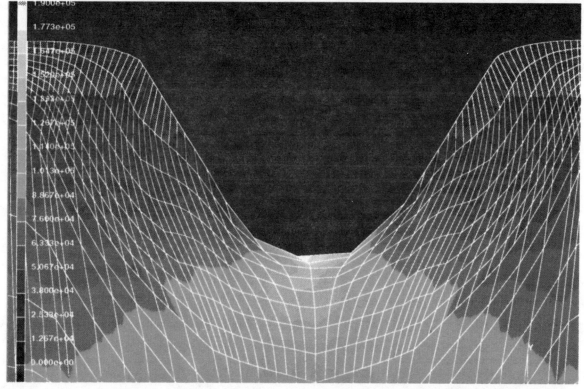

Figure 9: S_{11} Stress State w/o Residual Stress w/ Applied End Load to Von Mises Sy Averaged Thru Cross-Section - Tensile Stress Only

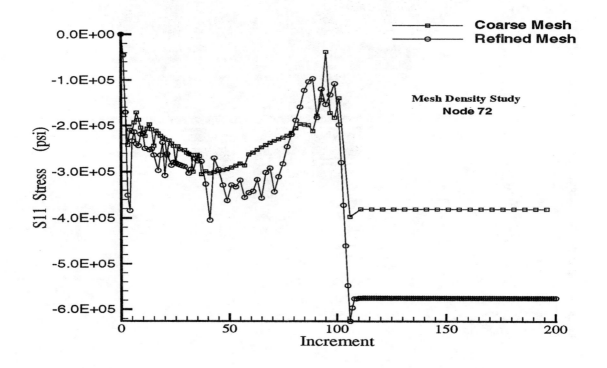

Figure 10a: Mesh Density Study Surface Node 72

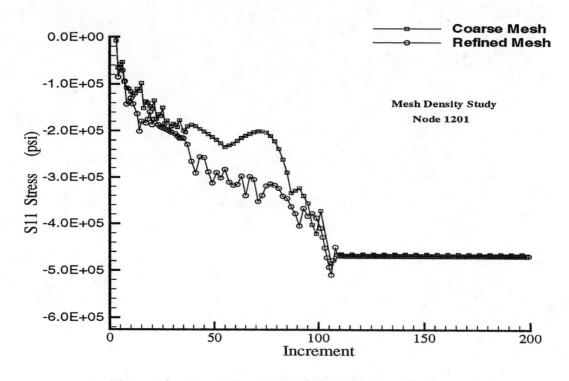

Figure 10b: Mesh Density Study Sub-Surface Node 1201

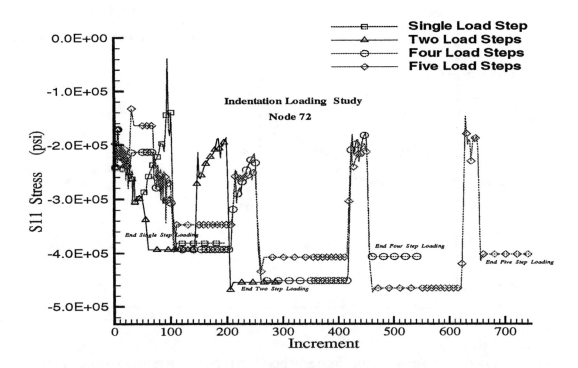

Figure 11: Incremental Loading Study w/ Varying Depths per Loading Series

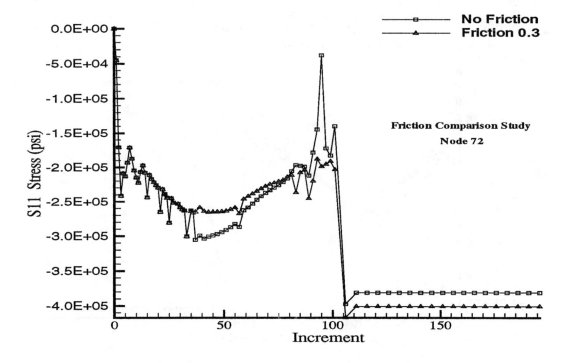

Figure 12: Comparison of Effects of Assumed Coefficient of Friction on Results

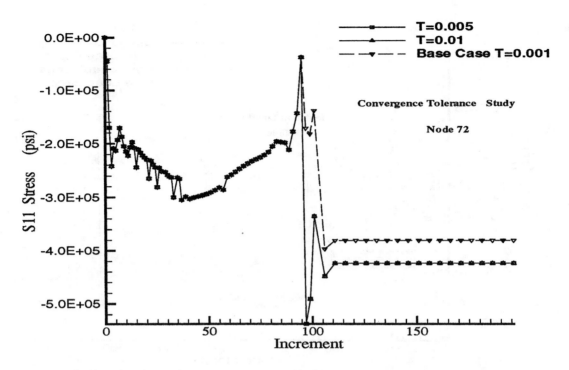

Figure 13: Variation of Convergence/Tolerance on Residual Stress State

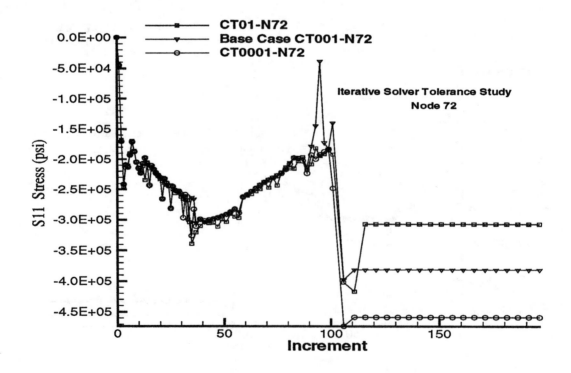

Figure 14: Variation of Iterative Solver Tolerance on Residual Stress State

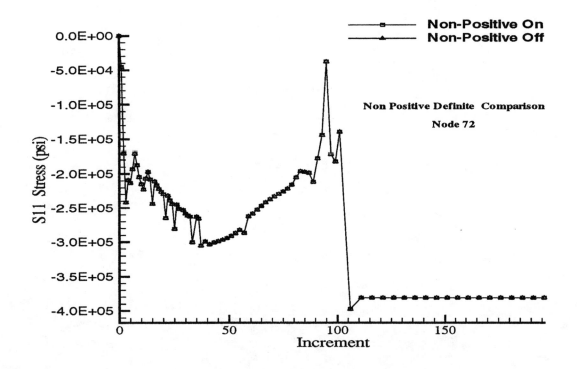

Figure 15: Non-Positive Definite Option Study

Residual Stresses in Inconel-718-Clad Tungsten Rods Subject to Hot Iso-Static Pressing

Ronnie B. Parker
Engineering Sciences and Applications Division (ESA-EA)
Los Alamos National Laboratory
Los Alamos, NM 87545

Partha Rangaswamy
Materials Science and Technology Division (MST-8)
Los Alamos National Laboratory
Los Alamos, NM 87545

ABSTRACT

The Accelerator Production of Tritium (APT) backup target design uses clad tungsten rods to generate neutrons via spallation reactions between high-energy protons and tungsten. Design requirements and safety considerations do not allow coolant to be in direct contact with the tungsten, necessitating cladding of the tungsten. Inconel 718 was chosen as the cladding material for its superior mechanical, thermal, and neutronic properties. Thermal-hydraulic considerations require the clad to remain in contact with the tungsten at all times during plant operations, necessitating bonding of the clad to the tungsten. The clad was diffusion bonded to the tungsten by hot-isostatic-pressing (HIP) at 1080 °C. The difference in the thermal expansion coefficients of the tungsten and inconel 718 generate large residual stresses after bonding. Finite element analyses (FEA) were used to predict residual stress levels from HIP for subsequent thermal stress analysis to determine operational stress levels for lifetime predictions. Residual stresses in the pressed specimens were measured using X-ray diffraction (XRD) (classical d vs. $\sin^2\psi$ technique) technique to benchmark FEA calculations. Residual stresses were obtained in the inconel 718 cladding only, and along the directions parallel to the orientation of the tungsten rod. This paper compares the residual stresses predicted from the FEA and XRD measurements in the inconel 718 cladding.

INTRODUCTION

The production of tritium within the APT facility will rely upon the preferential absorption of neutrons by ^{3}He, an isotope of Helium, to produce tritium. Neutrons are produced within the target region by spallation reactions between high-energy protons and tungsten. To optimize the production of neutrons within the target, the tungsten is split apart and spaced to allow neutrons to escape the target. To produce the required geometry and cool the target, a ladder-rung arrangement as shown in Fig. 1, has been proposed. Within the rungs of the ladder, inconel-718-clad concentric tungsten tubes will be placed to allow coolant to be circulated around them. A secondary backup design utilizes inconel 718 clad tungsten rods within the ladder-rung as a target. This paper focuses on the residual stresses after hot isostatic pressing of the clad tungsten rods.

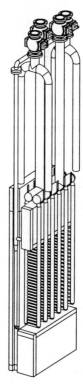

Figure 1. APT neutron spallation target ladder rung geometry.

Safety concerns for maintaining a geometry with the ability to be cooled required the cladding to remain in intimate contact with the tungsten at all times, allowing heat to be removed from the tungsten. The higher coefficient of thermal expansion (CTE) of inconel 718, coupled with the geometric layout of the rod, requires the cladding to be bonded to the tungsten during plant operation. Hot isostatic pressing of the clad rods has been proposed to diffusion bond the inconel 718 to the tungsten. However, the HIP process generates residual stresses within the rod that must be accounted for during plant operations. In this research the objective was to apply finite element modeling to simulate the evolution of stresses during the HIP process and benchmark these predictions with X-ray diffraction residual-stress measurements after the completion of the consolidation process.

NUMERICAL SIMULATION OF RESIDUAL STRESSES USING FINITE ELEMENT MODELING (FEM)

Physical Geometry

The clad rod studied by this analysis is depicted in Fig. 2. The rod is manufactured by sliding a 3.175 mm tungsten rod into an inconel 718 tube with a 0.125 mm wall thickness and a nominal 1.65 mm inside radius. This creates a nominal 0.0625-mm on-radius clearance between the cladding and tungsten. This gap is explicitly modeled within the finite element models to accurately predict processing stresses. End caps are then electron-beam welded on the ends of the tube to completely encase the tungsten rod in inconel 718.

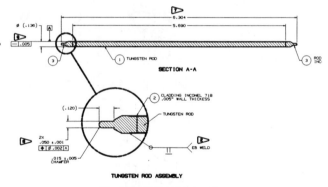

Figure 2. Clad rod geometry

HIP Process

The clad rod is then placed within a hot isostatic press and subjected to the pressure and time history, as shown in Fig. 3. Externally applying an appreciable pressure to the assembly forces the inconel 718 cladding to entirely contact the tungsten. At some undetermined time during the cycle, they become bonded together.

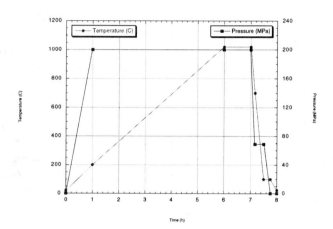

Figure 3. HIP cycle parameters showing pressure and temperature variation with time.

FEM Model Development

Finite Element Model

A two dimensional, axisymmetric FEM model with one-half symmetry was developed to model the HIP process. The general purpose, commercially available finite element package, ABAQUS/Standard, was chosen as the numerical solution to the problem for its capability to solve thermal-mechanically coupled problems, and its ability to explicitly model contact between bodies with friction. The model consists of 4,927 8-node axisymmetric elements, and 16,628 nodes to model the rod. The model is shown in Fig. 4.

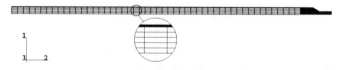

Figure 4. ABAQUS, 2-D, Axisymmetric Finite Element Model of One-half the Clad Rod.

FEM Constitutive Model

The tungsten target material has a minimum purity of 99.9%, and was obtained from a commercial vendor. An elastic constitutive model was used for the tungsten, as shown in Table 1. High-temperature data was unavailable for the tungsten, so it was assumed that these properties prevailed at all temperatures during the analysis. It should be noted that room temperature properties suggest the tungsten behaves as a brittle material, exhibiting no plasticity. (Machine Design 1995)

Table 1: Thermal and Mechanical Properties of Tungsten

Density	0.2 g/cm^3
Youngs Modulus	404 GPa
Yield Strength	1507 GPa
Ultimate Strength	1507 GPa
Coefficient of Therml Expansion	4.5 x 10^{-6} /K

The material for the cladding was chosen to be annealed inconel 718. This material was chosen for its superior thermal and neutronic properties. Inconel 718 is a nickel-based-super-alloy consisting of 51% nickel. The constitutive model for the inconel 718 clad consists of explicitly defined elastic-plastic true stress versus strain curves at various temperatures. The engineering stress versus strain curves used to develop the constitutive relationship are depicted in Fig. 5.(Dalder 1995) The temperature dependent elastic modulus, Poisson's ratio, and CTE are shown in Fig. 6.(Klopp 1995) Note that ABAQUS by default, uses a Von Mises yield surface and was used for all calculations. ABAQUS also requires isotropic hardening flow rule for coupled temperature-displacement analysis. Although isotropic hardening may not produce as accurate results as kinematic hardening for cyclic plastic loading, the coupled solution technique was chosen for its benefits and versatility thus limiting the plasticity model to isotropic hardening. ABAQUS also determines parameters by linearly interpolating to temperatures between curves, and assumes perfect plasticity beyond the last defined point on any curve for this material model.

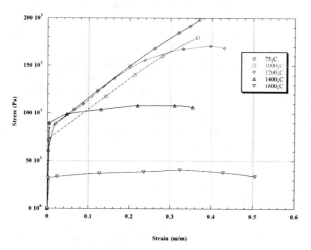

Figure 5. Engineering stress strain diagram for inconel 718 at various temperatures

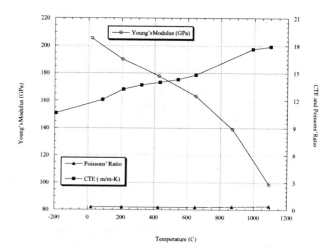

Figure 6. Elastic modulus, Poisson's ratio and coefficients of thermal expansion for inconel 718 at various temperatures

High-temperature creep of the clad was included in the analysis to include the stress relief effects during the HIP cycle. The ABAQUS power law model was used. The form of this material model is shown below: (Hibbit 1998)

$$\dot{\varepsilon}^{cr} = A\sigma^n t^m \qquad (1)$$

where:

$\dot{\varepsilon}^{cr}$ is the uniaxial equivalent creep strain rate,

σ is the uniaxial equivalent deviatoric stress,

t is the total time,

A, n, m are user defined material dependent coefficients.

The user defined coefficients for Inconel 718 were determined by a linear regression fit of available data, (Blatherwick 1966) and are listed in Table 2

Table 2: ABAQUS User Defined Power Law Creep Parameters for Inconel 718.

Temperature (°C)	A	n	m
24	8.3920 x 10^{-93}	17.170	-0.9342
538	1.645 x 10^{-72}	12.922	-0.3859
649	4.5579 x 10^{-59}	10.587	-0.3859
760	1.9145 x 10^{-31}	5.4141	-0.3054

FEM Analysis

The HIP process cycle, as shown in Fig. 3, was divided into numerical steps within the finite element analysis. All of the steps were coupled temperature-displacement, (carried to steady-state) implicitly integrated steps, except for the high-temperature explicit visco-plasticity (creep) step. During heat up, contact between the tungsten and inconel 718 was maintained by applying pressure, and modeled using a sliding interface with a friction coefficient of 0.95. (Mark's 1978). The 0.95 sliding friction was maintained through the visco-plasticity step. The friction was then changed, prior to cooling the rod back to room temperature. Various values of the friction coefficient were used to study the effect of incorporating friction between inconel 718 and tungsten on the final residual stress state. Changing the friction in this manner allows study of the effect of allowing compliance in the bond during the HIP cycle. In actual computations, sliding friction coefficients of 0, 0.95, 95.0, and "tied" were used at the interface between the inconel 718 and tungsten during cool down. Note, a sliding friction coefficient of "0" means there is no friction at the interface between the two materials, and a sliding friction of "∞" means the two materials cannot move independent of each other. A "tied" interface implies no separation and sliding coefficient of friction "∞". We used these conditions to emulate the diffusion bond between inconel 718 and tungsten. With this approach a comparison of the final residual stress state upon the completion of the HIP cycle for different interface conditions could be made with residual stress measurements.

Tied Interface FEM analysis

It was assumed that bonding is completed at the end of the visco-plasticity step prior to cooling the rod back to room temperature from the maximum process temperature of 1080 C. The bond was approximated using a "tied" interface condition between the inconel 718 and tungsten. The ABAQUS steps used to approximate the HIP cycle are summarized in Table 3. These steps were initially used to approximate the HIP process to diffusion bond the inconel 718 to the tungsten.

Table 3: Tied Interface FEM Analysis Step Summary.

Step No.	Temperature at End of Step (°C)	Pressure at end of Step (MPa)	Description of Steps
1	23 (Room)	198	Contact definition step. Sets initial contact parameters with a sliding friction coefficient of 0.95. Also applies pressure load.
2	1080	198	Steady-state temperature increase to Maximum HIP temperatures.
3	1080	198	Visco-plasticity step. Pressure and temperature held constant for 1-hour period.
4	1080	198	Beginning of contact pair definition change. All nodes are fixed with the exception of the master and slave nodes of the contact definition. This step in conjunction with steps 5 and 6 provide numerical stability while friction coefficients are redefined for contact pair.
5	1080	198	Intermediate step of contact pair definition change. Contact pair definition changed to "tied" contact. This change essentially creates a rigid bond.
6	1080	198	Final step of contact pair definition change. Original boundary conditions restored.
7	23 (Room)	198	Temperature ramped back to room temperature
8	23 (Room)	0.101	Pressure released.

Sliding Friction Interface FEM Analysis

Upon examination of the X-ray diffraction results (Fig. 11), it was discovered that the calculated stresses were higher than the measured values. The HIP analysis was repeated by allowing sliding friction between the inconel 718 and tungsten during the cooling process of the HIP cycle. By using sliding friction instead of tied contact between the clad and rod, the bond is allowed some compliance that is believed to exist within the real bond. This approach was used to possibly reduce the residual stresses to levels commensurate with X-ray diffraction measurements. The ABAQUS Coulomb friction model was used and the steps used to approximate this process are shown in Table 4.

Table 4: Friction Interface FEM Analysis Step Summary.

Step No.	Temperature at End of Step (°C)	Pressure at end of Step (MPa)	Description of Steps
1	23 (Room)	198	Contact definition step. Sets initial contact parameters with a sliding friction coefficient of 0.95. Also applies pressure load.
2	1080	198	Steady-state temperature increase to Maximum HIP temperatures.

Step No.	Temperature at End of Step (°C)	Pressure at end of Step (MPa)	Description of Steps
3	1080	198	Visco-plasticity step. Pressure and temperature held constant for 1-hour period.
4	1080	198	Beginning of contact pair definition change. All nodes are fixed with the exception of the master and slave nodes of the contact definition. This step in conjunction with steps 5 and 6 provide numerical stability while friction coefficients are redefined for contact pair.
5	1080	198	Intermediate step of contact pair definition change. Contact pair definition redefined with sliding friction contact. The friction value was changed to model compliance in the diffusion bond during cooling.
6	1080	198	Final step of contact pair definition change. Original boundary conditions restored.
7	23 (Room)	198	Temperature ramped back to room temperature
8	23 (Room)	198	Beginning of contact pair definition change. All nodes are fixed with the exception of the master and slave nodes of the contact definition. This step in conjunction with steps 9 and 10 provide numerical stability while friction coefficients are redefined for contact pair.
9	23 (Room)	198	Intermediate step of contact pair definition change. Contact pair definition changed to "tied" contact. This change essentially creates a rigid bond.
10	23 (Room)	198	Final step of contact pair definition change. Original boundary conditions restored.
11	23 (Room)	0.103	Pressure Released

FEM Results

Figure 7 shows a plot of axial stresses along the surface of the clad rod for the tied (∞), 95.0, and 0.95 friction values along the inconel 718, tungsten interface. The axial stress is tensile with a magnitude of 685 MPa for approximately 95% of the rod length. Only near the proximity of the end cap and towards the end of the clad rod do the stresses vary considerably, going from tension to compression, and then decreasing to zero towards the edges. A similar distribution follows for the hoop stresses shown in Fig. 8, except the tensile stresses are lower with a magnitude of 625 MPa.

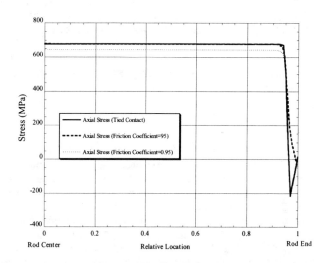

Figure 7. Calculated residual axial stresses for different interface conditions along clad rod surface

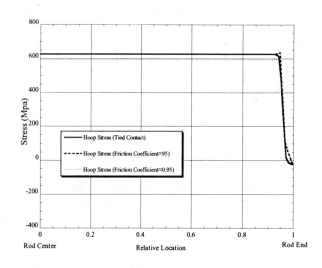

Figure 8. Calculated residual hoop stresses for different interface conditions along clad rod surface

A comparison of the stresses for the different interface friction values shows essentially no difference between the "tied" and 95.0 friction value for both axial and hoop directions. However, the 0.95 friction value shows an approximate 5% decrease in tensile stresses along the region where the stresses are uniform. In the region at the end of the clad rod, there is some disparity in the stress values for axial stresses. In contrast, the hoop stresses are all comparable. Since the interface conditions only apply to the region where inconel 718 and tungsten exist, it would be expected that they have no effects in the end-cap region, which is only inconel 718. However, the disparity in the axial stresses in the end cap region cannot be reasonably explained. Note that a value of "zero" friction predicted stresses no different than the 0.95 analysis. Also, the stresses in the tungsten (not shown) were compressive and mostly uniform for most of the tungsten rod. The only changes from this uniformity were at the end caps, where the radial stiffness of the end cap effected the final stress state in the tungsten rod.

EVALUATION OF RESIDUAL STRESSES USING X-RAY DIFFRACTION

X-ray diffraction

The measurement of residual stress by X-ray diffraction utilizes the spacing of the lattice planes as a gauge length for measuring strain (Noyan, 1987, Hilley, 1971). A change in stress results in a modification of the interplanar spacing, which alters the angular position of the diffraction peaks. The interplanar spacing of a specific set of planes is obtained from grains of different orientations to the surface normal. The variation of the interatomic spacing is determined as a function of ψ, the angle between the surface normal and the direction of the measured strain. This is determined by tilting (by an angle of ψ) and rotating the specimen (by an angle of ϕ) with respect to the incident X-ray beam. The change in interplanar spacing as a function of orientation (ϕ, ψ) is due to surface strains, this can be related to the surface stresses (Hilley, 1971). This is also known as the classical $d_{\phi\psi}$ versus $\sin^2\psi$ method of measuring near surface stresses using X-ray diffraction. Accordingly, for known values of $d_{\phi\psi}$ and d_0 (strain-free), the average strain along the $\phi-\psi$ direction with respect to the surface normal $(_{33})$ can be written as (Noyan, 1987)

$$(\varepsilon_{33})_{\phi,\psi} = \frac{d_{\phi\psi} - d_0}{d_0} \quad (2)$$

Where $(\varepsilon)_{\phi,\psi}$ can be related to the surface stresses by the conventional elasticity equations. For classical X-ray stress analysis, these strains are a linear function of the angle ψ, and the surface stress state in the sample can be determined by a linear relationship between $(\varepsilon_{33})_{\phi,\psi}$ and $\sin^2\psi$. (Noyan, 1987)

$$(\varepsilon_{33})_{\phi,\psi} = \frac{d_{\phi\psi} - d_0}{d_0} = \frac{(1+\nu)}{E}\left[\sigma_\phi \sin^2\psi - \frac{\nu}{E}(\sigma_{11}) + \sigma_{22}\right] \quad (3)$$

Where ν (Poisson's ratio), and E (Young's modulus) are elastic constants and $\sigma_{11}\cos^2\phi + \sigma_{11}\sin^2\phi = \sigma_\phi$ and the stresses are in the direction of σ_ϕ.

If (ε_{33}) is plotted versus $\sin^2\psi$, then equation (3) represents the equation of a straight line with the slope proportional to σ_ϕ. In this study, the assumption of a bi-axial plane stress state is generally valid since the average depth of penetration of Cu-K$_\alpha$ X-rays in this alloy has been computed to be between approximately 11 microns (at $\psi = 0°$) and 6 microns (at $\psi = 60°$).

The stress σ_ϕ measured by the X-ray method contains both the macrostress and microstress component. The raw data consists of plots of d versus $\sin^2\psi$ which were fairly "linear" also known as "regular". The slopes of the linear plots were used to determine the surface residual stress along the specific orientation of measurement. For a "regular" or linear behavior of d vs. $\sin^2\psi$ the stress σ_ϕ represent a uniform stress distribution in the measured area of the sample. This also implies that the microstress components cancel out due to equilibrium conditions within the sampling volume. Therefore, the remaining stress component is the macrostress. This macro-stress component can be used for validating the predictions from analytical or finite element models.

Measurement Procedure

Residual stresses were measured in the inconel 718 along the axial (parallel to the axis of the rod) direction for comparison with predictions from FEM models. The residual stresses in the rods were made using an X-ray (d versus $\sin^2\psi$) diffraction technique. These measurements were made using a Huber stress goniometer with X-rays generated from an 18 kW Scintag rotating anode located at the Lujan Center (LANSCE-12) at Los Alamos National Laboratory.

Specimens of size 50 mm long x 3.25 mm in diameter were sectioned in the central region of a long (300 mm) clad rod. The first sets of XRD (X-ray diffraction) stress measurements were performed in the as-received condition with minimal surface preparation (cleaning the surface of the specimen with acetone). The second set of measurements were performed after removing a layer of 5 microns by electro-chemical polishing using a mixture of perchloric acid, methanol and butyl cellusolve at temperatures less than –25 °C. This procedure is necessary to remove any ambiguities associated with surface measurement techniques. The specimens were periodically examined by optical microscope for proper polishing.

All measurements used Cu radiation (wavelength 1.54Å). Initial measurements involved recording 2θ scans from 20° to 160° for identifying the diffraction peaks in the high angle (> 135° 2θ) (Fig. 9). The diffraction pattern revealed a single-phase material having a face centered cubic structure (fm3m). Two diffraction planes (331) and (420) at 2θ approximately 137° and 145° respectively, were identified for residual stress measurements (Fig. 8). The X-ray beam was collimated using a 2 mm diameter collimator on the incident side to the sample for axial stress measurements. The distance from the sample surface to the tip of the collimator was 40 mm. Stress measurements were mapped at four positions across the length using an increment of 1 mm on either side from the approximate center of the clad rod (Fig. 10). Stress measurements in positive and negative ψ tilts were recorded to check for the presence of shear stresses, which were not present. The ψ angles ranged from 0 to 60° in increments of 10°.

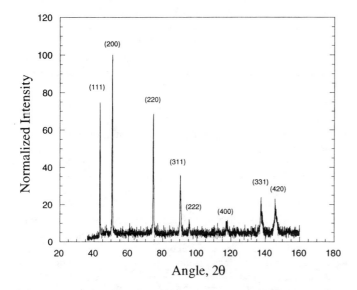

Figure 9. X-ray diffraction spectra for Inconel 718 showing the two diffraction peaks (331) and (420) used for stress analysis.

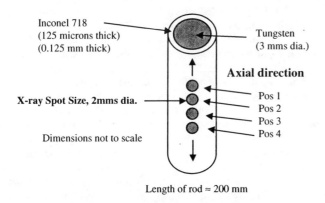

Figure 10. X-ray diffraction measurement locations and spot size on the surface of inconel 718 clad rod used for determining stresses.

Along with the measurements of stresses in the axial directions, a second study was conducted to determine the systematic errors that would effect the measurements. Preliminary stress measurments were made on flat specimens of Inconel 718 measuring 20 x 20 x 2 mms. They were measured in an "as-received", annealed, and hot isostatic pressed condition. These results of these measurements showed residual stress levels of each sample to be within 25 MPa of each other, a value corresponding to the resolution of this particular instrument. Stress measurements were made in Al_2O_3 powder mounted along with the clad rod on the same sample holder. The Al_2O_3 powder that measures zero stresses provides a check on the accuracy of the stress determination using the d versus $sin^2\psi$ technique. This approach also factors other possible effects which cause erroneous stress calculations, such as curvature effects, coarse grain size, texture, and sample displacement from the goniometer center. Fig. 11 shows the d versus $sin^2\psi$ plots for inconel 718 cladding (hkl 420) and alumina powder. The slope for the powder, which should ideally be zero, shows a negative slope, indicative of compressive stresses. However, the slopes for the inconel 718 cladding are positive and higher in absolute value than the powder, and despite some scatter in the data, it is clear that stresses in the cladding are in tension. Stresses were calculated using an X-ray elastic constant of 3.5×10^{-6} MPa. (Hilley, 1971)

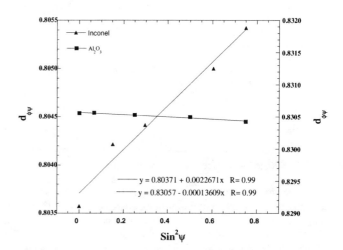

Figure 11. d vs. $Sin^2\psi$ plots at a position on the surface for inconel and in a stress free powder. Solid lines through the data symbols are least squares fit for calculating axial stresses from the slopes of inconel and Al_2O_3, respectively.

Experimental Results

Residual stresses in the as-received condition and after electropolishing are shown in Fig. 12. In the as-received condition, there is considerable scatter in the stresses for both the 331 and the 420 hkls. The stresses for the (331) peak range from 82 – 372 MPa for an average value of 273 ±134, and for the (420) hkl the stresses range from –211 to 423 MPa for an average value of 219 ±293 MPa. There are no clear reasons for the large scatter in the as-received conditions and for the negative stress value for the (420) peak. However, it can be stated based on previous experience that X-ray measurements are sensitive to the surface conditions and sometimes may not be representative of the stress region in the bulk of the material.(Rangaswamy, 1997) Electro-polishing away the surface layers has been shown to be an effective way of obtaining stresses representative of the bulk material in the absence of stress gradients.

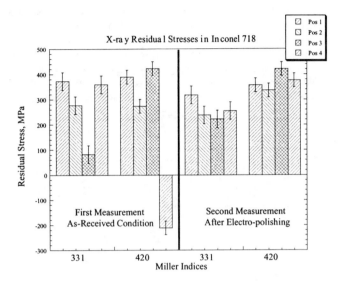

Figure 12. Axial stresses in inconel cladding for Miller Indices (331 and 420) in the as-received and electro-polished conditions.

The stresses recorded after electro-polishing (which resulted in a removal of 5 microns from the surface), are tensile and for (331) peak ranges from 222 – 318 MPa for an average value of 258 ±35 Mpa, and for (420) from 336 – 421 MPa for an average value of 373 ±27 MPa. The scatter is also reduced in comparison to the as-received condition. Note that stresses for peak (331) are lower than peak (420). Inconel 718, which is characterized by a nickel-based crystal structure, exhibits elastic anisotropy with an anisotropy factor of 2.5. (Hertzberg, 1996) The calculated single-crystal modulus for (331) is 249 GPa and for (420) is 189 GPa, respectively. The same X-ray elastic constant was used to calculate the stresses, so the results for the two peaks are not surprising. The more compliant of the two (in this case, 420) has a higher slope and therefore a higher stress. However, to compare with the finite element models, stresses calculated from peak (331) were used because the X-ray plane specific constant obtained from literature is for this peak. The error bars on the stresses are calculated based on the super-position of the stress in the "stress-free" powder, and the standard deviation of a set of five consecutive measurements repeated at one location on the surface.

DISCUSSION

The objective of this research was to determine and benchmark FEM predictions of residual stresses during the completion of the HIP process of an inconel 718 cladding onto a tungsten rod, against X-ray diffraction stress measurements. In this research, it was assumed that the X-ray technique was used to accurately measure stresses in the cladding. Tensile stresses within the inconel 718 were expected because of the differences in the coefficients of thermal expansion of tungsten (Table 2) and inconel 718 (Fig. 6). The finite element predictions overestimate the measurements by more than twice the magnitude for both tied and no friction conditions. Neither of these conditions brings an acceptable agreement with the experimental measurements.

Previous attempts to predict residual stresses in similar high-temperature manufacturing processes have also resulted in over prediction of stress levels. (Prime, 1997) To demonstrate the complexity in predicting residual stresses, the residual stresses generated based on a simple thermo-elastic solution will be examined (Chawla, 1987) If one assumes a temperature difference of 1057 °C (ΔT) (from the hot-press temperature to room temperature), and a difference in thermal expansion coefficients of 6.5 x 10^{-6} °K ($\Delta \alpha = \alpha_T - \alpha_I$), the misfit strain in the direction parallel to the axis of the rod between tungsten and inconel 718 should be about 6500 µe ($\Delta \alpha . \Delta T$). Furthermore, based on an inconel 718 elastic modulus of 200 GPa, the estimated stress within the inconel 718 would be 2400 MPa. Since the volume fraction of inconel 718 is only 15% of the clad rod, and the mechanical properties of tungsten are substantially higher, a large portion of the stresses are naturally imposed on the inconel 718 cladding and comparatively less stresses on the tungsten.

The finite element model process described within this paper uses a complex material behavior model, which includes both plasticity and creep, reducing the stresses to about one-quarter the value estimated from the thermo-elasticity solution. However, since varying the compliance did not greatly effect the residual stress levels at the end of the HIP cycle, the analyses still lacks correct physical understanding of the bonding process. By performing sensitivity studies with a finite element model we can gain insight into what parameters might effect the residual stress states after pressing. We have shown that by changing the nature of the interface (tied and sliding friction) between the inconel 718 cladding and the tungsten rod during cool-down, the predicted final residual stress state was changed, albeit by a small amount.

There are other parameters that could have been varied, which were not undertaken in the present analysis. As an example the time at which the bond was established during the HIP cycle could be changed, and how varying the bond time effects the final stress state would be studied. If this process were to greatly change the final stress state, one could postulate that the time of binding plays a more important role in the development of residual stresses when compared with varying compliance within the bond. Subsequent analysis will explore this avenue to better understand the bonding process, and to more accurately predict the residual stress state. In any case, the analysis performed to date suggests that a better understanding of the actual diffusion bonding process is needed to more accurately predict the residual stress states.

Although this technique does not accurately model the diffusion bond process, it does provide a representative stress state that can be used to perform other parametric studies that look into ways of reducing residual stresses. The parametric studies can also make sense of, or predict, trends of how the final stress state can change in subsequent analysis that explores the rods response to operational loading conditions.

SUMMARY AND CONCLUSIONS

FEA was used to predict residual stress levels from hot isostatic pressing of an inconel 718 clad-tungsten rod. Residual stresses in the hot pressed specimens were measured using the XRD technique to benchmark FEA calculations.

The FEM predictions of axial stress for the two interface conditions, namely "tied with no friction" and "sliding with zero friction" varied from 685 to 625 MPa, a reduction of 5%.

The XRD stresses in the axial direction were determined to be 285 MPa.

The FEM predictions of the axial stresses were more than twice the XRD measured stresses levels.

The disparity in the comparison of predicted and measured results indicate that a better understanding of the bonding process is needed to make more accurate computations.

ACKNOWLEDGMENT

The authors acknowledge Ron Barber and Mike Cappiello of the Los Alamos National Laboratory for their support of this research. Harlan Horner and Bob Johnson of General Atomics were instrumental in providing samples for measurement. We would like to thank Ed Dalder of Lawrence Livermore National Laboratory for his assistance in locating difficult to find material properties at elevated temperatures. This work was supported (in part) under the auspices of the United States Department of Energy, and the Accelerator Production of Tritium Technical Project Office. The Lujan Scattering Center is a national user facility funded by the United States Department of Energy, Office of the Basic Energy Sciences – Materials Science, under contract number W-7405-ENG-36 with the University of California.

REFERENCES

A. A. Blatherwick, A. Cers, "Fatigue, Creep and Stress-Rupture Properties of Nicrotung, Super A-286, And Inconel 718," AF Technical Report AFML-TR-65-447, Wright-Patterson Air Force Base, Ohio, 1966.

K. K. Chawla, "Composite Materials – Science and Engineering," Materials Research and Engineering, Edited by B. Ilschner and N. J. Grant, pp. 189-195, 1987.

Ed Dalder, Lawrence Livermore National Laboratory, facsimile communication of LLNL-SBIP Final Report, No Author Given, "Drop Hammer Formability of Inconel 718 and 718 SPF Alloys for Hughes Bros. Aircrafters, Inc., South Gate, CA 90280," 6/8/1995.

R. W. Hertzberg, "Deformation and Fracture Mechanics of Engineering Materials," Wiley, New York, 1996.

Hibbitt, Karlsson & Sorensen, Inc., ABAQUS/Standard User's Manual, Volume 1, Version 5.8, 1998

Residual Stress Measurement by X-ray diffraction", SAE Information Report J784a, M.E. Hilley, ed., Society of Automotive Engineers, New York, August, 1971.W. D. Klopp, "Aerospace Structural Metals Handbook," Volume No. 4, Code 4103, pp. 1-102, January 1995.

Machine Design 1995 Materials Selector Issue Part 2, Penton Publications, 1994, p. 89.

"Mark's Standard Handbook for Mechanical Engineers," Ninth Edition, McGraw-Hill Book Co., p. 3-26, 1978.

Mike Prime, Los Alamos National Laboratory, ESA-EA, private communication, 10/1997.

I.C. Noyan and J. B. Cohen, "Residual Stress: Measurement by Diffraction and Interpretation", Springer-Verlag, New York, 1987.

P. Rangaswamy, M.A.M. Bourke, P.K. Wright, E. Kartzmark, J. Roberts and N. Jayaraman, "Influence of residual stresses on the Thermo-mechanical processing of SCS-6/Ti-6-2-4-2 Titanium Metal Matrix Composites", Materials Science & Engineering A224, 200-209, 1997.

AN INVESTIGATION INTO THE RESIDUAL STRESSES IN AN ALUMINIUM 2024 TEST WELD

Robin V. Preston
Department of Engineering
University of Cambridge
Trumpington Street
Cambridge, CB2 – 1PZ, U.K.

Simon. D. Smith
TWI
Abington Hall
Abington
Cambridge, CB2 – 6AL, U.K.

Hugh R. Shercliff
Department of Engineering
University of Cambridge
Trumpington Street
Cambridge, CB2 – 1PZ, U.K.

Philip. J. Withers
Manchester Materials Science Centre
Grosvenor Street
Manchester, M1 – 7HS, U.K

ABSTRACT

This paper uses finite element (FE) analysis to examine the residual stresses generated during the TIG welding of aluminium aerospace alloys. It also looks at whether such an approach could be useful for evaluating the effectiveness of various residual stress control techniques.

However, such simulations cannot be founded in a vacuum. They require accurate measurements to refine and validate them. The unique aspect of this work is that two powerful engineering techniques are combined: FE modelling and neutron diffraction. Weld trials were performed and the direct measurement of residual strain made using the ENGIN neutron diffraction strain scanning facility. The predicted results show an excellent agreement with experimental values.

Finally this model is used to simulate a weld made using a "Low Stress No Distortion" (LSND) technique. Although the stress reduction predicted is only moderate, the study suggests the approach to be a quick and efficient means of optimising such techniques.

INTRODUCTION

"The only way this country can compete with foreign builders is with a design that is 10 to 15% lighter and 20% cheaper. To be credible, engineering must come in with a (joining) process that will work on the production line." These are the words of Charles Willets, Engineering Manager, National Aeronautics and Space Administration, discussing the role of welding in future aircraft design (Irving, 1997).

Mechanical fasteners are currently the primary method of joining aerospace structures. However, the need to reduce weight and cost is driving the development of welded aerospace structures; and as the commercial aircraft market becomes ever more competitive, so welded aerospace structures become increasingly attractive to commercial aircraft designers.

Riveted Aerospace Structures

Rivets and other mechanical fasteners have been used in aircraft fabrication since the 1920's. The major disadvantages of riveted aerospace structure are as follows:

- The riveting process is both complex and time-consuming when compared with other joining processes.
- The material in the vicinity of the rivet often requires strengthening through the use of thicker sections etc...
- The structure is often more complex as each joint must form a lap joint configuration.
- Much of the corrosion and fatigue damage seen in aerospace structures is generated around the rivet holes.

The wing set for an A340 Airbus currently contains some 80,000 mechanical fasteners, approximately two thirds of which are rivets. They are used to join relatively thin sheets for the fabrication of flight surfaces. This paper examines the use of welding, as an alternative to riveting, for these "thin sheet" applications.

Welded Aerospace Structures

A major problem with welding as a joining technique is caused by the residual stresses produced by the thermal contraction of the weld region upon cooling. If the structure is unable to support these stresses, it will either distort or crack. (Residual stresses will also combine with "in service" loads, to create stress fields often very different from those otherwise expected. This can cause problems for component lifing, etc...) Many of the most advanced aerospace alloys are considered "unweldable" by conventional techniques, due to their susceptibility to distortion and solidification cracking during welding. The welding technique chosen for the experimental program was Tungsten Inert Gas (TIG). Reasons for its selection include:

- It is a trusted, well-understood technique in many other safety critical applications (e.g. the nuclear industry (Weisman, 1973)).
- Its welds are of high quality and high reproducibility.
- The process is fast and easily automated.
- The process has the flexibility for use in repair welds during aircraft maintenance, etc...

Aerospace manufacturers are investigating many techniques to control the residual stresses generated by welding. One of the most promising is known as Low Stress No Distortion (LSND) welding.

Low Stress No Distortion Welding (LSND)

The thermally induced stresses in the region adjacent to the weld can cause excessive distortion of thin structures. The thinner the section the lower its resistance to distortion. If it became possible to reduce the residual stresses generated during welding, and hence control the distortion, TIG welding could become the preferred choice for aircraft fabrication.

One technique which has been suggested as a means of reducing residual stresses, and hence distortion, is LSND (Guan et al, 1988). LSND techniques combine one or more of the mechanisms listed below:

- The structure to be welded is constrained. This will restrict distortion on the time-scale required to reduce the temperature variations through the sheet thickness. This will reduce through thickness stress variations, and help to reduce out-of-plane distortion.
- Certain regions in the locality of the weld are subject to further heating to pre-tension the weld region. This will allow more of the thermal expansion of the weld pool to be accommodated elastically. To be practical for industrial applications, the heat sources would be required to travel along with the TIG torch. Suitable heating sources include flames, high intensity lamps and lasers.
- The area immediately adjacent to the weld is cooled so that the thermal expansions may be accommodated elastically. This is highly beneficial as only plastic deformations lead to the generation of residual stresses. Cooling techniques include water-cooled backing plates, high pressure air, etc...

In production terms, LSND is still in its infancy; the development of LSND from a concept into a commercially viable fabrication technique will require accurate weld models from which the optimum LSND parameters can be determined. This paper represents the first stage in the development of a commercial LSND technique.

THE WELDING TRIALS

All tests were conducted on the aluminium alloy 2024 T3. 2024 is a general-purpose aluminium alloy for aerospace applications, and is commonly used for thin section flight surfaces. It is a precipitation-hardened alloy, considered to be "unweldable" by conventional techniques, due to its susceptibility to both weld-induced distortion and solidification cracking (ASM, 1997a).

The material was in the T3 condition. This refers to the alloy's microstructure. T3 alloys are heat-treated, wrought, and then allowed to age naturally age. This processing modifies the microstructure, giving the alloy a higher strength than in its "as cast" form. The properties used during this paper are assumed to be those of 2024 T3 (ASM, 1997b). (Future models may consider the change in properties with non-optimal thermal histories, as would be found in the vicinity of the weld.)

The welding trials were based upon autogenous "bead on plate" welds, thus eliminating the added complication of variable constraint conditions. The test plates were 160mm by 150mm by 3.2mm. A single pass weld was made along the centre of each sheet in the long axis direction. The welds began and ended 10mm from the plate edges.

A summary of the welding parameters used for the trials is given in table 1.

Welding Current	215 Amps
Welding Voltage	~12 Volts*
Welding Speed	200 mm/min
Weld Time	42 seconds
Arc Length	~2.4 mm*
Electrode Diameter	2.4 mm
Shielding Gas	Argon (15 litre/minute)
Frequency / Balance	200 Hertz / 75%

Table 1 - Summary of Welding Parameters

* The welding voltage is related to the arc length. Changes in arc length (due to distortion of the plate etc) can cause slight (~1-2V) variations in welding Voltage.

Care was taken to ensure that all welds were produced under identical conditions. All welds were made using a heavy copper backing plate. The welding jig used an over centre toggle clamping arrangement positioned 20mm away from the weld centre line to ensure the sample remained flat. The backing plate was allowed to cool between welding runs. Contrary to the literature, none the tests with the welding conditions and plate geometry described above showed any indication of solidification cracking.

THE WELD MODEL

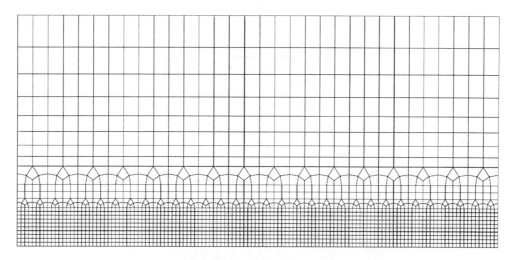

Figure 1 - Mesh used for FE Analyses

The weld modelling was carried out using ABAQUS (Version 5.7). In common with most authors, the model solved the thermal and mechanical analyses sequentially, to improve computational speeds. Due to the geometry of the problem, a shell element approach was chosen (NAFEMS 1992).

An example of the mesh used is shown in figure 1. It was generated using Patran (v 7.0) and consisted of 2000 elements with 5 section points through thickness. Due to the symmetry of the plate only one half was modelled.

THE THERMAL MODEL

Mechanical Model	Duration	Description
Step 1	54 sec	Welding / LSND
Step 2	600 sec	Cooling

Table 2 - Summary of the Steps used in the Thermal Analyses

The thermal model used linear heat transfer elements (DS4). A DFLUX user-subroutine was implemented to control the power input into the plate. No attempt was made to model the physics of the arc process; instead the various losses were incorporated into an arc efficiency term (η). The relationship is given below.

$$P = \eta I V \qquad eq.(1)$$

Arc efficiencies quoted in the literature vary from 60% (Gourd, 1990) to 90% (Michaleris et al, 1997) and above. Since the alternating current will increase the energy dissipated at the electrode, a lower 60% value for arc efficiency was chosen. The arc was simulated as a moveable body flux power source. Due to the high thermal conductivity of aluminium alloys (190 W m^{-2} °C^{-1}, over three times that of steel) and the relatively thin (3.2mm) plates, it was decided that the more complex heat source models favoured by some authors (Goldak et al, 1985) would offer insufficient benefit in this case, to justify the increase in complexity.

There was very little information in the literature on determining the heat losses from the conduction of heat into the backing plate. This was to be expected, as such values will be heavily dependent on surface finishes, clamping pressures, etc. A comparison of fusion zone widths with those observed experimentally was used to assess the extent of the heat loss into the backing plate. A constant value for the film co-efficient was used during the analysis (800 W m^{-2} °C^{-1}) in conjunction with a FILM subroutine to simulate the ambient temperature of the backing plate. A series of tests under different film coefficients and ambient temperature relationships were conducted to find the optimum parameters for this combination of welding and clamping.

Heat losses in to the atmosphere were ignored as the values quoted in the literature (~10W m^{-2} °C^{-1} (Michaleris and DeBiccari, 1997)) were significantly smaller than the conduction into the copper backing plate. The radiative losses were also ignored, as these only become significant at the high temperatures close to the arc, which can then be included in the arc efficiency factor η.

The LSND weld model was identical to that used for conventional welding except that an additional heat source was supplied to the plate to tension the vicinity of the weld. A schematic of the LSND arrangement is shown in Fig. 2.

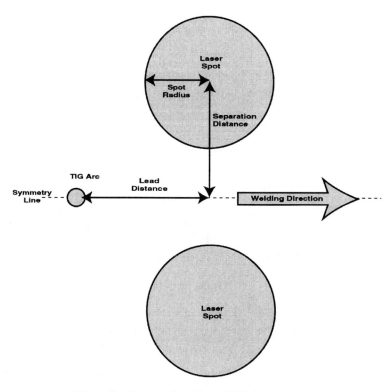

Figure 2 – Schematic of the LSND Arrangement

The LSND weld assumed heating from two defocused laser spots. The exact form of the heat source is not important, only the distribution of power. The laser spots may just as easily represent flame, lamp or resistive heat sources, etc... (Due to the symmetry, only one spot was actually required for the simulation). This was modelled by a 1200 Watt circular surface source. The LSND parameters were assigned by what seemed reasonable to achieve a thermal tensioning effect in these plates, but should not be considered optimal. A summary of the parameters is given in Table 3.

Parameter	Value Assigned
Laser Spot Radius	30mm
Lead Distance	20mm
Separation Distance	45mm
Laser Efficiency	0.5

Table 3 – Summary of LSND Parameters

All thermal models used temperature dependent material properties and included latent heat effects. Weld-stirring, which increases the transfer of heat from the weld pool, was simulated using an artificially elevated value for the thermal conductivity for temperatures above the melting point, as described by Goldak et al. (1985).

THE MECHANICAL MODEL

Mechanical Model	Duration	Description
Step 1	54 sec	Welding / LSND
Step 2	600 sec	Cooling
Step 3	1 sec	Removing Clamps

Table 4 - Summary of the Steps used in the Mechanical Analyses

The mechanical model used a von Mises yield surface, with an isotropic hardening behaviour (Hibbitt et al, 1996). S4R elements were chosen as these could represent the highly non-linear environment in the vicinity of the weld pool without the problems associated with shear-locking. No attempt was made to model the resetting of strains above the melting point.

The mechanical model was run using the ABAQUS non-linear geometry option. This was required as the dominant mode of distortion during the welding of thin plates is generally buckling (Masubuchi, 1980).

The clamping arrangement was modelled as boundary constraints in the direction perpendicular to the plane of the plate. It was assumed that the frictional forces exerted by the clamps would be insufficient to retard the thermal expansion of the plate, so no constraints were imposed in-plane. Further work will be required to confirm, or otherwise, the validity of this assumption.

The final step of the analysis was to remove the clamping boundary condition. This was done through a step of one second duration, during which a simple redefinition of the boundary conditions was applied.

The LSND mechanical model was identical to the conventional model, except that it used the thermal history for the LSND welding run.

NEUTRON DIFFRACTION

The results of FE analysis were compared with internal residual strain values measured directly using the ENGIN neutron diffraction beam-line at the Rutherford Appleton Laboratory (RAL).

Free neutrons are only produced in high energy environments, such as particle accelerators and nuclear reactors. They interact only weakly with matter enabling them to penetrate many centimetres of steel or many ten's of centimetres of aluminium.

The technique uses the Bragg diffraction (Cullity 1956) of a highly collimated neutron beam to determine the internal lattice spacings. Collimation is achieved using neutron-absorbent cadmium slits. The formula relating the wavelength of the neutrons (λ), the lattice spacing (d), and the diffraction angle (θ), is given in equation 2. The strain component measured will be perpendicular to the lattice planes.

$$\lambda = 2 d \sin\theta \qquad \text{eq.(2)}$$

The experimental arrangement at ISIS is shown in fig 3.

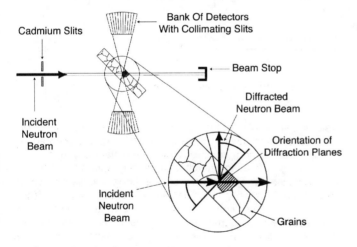

Figure 3 - Schematic of the ENGIN Beam Line

The position of the detectors is fixed, giving a constant diffraction angle.

The neutrons are generated in pulses. By measuring the time of flight for each neutron, the corresponding neutron wavelength can be calculated by use of the de Broglie relation (Beiser, 1987). This enables pulsed neutron sources to use a range of wavelengths simultaneously. This means that the lattice spacing for many peaks may be determined.

The peaks are analysed using a Reitweld fit (Reitweld 1969). This fits all peaks simultaneously thus reducing the effects of variations in modulus with crystallographic direction. The variations in lattice spacing are then used to determine the elastic strain (ε) through the relationship given below, where d_0 is the stress-free lattice parameter.

$$\varepsilon = (d - d_0) / d_0 \qquad \text{eq.(3)}$$

As can be seen, the derivation of elastic strain requires that the stress-free lattice spacing be known. However, in a welding problem, this is non-trivial, especially for precipitation hardened materials such as 2024, where the thermal history will affect the microstructure of the material, causing the composition of the phases being measured to vary.

A change in composition can change the stress free lattice parameter (Blackburn et al, 1996), which, if not taken into account, can introduce errors to the elastic strain measurements (a 0.1% increase in copper content will cause the measured stress to be underestimated by 8MPa (Preismeyer, 1991) in aluminium-based alloys).

To determine the variation in stress free lattice parameter with distance from the weld centre line, diffraction measurements were made on four thin, nominally stress free "match-sticks", electro-wire machined from four positions on a second plate welded under identical conditions. From this the variation in d_0 with distance was established (although only strictly for the central section of the plate from which the matchsticks were removed).

The measurement volume is defined by the intersection of the incident beam volume with the volume visible to the collimated detector bank. Since the gauge volume is spatially immobile, to make measurements at different positions, the sample is moved relative to the gauge volume by a computer controlled stage.

Line scans were taken outward from the weld centre line at five positions: 60mm, 30mm, 0mm, -30mm and -60mm, as shown in fig. 4.

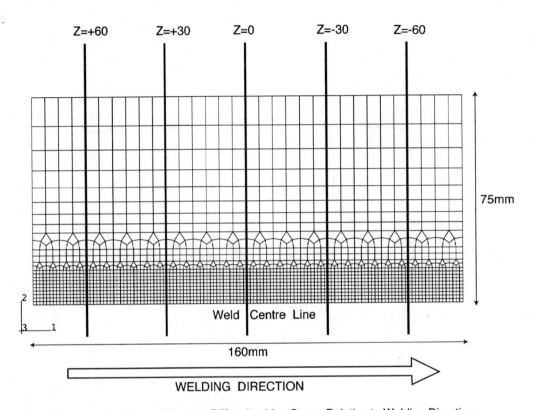

Figure 4 - Positions of Neutron Diffraction Line Scans Relative to Welding Direction

DISCUSSION
Results of Conventional Welding

This section looks first at the large-scale features predicted by the model. The second section compares the FE results with those obtained by diffraction measurements.

Model Direction	ABAQUS Axis
Longitudinal	Axis 1
Transverse	Axis 2
Through Thickness	Axis 3

Table 5 - Summary of Axis Notation

Figure 5 shows the temperature field for conventional welding during step one of the thermal analysis. During welding, the temperature contours are largely invariant with time. This is consistent with the use of high welding torch travel speeds and backing plates. Figure 6 shows the post weld longitudinal residual stress distribution. As would be expected, the narrow tensile region at the weld is supported by a much broader compressive region in the plate beyond.

Another interesting result is in the region in which the arc is extinguished. The model shows large localised tensile stresses in both the longitudinal and transverse directions (not shown) within this region. It was observed that if cracking of the welds ever occurred (cracking never occurred for the parameters listed in Table 1), it was initiated as radial solidification cracks in the zone where the arc is extinguished, further supporting the modelling predictions.

One of the plates welded conventionally was later examined using neutron diffraction to determine the accuracy of the models. Longitudinal, transverse, and through thickness elastic strain components are given below for the Z=30mm scan. The Z=30mm scan is of particular interest as it comes very close to the region of peak stress.

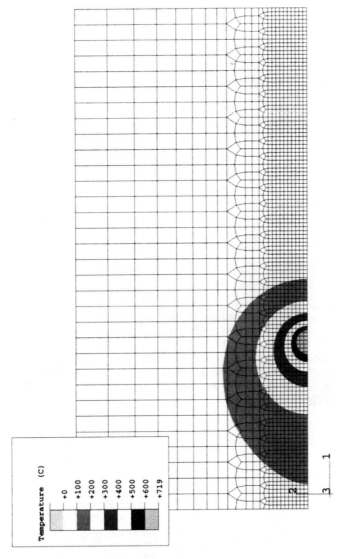

Figure 5 – Distribution of Temperature during Conventional Welding

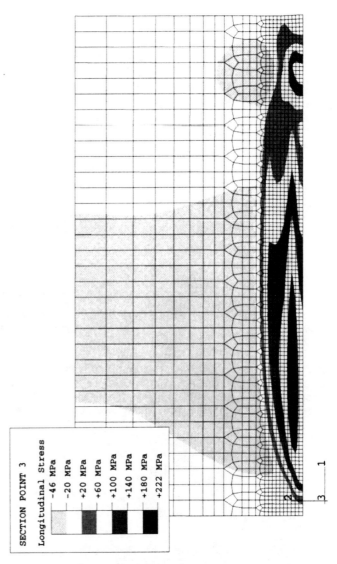

Figure 6 – Longitudinal Stress Distribution after Conventional Welding

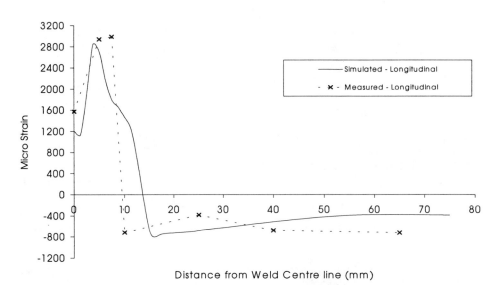

Figure 7 - The Longitudinal Strain Component - Measured at Z=30, as a Function of the Distance from the Weld Centre Line.

Figure 7 shows a comparison of predicted and measured strains for the longitudinal direction. Although there is a discrepancy in the far field, the overall agreement is good. It is interesting to note that the peak elastic strain, both measured and modelled, occurs on or just outside the fusion zone interface (5 to 10mm from the weld centre line). This supports the view that the material in the region immediately adjacent to the weld dominates the residual stress (Conrardy and Dull, 1995).

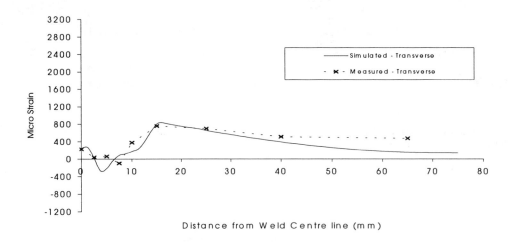

Figure 8 - The Transverse Strain Component - Measured at Z=30, as a Function of the Distance from the Weld Centre Line.

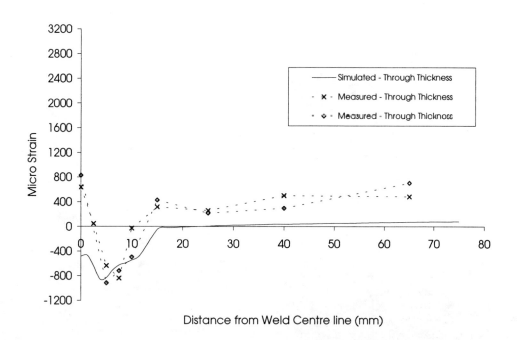

Figure 9 - The Through Thickness Strain Component - Measured at Z=30 as a Function of the Distance from the Weld Centre Line.

Figures 8 and 9 show comparisons of measured against predicted strain for the transverse and through thickness directions respectively. The through thickness strain predictions assumed all strains to be Poisson in nature, and could thus be calculated using:

$$\varepsilon_{Through\ Thickness} = -\upsilon\ (\sigma_{Longitudinal}\sigma_{Transverse})\ E^{-1} \quad eq.(4)$$

Where υ is Poisson's ratio and $\sigma_{Component}$ is the respective component of stress (Timoshenko 1955).

Again the agreement between measured and predicted results is satisfactory, although both plots show discrepancies in the far field. This maybe related to an error in a far field stress-free lattice parameter measurement. Especially pleasing is the degree of agreement within the weld region, as neutron diffraction measurements will often be erratic in such areas due to the effects of texturing (Webster, 1992). (Texturing is when the grains in a sample do not have a random alignment, as is desirable for this type of diffraction measurement.)

The other line scans show a similar level of agreement as that seen for Z=30mm.

Comparison of Conventional Welding with those Applying LSND Techniques

Figures 10 and 11 show the temperature distribution and longitudinal components of the residual stress for the LSND welds. Note the warping of the isotherms by the laser spot when compared to the conventional temperature distribution (fig. 5).

It is clear from the conventionally welded predictions (fig. 6 through 9) that the large longitudinal residual stress component will dominate the buckling response of the plate. The peak stress value in the plate welded conventionally is 222MPa. However, as can be seen in fig. 11, although the heat input to the plate has increased, the maximum stress has fallen to 195MPa. It is also important to remember that the parameters used for the LSND source represent guesses, so it is likely that far greater reductions in the levels of residual stress would be possible from optimised LSND conditions.

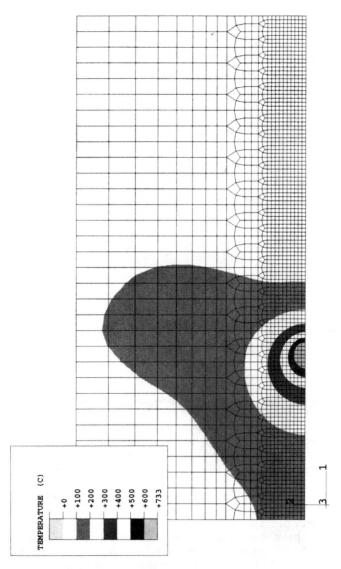

Figure 10 - Distribution of Temperature during LSND Welding

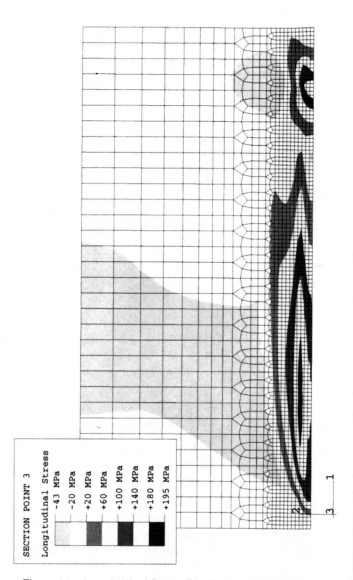

Figure 11 – Longitudinal Stress Distribution after LSND Welding

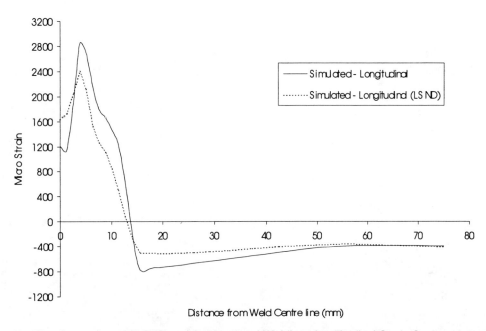

Figure 12 - The Comparison of LSND and Conventional Welding - Longitudinal Strain Components at Z=30 as a Function of the Distance from the Weld Centre Line.

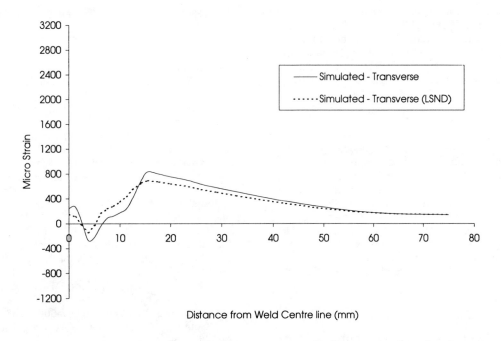

Figure 13 - The Comparison of LSND and Conventional Welding - Transverse Strain Component at Z=30 as a Function of the Distance from the Weld Centre Line.

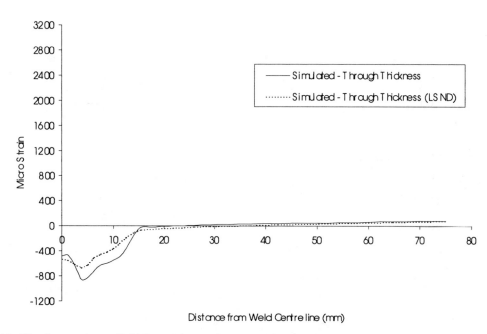

Figure 14 - The Comparison of LSND and Conventional Welding - Through Thickness Strain Component at Z=30 as a Function of the Distance from the Weld Centre Line.

Figs 12 through 14 show a comparison of strain prediction for conventional and LSND welding techniques for the Z=30mm line scan discussed earlier (figs 7, 8 and 9). At all points, and for every component of residual elastic strain, the thermally tensioned values are reduced from those obtained for conventional welding. Comparable effects are observed for the other line scans.

CONCLUSIONS

The results of the modelling are very encouraging. The agreement between the diffraction measurements and those predicted using ABAQUS is good. Validation data from a technique such as neutron diffraction gives a high degree of confidence in the modelling work to date.

The primary objective, to develop an efficient process optimisation tool, has been realised. The preliminary LSND predictions also seem very encouraging. It is likely that a much greater degree of residual stress reduction would be possible if the LSND conditions could be optimised, and this will be the subject of future work.

ACKNOWLEDGEMENTS

This work was jointly funded by the EPSRC and DTI (through the PTP studentship scheme) and by TWI. I would also like to thank EPSRC and the RAL for the allocation of beam time. Finally, I would like to thank David Dye, Mark Daymond and John Wright for their help with ENGIN, and Professor N. Fleck at CUED for the provision of computing facilities.

REFERENCES

ASM, - American Society of Metals Handbook Volume 2, ASM International, 1997.

ASM, - American Society of Metals Handbook Volume 6, ASM International, 1997

Beiser, A, - Concepts of Modern Physics, McGraw-Hill, 1987.

Blackburn, J M, Brand, P. C, Prask, H J & Fields, R J. – The Importance of D0 in the Determination of Weld Residual Stresses using Neutron Diffraction, Welding and Weld Automation in Shipbuilding, pp. 47-55, The Minerals, Metals and Materials Society, 1996

Conrardy, C & Dull, R, - Practical Techniques for Distortion Control, Edison Welding Institute - Report, 1995.

Cullity, B D, - Elements of X-ray Diffraction, Addison-Wesley, 1956.

Goldak, J, Bibby, J, Moore, R, House, R & Patel, B. - Computer Modelling of Heat Flow in Welds, Metallurgical Transactions B Vol. 17B, pp. 587-600, 1985.

Gourd, L M, - Principles of Welding Technology, Edward Arnold, 1980.

Guan, Q, Leggatt, R H & Brown, K W, - Low stress non-distortion (LSND) TIG welding of thin walled structural elements, TWI - Report, 1988.

Hibbitt, Karlsson & Sorensen, - ABAQUS/Standard User's Manual - Volume 2 (V5.6), Hibbitt, Karlsson & Sorenson Inc., 1996.

Irving, B, - Why Aren't Aircraft Welded ?, Welding Journal, Vol. 76, pp 31-41, 1997.

Masubuchi, K, - Analysis of Welded Structures, Pergamon Press, 1980.

Michaleris, P, Feng Z & Campbell, G, - Evaluation of 2D and 3D FEA models for predicting residual stress and distortion, ASME Pressure Vessels and Piping Conference, Orlando, FL, USA, 1997.

Michaleris, P & Debiccari, A, - Prediction of welding distortion Welding Journal (Miami) Vol. 76, pp 172-s-181-s, 1997.

NAFEMS, A Finite Element Primer, National Agency for Finite Element Methods and Standards, 1992.

Priesmeyer, H G, - The stress-free reference sample: alloy composition information from neutron capture, NATO Advanced Research Workshop on Measurement of Residual and Applied Stress using Neutron Diffraction, Oxford, UK, 1991.

Rietweld, H M, - A profile refinement method for nuclear and magnetic structures Journal of Applied Crystallography Vol. 65, pp. 65-71, 1969

Timoshenko, S P, - Strength of Materials, Van Nostrand, 1955.

Webster, P J, - Welding applications of neutron strain scanning, International Trends in Welding Science and Technology, Gatlinburg, TN, 1992.

Weisman, C, - Welding Handbook - Section 5 Applications of Welding, American Welding Society, 1973.

RESIDUAL STRESS IN TRANSIENT ZONES OF A HIGH VACUUM TUBE

Kent K. Leung

National Ignition Facility, Lawrence Livermore National Laboratory
P. O. Box 808, L447, Livermore, CA 94550
Email: Kkleung@Llnl.Gov

ABSTRACT

The Superconducting Supercollider (SSC) particle accelerator operates at ultra high vacuum. Evaluating the residual stress on a welded joint in a steel vacuum tube is necessary for the successful operation of the SSC. This paper presents the analysis of the residual stress on a single pass welded (TIG) joint on the tube. Thermal and structural interactions of the connected parts under the solidification process of the weldment are the main contributions to the residual stress.

INTRODUCTION

Welding induced residual stresses have been studied in the steel construction industry for years. A welded joint at flange to web intersection in an I-beam may reach 80% of the yield limit. (Johnson et al, 1968). Design codes employ safety factor to account for the residual stress. Research and development on residual stress using finite element method over the past decade and the recent advance in this field has been reported. (Dong et al, 1998). The availability of multi-field elements have enable engineers to perform residual stress since 1992. (ANSYS 5.0 1992).

Each beam tube is 53 feet long and is jointed together to form a 54 mile diameter oval shape particle accelerator. High vacuum must be maintained for the light speed charged particles to collide. A liquid helium leak in a high vacuum tube shall make a superconducting magnet inoperable for weeks. Therefore, a major test program was conducted to learn how to reduce the welded joint stress. Inspecting the surface of the welded joint by various constructions is used to define the best approach in reducing the residual stress.

Experience gained in operating the superconducting laboratory in Europe indicating that difference in thermal expansion between the vacuum tube and the dipole vessel (Figure 1) is the major contribution of the leaked joint. Materials of the same kind have small difference in the coefficient of expansion not only from manufacturing process but also from statically treatment of the reported material properties over many years. A SSC team went to inspect a used CERN's coaxial beam tube assembly by cutting the assembly in the axial direction. The inner beam tube is observed to expand few millimeters longer than the outer vessel tube after removing the circumferential weldment. Many coaxial jointed beam tube and vessel assemblies were constructed (Figure 1). Given the evidence, a bellow is added for all the beam tube in the SSC design (Figure 2). An analytical study of the residual stress on the welded joints of the beam tube with a bellow design was conducted. Software was needed to link the interaction between temperature from thermal analysis to a nonlinear structural analysis of the weldment under solidification process which is defined as the weldment at a temperature of 3000 degree F with a modulus of 7 psi to a room temperature with a modulus of 3E10 psi. Code was searched to link the thermal and structural interaction and a finite element code (ANSYS 1992) was used because the availability of a multi-field element and Lorenze force induced stress analysis. (Leung 1993, 94).

The result of using a multi-field element in predicting the residual stress in a beam tube is reported in this paper.

This paper presents a finite element model used to study of the residual stress induced by a single pass of vacuum weld (Fig. 5A). The radial movement of the interconnected parts during cooling process is a concern for the joint design (Fig. 5B). A maximum combined stress in the welded joint is specified as safety qualification and the stress profiles are shown in Fig, 3, 4, 8. The temperature profiles, Fig. 6, 7, on the joint at 300 seconds and 800 seconds are also reported.

The vacuum tube is subjected to additional stress under d added on top of the welded residual stress.

ASSUMPTIONS AND BOUNDARY CONDITIONS

Residual stress developed in the weldment depends on many factors. The thermal and structural properties of the weldment are different from the parent metal and were assumed to be identical in this analysis. Other assumptions are listed as follows:

(1) Stress and deformation on the vessel (Fig. 1, A) under assembling processes. The vessel is welded in two identical half cylinders in the longitudinal direction. These joints are designed to reach the yield limit of the weldment in developing maximum circumferential tension to confine motions of the superconducting coils under high field force.

(2) Ovality of the tube and the union tube at the joint.

(3) Gravity effect on the vessel which contains a solid block of steel laminations for confining the magnetic field and also used to transfer radial pressure to the superconducting coils. This gravity effect produces and bending to the beam tube at both end. The bending angle is measured but is not incorporated in the present joint residual stress analysis.

(4) For simplicity and conservatism, The axial restraint from the bellow support are not considered as a rigid axial boundary condition. Similarly, rigid radial supports are also used for both the union tube and the beam tube.

(6) Material property is simplified as 6 points for the state of the temperature with corresponding structural properties.

RESULTS OF ANALYSIS

Figure 3 to 7 show the residual stress, radial deflection and the temperature at the two different time after the welding starts.

Detailed stress (the von Mises stress) plot (Figure 8) is used as a design specification in a trade off study of various configurations in the welded connection design including the union tube and arrangement of the supports.

CONCLUDING REMARKS

(1) Maximum stress is found at the inner tip location of the welded joint. Maximum radial stress is 59.3 ksi in compression at the weldment (Fig. 3). Maximum hoop stresses are found in two different locations in tension at 40 ksi at the outer and inner circumferential areas on the outside diameter of the beam tube (Fig. 4).

(2) Maximum radial movements at 800 seconds after welding start is 0.0223" (Fig. 5B) at the union tube. This radial deflection as well as the radial stress of the welded joint may be reduced if the radial compliance of the bellow is selected to accommodate the thermal radial expansion of the assembly. Heat flow and temperature difference between the jointed parts may also improve in the connection design.

(3) The temperature drops from 419 degree F at 300 seconds to 183 degree F at 800 seconds after the welding process start (Fig. 6, 7). The temperature should be measured during welding process to provide information to improve the analytical model. The temperature profiles as shown in the thermal plots are conservative due to heat dissipation to ambient air is not used in the analysis.

REFERENCES

(1) Johnson, B., "Design of Steel Structures," John Wiley & Sons. 1968.

(2) Dong, P., "Introduction to Residual Stress," PVP-373, ASME 1998.

(3) Swanson Analysis Systems, "ANSYS 5.5" 1999.

(4) Leung, K., "Effective Stress at a 4 Degree K beam Tube," IEEE Particle Accelerator Conference, Washington, DC. 1993.

(5) Leung, K., "Quench Stress on SSC Collider Liner Tube," Cryogenic Engineering & International Cryogenic Material conference, NM. 1994.

(A) Shell   (B) Vacuum beam Tube

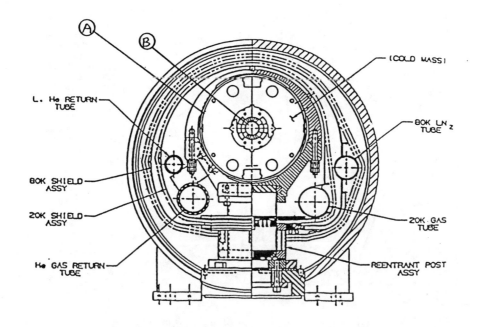

Figure 1. Cross Sectional Illustration of The Super Collider Beam Tube Within the Dipole Magnet Shell

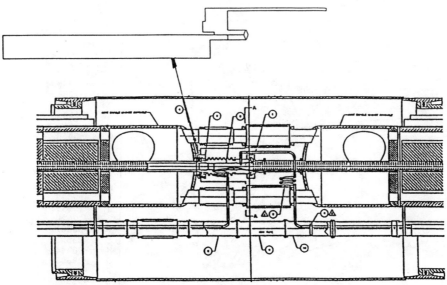

Figure 2. Longitudinal Sectional View of The Beam Tube (B) Within the Dipole Magnet Shell (A).

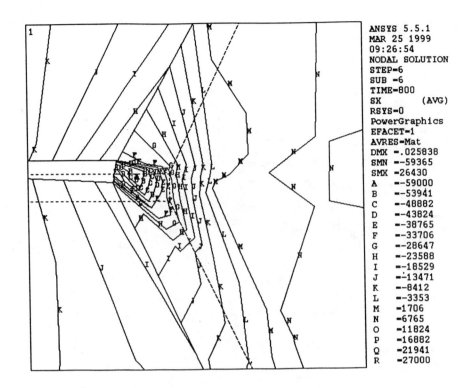

Fig. 3 Maximum radial stress is about 60 ksi in compression

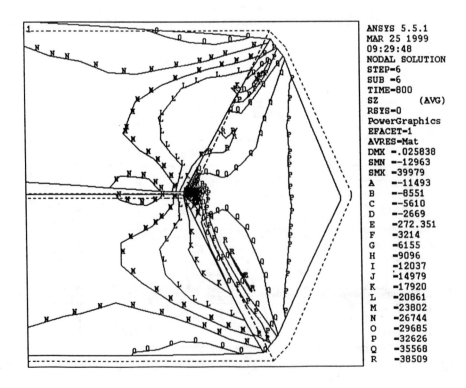

Fig. 4 Maximum hoop stresses in two locations in tension are about 40 ksi at outer and inner circumferential areas

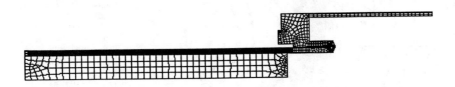

Fig. 5A FEA Model of the Union Tube Joints to the Beam Tube

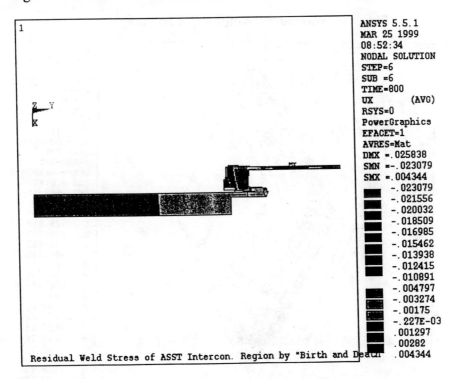

Fig. 5B Radial movements at 800 seconds after welding start is 0.0258 in

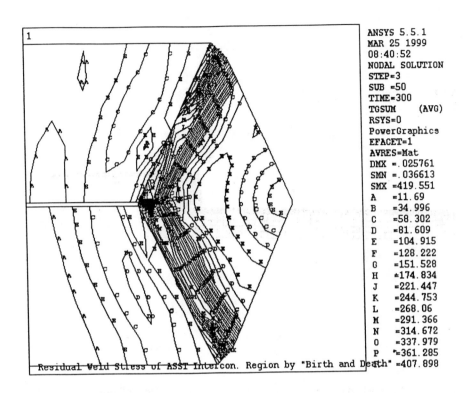

Fig. 6 Temperature (F) Profile at 300 Seconds

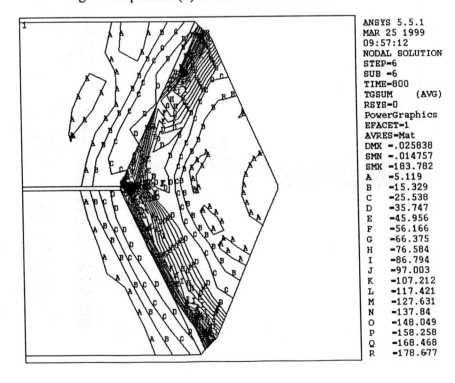

Fig. 7 Temperature (F) Profile at 800 Seconds

Detailed stress plots are shown in Figure 8 to indicate the von Mises stress that is used to qualify the exiting weldment design for the project. Design Specification for the joint is 60 ksi maximum.

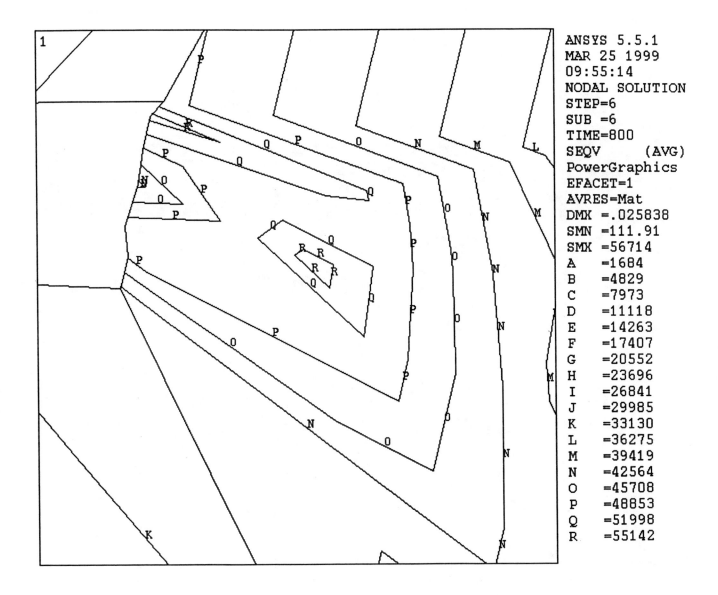

Figure 8 Detailed von Mises Stress Distribution on The Weldment

PVP-Vol. 393, Fracture, Fatigue and Weld Residual Stress
ASME 1999

INFLUENCE OF HEATING CONDITIONS ON TEMPERATURE DISTRIBUTION DURING LOCAL PWHT

Yukihiko Horii
Tsurumi R & D Center
Japan Power Engineering &
Inspection Corp.
14-1 Benten-cho, Tsurumi-ku
Yokohama, Japan Zip230-0044

Jinkichi Tanaka
Anan College of Technology
265 Aoki, Minobayashi-cho,
Tokushima, Japan,
Zip 774

Masanobu Sato
Inspection Division
Japan Power Engineering &
Inspection Corp.
Shin-Toranomon Bldg. 5-11,
1 Akasaka, Minato-ku, Tokyo, Japan
Zip 107-0052

Hidekazu Murakawa
Joining & Welding Resear Inst.,
Osaka University
11-1 Mihogaoka, Ibaraki-city
Osaka, Japan
Zip 567-0047

Jianhua Wang
Shanghai Jiao Tong Univ.
1954 Hua Shan Road
200030 Shanghai China

ABSTRACT

In a local band heating of pipe butt joint for post weld heat treatment (PWHT), temperature gradients occur inevitably and induce thermal stress depending on the heating conditions. To make clear the effect of heating conditions, temperature distributions during the local band heating PWHT were measured using a carbon steel pipe. An axis-symmetric model for finite-element-method (FEM) was adopted to calculate the temperature distributions under various heating band widths, heating rates, pipe wall thickness and heat transfer coefficients on the inner surface of pipe. From the results, a formula to estimate the temperature difference between the outer surface at the center of heating source and the inner surface at a distance of twice the wall thickness from the center from various heating parameters was obtained. In addition, a suitable heating source width was proposed from a view point of temperature difference.

INTRODUCTION

Post weld heat treatment (PWHT) is implemented mainly to reduce the welding residual stress and to improve the mechanical and metallurgical properties of weld heat affected zone (HAZ) and weld metal. Beneficial effects of PWHT are realized by the exposure of weldment at a relatively high temperature , for instance the temperature for carbon steel is 600℃ to 650℃.

PWHT is usually carried out using a furnace in which welded assemblies are installed and heated gradually up to the required temperature and held for some hours and cooled down slowly (a furnace PWHT). But for pipe butt joints on site, it is impossible to apply the furnace PWHT and a local PWHT method is adopted, in which a heating source is attached on the weldment circumferentially and a thermal insulation is positioned on the outer surface of pipe in some breadth.

The local PWHT of pipe with the circumferential heating source has two disadvantages in comparison with the furnace PWHT ; the first one is the occurrence of temperature gradient in the pipe axial direction which results in thermal stress and may change the degree of the reduction of welding residual stress during PWHT , the second is the existence of temperature gradient in the radial direction. In addition, the inner surface temperature is not usually measured due to the difficulty of an approach to the inner surface . So that it becomes important to ascertain the inner surface temperature to be in the required temperature range.

The above-mentioned subjects would be affected by many factors; pipe wall thickness, pipe diameter, heating source width, heating rate, insulation method and its width, heat loss by convection from the inner and outer surfaces etc. Although most of factors on the local PWHT above-mentioned are defined almost in the same way in various codes or standards, there appears an apparent different way in how to define the heating source width; one is that the heating source width shall be proportional to the wall thickness, the other is that the heating source width shall be proportional to the $\sqrt{Rd}$ (R is pipe inner radius, d is wall thickness). The heating source width supposedly plays the most important role in the formation of temperature gradient. Which is a more reasonable way?

In 1973, Shifrin and Rich have reported on a series of tests concerning the local stress relief of thick walled pipes (diameter:15-23 inch, thickness:3-4.5 inch) and suggested that the total heating source width should be " at least five times the wall thickness " to keep the maximum temperature gradient less than 100F(55℃). Recently, Bloch et al. (1997) have reported on a comprehensive experimental research work on the local PWHT of pipe (the wall thickness is up to one inch) and proposed that a new parameter " Hill Number Hi should be maintained at least 5.0" to keep the temperature difference between the top outer surface and the bottom inner surface at a distance of twice wall thickness from the weld center line less than 45 F (25℃). The " Hi " is defined as $Hi = Ac / (2Acs + Ai)$, where Ac is the exterior surface area of the pipe over which the heating source is positioned and Acs is the metal cross sectional area of the pipe and Ai is the interior pipe surface area in the pipe section defined by a width of twice the wall thickness on each side of the weld center line. He also suggested that the condition of $Hi \leq 2.5$ is not recommended for the usual process.

Shifrin and Rich(1973) and Bloch et al. (1973) have discussed about the temperature distribution in the local PWHT, but indicated no data on the change of welding residual stress due to the temperature gradient. Burdekin (1967) has analyzed theoretically the thermal stress due to the local PWHT of pipes and reported that a total heated band width of $5\sqrt{Rd}$ (R : pipe inner radius , d : thickness) seemed to be adequate to avoid an excessive thermal stress. But he did not make comment on the effect of thermal stress upon the change of welding residual stress during the local PWHT.

Although the local PWHT is adopted as a substitute of the furnace PWHT, the beneficial effects of PWHT should be fulfilled in almost the same degree as the furnace PWHT. In order to assure this expectation, the disadvantages, previously discussed, should be minimized as small as possible. Even if the issues on the formation of temperature distribution and thermal stress have been discussed as a function of the heating source width by above-mentioned authors, the totally combined discussions have not been presented. In other words, any proposal on the method to minimize the disadvantages of the local PWHT from the viewpoints of temperature distribution and thermal stress has not been presented.

JAPEIC (Japan power engineering and inspection corporation) has set up a working group to study the effects of local PWHT conditions of pipe butt joints. The working group has been studying mainly the effect of the circumferential heating source width on the temperature distribution as well as on the occurrence of thermal stress and its effect on the reduction of welding residual stress theoretically using Finite Element method and experimentally. This report concern with the temperature distribution and the thermal stress and its effect on the reduction of welding residual stress will be discussed in another paper by Murakawa et al. (1999).

In this study, the effects of heating source width, pipe size(diameter and thickness), heating rate, insulation condition and the heat loss by the convection from the pipe surfaces on the temperature distribution during the local PWHT were analyzed using Finite Element method, and the experimental equations were formulated to predict the temperature differences between the outer surface and inner surface of pipe at the center of heating source as well as between the outer surface at the center of heating source and the inner surface at a distance of twice wall thickness from the center of heating source. The prediction results were compared with those of experiments. Based on the prediction formulae, a table was proposed showing a suitable heating source width to maintain the temperature difference in the weld area within a limited range for various pipe sizes.

TEST MATERIAL AND EXPERIMENTAL METHOD

It is not always necessary to use a pipe with a girth weld in order to measure temperature distributions during local PWHT, and a simple straight C-Si-Mn type pipe of JIS STPT480 (340mm in diameter, 53mm in wall thickness, 1900mm in length) was used. Specification of STPT480 is shown in Table 1. Location of temperature measurement positions are shown in Fig. 1. To know temperature gradient toward a radial direction, temperatures were measured at the outer surface and 50mm below the outer surface in the same axial distance of the top (0 o'clock) and the bottom (6 o'clock) using thermocouples. (The wall thickness was 53mm and inner surface temperature was not measured.) The measuring axial distances from the center of heat source were 0, 1t, 1.5t, 2t and 3.5t (t : wall thickness). PWHT conditions tested are summarized in Table 1. Heating source width, heating rate and the end condition of pipe were changed. Test number 1 is a standard condition, namely half of heating source width B is three times the thickness from the toe ($\fallingdotseq 2.5\sqrt{Rt}$ (R : inner radius)) and heating rate 104℃/h is the maximum allowable rate based on the calculation of 220x25/t℃/h. In the test numbers 2 and 3 the heating

Table 1 Tested local PWHT conditions & specification of STPT480

Test No.	Half heat source width B (mm)	Heating rate Hv (℃/hr)	PWHT condition	Cooling rate (℃/hr)	Pipe ends codition	Reference
1	185	104	600℃、2hr	130	Free	3 t, ≒2.5√Rt
2	185	**220**		220	Free	High heating rate
3	**285**	104		130	Free	Wide heating width
4	185	104		130	**Capped**	Pipe edge conditin
Spec. of JIS STPT 480	C≦0.33% Si 0.10/0.35% Mn 0.30/1.00%, P≦0.035% TS≧480Mpa, YS≧275Mpa, EL25%, Seamless pipe					

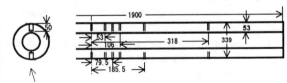

Thermocouple Top & Bottom outer surface , and
Top & Bottom inner side (50mm below the outer surface)

Fig.1 Temperature measurement positions

rate and the heating width was changed respectively. A hot air outflow was observed from the upper side of the pipe during PWHT in test numbers 1, 2 and 3 when those pipe ends were opened. Then about 90% of cross sections in pipe ends were capped with insulator in test number 4 to reduce the outflow by cap. Electric resistant heating was used as a heating source and a nichrome coil was wounded uniformly over the heating source width. Electric power was controlled by the temperature at the base point which was the center of the heating source in the top outer surface of the pipe (0 o'clock).

TEST RESULTS

Typical temperature changes at each point during local PWHT in test number 1 is shown in Fig. 2. The base point showed the highest temperature among other measured points. The thermal cycle itself almost coincided with the intended one. Temperatures at other positions were increased gradually during the holding period. Similar tendency was observed in other tests.

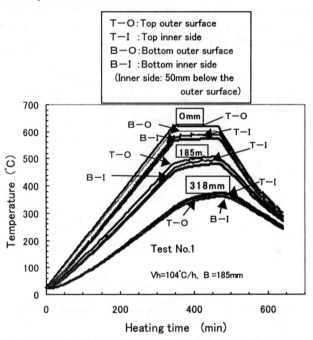

Fig.2 An example of temperature change during local PWHT

Temperature distribution.

Figure 3-a and 3-b show the axial distribution of temperature at 5min and 1h after the base point reached to the holding temperature and holding began respectively in the test number 1.

Figures 4 and 5 are axial distributions of temperature in test numbers 2 and 3 respectively. Temperature was decreased with increasing the distance from the center of heating source. The inner side temperature was lower than the outer surface one in both of the top and the bottom and a temperature gradient was observed in the axial and radial direction. In addition top side temperatures were higher than bottom side ones in both of the inner side and the outer

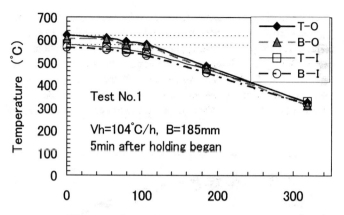

Fig.3-a Temperature distribution at 5min after the holding began in test No.1 (Pipe end: Open)

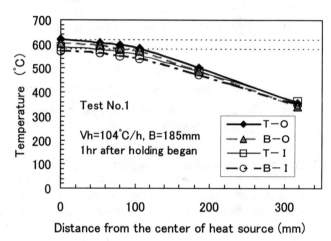

Fig.3-b Temperature distribution at 1h after the holding began in test No.1 (Pipe end: Open)

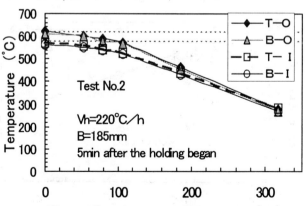

Fig.4 Temperature distribution at 5min after holding began in test No.2 (Pipe end: Open)

surface at the same axial distance. This means that temperature gradient exists in circumferential direction in test numbers 1, 2 and 3. The lowest temperature was appeared on the inner side of bottom in all tests. During the holding period temperature difference from the base point ΔTm(=To-Tm; To: temperature at the base point, Tm : temperature at the point in question) was reduced slightly. The temperature difference ΔTm was almost saturated when the holding was over 1 h .

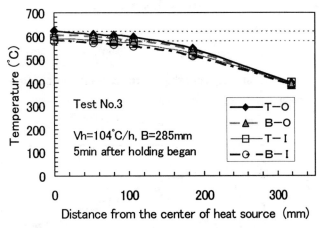

Fig.5 Temperature distribution at 5min after holding began in test No.3 (Pipe end: Open)

Temperature gradients and convection

In test numbers 1, 2 and 3, circumferential temperature gradients were observed as shown in Figs. 3, 4 and 5. On the other hand, in test number 4 which both pipe ends were closed partly with insulator, the bottom side temperature was close to the top side ones as shown in Fig. 6. Outflow of hot air from the upper side and inflow of air from them of pipe was observed during PWHT in test numbers 1, 2 and 3. Such airflow was largely reduced in number 4. It can be said that the convection of air in the inside of pipe influenced on the heat transfer at the inner surface and caused the above circumferential temperature gradient. Such convection may be small in site because pipe has long length to oppose airflow. Comparing the temperature distribution with and without cap at the end of pipe, the top inner side distribution without cap is similar to the inner side one with cap. Then results of test numbers 1, 2 and 3 can be evaluated by the top inner side distribution for practical use.

Effects of heating rate on temperature difference

Figure 7 shows the temperature difference ΔTm of the top inner side (T-I) and outer surface (T-O) points in the case of different heating rate. Temperature distributions in top outer surface were almost same in the heating rate of Vh104°C/h and 220°C/h. But the inner side temperature difference becomes larger in Vh220°C/h than in 104°C/h. The area showing less than 40°C of ΔTm is none at the 5min after the holding began, but after 1 h the area within about 45mm from the center showed less than 40°C of ΔTm on the inner side in the case of 104°C/h while it was about 10mm in the case of 220°C/h. Increasing of the heating rate can be said to increase the temperature gradients towards radius direction.

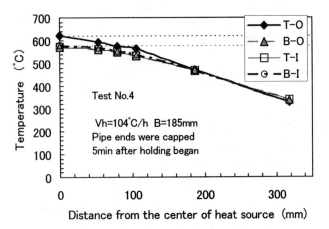

Fig.6 Temperature distribution at 5min after holding began in test No.4 (Pipe end: capped with insulator)

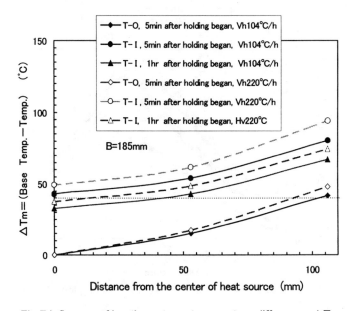

Fig.7 Influence of heating rate on temperature difference △Tm

Effects of heating width on temperature difference

Figure 8 shows the temperature difference ΔTm of the top inner side points in the case of different heating source width. In the case of half width of heating source B of 185mm there was no area where ΔTm was less than 40°C. By increasing the half heating width B to 285mm the area within 63mm from the center was heated to be less than 40°C of ΔTm in the top inner side. Then it may be said that in order to keep ΔTm of less than 40°C in more than a thickness area from the center of heat source, the heating width of 3 times the thickness from the toe is too small in the uniform heating case with the nichrom coil.

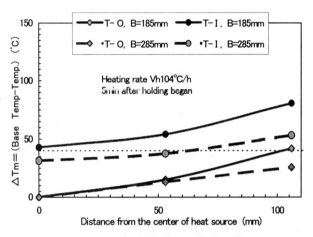

Fig.8 Influence of heating source width on temperature difference ΔTm

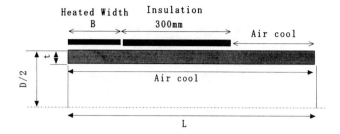

Fig.9 Model of analysis

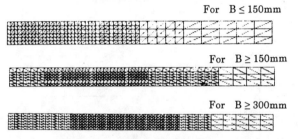

Fig 10 Finite Element Meshes

THEORETICAL ANALYSIS ON TEMPERATURE DISTRIBUTION
Method of analysis

The temperature distribution during the local PWHT of the pipe butt joint, in which a circumferential heating source was attached to the outer surface of pipe, was analyzed using an axis-symmetric model as shown in Fig.9 for the half length of the pipe.

The following assumptions were put in analysis.
1) Heat flux : Heat flowed into the outer surface of pipe from the heat source at a constant rate. Heat flux was adjusted to obtain various heating rates Vh on the outer surface of pipe under the center of heating source.
2) Thermal insulation : Thermal insulation was attached to the outer surface of pipe outside the heating source in a breadth of 300mm. Heat loss through the insulation was assumed to be estimated by Newton's law and the heat transfer coefficient β was set as 2×10^{-8} J/mm^{2}s°C. The ambient temperature was 0°C.
3) Heat loss from the outer surface: Heat loss from the outer surface of pipe without insulation was assumed to be estimated by The heat transfer coefficient β was set as 2.3×10^{-5} J/mm^{2}s°C. The ambient temperature was 0°C.
4) Heat loss from the inner surface : Heat loss was assumed to be estimated by Newton's law. The heat transfer coefficient β was set at various values between 0.1 ~ 5 $\times 10^{-5}$ J/mm^{2}s°C. The ambient temperature inside the pipe was assumed to be 0°C. In actual, the heat loss from the inner surface occurs by the combination of the convection and radiation. In order to adjust these effects, the heat transfer coefficient β was varied. Although it has been reported [2] that the temperature at the top position (0 o'clock) of pipe was higher than that at the bottom position (6 o'clock), this effect was not taken in mind.
5) Material constant : Material constants relating to the thermal conduction were set constant as follows ; Thermal conductivity : 0.05 J/mms°C, Specific heat : 0.063J/g°C, Density :0.0078 g/mm^3 .

Finite Element method (FEM) was adopted for calculation. Three types of finite element meshes were adopted depending on the heated width B as shown in Fig.10.

Series of computation

Computation was carried out for the combination of pipe size (D and t), heated width (B) and the heat transfer coefficient (β) as shown in Table 2. Heating rate was obtained at 550°C. The temperature distribution was analyzed at the time when the temperature on the outer surface of pipe under the center of heating source reached to 600°C to 660°C

Table 2 Series of computation factor

Pipe size D, t (mm)	Half width of heating source. B (mm)	Heating rate Vh(°C/h) at 550°C	Heat transfer coeff. $\beta \times 10^5$(J/mm^{2}s°C)
500, 25	80, 120, 150, 200	97~290	0.1, 2.3
500, 25	150, 230, 300, 500	60~230	0.1, 1.0, 2.3, 5.0
250, 25	120, 150, 200	80~230	0.1, 1.0, 3.0
340, 53	150, 190, 290, 300	78~90	0.1, 1.0, 3.0, 5.0

Notes 1. D : Outer diameter of pipe
2. t : Wall thickness of pipe
3. B : Half width of heating source
4. Vh: Heating rate at 550°C
5. β : Heat transfer coefficient on the inner surface of pipe

Examples of calculation.

Examples of the temperature distributions calculated at the time when temperature reached to a required one are shown in Fig.11. There existed a temperature difference in the radial direction under the heating source. The radial temperature difference showed a maximum at the center of heating source and almost disappeared at a distance of 150 mm outside the heating source. The temperature decreased gradually in the axial direction from the center of the heating source and dropped steeply outside the heating source. The axial length in which the temperature decreased gradually increased with increasing the heated width. These trends were observed in all calculated data.

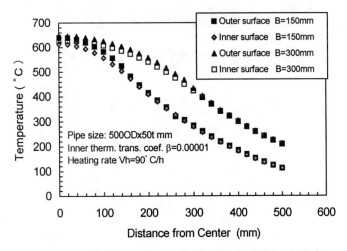

Fig.11 Examples of temperature distribution by FEM calculation

Prediction of temperature difference.

The analysis of temperature distribution during the local PWHT of pipe indicated the existence of temperature differences both in the radial and axial directions. In order to assure the beneficial effects of the PWHT, weld metal and HAZ shall be heated up to a required temperature, i.e. the holding temperature in the PWHT. This means that the temperature difference between the outer and inner surfaces of pipe shall be limited in a certain range throughout an area including weld metal, weld heat affected zone and some extended zone from HAZ. Considering that the width of weld metal is usually less than twice the wall thickness and the width of HAZ is less than one wall thickness in the usual welding process, the area where the temperature must satisfy the requirement for the holding temperature, was assumed to be twice the wall thickness in width in both directions from the center of heating source.

The temperature distribution below the heating source is schematically illustrated in Fig.12. To is the temperature on the outer surface at the center of heating source. Ti is the temperature on the inner surface at the center of heating source. T2 is the temperature on the inner surface at a distance of two times the wall thickness from the center of heating source.

Two kinds of temperature differences were defined in this report, a radial difference of $\Delta Tr = To - Ti$ and a combined difference of $\Delta T = To - T2 = (To - Ti) + (Ti - T2)$. From the viewpoint of the assurance of the beneficial effects owing to PWHT, ΔT would be more important.

ΔTr was extracted from the FEM calculated data and subjected to the regression analysis under the factors of heating rate at 550°C Vh, wall thickness t, heated width (= half width of heating source) B and the heat transfer coefficient β, resulting into the following equation;

$$\Delta T r (=To - Ti) = 0.001 \, Tr$$
$$Tr = Vh(10^5 \beta)^{1/3} t^2 / \sqrt{B} \quad \cdots\cdots(1)$$

The relation between the parameter Tr and ΔTr derived from FEM data is shown in Fig 13. The figure shows a fairly good correlation and a deviation of less than 5°C.

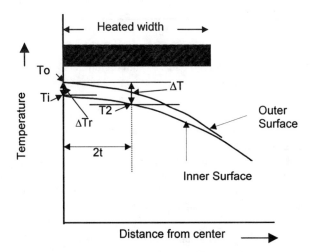

Fig.12 Schematic illustration of temperature distribution and notation of To,Ti,T2,ΔT and ΔTr

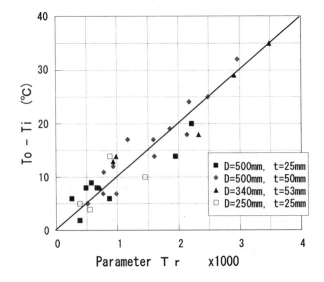

Fig.13 Relation between To-Ti and parameter Tr

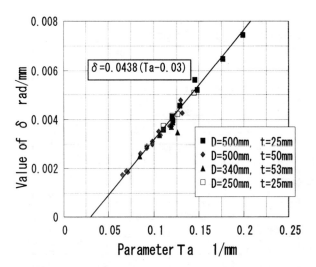

Fig.14 Relation between δ and parameter Ta

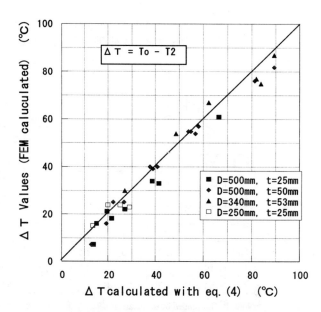

Fig.15 Comparison of estimated ΔT with calculated ΔT

In order to predict the temperature of T2, the axial temperature distribution curve was fit to a $\cos(\delta x)$ function, where x was a distance from the center of heating source. The value of δ was estimated from the temperature distribution on the inner surface calculated by FEM and subjected to the regression analysis, resulting in the following equation;

$$\delta = 0.0438 \, (Ta - 0.03)$$
$$Ta = Vh^{1/10}(10^5 \beta)^{1/30} / \sqrt{B} \quad \cdots\cdots\cdots(2)$$

The relation between δ from FEM calculated data and a parameter Ta is shown in Fig.14. Combining the equation (1) with (2) leads to a following equation;

$$\Delta T = 0.001 \, Tr + Ti \, \{1 - \cos(2 \delta t)\} \quad \cdots(3)$$
where x=2t

Using $Ti = To - 0.001 Tr$, equation (3) is rewritten as follow,
$$\Delta T = To \, \{1 - \cos(2 \delta t)\} + 0.001 Tr \cos(2 \delta t) \quad \cdots\cdots(4)$$

The comparison of the predicted value using the equation (4) with the estimated values from FEM analysis is shown in Fig.15. The figure indicates a good correlation and the deviation of less than 10℃.

Comparison with the experiments.

The comparisons of the estimated values with those of the experiments (pipe size, OD=340mm, wall thickness 53mm) were carried out about the temperature differences of ΔTr and ΔT on the top position of pipe, as shown in Table 3. The heat transfer coefficients were varied depending on the capping conditions in the experiments.

The estimated ΔTr values were in fair agreement with the experimental values except the test number 2 which was outside the calculation conditions. The estimated ΔT values were slightly larger than the experimental values.

Suitable heating source width and discussion

From the equation (4), it is possible to obtain a suitable heated width (= Half width of heating source) under a defined value of ΔT. ΔT is the temperature difference just after the time when the temperature reaches to the holding temperature and becomes smaller during a holding stage of PWHT and the maximum temperature gradient within soaking area is required to be below 50°C and 56°C by JIS and ASME VIII Div.2 respectively. Considering above, the authors assessed ΔT as 40℃ severely in the following discussions.

The maximum heating rate is defined as follows in many codes.
$Vh = 220 \times 25 / t$ ℃/h, where t is the wall thickness in mm.

This relation was inserted into equation (4). The required heated width(B) was calculated under the conditions of ΔT < 40℃ and To = 640℃ for three different values of the heat transfer coefficients(β)

Table 3 Comparison of estimated and experimental values of ΔTr and ΔT

Test No	Condition		Measured						Estimated		
	Pipe End	B	Vh	To	Ti	T2	ΔTr	ΔT	ΔTr	ΔT	β
1	Open	185	104	622	579	541	43	81	37	90	5x10⁻⁵
2	Open	185	220	620	571	526	49	94	77	138	5x10⁻⁵
3	Open	285	104	620	589	567	31	53	29	60	5x10⁻⁵
4	Capped	185	104	620	584	558	36	62	31	83	3x10⁻⁵

Table 4 Required heated width(=half width of heating source) to maintain $\Delta T \leqq 40°C$ and its ratio to wall thickness, inner radius for the commonly used pipe dimensions

Pipe Size		$ß=0.5\times10^{-5}$				$ß=1\times10^{-5}$				$ß=3\times10^{-5}$			
D	t	B	B/t	$B/\sqrt{Rt}$	Hi	B	B/t	$B/\sqrt{Rt}$	Hi	B	B/t	$B/\sqrt{Rt}$	Hi
267	15.1	44	2.9	1.0	1.1	50	3.3	1.2	1.2	58	3.8	1.4	1.4
"	21.4	76	3.6	1.6	1.4	84	3.9	1.7	1.5	102	4.8	2.1	1.8
"	28.6	118	4.1	2.2	1.7	132	4.6	2.4	1.9	160	5.6	2.9	2.3
355	19.0	64	3.4	1.2	1.2	70	3.7	1.3	1.3	86	4.5	1.6	1.6
"	27.8	114	4.1	1.8	1.6	126	4.5	1.9	1.7	154	5.5	2.4	2.1
"	35.7	164	4.6	2.3	1.8	182	5.1	2.6	2.0	222	6.2	3.1	2.5
508	26.2	104	3.9	1.3	1.4	116	4.4	1.5	1.6	140	5.3	1.8	1.9
"	38.1	178	4.7	2.0	1.8	200	5.2	2.2	2.0	244	6.4	2.7	2.4
"	50.0	258	5.1	2.6	2.1	288	5.8	2.9	2.3	356	7.1	3.5	2.8
660	34.0	152	4.5	1.5	1.6	170	5.0	1.7	1.8	208	6.1	2.1	2.2
"	49.1	252	5.1	2.1	2.0	282	5.7	2.4	2.2	348	7.1	3.0	2.7
"	64.2	352	5.5	2.7	2.2	394	6.1	3.1	2.4	492	7.7	3.8	3.0

Note. 1. The shaded boxes mean $B/\sqrt{Rt} > 2.5$, R : Inner radius 4. t: Wall thickness of pipe, mm
2. D: Outer diameter of pipe, mm
3. B: Heated width (=Half width of heating source), mm
5. ß: Heat transfer coefficient on the inner surface of pipe, $J/mm^2 s°C$

on the inner surface of pipe. The results are shown in Table 4. Table 4 indicates the pipe size (the outer diameter D and wall thickness t), required heated width B, ratio of B/t, ratio of $B/\sqrt{Rt}$ (R is the inner radius) and Hill Number Hi for three different conditions of convection on the inner surface.

The required heated width increased with increasing the wall thickness and the thermal coefficient. The same trend was observed for B/t, $B/\sqrt{Rt}$ and Hi. Referring to the ratio of B/t, it varied from 2.9 to 7.7 depending on the wall thickness and heat transfer coefficient. It is suggested that the required heated width should be proportional not to the wall thickness t but to the (wall thickness t)$^{1.4}$ approximately under author's assumptions. The proposal of $2B/t \geqq 5$ by Shifrin and Rich (1973) appears to be optimistic in comparison with table 4. This discrepancy would be brought about by the difference in assumptions, i.e. author's assumption is $\Delta T \leqq 40°C$ and Shifrin's one is $\Delta Tr \leqq 100F(55°C)$.

The heat transfer coefficient showed a comparatively large effect on the required heated width. The heat transfer coefficient should be changed depending on the pipe conditions, such as pipe position of horizontal or vertical, pipe length, pipe size etc. In our experiments, the pipe was positioned horizontally and the length was about 1.9m. A current of hot air flowing out from the ends of pipe was observed during PWHT operation. This phenomenon must have induced a larger heat loss from the inner surface of pipe by convection. The convection would be localized and become smaller for a longer pipe. For the local PWHT of a horizontal long pipe, the heat transfer coefficient β is supposed to be 1 to 3×10^{-5} $J/mm^2 s°C$.

The Hi number varied from 1.1 to 3.0 and was comparatively smaller than the value of 5 recommended by Bloch et al.(1997). This discrepancy maybe depends on the neglect of the temperature difference between the top and bottom of pipe and on a value of ΔT.
As illustrated in our experimental data in chapter 2, the temperature difference between the top and bottom amounted to 15°C. This temperature difference could be reduced by use of a multiple heating and regulation system, in which the temperature would be controlled both at the top and at the bottom of pipe or at three ~ four positions around the pipe. Hi numbers calculated by equation (4) would vary from 1.8 to 5.0 in the case of $\Delta T \leqq 25°C$ as assumption of Bloch et al. (1997). Assuming that the temperature difference between the top and bottom of pipe were 10°C, Bloch's assumption would become $\Delta T \leqq 15°C$ and Hi numbers would be between 3.0 and 9.0.

The ratio of $B/\sqrt{Rt}$ increased from 1.0 to 3.8 with increasing the wall thickness and heat transfer coefficient. The shaded box in table 4 means that the condition of $B/\sqrt{Rt} > 2.5$ stands. As the report 3 would indicate, the condition of $B/\sqrt{Rt} \geqq 2.5$ would be recommended to make sure the reduction of welding residual stress during the local PWHT. This recommendation suggests that the heated width for the pipe size corresponding to the shaded box should be decided from the viewpoint of the temperature distribution. In addition, the heated width for the pipe size corresponding to the no-shaded cell should be decided from the viewpoint of the reduction in the welding residual stress. Table 4 suggests that the suitable heated width should be decided in considerations for the effect of thermal stress on the reduction of welding residual stress during the local PWHT for the above 2/3 of the commonly used pipe sizes. In other words, the heating width should be decided on the condition of $B/\sqrt{Rt} \geqq 2.5$, instead of the form of $B \geq \alpha t$ (α is constant such as 2.5).

CONCLUSIONS

Effects of heating source width, heating rate and heat transfer from the surfaces upon the temperature distribution during a local PWHT of girth weld in carbon steel pipes were investigated experimentally and theoretically using the Finite Element Method. The heat source is an electric band heater. A suitable heating condition for local PWHT of girth weld was proposed based on investigations in view of obtaining the beneficial effect of tempering by PWHT. The main results are as follows;

1. Experimental investigations revealed an existence of temperature gradients in the radial and axial directions. In addition, a temperature difference between the top and bottom sides was observed. These temperature gradients decreased with an increasing heating source width and a decreasing heating rate and also with an adoption of pipe end cap that restricted an occurrence of air convection. Probably temperature difference between top and bottom may be caused by the occurrence of convection of air inside the pipe.

2. Theoretical studies indicated ;

a). The temperature difference between the outer and inner surfaces at the center of the heating source (ΔTr) could be predicted by the following formula;

$$\Delta Tr = 0.001 Vh (10^5 \beta)^{1/5} t^2 / \sqrt{B}$$

Vh : heating rate, β : heat transfer coefficient on the inner surface,
t : wall thickness, B : heated width (=half of heating source width)

b). The temperature distribution in the axial direction could be fit in a form of $\cos(\delta x)$ function (x is a distance from the center) and δ could be expressed as follows;

$$\delta = 0.0438 \{ Vh^{1/10} (10^5 \beta)^{1/30} / \sqrt{B} - 0.03 \}$$

c). The temperature difference (ΔT) between the outer surface at the center of heating source and the inner surface at a distance of twice the wall thickness from the center could be predicted as follows:

$$\Delta T = \Delta Tr + (To - \Delta Tr)(1 - \cos(2 \delta t))$$

To is the temperature on the outer surface at the center of heating source.

3. Under the condition of $\Delta T \leq 40°C$, a required heating source width was calculated for the commonly used pipe dimensions. The results suggested that the heating source width should be proportional not to the wall thickness but also to the (wall thickness)$^{1.4}$ approximately.

4. The calculated B values were smaller than that calculated from $B/\sqrt{Rt} \geq 2.5$ (R is inner radius of pipe) for the 2/3 of calculated pipe dimensions. The equation of $B/\sqrt{Rt} \geq 2.5$ indicated the condition to assure a fully decrease in welding residual stress by local PWHT.

ACKNOWLEDGEMENTS

The authors would like to thank to the JAPEIC working group members for useful discussion and advice and Dr. H. Nakamura for helpful comments to improve this study. The Authors would also like to thank NKK Corporation for supplying the test pipe and Mr. S. Nishikawa and Keihin Corporation for supporting the local PWHT experiments.

REFERENCES

Bloch, C., Hill, J., and Connel, D., 1997, "Proper PWHT Can Stop Stress-Induced Corrosion," Vol.76, Welding Journal, May, pp31

Burdekin, F.M., 1963, "Local Stress Relief of Circumferential Butt Welds in Cylinders," British Welding Journal, Vol.10, September, pp483

Murakawa. H., Wang. J., Tanaka. J., Horii. Y., and Sato. M.,1999, "Determination of Critical Heated Band Width during Local PWHT by Creep FE Analysis," Accepted for presentation in PVP meeting, Boston, August

Shifrin E.G., and Rich. M.I., 1973, "Effect of Heat Source Width in Local Heat Treatment of Piping," Welding Journal, Vol.52, December, pp792

DETERMINATION OF CRITICAL HEATED BAND WIDTH DURING LOCAL PWHT BY CREEP ANALYSIS

Hidekazu Murakawa
Osaka University
Joining and Welding Research Institute
11-1 Mihogaoka, Ibaraki-city
Osaka 567-0047, Japan

Jianhua Wang
Shanghai Jiao Tong University
Material School
1954 Hua Shan Road
Shanghai 200030, China

Jinkichi Tanaka
Anan College of Technology
Tokushima
265 Aoki, Minobayashi-cho
Tokushima 774, Japan

Yukihiko Horii
Japan Power Engineering & Inspection Corp.
Tsurumi R & D center
14-1 Benten-cho, Tsurumi-ku
Yokohama 230-0044, Japan

Masanobu Sato
Japan Power Engineering & Inspection Corp.
Inspection Division
Shin-Toranomon Bldg. 5-11
1 Akasaka, Minato-ku, Tokyo 107-0052, Japan

ABSTRACT

In the related report by authors (Horii et al., 1999), suitable heating conditions are discussed from the aspect of the temperature distributions. In this report, optimum heating conditions are discussed from the residual stress point of view using the thermal-visco-elastic-plastic finite element method. The whole history of the stress relaxation during the local PWHT can be clarified through the FEM analysis. Present investigations show that the transient and the residual stresses induced by the local PWHT are significantly affected by the creep behavior and the temperature dependency of Young's modulus. Stresses decrease quickly in the heating stage. They continue to decrease slowly according to the creep law in the holding stage and suddenly increase due to the recovery of Young's modulus when the cooling stage starts. The maximum residual stress after PWHT decreases with the increase of heated band width. When the heated band width is large enough, the residual stress after the local PWHT becomes almost the same as that for the uniform PWHT. Thus, the critical heated band width based on residual stress can be defined as the heated width which gives the residual stress close to that of the uniform PWHT. Through the comparison with the existing codes, it is found that the heated band width of $2.5\sqrt{Rt}$ on either side of the weld seems reasonable from the residual stress point of view. The H_i criterion based on the through thickness temperature gradient is also discussed in this study.

NOMENCLATURE

A_c = area of heat source on outside surface
A_{cs} = cross sectional area of pipe wall
A_i = inside surface area of soak band

B = half heated band width
b = coefficient of creep law
D = outer diameter
DT = temperature gradient within soak band
Hi = Hill number
n = coefficient of creep law
L = half length of pipe
SB = soak band width
SB_o = uniform heat width on outer surface
SB_i = uniform heat width on inner surface
Sr = residual stresses after stress relief by local PWHT
t = thickness
t_H = hold time
V_H = heating rate
W_{in} = insulation width
σ = stress
$\bar{\sigma}$ = equivalent stress
$\{\sigma'\}$ = deviatric stress
$\dot{\varepsilon}^c$ = creep strain rate
$\dot{\bar{\varepsilon}}^c$ = equivalent creep strain rate
β = heat transfer coefficient on inner surface

INTRODUCTION

Local PWHT is usually performed when it is impractical to heat treat the whole vessel in a furnace. PWHT can have both beneficial and detrimental effects. Two primary benefits of PWHT are the tempering and the relaxation of residual stresses. Benefits, such as improved ductility, toughness and corrosion resistance, are the results from the primary benefits. For the successful PWHT, the process must be optimized by selecting proper parameters. Many factors influence PWHT process such as size of the pipe, heated widths, insulation conditions, heating rates, soak temperatures, hold times, material composition, etc. However up to now the influences these factors have on PWHT are not clearly understood and differences in criteria for sizing the parameters, such as the heated band, are found among the existing codes.

This study provides a direct method to assess the effectiveness of the local PWHT. An axisymmetric model based on the thermal-visco-elastic-plastic finite element method is developed for this purpose. By using this model, both temperature and stress distributions during the whole local PWHT history can be obtained. Investigations show that the thermal stresses induced by local PWHT are strongly affected by the creep behavior and the temperature dependency of Young's mdulus. The study also demonstrates the possibility that through a series of computations the effects of various factors can be assessed and the optimum parameters can be found.

This report is separated into two parts. The method of computation and the characteristics of the mechanical behaviors in the local PWHT are discussed in the first part. The discussions on the critical heated band width and the influential factors are presented as the second part.

METHODS OF ANALYSIS
Method of Analysis and Model to be Analyzed

The local PWHT of a butt-welded pipe joint as shown in Fig. 1 is idealized as an axisymmetric problem. The half of the pipe is analyzed using the thermal and the mechanical axisymmetric codes developed by the authors (Wang et al., 1998). Since the local PWHT is performed only when the constraint is small enough, the pipe is assumed to be mechanically free in the axial direction. The heated band width B is defined as half width of heating source. The mesh division was formed by rectangular elements. The PWHT under given heating and cooling rates, holding temperatures and times can be easily simulated by automatic control of heat input using PID method implemented in the program.

Creep Law

In this study, the power creep law in the following form is used.

$$\dot{\varepsilon}^c = b\sigma^n \qquad (1)$$

where, σ : stress, $\dot{\varepsilon}^c$: creep strain rate, b and n : constants of the creep properties dependent on temperature.

In three dimensional problems, the component of the creep strain rate can be written as

$$\{\dot{\varepsilon}^c\} = 1.5\overline{\dot{\varepsilon}}^c \{\sigma'\}/\overline{\sigma} \qquad (2)$$

where, $\overline{\dot{\varepsilon}}^c$: equivalent creep strain rate, $\overline{\sigma}$: equivalent stress, $\{\sigma'\}$: deviatric stress. The details of the formulation are presented in the references (Ueda and Fukuda, 1975, Wang et al., 1998).

EFFECT OF CREEP ON TRANSIENT AND RESIDUAL STRESSES DURING LOCAL PWHT

Thermal stresses are not produced during the ideal uniform heating and cooling cycle. However the thermal stresses and the residual stresses are produced due to the temperature gradient and the constraint in the process of local PWHT. Because of the high temperature and the long hold time in local PWHT, both the creep phenomenon and the temperature dependency of the material properties must be taken into consideration. The material properties used in this part of the analysis are presented in Table 1.

Table 1 Material properties

Temperature (°C)	0	200	400	600
Heat conductivity (W/cm °C)	.546	.469	.385	.268
Thermal capacity (J/g °C)	.410			.913
Density (g/cm^3)	7.82			7.61
Yield stress (MPa)	275	275	200	100
Young's modulus (GPa)	210	210	170	94
Heat transfer coefficient (W/cm^2°C)	0.0033			
Thermal expansion coefficient (°C^{-1})	0.000013			
Poisson's ratio	0.3			

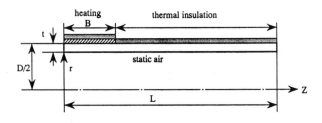

Fig. 1 Axisymmetrical model for analysis

The diameter D, the thickness t and the length of the pipe L are assumed to be 1000 mm, 25 mm and 1000 mm, respectively. The maximum temperature of local PWHT and the heating rate are 600 ℃ and 220 ℃/h. The hold time is one hour.

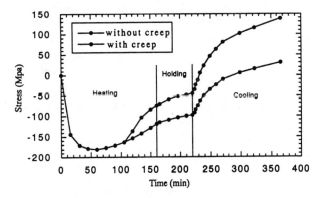

Fig. 2 Effect of creep on thermal stress (B=80 mm)

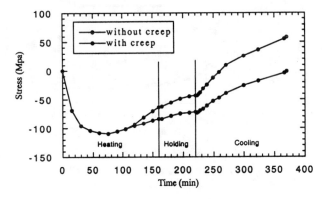

Fig. 3 Effect of creep on thermal stress (B=175 mm)

The effect of the creep on the transient and the residual stresses induced by the local PWHT is clarified using the two models with and without considering the creep. The creep phenomenon is considered not only during the holding stage but also in the heating and the cooling stages when the temperature is higher than 400 ℃. In these computations, the local PWHT is assumed to be performed on the stress free specimens. Figures 2 and 3 show the history of the axial component of the stress during the PWHT for two cases with relatively short and long heated band widths, namely B=80 mm and 175 mm. The stresses at the center on the internal surface of the pipe are plotted. The transient stress is produced by the thermal, the plastic and the creep strains. While, the residual stress is produced by the plastic and the creep strain. Thus, the difference between the two curves appearing after 100 min. in both Figs. 2 and 3 are influence of the creep. It can be seen that the residual stress induced by local PWHT considering creep is larger than that without creep. Comparing curves without considering creep in Figs. 2 and 3, it can be seen that the formation of residual stress due to plastic deformation during PWHT can be avoided when the heated band width is large enough. However, when the creep is considered, small amount of residual stress is produced even when the heated band width is long as seen in Fig.3. This means that the parameter of PWHT must be determined carefully considering the effect of the creep.

ANALYSIS OF STRESS RELIEF BY LOCAL PWHT
Welding Residual Stresses

To analyze the stress relief by the local PWHT, a simple welding residual stress generated by the thermal-elastic-plastic FEM is introduced as an initial residual stress. The size of the model is 500 mm, 25 mm, 600 mm in the diameter, the thickness and the length, respectively. One circumferential bead is applied on the pipe. The effective heat input per unit length is 11 kJ/cm and it is distributed on the four elements near the outer surface (2.5 × 10 mm) as shown in Fig. 4. Figure 5 shows the distributions of the residual stresses along the line which is about 8 mm from the outer surface of the pipe where the maximum residual stress is induced.

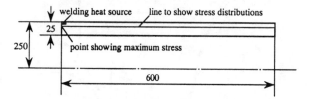

Fig. 4 Welding model of a pipe

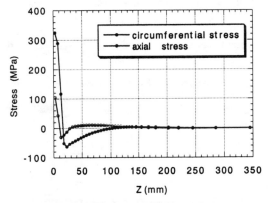

Fig. 5 Distributions of welding residual stresses

Analysis of Local PWHT

The local PWHT of a pipe with the same residual stresses discussed in the preceding section is analyzed under four different heated band widths. The four cases are,

(1) local PWHT with headed band width B= 25 mm,
(2) local PWHT with headed band width B= 65 mm,
(3) local PWHT with headed band width B= 125 mm,
(4) uniform heat treatment.

The heating rate is 220 °C/hr and the hold time is 4 hours. Figure 6 shows the history of the circumferential component of stress under four conditions. The stress at the point where the circumferential welding residual stress is the maximum is plotted. As seen in Fig. 6, the stresses decrease quickly in the heating stage because of the low values of yield stresses of the material at high temperature. The changes of stresses under four conditions are almost the same in the holding stage and they decrease slowly according to the creep law. However the stresses suddenly increase and are separated when the cooling stage starts. From Fig. 6, it can be seen that the residual stress after the local PWHT is almost the same as that for the uniform heat treatment when the heated band width is 125 mm. Figure 7 shows the distributions of circumferential component of the stresses after PWHT with different heated band width. The distributions of the residual stress along the same location as in Fig. 5 are shown.

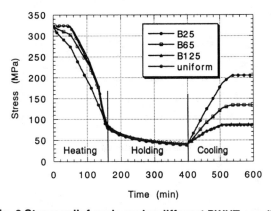

Fig. 6 Stress relief cycle under different PWHT conditions

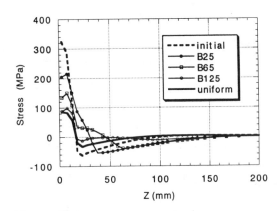

Fig. 7 Distributions of welding residual stresses with different PWHT conditions

Effect of Temperature Dependency of Young's Modulus

As it has been generally observed, the stress increases rapidly at the beginning of cooling stage. Two reasons can be given to explain this phenomenon. One is the thermal stress induced by the temperature gradient due to the sudden change of heat flow. The other reason is the recovery of the Young's modulus with cooling. To demonstrate these, two models with constant (E=130 GPa) and temperature dependent Young's moduli are analyzed under both uniform and local PWHT conditions. Figures 8 and 9 show the effect of the Young's modulus and the temperature gradient on the stress cycle. As expected, the phenomenon is not observed in the case of uniform PWHT with constant Young's modulus. The largest increase of stress is observed in the case of the local PWHT with temperature dependent Young's modulus as is shown in Fig. 9.

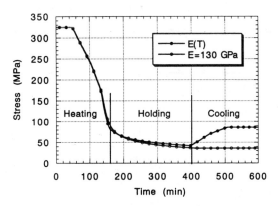

Fig. 8 Effect of temperature dependency of Young's modulus on stress cycle (uniform PWHT)

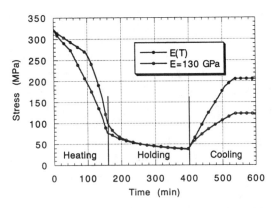

Fig. 9 Effect of temperature dependency of Young's modulus on stress cycle (B=25 mm)

Effect of Yield Stress

The temperature dependency of the yield stress may influence the stress relaxation by the local PWHT. Figure 10 show the time history of stress for cases with creep and temperature dependent yield stress, with creep and constant yield stress and without creep.

Significant difference due to the temperature dependency of yield stress is not observed in the residual stress shown in Fig. 10. In general, the residual stress is produced by both the plastic and creep strains. The time histories of these strain components are shown in Fig. 11 using the case of B=125 mm as an example. As seen from Fig. 11, the process can be divided into four stages, namely the first stage of heating process up to 400 ℃, the second stage of heating above 400 ℃, the holding stage and the cooling stage. In the first stage of the heating process only plastic strain is produced. While the creep strain is dominantly produced during the second stage of heating. The creep strain continues to increase slowly during the holding stage. In the cooling stage no change is observed in both the plastic and the creep strain. As observed from Fig. 11, the creep strain is dominant relative to the plastic strain. This is the reason why the temperature dependency of the yield stress has small influence on the stress relaxation as shown in Fig. 10. However, the situation may change depending on whether the plastic strain or the creep strain is dominant. In other word it may change with the dimension of the pipe and the heating condition. Further study is necessary from this aspect.

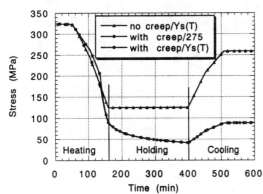

Fig. 10 Effects of temperature dependency of yield stress

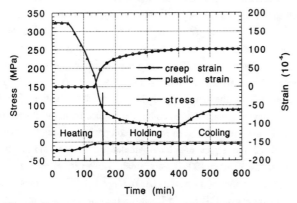

Fig. 11 Creep and plastic strains during local PWHT

CRITERIA FOR HEATED BAND WIDTH
Existing Code

Stress corrosion cracking due to the combined effects of service environment and localized weld-induced stress has been recognized as a serious problem in the industry for a number of years. In order to improve the ductility of welds and to relieve the residual stresses, the local PWHT is usually performed when it is impractical to treat the whole vessel in a furnace. The soak band, the heated band and the gradient control band are the main control parameters during local PWHT. The soak band consists of the through-thickness volume of the metal, which must be heated to the minimum but not exceeding the maximum required temperature. As a minimum, soak band should contain the weld metal, HAZ, and a portion of the base metal adjacent to the weld. The heated band consists of the surface area over which the heat source is applied to achieve the required temperature in the soak band and limit induced stresses in the vicinity of the weld. The gradient control band consists of the surface area over which insulation and/or supplementary heat sources are placed.

The heated band width is the single most important parameter determining the effectiveness of the local PWHT. However the criteria to determine the heated band widths are very different among existing codes. ASME Sections III, B31.1 and B31.3 do not provide specific guidance regarding the width of the PWHT heated band. BS 2633 provides a minimum recommended PWHT heated band width of five times pipe thickness (5t). However, one figure in BS 2633 recommends the use of a heated area of $2.5\sqrt{Rt}$ on either side of a branch connection, where R is the inside radius and t is the wall thickness. BS 5500 (BS 5500, 1997) and AS 1210 (AS 1210, 1989) also provide a minimum recommended PWHT heated band width of $5\sqrt{Rt}$ centered on the weld for circumferential welds and $2.5\sqrt{Rt}$ on either side of welds which connect nozzles or attachments to the shell. An American National Standard (ANSI/AWS D10.10-9X) (McEnereney, 1998) suggests that the size of the heated band is determined by two considerations. One is the through-thickness temperature gradient, and the other is induced stresses and distortion. The minimum heated width is determined by the larger one of the through-thickness temperature (with $H_i = 5$) and induced stress (SB plus $4\sqrt{Rt}$) criteria. H_i is an empirically derived ratio described as follows (Bloch, Hill and Connell, 1997),

$$H_i = A_c/(2A_{cs} + A_i) \qquad (1)$$

where, A_c = area of heat source on outside surface,

A_{cs} = cross sectional area of pipe wall,

A_i = inside surface area of soak band (assumed 4t wide, centered on weld).

All of the above criteria are not clear enough to specify the effectiveness of stress relief in local PWHT. The best way to evaluate the effectiveness is to know the residual stresses after PWHT directly.

In this study, a direct criterion for heated band width and the creep analysis are employed to assess the effectiveness of stress relief during PWHT. This method is clear enough to show the real situations of stress relief in PWHT and a critical heated band width can be determined. The pipes with the standard welding residual stresses are analyzed under different conditions of PWHT. A series of computations have been done to find the critical heated band widths using this method.

Direct Criteria to Assess Heated Band Width

A direct criteria based on the residual stress after local PWHT is to find a wide enough heated band width in local PWHT to produce an equivalent residual stress distribution to that for uniform PWHT with the same maximum temperature, the heating rate and the hold time as in the local PWHT. Figure 12 shows the method to define the critical width of the heated band. The heated width B is defined as a half width of heating source. As illustrated in Fig. 12, the maximum residual stress after PWHT decreases as the heated width increases. When the heated band width is large enough, the residual stress after PWHT becomes close to that for the uniform PWHT. Thus, the heated band width for such point can be defined as a critical heated band width.

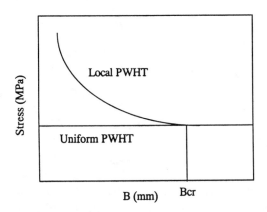

Fig. 12 Definition of the critical heated band width

NUMERICAL STUDY ON CRITICAL HEATED BAND
Numerical Model and Series of Computations

An axisymmetric model is used and the following considerations are made in the serial computations.

(1) use the material properties representing the low carbon steel as shown in Fig. 13
(2) consider a welding residual stresses distribution occurring in idealized through thickness welds
(3) consider the convection heat transfer on the inside surface
(4) use an average stress in high stress zones, such as the weld and the HAZ, to assess the effectiveness of the local PWHT

Table 2 Series of computations

No.	D (mm)	t (mm)	L (mm)	V_H (°C/hr)	T_H (hr)	W (mm)	B (mm)
A1–A11	500	25	750	220	1.0	338	Varied
B1–B10	250	25	750	220	1.0	238	Varied
C1–C9	1000	25	100	220	1.0	470	Varied
D1–D9	500	50	1000	220	1.0	475	Varied
E1–E9	500	50	1000	110	1.0	475	Varied

A series of local PWHT under different conditions listed in Table 2 are computed to assess their effects. In Table 2, D : outer diameter, t : thickness, L : half length of the pipe, V_H : heating rate, T_H : hold time, W_{in} : insulation width, B : heated band width. The series A, B and C are the cases to clarify the effect of the diameter D. The series D and E are the cases to clarify the effect of the thickness t and the heating rate V_H, respectively. The power creep law with n=5.0, b=0.15×10^{-9} (at 600°C) is used. The convection heat transfer is assumed on the internal surface of the pipe and the heat transfer coefficient β is assumed to be 0.0017 W/cm²°C.

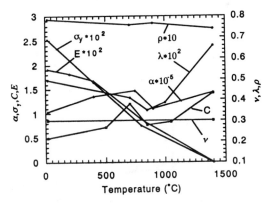

Fig. 13 Temperature dependent material properties

Through Thickness Temperature Gradient

The computed results for the thermal analysis are presented before those for stress. In this report, the through thickness temperature gradient DT is defined as the temperature difference between the exterior center of the pipe and the interior surface at a distance of two thickness from the weld centerline. Also the uniform heat width in which the temperature difference ΔT is maintained within 25°C at the start of the hold time on the outside surface SB_o and the inside surface SB_i are computed. Figure 14 shows the relations between DT and Hi. As seen from the comparison among the series A, B and C, the effect of pipe diameter on the through thickness temperature gradient DT is very small. However, as seen from the comparison between the groups (A, B, C) and (D, E), the thickness has a strong effect on DT. Figure 15 shows the effect of the thickness t on the relation between the uniform heat widths and Hi by comparing the cases A and D. The uniform heat widths on outside surfaces SB_o and inside surfaces SB_i are very different in thick pipe as the case of t=50 mm.

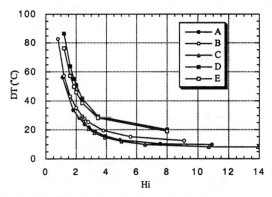

Fig. 14 Relations between through-thickness temperature gradient DT and Hi

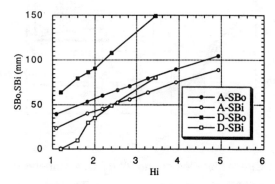

Fig. 15 Relations between uniform heat widths and Hi

Axial Temperature Gradient

Some of the existing codes recommend controlling the axial temperature gradient by limiting the temperature drop at the edge of the heated band to one-half of the temperature at the edge of the soak band. In this section, the effect of the heated band width on the axial temperature gradient is examined. Series A in Table 2 are computed with considering the convection heat transfer on the inside surface (β =0.0017). The temperature at the edge of the soak band is 600 ℃. Table 3 shows the temperature at the edge of the heated band T_{HB} and the temperature drop between these edges of soak band and heated band at the start of the holding for different value of heated band width. From Table 3 it can be seen that the temperature drop increases when the heated width increases and all are within the limit of one-half of the temperature at the edge of the soak band. In the strict sense, the temperature drop should be divided by the distance between above two edges to show the axial temperature gradient. If the temperature gradient is defined in this way, it decreases with the increases of the heated width.

Table 3 Axial temperature gradient

Heated Width (mm)	75	125	150	175	195	225	270	338
THB(°C)	530	483	462	442	427	406	374	319
Temp.drop(°C)	70	117	138	158	173	194	226	281

Influence of Welding Residual Stress Distributions

Three types of welds are analyzed by FEM for comparison. Namely, the heat sources are applied near the outside surface, near the inside surface and uniformly through thickness, respectively, as shown in Figs. 16 a), b) and c). The distribution of the circumferential component of the welding residual stresses along the thickness direction is shown in Fig.17. It can be seen that the locations of maximum stress in welds and the stress level in outside parts of the pipe are very different among three types. These may have an influence on stress relief during local PWHT. Figure 18 shows the effects of heated width on stress relief under three types of weld conditions. As seen from the figure, the residual stresses of type (a) can be reduced by local PWHT even more than by uniform PWHT when the heated width is large enough. However it is difficult to get the same result for type (c). The type (b) is in the intermediate situation. These phenomena may be explained by the interaction between welding residual stresses and the thermal stresses induced by the local PWHT. The bending stresses induced by the local PWHT are positive on outside surfaces and negative on inside surfaces and these may be favorable to reduce the welding residual stresses of type (a).

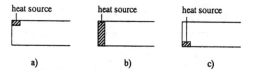

Fig. 16 Three types of weld models

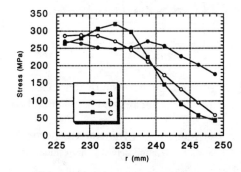

Fig. 17 Distribution of welding residual stresses through thickness for three types of welds

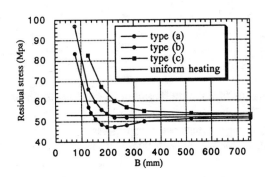

Fig. 18 Relation between residual stresses and heated band width for three types of welds

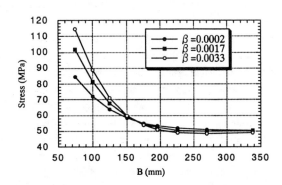

Fig. 19 Effect of heat transfer conditions on residual stresses

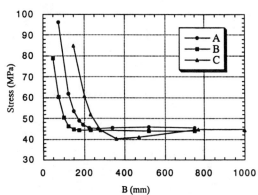

Fig. 20 Effect of diameter on relation between residual stresses and heated band width

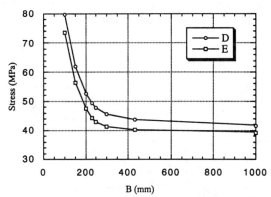

Fig. 21 Effect of heating rate on relation between residual stresses and heated band width

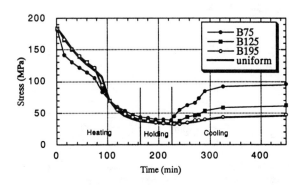

Fig. 22 Stress histories during PWHT under different heated band widths

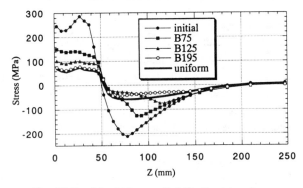

Fig. 23 Residual stress distributions under different heated band widths

Influence of Convection Heat Transfer on Internal Surface

Figure 19 shows the relation between the residual stress and the heated band width of series A with different convection heat transfer on the inside surface. It can be noticed that the residual stress increases with an increase of convection heat transfer coefficient only in the case of narrow heated band widths. When the heated band width is large enough, significant difference is not observed in the residual stress and the critical heated widths becomes almost the same under different conditions of convection heat transfer. Of course, it is necessary to increase the heating power when convection heat transfer increases.

Table 4 Critical heated band widths and related computed results

No.	D (mm)	t (mm)	DT (°C)	SB$_o$ (mm)	SB$_i$ (mm)	Sr (MPa)	Bcr (mm)	Bcr/t	Bcr/$\sqrt{Rt}$	H$_i$
A	500	25	23	67	53	48.5	180	7.2	2.4	2.62
B	250	25	35	51	40	46.4	125	5	2.5	2.0
C	1000	25	16	92	74	47.5	260	10.4	2.5	3.62
D	500	50	51	91	35	47.8	250	5	2.5	2.0
E	500	50	46	93	49	44.3	230	4.6	2.3	1.84

It is also generally observed in Figs. 18 and 19 that it is impossible to reduce the residual stresses by simply increasing the heated width unnecessarily.

Residual Stresses after Stress Relief by PWHT

The residual stresses after stress relief in local PWHT under different conditions shown in Table 2 are computed by FEM creep analysis. The through thickness heat source is used to generate the welding residual stresses (type b). The convection heat transfer on the inside surface is assumed to be $\beta = 0.0017$.

Figures 20 and 21 show the relation between residual stresses and heated band width for series A through E. The average value of circumferential stress in weld and HAZ (high stress zone) is used to assess the effectiveness of the PWHT since it is more stable than the maximum stress at single point. The critical heated width is defined as the width which reduces the residual stress after PWHT close enough to the value obtained from the uniform PWHT. Figures 22 and 23 show the histories and the axial distributions of the circumferential component of the stress during the local PWHT for series A, respectively.

Table 4 summarizes the critical heated band widths and other computed results. Sr is the residual stresses after stress relief by local PWHT for the pipe treated with the proposed critical heated band widths.

DISCUSSIONS
Criteria for Sizing Heated Band Width

Present study shows that direct criterion based on the residual stress after local PWHT is very useful for sizing heated band widths.

It is found that the diameter and the thickness of the pipe have a large influence on the residual stresses after local PWHT. As seen in Table 4, the criterion of five times pipe thickness (5t) seems not suitable. A heated area of $2.5\sqrt{Rt}$ on either side of a branch connection, and SB plus $4\sqrt{Rt}$ (centered on weld, SB=3t), are two criteria that seem reasonable judged from the computed results in this study. The Hi=5 criterion is too large from the residual stress point of view even under the case of convection heat transfer on inside surface. It is necessary to make the H$_i$ concept clearer by further investigations.

This criterion is based on the measured through thickness temperature gradients considering the temperature differences between the 0 o'clock and the 6 o'clock position due to the convection air flow inside the pipe. According to the proposal by Bloch et al. (1998), H$_i$ number should have a value of at least 5 to maintain a maximum temperature gradient within 25°C. However through our computations it is found that the relations between DT and H$_i$ are strongly influenced by the thickness as shown in Fig. 14. Noting that the criteria Hi=5 is proposed for the pipe with thickness of one inch (25mm), it may be too large for small thickness and may not be large enough for large thickness.

Heating and Cooling Rates and Hold Time

According to the creep law the degree of stress relief is primarily related to the hold temperature. The creep strain rate at 650°C is much greater than that at 600°C. Rate of stress relief becomes slow when the hold time increases, so too long hold time is not necessary. As seen from the comparison between the series D and E in Fig. 14, the heating and cooling rates have some influence on residual stress when the heated band width is not sufficient. When the heated band width is wide enough, its influence becomes small.

SUMMARY AND CONCLUSIONS

The main conclusions drawn from the present study are summarized as follows.

(1) The transient and the residual stresses induced by the local PWHT are strongly affected by creep behavior and the temperature dependency of Young's modulus.

(2) The stress cycles under the local PWHT are studied using the FEM. The results show that the stresses decrease quickly in the heating stage. In the holding stage, they decrease slowly according to the creep law. Then they increase suddenly when the cooling stage starts. The reasons for such sudden increase are the temperature gradient produced by the change of heat flow and the recovery of Young's modulus with cooling.

(3) Three types of welding residual stress distributions are used to show their effects on stress relief during the local PWHT. That for the through thickness weld is employed as the standard case in this study.

(4) The residual stress increases with an increase of the convection

heat transfer coefficient on the internal surface only in the cases of insufficient heated widths.

(5) From the point of view of stress relief, the criterion of five times pipe thickness (5t) seems not suitable. A heated area of $2.5\sqrt{Rt}$ on either side of a branch connection, and SB plus $4\sqrt{Rt}$ (centered on weld, SB=3t), seem reasonable.

(6) The relations between the through thickness temperature gradient DT and Hi are influenced by the thickness. The Hi concept may not be suitable for pipes with arbitrary thickness.

(7) Improvement of material property and stress relief are the two main considerations for sizing the heated band width. The heated band width should be determined by the larger one given from the two considerations.

ACKNOWLEDGEMENT

The authors would like to acknowledge that part of the present research program is supported by Pressure Vessel Research Council.

REFERENCES

Bloch, C., Hill, J. and Connell, D., 1997, "Proper PWHT Can Stop Stress-Induced Corrosion," Welding Journal, May 1997, pp.31-41.

Horii, Y., Tanaka, J., Sato, M., Murakawa, H. and Wang, J., 1999, "Influence of Heating Conditions on Temperature Distribution during Local PWHT," to appear in Proceedings of the ASME/PVP Conference (PVP '99), July 1999, Boston, Massachusetts.

McEnerney, J.W., 1998, "Recommended Practices for Local Heating of Welds in Pipe and Tubing", ANSI/AWS D10.10-9X.

Ueda, Y. and Fukuda, K., 1975, "Analysis of Welding Stress Relieving by Annealing Based on Finite Element Method", Transactions of JWRI, Vol.4, No.1, pp.34-45.

Wang, J., Lu, H. and Murakawa, H., 1998, "Mechanical Behavior in Local Post Weld Heat Treatment (Report I) - Visco-Elastic-Plastic FEM Analysis of Local PWHT -", Transactions of JWRI, Vol.27, No.1, pp.83-88.

AS 1210, 1998, Australian Standard Unfired Pressure Vessels Code (AS 1210).

BS 5500, 1997, British Standard Specification for Unfired fusion Welded Pressure Vessels (BS5500).

JAPEIC, 1998, "Examination of Effective Heated Band for Local PWHT," Research Report of Japan Power Engineering and Inspection Corporation, March 1998.

CLASSICAL AND EMERGING FRACTURE MECHANICS PARAMETERS FOR HISTORY DEPENDENT FRACTURE WITH APPLICATION TO WELD FRACTURE

Frederick W. Brust

Battelle Memorial Institute
Columbus, Ohio

ABSTRACT

This paper summarizes current and emerging mechanics based methods for predicting crack growth and fracture. In particular, history dependent fracture, where a material has undergone one or more sources of nonlinear deformation prior to crack initiation and growth, is discussed. A number of classical methods of fracture prediction are overviewed, and examples are presented which illustrate performance for history dependent situations. In addition, emerging fracture parameters are discussed which have the potential to characterize failure for materials that have undergone important prior fabrication history. Again, some examples are illustrated. One main focus of this paper is to illustrate the effect of prior history on fracture response, and to point out the areas where the current methods need improvement.

INTRODUCTION

Fracture mechanics methods are now widely used by nearly all industries in both the public and private sectors to help ensure that fabricated structures do not fail prematurely. A (subjective) categorical list of fracture methods used today for static failure assessments are:

1. K-Based (Including Plastic Zone Correction Methods: Irwin, Dugdale, etc.)
2. R6 or FAD (Interpolate between Elastic and Fully Plastic)
3. J-Integral (History Effects are not properly accounted for)
4. C*-Integral and C_t (High Temperature)
5. 'Energy Release' Type
6. Damage Based ('Lemaitre Type', Coupled or Uncoupled)
7. CTOA (Limited to Plastic Stable Growth)
8. Integral Parameters (T*, McClintock, Watanabe, etc.)

The first five of these might be termed classical fracture parameters, are simple to use, but have questionable accuracy when applied to situations where material history affects the fracture response. The latter three are emerging methods that are more complicated to use, but are more fundamental for use with history dependent fracture including weld fracture.

Classical Methods

K-based methods are used in brittle materials. For moderately ductile materials, plastic zone corrections of the K-based solutions (of the Irwin or Dugdale type) are often used in the steel fabrication industries. The *R6 or FAD* (Failure Assessment Diagram) approach, which was developed in Great Briton for the nuclear industry and is used extensively today in Europe, consists of interpolating between the K-based solution and a rigid plastic solution. *J-Integral/Tearing* theory, developed in the US and used extensively in the energy production fields, is used under fully plastic conditions. High temperature fracture of weldments will not be considered here and method 5 has theoretical limitations. All of these methods are useful under certain circumstances, but none of these methods are satisfactory for use under general history dependent fracture conditions. Examples are provided to illustrate this.

Emerging Methods

The *CTOA parameter* has seen use in predicting welded pipe fracture and, in recent years, fracture in thin structures (aging aircraft fuselages). CTOA does not appear useful for 3D fracture predictions and for situations outside stable growth (i.e., history dependent cyclic loading appears unworkable). *Damage based approaches* have seen increasing use for two-dimensional fracture predictions and some limited 3D calculations. The methods are particularly useful for crack initiation predictions, but are less reliable for crack growth predictions because the predictions are mesh size dependent. *Integral Parameters*, especially the T*-Integral, are the most general crack growth parameter for use in complicated loading situations and have seen increasing use in characterizing fracture (both time dependent and time independent) in recent years, especially in the energy production field and the aging aircraft regime.

These three emerging methods have not been used extensively in industry because the computational power required to perform such detailed fracture and failure analyses has not available. With the beginning of the emergence of the massively parallel machines and the corresponding code vectorization in progress at present, the use of these resource demanding fracture parameters will soon be a reality. Examples of the use of most of these parameters are provided in this paper, with discussion of their usefulness in weld fracture.

MATERIAL HISTORY DEPENDENCE

Structures continually degrade during the service performance years. Initial fabrication flaws (especially for welded fabrication) may be present, fatigue and corrosion cracks may initiate and grow, etc. Assuring that these defects do not lead to service failures is an important goal for the product manufacturer and the structural fabricator. Structural integrity can be ensured by assuming the presence of flaws at highly stressed regions of the structure and performing fatigue, corrosion, and fracture analyses of the structure under service conditions. In addition, when flaws are found in service, the repair schedule (timing, type of repair, etc.) can be determined by performing such a fracture assessment.

Usually damage assessments are made by neglecting prior history in the components that make up fabricated structures. Consider the cycle represented by Figure 1 for a fabricated steel structure. Start from the upper left and move clockwise. The different components may be cast, rolled, forged, etc. from the material supplier, and often combinations are used in the service structure. An initial residual stress and distortion pattern develops depending on which process is used. For instance, in rolled plate, tensile stresses usually develop near the plate edges and compressive stresses develop in the interior.

The rough material stock is then transported from the steel mill to the fabrication shop. The transportation, handling, and storage may alter the original stress and deformation history inherent in the rough stock. The material may then be surface treated, depending on the application, further altering the original history in the rough stock. The material is then cut into the desired shape required for the fabrication (for instance, consider a truck frame consisting of a number of components (forged, cast, rolled) and welded together). This cutting process alters the original residual stress history (and thus distortion history). Moreover, the cutting process may induce additional residual stresses near the cut edge (particularly for thermal cutting since the thermo-mechanical flame cutting process induces nonlinear strains near the cut edges).

The component might then be bent and formed into the required shape for the fabrication. Forming again alters the original history, and can add additional significant stresses and strains in the bent regions of the component. Connecting the parts with welds, bolts, adhesives, etc assembles the different components into the desired structure. The connection process further alters the residual history in the component parts. In particular, the welding process induces significant residual stresses and distortions into the assembled structure. Finally, the assembled component or structure might then be straightened to achieve certain desired tolerance requirements. This process further alters the history in the structure.

As seen, by the time the material from the steel plant makes its way into the service structure, each component has already seen a history of stresses, nonlinear strains, and corresponding displacements. As an example, consider Figure 2. This is an example of the effect that bend forming and welding on the service residual stress state. As illustrated on the left of Figure 2, a two-dimensional nonlinear finite element analysis was performed of a plate being bent in a die. Next, a weld bead was deposited on the exterior of the bend curvature. Both the bending and welding processes were modeled via the finite element method (the weld models used are described in References [1-8]). The right side shows black and white contour plots of the 'X' component of stress (this component will likely lead to crack growth at the weld toe). The material was a standard structural steel. The stress magnitudes vary between + and – 300 Mpa (the magnitudes are not important for our purposes). It is clearly seen that stresses at the weld toe are markedly different between the two cases.

This history can have an important effect on the service life of the structure. Fatigue, corrosion cracking, fracture, etc.

can all be affected by prior history. However, in most cases this prior history is neglected in making a damage assessment of the structure. *Fatigue and fracture assessments are often made by pretending that the service material is pristine, free of history.* The fatigue life predicted assuming a pristine structure is different from that which would be predicted by including the prior history (welding alone (Figure 2 (a) or bending and welding 2(b)). References [9-12] further illustrate this point and a examples will be shown later that clearly illustrate this effect.

Of course, for some critical structures, stress relief of many of the components is performed before the structure goes into service. However, the costs associated with stress relief are not practical for the vast majority of fabricated structures. In addition, 'stress relief', annealing, and other processes do not eliminate all residual stresses and strains leading to a pristine structural component.

The remainder of this paper illustrates other examples that show material history effects are important in evaluating damage.

FRACTURE PREDICTIONS OF WELDS USING CLASSICAL PARAMETERS

Welded Pipe

In excess of seventy extensive full scale tests of pipe have been performed at Battelle over the last 20 years (References [13-16]) and numerous references cited therein. These tests have included circumferential through-wall and surface cracked pipe subjected to bending and combined bending and tension. Base metal, weld metal, and fusion zone cracks have been considered. The first five of the eight general types of fracture parameters listed in the introduction were extensively studied by comparing predictions to experiments.

Let us first consider the use of the R-6 (Revision 3) method (Reference [17]). The R-6 methodology for assessing the integrity of cracked structures was developed and validated by CEGB, UK (References [17-18]) and will not be detailed here. The assessment procedure consists of interpolating between the linear elastic fracture mechanics criterion for failure and the fully plastic limit-load solution using a failure assessment diagram. The failure assessment diagram (FAD) is a prescribed function of L_r (i.e. $K_r = K_r (L_r)$), where $K_r = K_I / K_c(\Delta a)$ and $L_r = \sigma/\sigma_{net}$. Here $K_c(\Delta a)$ is the material resistance (obtained from a J-R curve and converted to K space), K_I is the applied stress intensity factor, and σ and σ_{net} applied stress and net section (limit load) stress, respectively. As such, the R-6 method (like the ASME Section XI criteria) is basically a design or failure avoidance procedure with an inherent safety margin already included in the analysis, i.e., no safety factor is explicitly applied. Crack initiation and maximum load predictions using this scheme are expected to be conservative. Three possible options for the FAD equation exist: here we use option 1 only (see Reference [17]). For the results shown below, the J-R curve was obtained from a compact tension (CT) specimen cut from a welded pipe of the same size (diameter and thickness) as the test result (See Reference [20, 21]). The crack was placed in the weld itself for both the CT specimen and the pipe.

Figure 3 illustrates the comparisons between predictions of maximum load using the R-6 option 1 method and experimental data. The different materials tested and the mean radius divided by the thickness (R_m/t) is also shown. The predictions are conservative since they are all larger than 1.0 (numbers above 1.0 represent the amount of conservatism). These fracture predictions did not include the effect of the residual stresses in the analysis. Indeed, obtaining the residual stresses and including this effect in the analysis is possible by using a weld modeling strategy such as that of References [1-8]. Including the residual stresses in the R-6 analysis procedure would have clearly reduced the conservatism in the predictions.

For these seven large full-scale tests, the mean ratio of experimental divided by predicted load was 1.50 with a coefficient of variation of 12.8 percent. In all, thirty-two surface cracked pipes were tested with cracks in either base or weld metal and analyzed via the R-6 method. The mean ratio of experimental divided by predicted load was 1.3. In addition, a number of other methods were evaluated in Reference [20] including six J-estimation methods, ASME section XI methods, plastic zone correction methods, and net section collapse and screening criteria approaches. In all cases the predictions for the welded pipe were much poorer compared with predictions for the base metal pipes. Similar results were obtained for through-wall cracked pipes where upwards of forty tests were performed (Reference [19]). It should be clear that fracture predictions of structures that experience prior history of the type illustrated in Figure 1 need improvement. Many more details regarding these results and many others cam be found in References [13-16, 19-21].

Fracture of Welded Beams

Following the 1994 Northridge earthquake, wide spread damage was discovered in the pre-qualified welded steel moment frames. Detailed inspections indicated that most of the structural damage occurred at the weld connections between the beam and column flanges. In particular, the weld joints between the bottom beam flange and the column face suffered the most severe damage. Cracks mostly initiated at the weld root and propagated with very little indication of plastic deformation. The desired plastic hinges assumed in the structural building codes 'plastic design' was not formed in the weaker beam away from the weld joint. Instead, brittle weld fracture was identified as the dominant failure mechanism.

Welding-induced residual stresses are believed to be one of the factors contributing to the brittle fracture. Indeed, there exist ample evidence that residual stresses can play dominant role in the fracture process of highly restrained welded joints (References [22,23]). The design of welded moment resistant frame connections presents perhaps the most severe mechanical restraint conditions both during welding and in service. Consequently, the presence of high weld residual stress is expected. In addition, the triaxiality of the residual stress state in these joints can be significant. As such, the anticipated plastic deformation cannot develop before the fracture driving force reaches its critical value, resulting in brittle fracture.

Figure 4 illustrates an analysis of a typical beam-column connection (A36 beam, A572 Gr. 50 column, and E70T-4 weld material. The insert illustrates the finite element mesh that was used to model the welds in the beam-column. The residual stresses (not shown here – see Reference [12]) showed a very high degree of triaxiality and were directly caused by constraints induced by the welding process and the geometry. The two-dimensional mesh illustrated below the beam column (Figure 4) was used to study the fracture response of a typical lower flange in a beam column connection. Nine passes were deposited (see inset mesh of Figure 4). The stress intensity factor is plotted as a function of the ratio of applied tensile stress to weld yield stress in the right of Figure 4. It is seen that including the history of residual stresses in the analysis procedure has a marked effect on the stress intensity factor. Indeed, for fracture in the brittle or transition regime, including prior history in the analysis is critical to obtain correct results. References [11-13] show further evidence that prior history in the form of weld residual stresses and strains has an important effect on fatigue and stress corrosion cracking, as well as ductile fracture characterized by the crack tip opening fracture parameter.

FRACTURE PREDICTIONS USING EMERGING FAILURE PARAMETERS
Composites Processing and Failure.

The response of composites is another area where the constitutive behavior, damage development and failure is affected by the history of fabrication. During fabrication of composite components, whether polymer, metal, or ceramic matrix with continuos or chopped fibers, significant residual stresses develop. These stresses can be used to advantage to improve service performance. Properly including the effects of processing induced residual stresses in the analysis model has a very important effect on the predicted failure response.

The material considered was an eight-ply unidirectional Ti-15-3/SCS6 composite, approximately 1.99 mm thick, fabricated by Textron Specialty Materials (serial number 890556), using a Hot Isostatic Press (HIP) consolidation technique. The SCS6 fiber is approximately 0.14 mm in diameter, and contains alternating outer layers of Carbon and non-stoichiometric SiC, which protect the fiber during handling. The alloy is a metastable b.c.c. β-Ti alloy, the b.c.c. phase being stabilized by vanadium. The thickness of the fabricated 8-ply unidirectional system is about 1.99 mm and the fiber volume fraction is approximately 0.34. The approach taken to produce experimental data with which to compare predictions is described in detail in References [24, 25].

Two steps are involved in the processing model: modeling the Hot Isostatic Press (HIP) process, and then cooling to room temperature. As schematically illustrated in the top portion of Figure 5, a HIP pressure is assumed to be applied to the thin composite at near the consolidation temperature, where the matrix begins to attain observable stiffness. The actual fabrication process involves the large deformation flow behavior of very soft matrix material as it is pressed around the rather rigid fibers. This complex process is neglected, assuming that the matrix material behaves as a fluid (with no memory of deformation history), until the titanium matrix begins to obtain stiffness (at about 815 C). The HIP pressure coupled with the high temperature application leads to oriented thermal plastic deformation in the matrix and the neat packing arrangement shown in Figure 5 is attained. After the HIP pressure is applied, the system is cooled to room temperature while the pressure is linearly reduced to zero. After cooling to room temperature, significant residual stresses develop in the system. The residual stress and strain state is not shown here but can be observed in References [26]. The system is then loaded.

To predict the point of failure in this system caused by crack nucleation and growth, a phenomenological classical damage mechanics approach was used. The damage relationship/model, which is a modification of Lemaitre's [27] model by Tai and Yang (Reference [28]) is written as:

$$D = \int_0^{\varepsilon_p} (\sigma^* \, d\varepsilon_p)/\varepsilon_o^p$$

where,

$$\sigma^* = 2/3(1+v) + 3(1-2v)(\sigma_m/\sigma))$$

and v is Poisson's ratio, σ_m is 1/3 the trace of the stress tensor, and ε_p is the classical Von Mises equivalent stress. The only material constant needed is the uniaxial plastic strain at failure; ε_o^p, at each temperature (from 815 C to room temperature since damage is accumulated within the fabrication process as well). $D = 0$ represents no damage, and $D = 1$ represents local failure. D is integrated throughout the strain history, including processing. This parameter predicts failure rapidly at regions of high constraint (i.e., high σ_m), and predicts failure conditions much slower at conditions of low constraint. This type of classical local approach appears to perform quite well in composite materials where a dominant crack is not present. However, as discussed in the introduction, for monolithic materials with a dominant crack present, performance is poor because damage localizes at the crack tip and failure predictions depend on the mesh size near the crack tip.

The unit cell model, illustrated in the lower left of Figure 5, represents a ¼ unit cell within the composite. Periodic boundary conditions and symmetry conditions are imposed, as appropriate (Figure 5). The fiber/matrix interface is modeled as a contact/friction surface. As illustrated, this permits fiber matrix debonding to occur – something clearly observed in the experiments.

The damage development up to impending failure is observed at the bottom left of Figure 5. The 'black' is a prediction of material failure location and the prediction indicates failure at an applied stress of about 403 Mpa. The experimental failure development via the development of slip bands is seen at the lower right of Figure 5. Failure is observed at an applied stress of 414 Mpa. Note from Figure 5 that the failure predictions (both location and magnitude) compare very well with the experimental observations. Similar analyses neglecting the fabrication history produce very poor comparisons. Moreover, the predicted stress versus strain response is very poor without including fabrication history. More details of the importance of properly including the prior fabrication history in the models for other loading conditions (including creep and creep damage development) are illustrated in References [26,29]. It is noted that in practice, fabrication history is usually not included in failure predictions of composite systems. This could lead to erroneous failure predictions.

Revisiting the J-Integral For Weld Fracture

A welded structure consisting of more than one ductile material undergoes a complicated local deformation pattern when a crack is located in the vicinity of the material interface. At an interface between two distinctively different materials, displacements, u_i, and stresses, σ_{ij}, generally are continuous, whereas their first derivatives can be discontinuous and the higher derivatives do not exist. The same statement can be made for weld-induced residual stresses across such a material interface. Hence, strain, ε_{ij}, and stress working density, W, are not differentiable, where

$$W = \int \sigma_{ij}\, d\varepsilon_{ij} \quad (1)$$

In such a situation, the applicability of J, extensively used for characterizing and predicting the stability of a crack in a homogeneous structure, must be examined.

Suppose a contour path, Γ, of the J integral traverses the material interface of the weld-base metal as depicted in Figure 6. Section A-A denotes the material interface and the interface local coordinate system X_a-X_b is tilted from the crack-tip local coordinate system X_1-X_2 by the angle θ. The J-integral on Γ can be decomposed into three similar integrals, i.e.:

$$J = \int_\Gamma (W n_1 - n_j \sigma_{ij} u_{i,1}) = J_{\Gamma_1} + J_{\Gamma_2} + J_{\Gamma_3}. \quad (2)$$

In Equation (2) Γ_1, Γ_2, and Γ_3 are the paths illustrated in Figure 6, $u_{i,1}$ denotes the partial derivative of u_i with respect to X_1. If there are no stress or strain singularities, and no defects nor discontinuities within Γ_2, then J_{Γ_2} is evidently zero under the condition where deformation theory prevails.

Now, applying Gauss' theorem to the closed path, J_{Γ_3}

$$J_{\Gamma_3} = \int_\Gamma (W n_1 - n_j \sigma_{ij} u_{i,1})$$

$$J_{\Gamma_3} = \iint \left[\left(\frac{\partial W}{\partial X_a} - \sigma_{ij}\frac{\partial \varepsilon_{ij}}{\partial X_a}\right)\cos(\theta) + \left(\frac{\partial W}{\partial X_b} - \sigma_{ij}\frac{\partial \varepsilon_{ij}}{\partial X_b}\right)\sin(\theta) \right] dV$$

$$(3)$$

where n_j is the X_j-component of the outer normal vector to the contour Γ_3 enclosing the area V_3. In deriving Equation (3), the linear and the angular momentum balance equations and the small strain relationship ($\varepsilon_{ij} = 0.5\,(u_{i,j} + u_{j,i})$) is used. Equation (3) holds if W and $u_{i,1}$ are continuously differentiable everywhere in V_3. Note that the values for $\partial W/\partial X_a$ and $\partial \varepsilon_{ij}/\partial X_a$ exist in the domain (note that X_a parallel to the weld interface); however, the values for $\partial W/\partial X_b$ and $\partial \varepsilon_{ij}/\partial X_b$ (perpendicular to the interface) do not exist at the material interface. This means that, for the general case, J_{Γ_3} is nonzero. Therefore, the contour J-integral does not furnish path independence for a multimaterial body even under the assumption of the deformation theory of plasticity, if a contour is considered to traverse the material interface. Similar concepts have been shown by Smelser and Gurtin [30], Kikuchi and Miyamoto [31]. Nakagaki, et al [32] discuss these concepts in more detail with examples.

In the special case when the material interface A-A is parallel to the crack plane, i.e., $\theta = 0$, Equation (3) holds in the absence of those non-differentiable terms in the integrand. Furthermore, if W is a single valued function of ε_{ij}, such as occurs with proportional loading, it is clear that the right hand side of Equation (3) will be zero through

$$\frac{\partial W}{\partial X_a} = \sigma_{ij}\frac{\partial \varepsilon_{ij}}{\partial X_a} \quad (4)$$

$$J_\Gamma = J_{\Gamma_1}. \quad (5)$$

i.e., the path independence of the J-integral across the material interface is maintained provided that (a) the loading is proportional throughout the body, and (b) the material interface is parallel to the crack plane. Condition (b) is satisfied in the analysis of the test geometry's considered in the present work. Condition (a) is satisfied under a monotonic loading up to crack initiation. If one can assume proportional loading, then the J-contour integral, as well as simple estimation formulae to determine J from applied load-displacement records can be used for the analysis of experiments discussed in [32].

SIGNIFICANCE OF THE J-INTEGRAL FOR WELD FRACTURE PROBLEMS
General Path Independence.

Many investigators over the years have attempted to find forms of the J-Integral that remain path independent for material interface problems. As shown above, J of equation (2) is indeed path independent when the crack is parallel to the material interface. References [30] and [32-35] illustrate this with a number of example problems. This is important since many cracks that develop in service are nearly perpendicular to the weld interface. Moreover, as illustrated in References [32, 33], J was found to be nearly path independent for many cases where the crack was not perpendicular to the interface even though the amount of plasticity across the interface was markedly different. The reasons for this are not entirely clear based on the arguments presented above via Equations (1-5).

Ma [36, 37] presented a form for the J-Integral that is path independent for general cracks at arbitrary angles to the material interface as in Figure 6. This was done by performing an analysis analogous to that above, and converting the integral along the interface (i.e. the contribution along $J_{\Gamma 3}$) into an additional line integral contribution along the interface. It appears that this new integral is not convenient since the mesh refinement required along the interface permits one to calculate J along $J_{\Gamma 1}$ directly instead of using Ma's form. In other words there is no advantage in using Ma's form for J since one can easily calculate J along $J_{\Gamma 1}$ since the mesh must be adequately refined near the crack and interface to render Ma's form to be accurate.

In the next section we discuss the effect that fabrication history has on the path independence of J-Integral and its modifications. It is shown that the search for path independence of J type parameters for fabrication history dependent (such as welds) fracture problems such as those done in References [36, 37] is probably fruitless. This is because the residual stresses and strains induced from the fabrication history renders all integrals path dependent in the strict sense (path independence can be shown via the divergence theorem; however this leads to a volume term in the integral definition along with the line integral term.

Weld Residual Stress/Strain and History Dependence

Under arbitrary load histories, where global or local stress unloading may occur, such as under cyclic loading or during large crack growth in ductile materials, the stress and the stress work are multi-valued functions of the strain, and hence dependent on the stress/strain history. As is well known, J becomes path dependent in this regime since it is outside the assumption of deformation theory plasticity. This path dependence occurs with or without the presence of a weld.

As illustrated in Figures 1 and 2, this history is often introduced before service via the fabrication process. Consider Figure 7, which is a blowup of the bottom portion of Figure 2. After bending and welding, a crack is introduced.

There are two paths illustrated in Figure 7a, Γ_A and Γ_B both of which cross the weld fusion line interface. After the crack is introduced, these residuals stress and strain patterns redistributes (typical redistribution can be seen in References [6, 9, 11, 38]). In Figure 7 the crack is drawn so that it intersects the interface at an angle other than zero, i.e., the crack in not parallel to the interface. As such, the value of J calculated along either path is expected to be different. In fact, J is expected to be path dependent even for monotonic loading, stationary crack and no initial fabrication history as summarized above. When there is prior history, for example from bending and welding as illustrated in Figure 7, J is never path independent. Considering Figure 7b, where both paths remain in the cracked material, J would be path independent for all paths that remain in the base metal if prior fabrication history was neglected. However, because of prior history (bending and welding in this case) the J-integral for both paths is not path independent, and the meaning of J as a fracture parameter is obscured.

Ductile Materials

For low toughness materials and moderately tough materials, the effect of residual stresses and strains caused by fabrication history on the fracture response is very important (see References [5-12, 39]). For low toughness materials, where the crack tip plastic zone is small, the driving force value of K is strongly affected by residual history. For moderately tough materials, where K is not valid and J is no longer path independent and is thus invalid, predicting fracture is quite difficult since current fracture methods are not entirely applicable. This will be discussed in more detail later in the context of emerging fracture parameters. However, prior history has a very important effect on the fatigue and stress corrosion crack growth response of all materials (from brittle to ductile).

For tough materials the J-integral approaches path independence at high service loads. Reference [9] provided a careful study of the effects of weld residual stresses on fracture response of a tough 304 stainless steel pipe welds. After welding, residual stresses were first modeled and developed of the form illustrated at the top of Figure 2. A crack was then introduced into the model. The value of J evaluated along numerous paths was plotted as a function of applied load. At low loads, the effect of the weld residual stresses and strains were very important and J varied significantly from path to path. At high loading, approaching the net section plastic collapse loads, J begins to approach path independence. *Moreover, the value of J that was calculated by neglecting the fabrication history was not much different from that calculated when including the history (Reference [9]) at high applied loads.* The apparent reason for this is that, at high loads in tough materials, the plastic strains that result from loading begin to dominate the strains that resulted from the fabrication history. As such, J can be used as a fracture parameter, at least to predict initiation loads.

However, the J_R (or material resistance) curve is affected by the history and must be obtained from specimens with the crack in the weld or heat affected zone. Indeed, References [14-16- 40-42] clearly show J can be used to predict fracture of tough ductile pipe welds by using a resistance curve that includes a weld and a J estimation scheme that adjusts for the effect of the weld (without including residual stresses). The above comments regarding tough materials and the applicability of J appears valid for cases tested to date. However, more work is necessary to generalize these comments for all materials and all crack geometry's.

NEW FRACTURE PARAMETERS

Local or Damage methods, summarized in the recent book by Lemaitre and Chaboche [27], attempt to take into account the microscopic breaking processes such as void nucleation, microcracking, cavitation, etc., while staying within the framework of continuum mechanics. The models range from simple phenomenological approaches where the necessary parameters are fit based on what 'works' to complicated micro-based approaches which require extensive micrographic work to develop the necessary parameters. The 'Gurson' type models may be considered to be intermediate to these two extremes. While these approaches show promise, they still must overcome one critical obstacle: predictions depend on the mesh size chosen if damage at a dominant crack is attempted. This is because the damage parameters depend on integrals of stress and strain type quantities at the all material points which accumulate damage. Since, at a crack tip these quantities are mesh size dependent, the accumulated damage, and hence, fracture predictions, depend on mesh refinement. Hence, one must define an element size at the crack tip that is 'fit' based on calculations and this size must be used for all future analyses. Otherwise, the finer the mesh, the more rapid the crack growth is predicted. For crack nucleation predictions this is not a problem as the example shown above in Figure 5 illustrates. While damage based methods are used to predict fracture of weldments (see Reference [39], which successfully used a 'Gurson' type model), the mesh size requirements continue to be a hurdle.

Integral or Energetic methods consist of evaluating an integral of stress and strain type quantities which encompasses the growing crack tip. For elastic-plastic, creep, or dynamic fracture a resistance curve is developed by analyzing a laboratory specimen, analogous to classical J-Tearing procedures. This resistance curve is assumed to be an intrinsic material property. For high cycle fatigue, the range of the integral parameter is fit to a simple equation, completely analogous to classical fatigue crack growth analyses that are routinely performed based on ΔK. With the emergence of advanced computational methods for fracture based on inelastic finite element alternating methods, practical fracture analyses based on these integral parameters are now possible (see Wang et al., References [44-46]). Hence, the historical stigma regarding these integral methods that they are 'nice but not practical for real structures' are no longer valid.

To use an integral parameter to characterize fracture, a material resistance curve is developed by performing an experiment on a laboratory fracture specimen. The experiment is then modeled via the finite element method where the integral parameter(s) of interest are calculated along a small finite-sized path. This numerical experiment then produces the material resistance curve(s). The behavior of other arbitrarily loaded and cracked structures can then be predicted by modeling the time history of loading and using this generated resistance curve as a crack growth criterion. The resistance curve is assumed to be an intrinsic material property. The time history of crack initiation and growth may thus be predicted. The analogy to elastic-plastic fracture mechanics based upon the J-integral may be made.

Integral parameters have shown promise in a number of non-linear applications in recent years and applications are rising rapidly. A number of crack tip parameters expressed in integral forms have appeared in the literature. These include Blackburn [47] (J_B); Kishimoto, Aoki, and Sakata [48] $\hat{J}$; McClintock [49] (J_M); Watanabe [50] (J_W); Goldman and Hutchinson [51] (C^*); Brust and Atluri [52], and Cherepanov [53] (T^*); and others.

The physical interpretation of many of these integral parameters is not entirely clear except for the case of T^*. The physical interpretation of the T^* integral for a growing crack is that of the energy release rate to a *finite*-sized material volume in the vicinity of the crack tip. The mathematical details of this interpretation are completely overviewed in the recent book by Atluri [54] (see also Reference [55]) and numerous references cited therein. Applications of T^* to problems of cyclic creep fracture, multiple site damage in aging aircraft, history dependent fracture in elastic plastic fracture, and others applications are presented in Reference [55] and references cited therein. A recent application of T^* to dynamic fracture is illustrated in Reference [56]. Applications of T^* to characterize weld fracture, where residual stresses are accounted for, is provided in Reference [57]. Reference [57] also provides the results of a recent large study in Japan to study elastic-plastic weld fracture. Due to paper length restrictions, an example is not shown here.

The T^*-Integral is beginning to be used to characterize fracture more and more in situations where classical methods break down. A number of new applications of T^* are being reported in Europe, Russia, Japan, and the US, some of which are discussed in References [5, 56]. As such, it appears reasonable to examine the performance of T^* in situations where fabrication history has an important effect on the fracture response of structures. This work is ongoing and References [58, 59] will provide further studies of history dependent fracture and characterizing weld fracture using T^* and the new parameter proposed by Pan discussed in the next subsection.

New Path Independent Fracture Parameter for Weld Fracture. Pan et al in Reference [60, 61] propose a new fracture parameter for characterizing weld fracture. In [60], the parameter is referred to as J_{1d}, and represents the energy difference of two cracked plates with slightly different crack sizes with consideration of residual stresses. The parameter is calculated along different paths and is somewhat analogous to the stiffness derivative approach of Hellen [62] and Parks [63]. Examples are illustrated in References [60, 61] which show that J_{1d} is path independent when calculated along numerous paths when weld residual stresses are accounted for in the analysis. The physical interpretation of J_{1d} is not clear at present and it has not been used as yet to characterize weld fracture. However, the response of J_{1d} appears to correspond to that of the crack tip opening angle. This parameter is studied in References [58, 59] as well.

SUMMARY OF HISTORY DEPENDENCE

Here we have summarized the problems associated with characterizing fracture in structural systems that are affected by fabrication history. Classical methods were overviewed and their ranges of applicability were discussed. New emerging fracture parameters were reviewed and the potential for characterizing history dependent fracture, especially in weld fracture, were overviewed. However, while these new parameters show promise for adequately characterizing weld fracture, much more verification is required.

In our view, the question of fabrication history dependent fracture, including weld fracture where residual stresses and strains are important, is very much open. New fracture parameters that perform and are verified in this regime are sorely needed.

It is clear that the current fracture procedures are inadequate and no magic parameter has been clearly shown to be generally applicable for fabrication history dependent fracture. In addition, from Figures 1 and 2, the effect of prior fabrication history on field failure response cannot be neglected. The current procedure of assuming a pristine cracked structure when doing a fracture assessment of a weld crack, for instance, is not adequate. Ongoing work, to appear in References [58, 59] will address these issues as well as further study the appropriateness of the emerging fracture parameters overviewed in the last section.

REFERENCES

1. F. W. Brust, J. Zhang, P. Dong, *Pipe and Pressure Vessel Cracking: The Role of Weld Induced Residual Stresses and Creep Damage during Repair.* Transactions of the 14th International Conference on Structural Mechanics in Reactor Technology (SMiRT 14), Lyon, France, Vol. 1, pp. 297-306 (1997).

2. Dong, P., Hong, J. K., and Tsai, C.L., "Finite Element Simulation of Residual Stresses in Multi-Pass Welds", Proceedings of International Conference on Modeling and Control of Joining Processes, December 8-10, 1993, Orlando, Florida.

3. Dong, P., et al, *Effects of Weld Residual Stresses on Crack-Opening Area Analysis of Pipes for LBB Applications*, Proceedings of LBB-95 International conference, Lyon, France, October, 1995, pp. 1171-1195.

4. F. W. Brust, P. Dong, J. Zhang, *A Constitutive Model for Welding Process Simulation using Finite Element Methods.* Advances in Computational Engineering Science, S. N. Atluri and G. Yagawa, Eds., pp. 51-56 (1997).

5. F. W. Brust, P. Dong, J. Zhang, *Pipe and Pressure Vessel Cracking: The Role of Weld Induced Residual Stresses during Fabrication and Repair.* Advances in Computational Engineering Science, S. N. Atluri and G. Yagawa, Eds., pp. 177-182 (1997).

6. F. W. Brust, P. Dong, J. Zhang, *Crack Growth Behavior in Residual Stress Fields of a Core Shroud Girth Weld.* Fracture and Fatigue, H. S. Mehta, Ed., PVP-Vol. 350, pp. 391-406 (1997).

7. J. Zhang, P. Dong, F. W. Brust, *Analysis of Residual Stresses in a Girth Weld of a BWR Core Shroud.* Approximate Methods in the Design and Analysis of Pressure Vessels and Piping Components, W. J. Bees, Ed., PVP-Vol. 347, pp. 141-156 (1997).

8. F. W. Brust, P. Dong, J. Zhang, *Influence of Residual Stresses and Weld Repairs on Pipe Fracture.* Approximate Methods in the Design and Analysis of Pressure Vessels and Piping Components, W. J. Bees, Ed., PVP-Vol. 347, pp. 173-191 (1997).

9. C. Hou, M. Kim, F. W. Brust and J. Pan, "Effects of Residual Stresses on Fracture of Welded Pipes," in Residual Stresses in Design, Fabrication, Assessment and Repair, ASME PVP-Vol. 327, ASME PVP Conference, Montreal, Quebec, Canada, July 21-26, 1996, pp. 67-75.

10. I. S. Abou-Sayed, J. Ahmad, F. W. Brust, and M. F. Kanninen, "An Elastic-Plastic Fracture Mechanics Prediction of Stress Corrosion Cracking in a Girth Welded Pipe," in Fracture Mechanics: 14th Symposium ASTM-STP 791, (1983).

11a. M. F. Kanninen, F. W. Brust, J. Ahmad, and V. Papaspyropoulos, "An Elastic-Plastic Fracture Mechanics Prediction of Fatigue Crack Growth in the Heat Affect Zone of a Butt Welded Plate," in <u>Fracture Tolerance Evaluation</u>, Edited by P. Kanazawa, A. S. Kobya, and K. Iida, published by Poyo Press, Tokyo, 1982, pp. 113-120. Also;

11b. M. F. Kanninen, F. W. Brust, J. Ahmad, and I. Abou-Sayed, "The Numerical Simulation of Crack Growth in Weld Induced Residual Stress Fields," <u>Residual Stress and Stress Relaxation</u>, E. Kula and V. Weiss, Editors, Proceedings of 28th Army Sagamore Research Conference, Lake Placid, New York, July 13-17, 1981, Plenum Press, 1982.

12. Zhang, J., Dong, P., and Brust, F. W., "Residual Stress Analysis and Fracture Assesment of Welded Joints in Moment Resistant Frames", in Modeling and Simulation Based Engineering, Ed. S. N. Atluri and P. E. O'Donoghue, Tech Science Press, pp 1894-1902, 1998.

13. G. M. Wilkowski, et al., "Analysis of Low-Energy Test Results of Degraded Piping," in <u>Nuclear Engineering and Design</u>, 89, 1985, pp. 257-269.

14. G. M. Wilkowski, et al., "Analysis of Experiments on Stainless Steel Flux Welds," NUREG/CR-4878, NRC Topical Report, November, 1988.

15. G. M. Wilkowski, et al., "Degraded Piping Program - Phase II Progress," <u>Nuclear Engineering and Design</u>, 98 (1987), pp. 195-217.

16. G. M. Wilkowski, et al, "Short Cracks in Piping and Piping Welds – Final Report, NUREG/CR-4599, Vol. 4, No. 2, 1993.

17. Milne, I., Ainsworth, R. A., Dowling, A. R., and Stewart, A. T. "Background To And Validation of CEGB Report R/H/R6 - Revision 3," January 1987.

18. Milne, I, Ainsworth, R. A., Dowling, A. R., and Stewart, A. T. "Assessment of the Integrity of Structures Containing Defects, CEGB Report R/H/R6 - Revision 3," 1986.

19. Brust, F. W., et al, "Analysis of Pipe With Short Through-wall Cracks", NUREG-CR-6235, April, 1995.

20. Krishnaswamy, P. K., et al., "Fracture Behavior of Short Circumferentially Surface-Cracked Pipe", NUREG/CR-6298, November, 1995.

21. Rahman, S., Ghadiali, N., Paul, D., and Wilkowski, G. M., "Probabilistic Pipe Fracture Evaluations for Leak-Rate-Detection", NUREG/CR-6004, April, 1995.

22. Gates, W. E.; and Morden M. 1995. "Lessons from Inspection, Evaluation, Repair and Construction of Welded Steel Moment Frames Following the Northridge Earthquake", <u>Technical Report: Surveys and Assessment of Damage to Buildings Affected by the Northridge Earthquake of January 17, 1994.</u> Report No. SAC-95-06: 3-1 to 3-79.

23. Naeim F.; DiJulio, R. Jr.; Benuska K.; Reinhorm, A. M.; and Li C. 1995. "Evaluation of Seismic Performance of an 11-Story Steel Moment Frame Building during the 1994 Northridge Earthquake", <u>Technical Report: Analytical and Field Investigations of Buildings Affected by the Northridge Earthquake of January 17, 1994.</u> Report No. SAC-95-04, Part 2: 6-1 to 6-109.

24. Majumdar, B. S., and Newaz, G. M., Inelastic Deformation in Metal Matrix Composites", Philosophical Magazine, Vol. 66, No. 2, pp. 187-2012, London, 1992.

25. G. Newaz, B. Majumdar, and F. W. Brust, "Thermo-Cycling Response of Quasi-Isotropic Metal Matrix Composites,", <u>Journal of Enginering Materials and Technology</u>,Vol. 114, pp 156-161, April, 1992.

26. Brust, F. W., Majumdar, B. S., and Newaz, G. M., "Analysis of Damage and Failure in Metal Matrix Composites", Fracture Mechanics: Twenty-Sixth Symposium, ASTM STP 1256, W. G. Reuter, J. H. Underwood, and J. C. Newman, Jr., Eds., 1995.

27. Lemaitre, J. and Chaboche, J. L., "Mechanics of Solids Materials," Cambridge University Press, 1990.

28. Tai, W. H., and Yang, B. X., "A New Damage Mechanics Criterion for Ductile Fracture", Engineering Fracture Mechanics, Vol. 27, No. 4, pp. 371-378, 1987.

29. Brust, F. W., and Newaz, G. M., "A Study of Tension, Compression, and Creep Deformation of Metal Matrix Composites", Proc. Of the American Society for Composites, 8th Technical Conference, October, 1993.

30. Smelser, R. E., and Gurtin, M. e., "On the J-Integral for Bi-Material Bodies", International Journal of Fracture, Vol. 13, pp 382-384, 1977.

31. Kikuchi, M., and Miyamoto, H., "Evaluation of J_k Integrals for a Crack in Multi-Phase Materials", Recent Research on Mechanical Behavior of Solids, Vol. 1, pp. 74-92, 1982.

32. Nakagaki, M., C. W. Marschall, and F. W. Brust, "Elastic-Plastic Fracture Mechanics Evaluations of Stainless Steel TIG Welds," <u>ASTM STP, 995</u>, Landes, Saxena, and Merkle, Editors, pp. 214-243, 1989.

33. Nakagaki, M, Marschall, C. W., and Brust, F. W., "Analysis of Cracks in Stainless Steel TIG Welds", NuREG/CR-4806, December, 1996.

34. Wilkowski, G. M., et al, "Analysis if Experiments on Stainless Steel Flux Welds", NUREG/CR-4878, April, 1987.

35. Rosenfield, A. R., Held, P. R., and Wilkowski, G. M., "Stainless Steel Submerged Arc Weld Fusion Line Toughness", NUREG/CR-6251, April, 1995.

36. Ma, W., Xu, X., Tian, X, and Goldak, J. A., "Influence of the Material Inhomogeneity of Overmatched Welded Joints on the Crack Driving force", Proc. International Conference on Fracture, Houston, TX, 1991.

37. Ma, W. et al., "Application of the J-Integral to elastoplastic Fracture Study of Non-homogeneous Welded Joints", Chinese Journal of Mechanical Engineering, Vol. 23, pp 43-50, 1987.

38. J. Zhang, P. Dong, F. W. Brust, W. J. Shack, M. Mayfield, M. McNeil, *Modeling of Residual Stresses in Core Shroud Structures*. To appear in International Journal for Nuclear Engineering and Design (1999).

39. Panontin, T. L., Nishioka, O., and Hill, M. R., "Fracture Assessments of Welded Structures", in Modeling and Simulation Based Engineering, Ed. S. N. Atluri and P. E. O'Donoghue, Tech Science Press, pp 1144-1149, 1998.

40. Rahman, S., Brust, F., Nakagaki, M., and Gilles, P., "An Approximate Method for Estimating Energy Release Rates of Through-Wall Cracked Pipe Weldments," Proceedings of the 1991 ASME Pressure Vessels and Piping Conference, Vol. 215, San Diego, California, 1991.

41. Rahman, S. and Brust, F., "An Estimation Method for Evaluating Energy Release Rates of Circumferential Through-Wall Cracked Pipe Welds," Engineering Fracture Mechanics, Vol. 43, No. 3, pp. 417-430, 1992.

42. Rahrnan, S. and Brust, F., "Elastic-Plastic Fractme of Circumferential Through-Wall Cracked Pipe Welds Subject to Bending," Journal of Pressure Vessel Technology, Vol. 114, No. 4, pp. 410-416, November 1992.

43. Lemaitre, J. and Chaboche, J. L., "Mechanics of Solids Materials," Cambridge University Press, 1990.

44. L. Wang, F. W. Brust, S. N. Atluri, "The Elastic-Plastic Finite Element Alternating Method (EPFEAM) and the prediction of fracture under WFD conditions in aircraft structure; Part II: Fracture and the T^*-Integral Parameter", Computational Mechanics, 19(5):370-379, 4/1997.

45. L. Wang, F. W. Brust, S. N. Atluri, "The Elastic-Plastic Finite Element Alternating Method (EPFEAM) and the prediction of fracture under WFD conditions in aircraft structure; Part III: Computational predictions of the NIST multiple site damage experimental results", Computational Mechanics, 20(3):199-212, 8/1997.

46. L. Wang, F. W. Brust, S. N. Atluri, "The Elastic-Plastic Finite Element Alternating Method (EPFEAM) and the prediction of fracture under WFD conditions in aircraft structure; Part I: EPFEAM Theory", Computational Mechanics, 19(5):356-369, 4/1997.

47. Blackburn, W. S., "Path Independent Integrals to Predict Onset of Crack Instability in an Elastic Plastic Material", *Intl J of Fracture Mechanics*, **8**, p. 343-346, 1972.

48. Kishimoto, K., Aoki, S., and Sakata, M., "On the Path Independent Integral-J", *Engineering Fracture Mechanics*, **13**, p. 841-850, 1980.

49. McClintock, F. A., in *Fracture 3*, Ed. H. Liebowitz, Academic Press, 1971.

50. Watanabe, K., "The Conservation Law Related to Path Independent Integral and Expression of Crack Energy Density by Path Independent Integral", *Bul of JSME*, **28** (235), January 1985.

51. Goldman, N. L. and Hutchinson, J. W., "International Journal of Solids and Structures" Vol. 11, No. 5, pp 575-592, 1975.

52. Brust, F. W., and Atluri, S. N., "Studies on Creep Crack Growth Using the T^* Integral", *Engineering Fracture Mechanics*, **23** (3), p. 551-574, 1986.

53. Cherepanov, G. P., "A Remark on the Dynamic Invariant or Path-Independent Integral", *Intl J Solids and Structures*, **25** (11), p. 1267-1269, 1989.

54. Atluri, S. N., "Structural Integrity and Durability", Tech Science Press, 1997.

55. Brust, F. W., "The T*-Integral: Definition and Use for predicting Damage Accumulation and Fracture", invited presentation at the *A . C. Eringen Medal of the Society of Engineering Science Award meeting* at the 32nd Annual Meeting (S. N. Atluri is Recipient), New Orleans, La, Oct. 1995. In *book Contemporary Research in Engineering Science*, ed. R. C. Batra, pp. 118-140, Springer, November, 1995.

56. Beissel, S. R., Johnson, G. R., and Popelar, C. H., "An Element Failure Algorithm for Dynamic Crack Propagation in General Directions", Engineering Fracture Mechanics, Vol. 61, pp. 407-425, 1998.

57. Yagawa, G., et al., "Study on Elastic-Plastic Fracture Mechanics in Inhomogeneous Materials and Structures", Prepared by the Nuclear Engineering Research Committee of the Japan Welding Engineering Society, March, 1994.

58. Brust, F. W., Dong, P., and Zhang, J., "Fracture Analysis of Welds Including Weld Fabrication Effects", Under preparation, to be submitted 1999.

59. Brust, F. W., Dong, P., and Zhang, J., "History Effects In Fracture and the T*-Integral", Under preparation, to be submitted 1999.

60. Y.-C. Hou, J. Pan, "A Finite Element Procedure to Determine a Fracture Parameter for Welded Structures with Residual Stresses", Fracture and Fatigue – 1997-, Vol. 2, PVP-VOL. 346, ASME, New York, pp. 309-315, 1997.

61. Y.-C. Hou, J. Pan and F. W. Brust, "A Fracture Analysis of Welded Pipes with Consideration of Residual Stresses," in PVP-Vol. 373, Fatigue, Fracture, and Residual Stresses -1998-, ASME PVP Conference, San Diego, July 26-30, 1998, pp. 433-437.

62. Hellen, T. K., "On the Method of Virtual Crack Extensions", International Journal for Numerical Methods in Engineering, Vol. 9, pp. 182-207, 1975.

63. Parks, D. M., "A Stiffness Derivative Finite Element Technique for Determination of Crack Tip Stress Intensity Factors", International Journal of Fracture, Vol. 10, pp. 487-502, 1974.

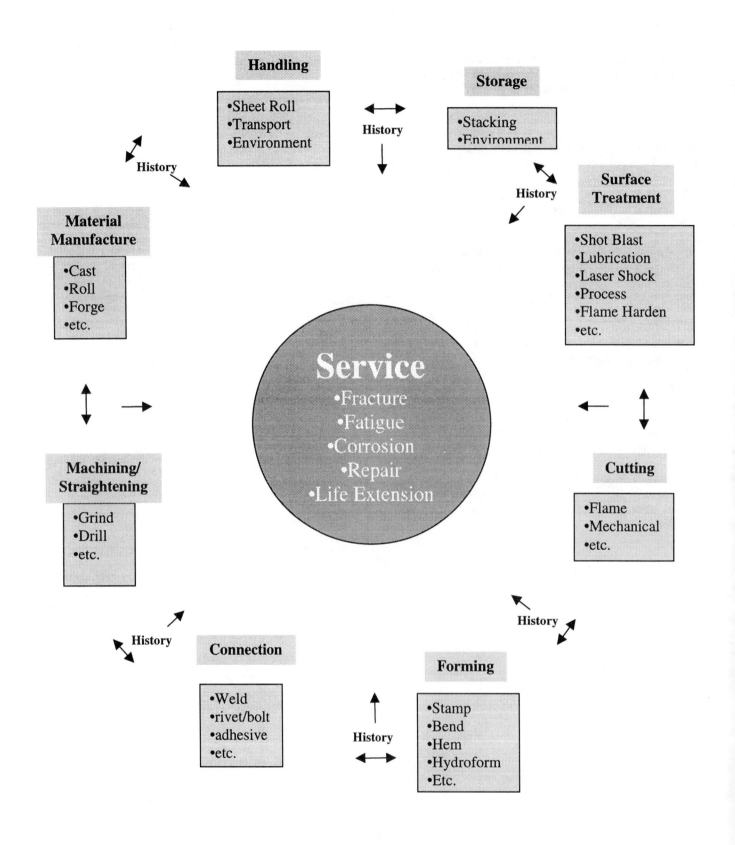

Figure 1. Operations History Leading to Structural Fabrication.

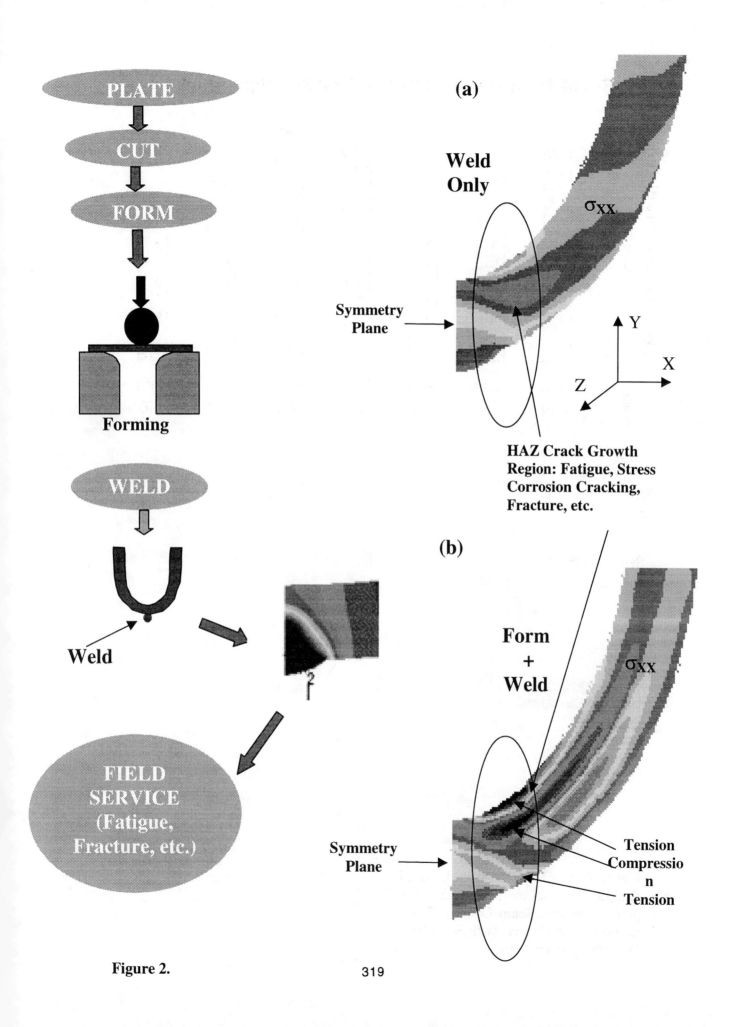

Figure 2.

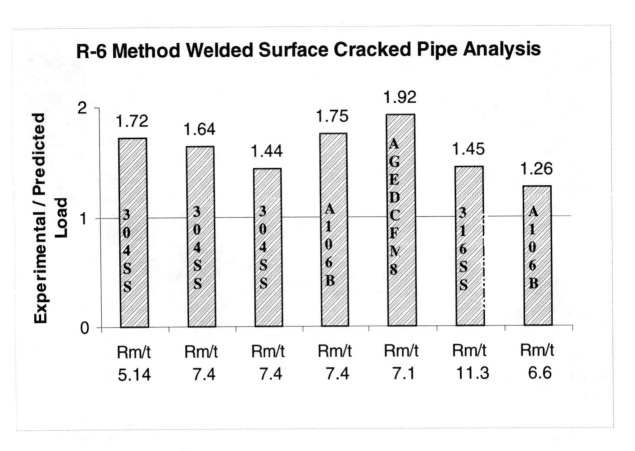

Figure 3.

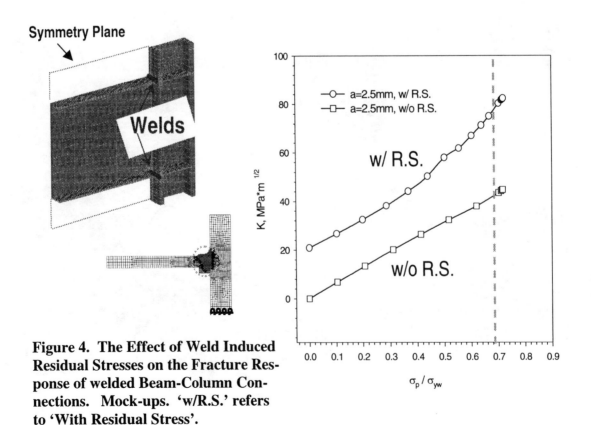

Figure 4. The Effect of Weld Induced Residual Stresses on the Fracture Response of welded Beam-Column Connections. Mock-ups. 'w/R.S.' refers to 'With Residual Stress'.

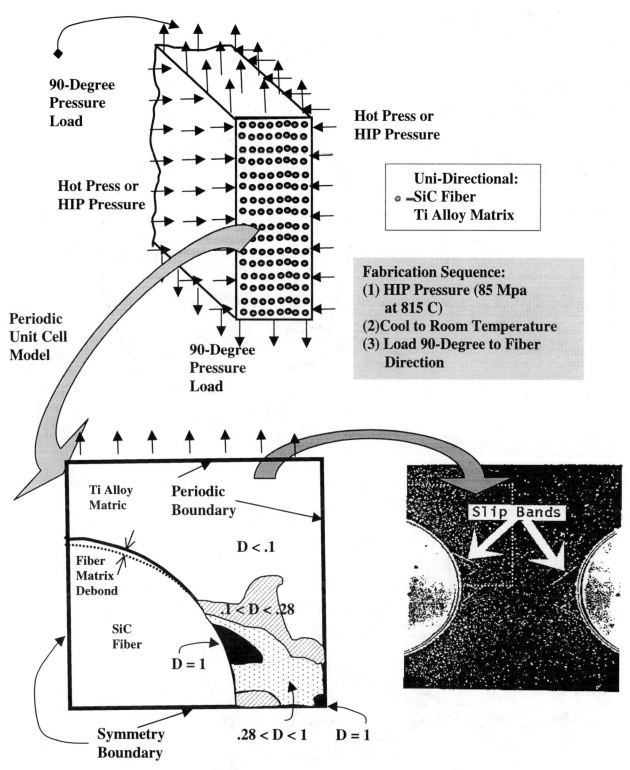

Figure 5. Metal Matrix Composite Damage Evolution Illustrating the Importance Of Fabrication History On Failure. D = 1 Represents Failure.

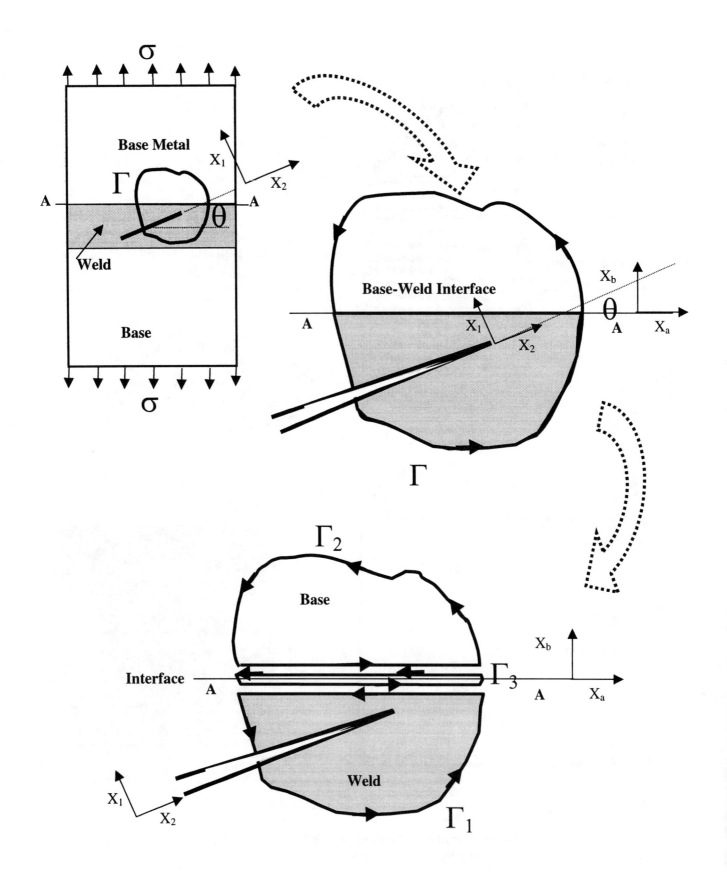

Figure 6. Crack Definition and Contour Path For Crack In Weld.

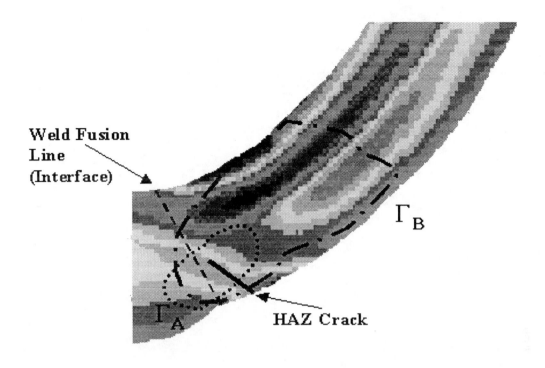

Figure 7a. Introduction of Crack in Bent and Welded Plate. Integral Paths Both Cross the Interface.

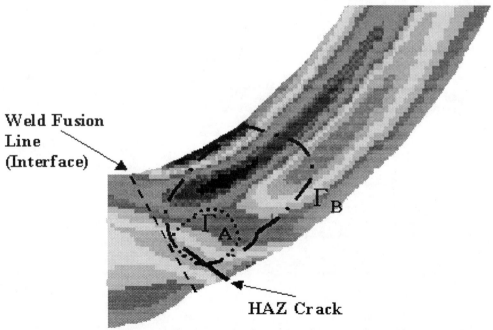

Figure 7b. Integral Paths All Within 'Cracked' Material.

MICRO- AND MACROSCOPIC FRACTURE BEHAVIOR OF HEAT-AFFECTED ZONE IN MULTI-PASS WELDED CRYOGENIC STEEL

Jae-il Jang
School of Materials Science and Engineering,
Seoul National University,
Seoul 151-742, Korea

Young-chul Yang
Research and Development Center,
Korea Gas Corporation,
Ansan 425-150, Korea

Woo-sik Kim
Research and Development Center,
Korea Gas Corporation,
Ansan 425-150, Korea

Dongil Kwon
School of Materials Science and Engineering,
Seoul National University,
Seoul 151-742, Korea

ABSTRACT

Evaluation of soundness of welded structures by mechanical testing for failure in the weld heat-affected zones (HAZs) has become general practice throughout the world. HAZs of steel welded joints show a gradient of microstructure from fusion line to the unaffected base metal. This study is concerned with a correlation between the microstructural change and the fracture characteristics in HAZ of multi-pass welded QLT (quenching, lamellarizing & tempering)-processed 9% Ni steel which is used for LNG (liquefied natural gas) storage tank in Korea due to the advanced cryogenic fracture toughness. The focus in present study was the investigation of micro- and macroscopic fracture behaviors of the various micro-zones within HAZ from viewpoint of the effects of microstructure. The changes in microstructures and impact toughness were observed using synthetic HAZ specimens. Also the microscopic fracture mechanism and apparent fracture toughness at local region were identified by *in situ* scanning electron microscopy (SEM) observations using miniaturized SENT (single edge notched tension) specimens. As results from the tests, it was found that the inter-critically reheated coarse grained HAZ (IC CGHAZ) was local brittle zone (LBZ) of this steel HAZ only at low temperature. Finally, these cryogenic LBZ phenomena were analyzed and discussed.

INTRODUCTION

Due to the fact that LNG (liquefied natural gas) is stored at, or below, its 111K boiling temperature, the inner walls of LNG storage tanks must be constructed with a material which possesses high strength and suitable fracture toughness at cryogenic temperature. Therefore, 9% Ni steel is widely used around the world as inner walls of LNG storage tanks because of its excellent fracture toughness at cryogenic temperature. For LNG storage tank in Korea, QLT (quenching, larmellarizing and tempering) heat-treated 9% Ni steel is generally used unlike the QT (quenching and tempering), NNT (double normalizing and tempering) and DQT (direct quenching and tempering) treated steel used in other countries. The QLT process, originally developed for lower Ni steel such as 5.5% Ni steel, includes lamellarizing (quenching from the temperature between A_{C1} and A_{C3}) treatment between quenching and tempering. This process enhances cryogenic toughness considerably because of the larger amount of retained austenite and the refinement of effective grain size. For the construction of LNG storage tanks, the QLT-treated steel is welded by SMAW (shielded metal arc welding) and SAW (submerged arc welding) processes with 70% Ni superalloys (Inconel type or Hastelloy type) as welding consumables.

It is well known that welding can seriously alter the metallurgical and mechanical properties of materials, and generally cause degradations of the properties. Therefore, the variation of the cryogenic fracture toughness within the weld HAZ (heat-affected zone) of this steel must be properly evaluated for the safety performance of LNG storage tanks. In authors' previous studies (Jang et al., a 1997, b 1998), the fracture toughness of multi-pass welded QLT-9% Ni steel was estimated through the modified CTOD test, which was newly proposed for thick weldments with X-groove. Now, the present study was focused on the microscopic fracture characteristics of various micro-zones within QLT-9% Ni steel HAZ including LBZ (local brittle zone) phenomena.

The structure of the HAZ consists of several zones with significantly different microstructures. Generally there is a wide range of microstructures in the HAZ depending on the peak temperature of single pass welding thermal cycle : coarse grained HAZ (CGHAZ), fine grained HAZ (FGHAZ), inter-critical HAZ (ICHAZ), and sub-critical HAZ (SCHAZ). It is well accepted that the CGHAZ adjacent to the weld metal has the lowest toughness among them. This CGHAZ can be roughly categorized into four micro-zones again according to the peak temperature of subsequent thermal cycle in multi-pass welding procedure : unaltered CGHAZ (UA CGHAZ), the region not reheated or reheated above temperature of grain growth, super-critically reheated CGHAZ (SCR CGHAZ), the region reheated above A_{C3}, inter-critically reheated CGHAZ (IC CGHAZ), the region reheated between A_{C1} and A_{C3}, and sub-critically reheated CGHAZ (SC CGHAZ), the region reheated below A_{C1}. Among them, SCR CGHAZ is also called FGHAZ because the second thermal cycle above A_{C3} changes the coarse grained microstructure to the fine grained microstructure through recrystallization.

Although many studies, for examples by Kim et al. (1991) and Okada et al. (1994), have been conducted on microscopic fracture behaviors in these regions of structural steels HAZ, the existence and fracture behaviors of microscopically embrittled zones, termed as LBZ, in cryogenic steel HAZ have not been studied yet. So, in the present work, main attention was paid to the evaluation of fracture characteristics in the various micro-zones, including LBZ, within HAZ of multi-pass welded QLT-treated 9% Ni steel. The changes in toughness due to the microstructural variation were observed using synthetic HAZ specimens. In addition, the microscopic fracture mechanism and apparent fracture toughness at local region were identified by *in situ* scanning electron microscopy (SEM) observations using miniaturized SENT (single edge notched tension) specimens.

EXPERIMENTAL PROCEDURES

Used Material

A commercial grade of 9% Ni steel with QLT heat treatment produced by Pohang Iron & Steel Company was used in this study. Table 1 and Table 2 list chemical compositions and mechanical properties of the used material.

Weld Simulation Test

Weld simulation test was performed for the systematic evaluation of toughness variation in several micro-zones. Specimens with size of 11×11×60 mm were prepared using Fuji's metal thermal cycle simulator (MTCS).

The thermal cycle for weld simulation is generally characterized by the peak temperature and the cooling time from 1073K to 773K. After reaching the peak temperature concerning the various micro-zones, the specimens were cooled down with the cooling time of 13.5 and 19.4 seconds. The cooling times of 13.5 and 19.4 seconds are approximately equivalent to those of a SAW and a SMAW of a 22-mm-thick plate with heat inputs of 23 kJ/cm and 28 kJ/cm. These simulated welding conditions are the same as the conditions generally used for the construction of LNG storage tanks. For the calculation of the time, the equation for welding thermal cycle (Masubushi, 1980), which was modified on the basis of Rosenthal's heat-flow equation (1946), was used.

The peak temperature of the first thermal cycle simulating CGHAZ, generally known as the most embrittled region within the single-pass welded HAZ, was 1623K and the peak temperature of the second thermal cycle was varied. For some cases, the peak temperature of the first thermal cycle was varied. In addition, a dilatation test was conducted to determine the A_{C1} and A_{C3} temperatures before the weld simulation test.

Fracture Test : Impact test and *in situ* fracture test

The Charpy V-notch impact tests using the simulated specimens were performed at room temperature and 77K, liquid nitrogen temperature, according to ASTM E-23. In addition, the fracture morphology of the specimens was observed by SEM.

To investigate micro-fracture behaviors, *in situ* fracture observations of the simulated specimens were conducted using tensile loading stage within the vacuum chamber of a SEM. For the test, miniaturized SENT specimens were prepared as shown in Fig. 1. The central groove part of the specimen was made thin so that it could be easily fractured under a light load, and sharp notch with the width of about 30 μm was introduced through electric wire cutter so as to approach the fatigue pre-crack. The loading rate was 0.032 mm/sec. From the tests, the apparent fracture toughness under plane stress condition could be attained using the stress intensity factor for SENT specimen. The related equations were as follows.

$$K_C = \frac{P_{max}}{B\sqrt{W}} f(\frac{a}{W})$$

where

$$f(\frac{a}{W}) = \frac{\sqrt{2\tan\frac{\pi a}{2W}}}{\cos\frac{\pi a}{2W}} [0.752 + 2.02(\frac{a}{W}) + 0.37(1-\sin\frac{\pi a}{2W})^3].$$

Metallographic Examination of Microstructures

All samples in this investigation were prepared for metallographic examination using standard techniques. 2% Nital was used as chemical etchant for viewing under an OM (optical microscope).

X-ray diffractometry was used to determine the amount of retained austenite. The amount of retained austenite was estimated by comparing the integrated peak intensities of $(200)_\alpha$ and $(220)_\gamma$ plane, using CrK_α.

For the observation of M-A constituents, the two stage electrolytic etching method, developed by Ikawa et al. (1980), was used before SEM observation. At the first stage the aqueous solution of EDTA and NaF preferred to etch the matrix. At the second stage the aqueous solution of picric acid and NaOH preferred to etch carbide. The obtained metallography could be characterized by a light black matrix with white island-like M-A constituent and dark black particles of carbide, if they existed.

The area fraction of M-A constituent was measured by image analyzer and the carbon contents in M-A constituent were analyzed through EPMA (electron probe X-ray micro-analysis).

RESULTS AND DISCUSSION

Variation of Toughness in Simulated HAZ

The dilatation test, for determining the transformation temperature of QLT-treated 9% Ni steel, showed that A_{C1} was about 843K and A_{C3} was about 963K.

As mentioned before, the CGHAZ exhibits the lowest toughness among the various zones within the HAZ under a single thermal cycle. So, in the present study, the CGHAZ was the target region to find LBZ and the effects of the second thermal cycle on the toughness of multi-pass welded HAZ was estimated. After applying the first thermal cycle with a peak temperature of 1623K, second thermal cycles with peak temperatures between 1473K and 823K were applied. The authors supposed that the second thermal cycles with peak temperatures between 1473K and 1373K simulated UA CGHAZ while the cycles between 1273K and 1073K simulated SCR CGHAZ, and the cycles between 923K and 823K simulated IC CGHAZ, respectively.

Before the impact tests using synthetic HAZ specimens, Charpy V-notch impact tests using base metal specimens were performed. The average impact value at room temperature was 293 J and that at 77K, liquefied nitrogen temperature, was 227 J. These values were similar to the general values in Table 2.

Figure 2 shows the results from the Charpy V-notch impact tests at room temperature using specimens simulating both SMAW and SAW conditions. Although the impact values of the specimens were smaller than that of base metal, all the specimens exhibited moderate impact toughness near 200J at room temperature. Consequently, there was no LBZ at room temperature. Also, a clear tendency of change in toughness with the peak temperature of the second thermal cycle did not exist.

For the 9% Ni steel the cryogenic toughness is more important than the toughness at room temperature because LNG storage tanks are operated at 111K. Figure 3 shows the results from the Charpy V-notch impact tests at 77K using the specimens for SMAW and SAW simulations. In both cases of SMAW and SAW simulation, the impact values of the synthetic HAZ largely decreased from that of base metal. In addition, the figure shows that CGHAZ exhibits low Charpy impact energies in two cases. The first is the one simulated by a single cycle with a peak temperature of 1623K(UA CGHAZ) and the second is the one simulated by double thermal cycle with the second peak temperatures in the range of IC CGHAZ. The specimen simulating SCR CGHAZ showed the highest value. The values of the synthetic IC CGHAZ for both SMAW and SAW were lower than required minimum impact value of improved 9% Ni steel at 77K, 35 J, in the several standards like EEMUA No.147 (1986) and British Standard 7777 (1993). The study for synthetic HAZ of QT-9% Ni steel by Tamura et al. showed a similar tendency of the change in toughness. But, in their study, only the change in toughness was shown and the details of the cause were not analyzed. Because the trends of toughness variation were nearly the same on the specimens for both SMAW and SAW simulation, the SMAW case was mainly analyzed in the following study.

The fractographs of the tested specimens at 77K are shown in Fig. 4. From the fractographs, the UA CGHAZ specimens showed the transgranular cleavage fracture mode and the IC CGHAZ specimens showed the intergranular cleavage fracture mode. On the other hand, the tested surfaces of SCR CGHAZ specimens with the highest impact value indicated that failure occurred by the mixed mode of localized quasi-cleavage and ductile failure which was proved by dimpled area. This change in fracture mode was quite consistent with the change in impact toughness.

To verify that the IC CGHAZ was the weakest region within HAZ of this steel, the peak temperature of the second thermal cycle was fixed at 923K, an intercritical temperature between A_{C1} and A_{C3}, and first thermal cycles with various peak temperatures were applied. The results from Charpy tests at 77K are shown in Fig. 5. The figure also certifies that IC CGHAZ is the most embrittled region within the steel HAZ at low temperature. The high impact value at 1173K and 1273K was due to the refinement of grain size through recrystallization during the first thermal cycle.

Consequently, IC CGHAZ was the LBZ of QLT- 9% Ni steel at low temperature, but not at room temperature. This is quite consistent with the fact that the load-displacement curve from CTOD tests using actual HAZ specimens showed many more pop-ins at LNG temperature than at room temperature in authors' previous study (1997). Therefore, we called embrittlement of the IC CGHAZ just at low temperature the "cryogenic LBZ phenomena" and focused on the analysis of these phenomena.

Microstructural Analyses

To find the cause of the cryogenic LBZ phenomena, microstructural analyses were performed using the synthetic HAZ specimens. At first, the amount of retained austenite was measured from the samples attained from the tested specimens concerning Fig. 2 and Fig. 3. This was due to the fact that the retained austenite was well known as the most effective factor on the cryogenic toughening mechanism of 9% Ni steel from many studies, for examples by Fultz et al. (1985). Figure 6 shows the examples of the retained austenite measurements using X-ray diffractometry. At $2\theta=130.17°$ where a peak for austenite might rise, if it existed, there was no peak and only the α-Fe peak existed at $2\theta=106°$. All the samples showed the same results. Therefore, it was concluded that the retained austenite did not change the toughness because retained austenite almost did not exist in CGHAZ due to the high peak temperature of the first thermal cycle. This result was compatible with the results in our previous study (1997) using actual HAZ specimens. In the study, the average amount of retained austenite in the HAZ near the fusion line, the region mostly consisted of CGHAZ, was lower than 1%.

For the second effort to find the cause of the cryogenic LBZ phenomena, authors tried to observe the martensite-austenite constituent (M-A constituent), sometimes called martensite islands, which is a microstructure with high carbon and high hardness. The M-A constituent was generally known as the one of the main factors that deteriorated the HAZ toughness of high strength structural steel welded with high heat input (for example, Okada et al. 1994). But until now no studies on M-A constituents in 9% Ni Steel HAZ have been performed yet.

In this study, through the two stage electrolytic etching method, M-A constituents were observed in all the specimens simulated by double thermal cycles with the first peak temperature fixed at 1623K for simulating the micro-zones within CGHAZ. Figure 7 shows the examples of the observations of M-A constituents in this steel HAZ. M-A constituents were present in two kinds of elongated types in the figure, i.e. long with large width and short with small width. The M-A constituents were known to be formed along the lath boundaries and prior austenite grain boundaries as shown in the figure. So, the difference of the M-A constituent size between the two types was due to the differences of grain size among the regions concerning the fine grain in SCR CGHAZ. The average width of M-A constituents was 0.2-0.3 μm and the length was about several μm.

For all the synthetic micro-zones within CGHAZ, the area fraction of M-A constituents was measured by image analyzer using the specimens which were sub-zero treated at 77K or not treated. Table 3 shows the evaluated area fraction for synthetic UA CGHAZ, SCR CGHAZ and IC CGHAZ. As listed in the table, the amount of M-A constituents showed the minimum value for SCR CGHAZ and similar values for UA CGHAZ and IC CGHAZ. In addition, the difference from sub-zero treatment was very small.

By the change in the area fraction of M-A constituents alone, the cryogenic LBZ phenomena of IC CGHAZ could not be explained. Therefore, authors tried to estimate the change in hardness of M-A constituents in every specimen. However, although a very small load of 0.1g was applied, hardness of only M-A constituents was not able to be measured because M-A constituent was too small. To overcome the problem, EPMA was used for the estimation of the carbon contents in M-A constituents, due to the fact that carbon contents were generally known as very effective factor on the increase in the hardness of the second phase particles. Although carbon contents in the matrix were estimated quantitatively through ZAF method of EPMA, carbon contents in M-A constituents could not be estimated quantitatively because the size of M-A constituents were smaller than the effective beam radius, about 2-3 μm, of EPMA. Therefore, in case of M-A constituents only approximate values were evaluated through mapping method of EPMA. Figure 8 and Table 4 show the results of EPMA. Carbon contents in the matrix near M-A constituents decreased in order of SCR CGHAZ, UA CGHAZ, IC CGHAZ. Carbon contents in M-A constituent showed the maximum value in IC CGHAZ, about 2.2%, and the minimum value in UA CGHAZ, about 0.75%. From the results, it was concluded that the M-A constituents in IC CGHAZ were the most stiff among the M-A constituents in CGHAZ. Oppositely, M-A constituents in UA CGHAZ had the lowest hardness among them.

In Situ Observation of Fracture Behavior

Microscopic fracture procedures were observed by SEM with tensile loading stage. Because the fracture behavior at 77K could not be observed, sub-zero treated specimens were used instead of direct observation of fracture at 77K. In these tests, it was considered that the specimens were under plane stress condition because of the existence of central groove as shown in Fig. 1 and the fractures of the specimens indicated higher apparent toughness and larger tendency toward ductile fracture than that of the specimens under plane strain condition. Unfortunately, the fracture and deformation behaviors near M-A constituents could not be observed *in situ* microscopically due to the very small scale of M-A constituents and the fluctuation from loading.

Figure 9 shows a series of SEM micrographs for the synthetic IC CGHAZ specimen which were not sub-zero treated. As shown in the figure, the fracture was accompanied with the serious deformation. Figure 10 shows the locally magnified micrograraphs. It could be shown that the heavy deformation occurred in the matrix between neighboring M-A constituents. Since the specimens tested were well away from plane strain condition, valid K_{IC} values were not obtained. The apparent fracture toughness, K_C, calculated from maximum load was 162.19 MPa(m)$^{1/2}$.

Figure 11 shows a series of SEM micrographs for the synthetic IC CGHAZ specimens which was sub-zero treated instead of the test at 77K. As shown in the figure, although the specimen was under plane stress condition, complete cleavage fracture was observed. From the magnified micrographs in Fig. 12, it was verified that there was no deformation. Additionally, as shown in the figure, cleavage occurred across the M-A constituents and the broken M-A constituents were observed. The apparent fracture toughness, K_C, calculated from maximum load was 108.78 MPa(m)$^{1/2}$. As expected from cleavage fracture, the toughness value of the treated specimen was much lower than that of untreated specimen.

Other specimens simulating UA CGHAZ and SCR CGHAZ exhibited similar behaviors with the synthetic IC CGHAZ specimen which was not sub-zero treated, i.e. ductile fracture accompanied with heavy deformation occurred without any brittle fracture.

Figure 13 shows examples of the load-displacement curves from the *in situ* fracture test. The curve for sub-zero treated IC CGHAZ indicated not only the lowest apparent fracture toughness but also the absence of stable crack growth which was representative of the ductile fracture. It is also noted in the figure that both the maximum load and the displacement value at the maximum load point in the treated IC

CGHAZ specimen were much lower than those in other specimens. As a result, it was verified in this cryogenic steel HAZ that, although the specimen was under plane stress condition, the fracture mode at low temperature of IC CGHAZ was cleavage without any deformation while those of other micro-zones at any temperature and IC CGHAZ at room temperature were ductile failures with heavy deformations in matrix between the neighboring M-A constituents.

Mechanism for Cryogenic LBZ

Although the presence of M-A constituents in the IC CGHAZ caused the reduction in toughness of high strength structural steels, the mechanism by which it occurred has not been clear yet. Possible mechanisms from many studies, which were well reviewed by Toyoda (1993), were mainly focused on the fact that the relationship between brittle (hard) M-A constituents ($Hv = \sim 700$) and ductile (soft) ferrite matrix ($Hv = \sim 200$) increased the brittleness of the interfaces or M-A constituents themselves.

Because the matrix of QLT-9% Ni steel CGHAZ was a lath structure of tempered martensite with the hardness of about 400 Hv (Jang et al., 1997), the difference in hardness between the matrix and the M-A constituents was not enough to make the interface debonding by the same mechanisms. On the other hand, Chen et al. (1984) proposed that the stress concentration and triaxiality of the neighboring matrix were increased by the stiff phase particle and the raised stress reached several times the average stress in a uniform matrix. Including this suggestion, the cryogenic LBZ phenomena and the change in toughness of several micro-zones within QLT-9% Ni steel HAZ could be explained by the following mechanisms.

In the case of IC CGHAZ, because the large amount of M-A constituents containing high carbon contents of about 2.2% existed, the stress in the matrix near M-A constituents increased to several times the average stress. At room temperature, it was so easy for the matrix to deform plastically that the raised stress was relieved. So, heavy deformation between neighboring M-A constituents was observed as shown in Fig. 9 and Fig. 10. Oppositely at low temperature, the plastic deformation of matrix was not easy and it was not able to relieve the high stress concentration. So, if a cleavage plane of the matrix was perpendicularly oriented to the applied load, the matrix cleaved. However, interfacial debonding could not occur due to the insufficient difference in hardness between M-A constituents and the neighboring matrix, which consisted of tempered martensite.

On the other hand, in the case of UA CGHAZ, although the fraction of M-A constituents was a little larger than that in IC CGHAZ, a serious increase in the stress of matrix did not exist. This was caused by the fact that the carbon contents in M-A constituents were low, 0.75%, in comparison with M-A constituents in other regions as reported in Table 4. Also, the relatively high carbon contents in matrix near the M-A constituents more greatly reduced the differences in hardness between M-A constituents and neighboring matrix. Finally, in the case of SCR CGHAZ, the moderate toughness was mainly due to the fine-grained microstructure which was made by the second thermal cycle with a peak temperature at recrystallization range. In addition, the relatively small fraction of M-A constituents assisted the relatively high toughness.

In this work, the toughness change within the several micro-zones of QLT-9% Ni steel HAZ was analyzed. But this metallurgical analysis using synthetic HAZ specimen cannot be directly applied to the analysis of actual HAZ toughness because there are several mechanical factors, for examples, residual stress and strength mismatch between HAZ and weld metal. Now, analyses of the mechanical factors in QLT-9% Ni steel HAZ are going to be studied in authors' next work.

CONCLUSIONS

Weld thermal cycle simulation specimens were used to evaluate the toughness variation of the several micro-zones within multi-pass welded QLT-9% Ni cryogenic steel HAZ. The primary studies of this investigation were:

(1) Through Charpy V-notch impact test using synthetic HAZ specimens, the toughness change was estimated within the CGHAZ which was generally expected to be the weakest region in HAZ. From the tests at room temperature, it was demonstrated that there was no LBZ. But, from the tests at 77K, it was found that the IC CGHAZ had the lowest toughness among the several micro-zones within the CGHAZ. So the IC CGHAZ of this steel was called the "cryogenic LBZ" because the region showed the tendency of LBZ only at low temperature.

(2) To find the effective factors on the cryogenic LBZ phenomena, analyses of microstructures were performed using XRD, OM, SEM and EPMA. Unexpectedly, it was found that retained austenite was not an effective factor. On the other hand, M-A constituents were observed in every CGHAZ specimens. The area fraction of the M-A constituents and the carbon contents in M-A constituents for each specimens were analyzed. Although UA CGHAZ had a little larger amount of M-A constituents than IC CGHAZ, IC CGHAZ showed much higher carbon contents in M-A constituents than UA CGHAZ.

(3) Microscopic fracture behavior was observed *in situ* using a tensile loading stage in a SEM. The miniaturized SENT type specimens were used for the estimation of apparent fracture toughness under plane stress condition. To simulate the cryogenic test, sub-zero treatment was applied. From the results, the sub-zero treated IC CGHAZ specimen showed cleavage fracture including breaking of the M-A constituents while untreated IC CGHAZ specimen showed ductile failure accompanied with heavy deformation between the neighboring M-A constituents.

(4) A mechanism for the toughness change within the CGHAZ was proposed. In the case of IC CGHAZ, because of the large amount of M-A constituents containing high carbon contents, the stress of matrix increased. At low temperature, the plastic deformation of matrix for relieving the high stress was not easy and therefore the matrix cleaved. However at room temperature, plastic

deformation of the matrix easily occurred and relieved the high stress. On the other hand, similar behaviors were not observed in UA CGHAZ and SCR CGHAZ, mainly due to the small carbon contents in M-A constituents and fine-grained microstructure.

REFERENCES

British Standard 7777, 1993, "Flat-bottomed, Vertical, Cylindrical Storage Tanks for Low Temperature Service," British Standards Institution.

Chen, J.H., Kikuta, Y., Araki, T., Yoneda, M., and Matsuda, Y., 1984, "Micro-Fracture Behavior Induced by M-A Constituent (Island Martensite) in Simulated Welding Heat Affected Zone of HT80 High Strength Low Alloy Steel," Acta Metallurgica, Vol.32, pp.1779-1788.

EEMUA Standard No. 147, 1986, "Recommendation for the Design and Construction of Refrigerated Liquefied Gas Storage Tanks," The Engineering Equipment and Materials Users Association.

Fultz, B., Kim, J.I., Kim, Y.H., Kim, H.J., Fior, G.O., and Morris, Jr., J.W., 1985, "The Stability of Precipitated Austenite and the Toughness of 9Ni Steel," Metallurgical Transactions, Vol.16A, pp.2237-2249.

Ikawa, H., Oshige, H., and Tanoue, T., 1980, "Study on the Martensite-Austenite Constituent in Weld-Heat Affected Zone of High Strength Steel," Journal of Japanese Welding Society, Vol.49, pp.467-472.

Jang, J.-i., Yang, Y.-c., Kim, W.-s., and Kwon, D., 1997, "Evaluatoin of Cryogenic Fracture Toughness in SMA-Welded 9% Ni Steel through Modified CTOD Test," Metals and Materials, Vol. 3, pp.230-238.

Jang, J.-i., Yang, Y.-c., Kim, W.-s., and Kwon, D., 1998, "A Study of Fracture Toughness and Microstructures in the Weld Heat-Affected Zones of QLT-Processed 9% Ni Steel," Advances in Cryogenic Engineering(Materials), Vol. 44, pp.41-48.

Kim, B.C., Lee, S., Kim, N.J., and Lee, D.Y., 1991, "Microstructure and Local Brittle Zone Phenomena in High-Strength Low-Alloy Steel Welds," Metallurgical Transactions, Vol. 22A, pp. 139-149.

Masubuchi, K., 1980, "Analysis of Welded Structures," Pergamon Press, New York, NY, ch.2.

Okada, H., Matsuda, F., and Li, Z., 1994, "The Behaviors of M-A Constituent in Simulated HAZ with Single and Multiple Thermal Cycles," Quarterly Journal of Japanese Welding Society, Vol.12, pp.126-131.

Rosenthal, D., 1946, "The Theory of Moving Sources of Heat and Its Application to Metal Treatments," Transaction ASME, Vol. 68, pp.849-860.

Tamura, H., Onzawa, T., and Uematsu, S., 1980, "Retained Austenite and Notch Toughness in Synthetic HAZ of 9% Ni Steel," Journal of Japanese Welding Society, Vol.49, pp.854-860.

Toyoda, M., 1993, "Brittle Fracture Strength and Fracture Toughness Evaluation of Welds," Journal of Japanese Welding Society, Vol.62, pp.603-616.

Table 1. Chemical compositions and mechanical properties of QLT-9% Ni steel.

Chemical Compositions (wt%)					
C	Si	Mn	P	S	Ni
0.066	0.24	0.65	0.005	0.005	9.28

Table 2. Mechanical properties of QLT-9% Ni steel.

Properties at R.T. (at 77K)

YP (MPa)	TS (MPa)	EL (%)	Charpy vE (J)
640 (910)	710 (1140)	36 (34)	290 (210)

Table 3. Change in area fraction of M-A constituents.

	R.T. Specimen	77K Specimen
UACGHAZ	14.5%	13.3%
SCR CGHAZ	9.1%	8.8%
ICCGHAZ	13.1%	12.9%

Table 4. Change in carbon contents in M-A constituents and neighboring Matrix.

	UACGHAZ	SCR CGHAZ	ICCGHAZ
M-A	0.75 wt%	1.1 wt%	2.2 wt%
Matrix	0.207 wt%	0.242 wt%	0.175 wt%

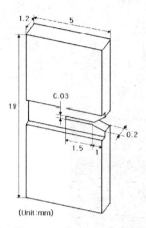

Fig. 1. Schematic diagram of miniaturized SENT specimen.

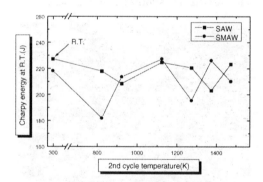

Fig. 2. Relation between Charpy impact energy at room temperature and the second peak temperature.

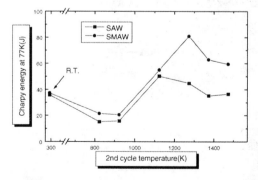

Fig. 3. Relation between Charpy impact energy at 77K and the second peak temperature.

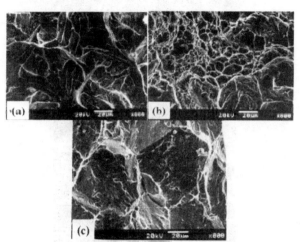

Fig. 4. SEM fractographs of Charpy tested specimens at 77K ; (a) synthetic UA CGHAZ, (b) synthetic SCR CGHAZ, (c) synthetic IC CGHAZ.

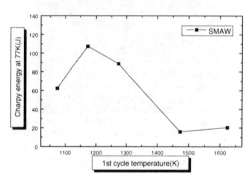

Fig. 5. Relation between Charpy impact energy at 77K and the first peak temperature.

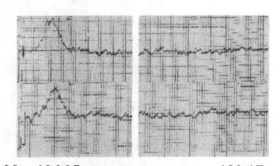

Fig. 6. Examples of XRD test for the estimation of retained austenite using samples attained from fractured specimens ; synthetic UA CGHAZ specimens tested at room temperature (upper) and at 77K (lower).

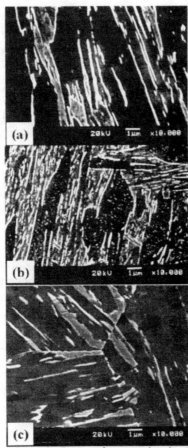

Fig. 7. Scanning electron micrographs in the synthetic CGHAZ prepared by 2 stages electrolytic etching technique to observe M-A constituents ; (a) synthetic UA CGHAZ, (b) synthetic SCR CGHAZ, (c) synthetic IC CGHAZ.

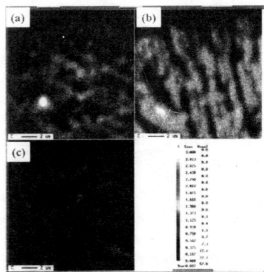

Fig. 8. Distribution of carbon contents in M-A constituents of synthetic CGHAZ ; (a) synthetic SCR CGHAZ, (b) synthetic IC CGHAZ, (c) synthetic UA CGHAZ.

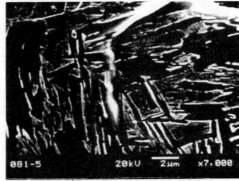

Fig. 10. Magnified SEM micrograph of a fracture sequence in Fig. 9.

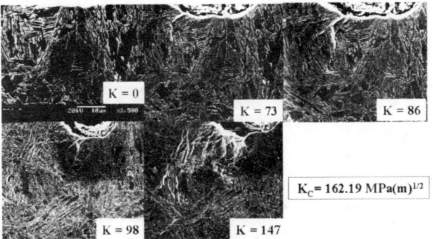

Fig. 9. A series of SEM micrographs near notch tip region of the synthetic IC CGHAZ specimen which was not sub-zero treated.

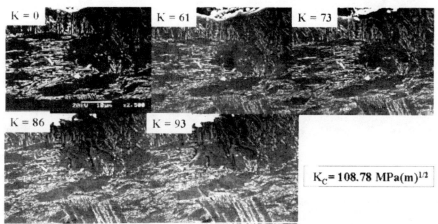

Fig. 11. A series of SEM micrographs near notch tip region of the synthetic IC CGHAZ specimen which was sub-zero treated at 77K.

Fig. 12. Magnified SEM micrograph of a fracture sequence in Fig. 11.

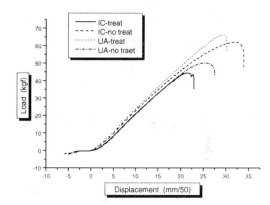

Fig. 13. Load vs. displacement curves from in situ fracture test using miniaturized SENT specimens (IC : synthetic IC CGHAZ, UA : synthetic UA CGHAZ, treat : sub-zero treated specimen, no treat : not sub-zero treated specimen).

WELDMENT RESIDUAL STRESSES IN A HYDROCARBON OUTLET VALVE OF A CATOFIN REACTOR

Germán Crespo

Zulay Cassier

Universidad Simón Bolívar
Departamento de Mecánica
Apdo. 89000, Caracas, 1080-A, Venezuela
Tlf: 58-2-906-406068, Fax: 58-2-906-4062
e-mail: gcrespo@usb.ve

ABSTRACT

The effect on fatigue of weld residual stresses in a hydrocarbon outlet valve of a catofin reactor was analyzed. The guide bars consisting of 5Cr-1/2Mo (ASTM A182-F5) and cobalt base abrasive resistance were welded with a A217-C5 steel using the electrode E-NiCrFe-3 for SMAW process following E502-15 AWS specification. The preheat temperature was 250 °C minimum up to 400 °C and conditions with and without postheating treatment was studied. Shear stress was applied on the fillet weld with cyclically full open-close process at 650 °C. A comparative analysis of the fatigue life and corrosion fatigue with several conditions of the weld was developed. The study show that both length of the guide bars and delay in postheating application are the principal effects on fatigue life and corrosion fatigue of the valve. A new design is presented.

INTRODUCTION

The selection of martensitic and austenitic chromium-molybdenum and chromium-nickel alloy steels for high temperature services were accepted and normalized (ASTM A182, ASTM A217). The heat treating and repair welding requirements, specially for martensitic ones, are specified detailed in the standards. However, some welding can not be qualified in accordance with Section IX of the ASME Code or ASTM A488, because it is impossible to consider the geometry and estimate the behavior of composite materials together. This is the case of guide bars of hydrocarbon outlet valves.

This study is the result of six years of trial-error analysis with the repairs to improve the design of the valves. It is not a strictly research because, obviously, the business of the plant where the valves work is to produce metil-ter-butil-eter (MTBE).

The fracture of the welded zone is showed in Figure 1. The analysis suggest that the fracture mechanism is a combination of fatigue and corrosion stress due to an intergranular attack of the hydrocarbon at high temperatures (Figure 2) that includes the following steps (Grabke et al. 1993):
- Over saturation of the metal matrix with carbon,
- Formation of an unstable carbide as M_3C,
- Decomposition of the carbide in metal plus carbon.

Fig 1. Cracks on welded zone. Body of the valve above; guide bars below.

In the other cases, the fracture appear in the interface between fusion zone and the heat affected zone as showed in Figure 3.

An analysis of the influence of postheating treatments, delay in their application and the length of the guide bars to improve the design is one of the principal objectives of this study. It is demonstrated in this paper that the best control of these parameters produce longest life of the valve.

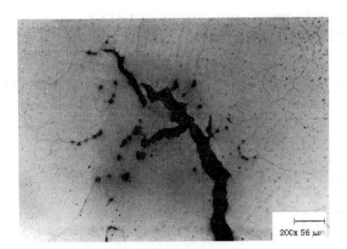

Fig 2. Corrosion intergranular with carbide formation in the welded zone.

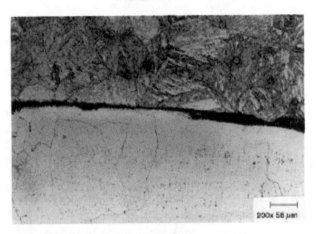

Fig. 3 Interface between fusion zone and heat affected zone.

EXPERIMENTAL STUDY

The valves used in this study are show in Figure 4, where the arrows indicate the guide bars. There were only 8 valves of 30" diameter and 8 valves of 18" diameter available for analysis. The body of the valves was made with A217-C5 steel and the guide bars with A182-F5. The original length of the guide bars was 48" and 32" for 30" and 18" diameter respectively. One of the sides of the guides was coated with plasma spraying (cobalt-chromium base) to improve the wear resistance.

The principal specifications of the existing welding procedure are the following:
- Filler metal: 3/32" – 1/8" diameter AWS E-NiCrFe-3, with "F" N° 43 and "A" N° 4, in accordance with SFA N° 5.11, SMAW process in 2F-weld position.
- Preheating and interpass temperature: 400 °F (205 °C) minimum, 700 °F (370 °C) maximum.
- Postheating: Not requiere per ANSI B31.3 (see Note 1) and 1250 °F (625 °C) per ASTM A 182.

Fig. 4. Body valves of 30" diameter. The arrows indicate the welded guide bars.

Since there are only 16 valves available for this analysis and the fatigue life is around one year, the changes on the design were made during each stop of the plant or when the valves failed and the section had to be closed. As a reference, a resume of the new parameters is listed in Table 1.

Table 1. Parameters for the new design.

Valve N°	Length of the guide bar (")	Postheating temperature (°F)	Delay postheating temperature (°F)
30"- 01	48 (a)	None (a)	(b)
30"- 02	48	1250	2
30"- 03	24	None	(b)
30"- 04	16	None	(b)
30"- 05	24	None	(b)
30"- 06	24	900	2
30"- 07	48	1250	24
30"- 08	16	900	24
18"- 01	32 (a)	None (a)	(b)
18"- 02	32 (a)	None (a)	(b)
18"- 03	32	1250	2
18"- 04	32	900	24
18"- 05	16	None	(b)
18"- 06	32	900	2
18"- 07	16	1250	24
18"- 08	16	900	24

(a) Original parameters
(b) At least one week

RESULTS

The fatigue life of the valves for the original conditions (see strings remark in Table 1) was about six month plant operation. In 1992, the first modification was introduced and since that year new conditions were analyzed. Although it is not possible to establish mathematical solutions for determine the behavior of fatigue life of the valves, it increased more than four times the original life thanks to the new designs as was demonstrated by experimental study.

In Figure 5 and 6, the increase of fatigue life as a function of delay in the application of postheating treatment is showed for three different lengths of guide bars in 30" diameter valves and two lengths for 18" diameter valves, respectively. As observed, the shorter guide bars with postheating treatment applied after two hours of welding process produce the longest life.

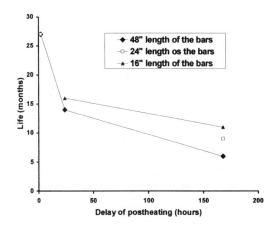

Fig 5. Fatigue life as a function of delay of postheating for the 30" diameter valves.

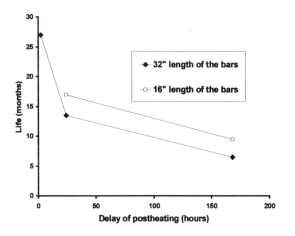

Fig 6. Fatigue life as a function of delay of postheating for the 18" diameter valves.

Apparently, it makes no difference when the postheating temperature is 900 °F or 1250°F. The fracture mechanism is the same in the valves with a new design but cracks appear in the interface only between the heat affected zone and fusion zone.

The effect of wear resistance layer may be ignored. Additional residual stress is present due a difference in thermal expansion coefficients. In all the parts where this surface treatment is applied, an open crack is produced only for the hard layer as it is observed both in guide bars and the discs of the seals (Figure 7). For the shorter guide bars these cracks not appear again.

Fig 7. Cracks in wear resistance layer applied on the discs of the valves.

DISCUSSIONS

On the valves used in this plant, a steam purge is used to form an maintaining an oxide layer on the metal surface for protection. Whatever the steam purge is insufficient or the oxide layer is very thin, it is impossible to avoid some hydrocarbon leaks through the discs when the reactor is in reaction mode, and the carburization porcess will begin early or later, but it is strongly acelerated due to residual stress.

CONCLUSIONS

Conclusion from this analysis is that the fail of weld guide bars is produced by a fatigue corrosion mechanism, long bar length and no postheating treatment and/or delay after the welding process induce residual stresses that reduce the life of the valves.

REFERENCES

ASME *Boiler and Pressure Vessel Codes, Section IX*, New York, USA.

ASTM A182-91, *Forged or Rolled Alloy-Steel Pipe Flanges, Forged Fittings and Valves and Parts for High-Temperature Service*, Pennsylvania, USA.

ASTM A217-91, *Steel Castings, Martensitic Stainless and Alloy, for Pressure-Containing Parts, Suitable for High-Temperature Service*, Pennsylvania, USA.

ASTM A488-91, *Practice for Steel Castings, Welding, Qualifications of Procedures and Personal*, Pennsylvania, USA.

AWS SFA-5.11, *Specification for Nickel and Nickel-Alloy Covered Welding Electrodes*, New York, USA.

Grabke H.J, Krajak R. And Müller-Lorenz, 1993, Metal dusting of high temperature alloys, *Werkstoffe un Korrosion V 44, pp. 89-97*.

INFLUENCE OF THE WELDING VARIABLES ON THE MECHANICAL PROPERTIES IN BUTT JOINTS FOR ALUMINUM 6063-T5

María Carolina Payares

Minerva Dorta

Patricia Muñoz-Escalona

Universidad Simón Bolívar
Departamento de Mecánica
Apdo. 89000, Caracas, 1080-A, Venezuela
Tlf: 58-2-906-4060, Fax: 58-2-906-4062
e-mail: pmunoz@usb.ve

ABSTRACT

The study of the mechanical properties in butt joints for aluminum 6063-T5 under the influence of variables such as arc current, arc voltage and welding speed is important for determining the optimums conditions for an efficient joint.

Tests were design and performed using Gas Metal Arc Welding process (GMAW), with a constant power source using a direct current reverse polarity. An aluminum plate was used as the base material and an ER electrode wire as filler metal as recommended by the AWS A 5.10.

The results showed that the arc current and the welding speed are the variables with the most influence on the tensile strength, the ultimate tensile stress and welding efficiency. Respect the arc voltage a defined behavior did not show up.

NOMENCLATURE

GMAW: Gas Metal Arc Welding
AWS: American Welding Society
BM: Base Metal
d: Height of welding reinforcement (mm.)
w: Weld bead width (mm.)
p: Penetration (mm.)
I: Arc current (A)
E: Arc voltage (V)
v: Welding speed (cm/min.)
h1: Welding Efficiency for welds with weld reinforcement (%)
h2: Welding Efficiency for welds without weld reinforcement (%)
Sy: Yield strength (Kg/mm^2)
Su: Ultimate tensile strength (Kg/mm^2)
Sy 1: Yield strength for welds with weld reinforcement. (Kg/mm^2)
Sy 2: strength for welds without weld reinforcement (Kg/mm^2)
Su 1: Ultimate tensile strength with weld reinforcement (Kg/mm^2)
Su 2: Ultimate tensile strength welds without weld reinforcement (Kg/mm^2).
Fu/Xbm: Ultimate load per unit of base metal length. (Kg/mm.)
Fu/Xweld: Ultimate load per unit of weld length (Kg/mm.)
t: Thickness (mm.)

INTRODUCTION

A wide variety of weldments made of aluminum 6063-T5 find numerous engineering application such as cryogenics, LNG tank, unfired pressure vessels, marine components, pipes and irrigation pipes. However, it has long been known that the heat of welding deteriorates mechanical properties of welded joints in heat treatable aluminum alloys; this includes loss of hardness and tensile strength in the heat affected zone (HAZ) of the alloys.

The amount of heat which can be inserted in the aluminum when it is been welded, depends on the welding parameters used. These welding parameters could change the geometry, the mechanical properties, failure location of both the fusion zone and the heat affected zone.

As a result, some researchers have attempted to investigate the effects of various process variables on the weld geometry and penetration. Kiyohara, et al (1977), Datsko (1977), Yasuda (1991) and Payares, et al (1998).

Some authors have studied the effects of welding variables on tensile strength failure location and hardness of aluminum alloy weldments. Burch (1958) studied the effect of welding speed on strength of 6061-T4 aluminum joints and showed that as the welding speed increased, the joint strength improved rapidly up to a speed of about 15 inch/min., at higher speeds, the improvement was less pronounced.

In this research, an attempt has been made to present the effects of welding variables (arc current, arc voltage and welding speed) on the mechanical strength and welding efficiency of 6063-T5 Aluminum welded joints.

EXPERIMENTAL PROCEDURES

Tests were designed and performed to evaluate the influence of welding variables in welding efficiency and both the yield strength and the ultimate tensile strength of a butt joint of aluminum 6063 - T5. Automatic GMAW equipment with a constant power source, using a direct-current reverse polarity to make weld reinforcement was employed.

A 6063-T5 aluminum plate (6.35 mm. thick) was used as the base material, with a chemical composition as shown in Table 1. The test plate was cut into 50.8 * 200 mm. An electrode wire of 1.19 mm. diameter was used as recommended by AWS A 5.10.

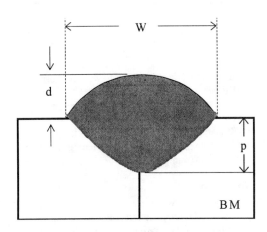

Figure 1. Scheme of the weld reinforcement geometry and the penetration in a butt joint.

Table 1. Chemical composition of Al 6063-T5 and ER 5356.

El*	Si	Fe	Cu	Mn	Mg	Cr	Zn	Ti	Al
BM	0.38-0.49	1.24	a**	a	0.45-0.59	a	a	1.3	Bal
FM	b***	b	0.02	0.07	4.78	0.07	0.01	0.12	Bal

* El (Wt % +/- 0.01)
** a < 0.03 %
*** b = Si + Fe = 0.33 %

The weld reinforcements of butt joints were produced using the welding parameters which had given best results for weld penetration and weld reinforcement in previous research, Payares, et al.(1998), Payares and De Barros (1997).

Table 2. Tested welding parameters.

Sample	I (A)	E (V)	V(cm/min.)
1	210	24	400
2	210	24	500
3	210	24	600
4	210	24	700
5	210	24	800
6	210	24	900
7	210	24	1000
8	210	21	600
9	150	24	600
10	180	24	600
11	150	18	600
12	150	21	600

For metallographic studies two (2) specimens of 15 mm. of length were fabricated from the uniform weld area from the piece of each condition, giving a total of 24 specimens. The weld specimens were treated according to standard metallographic techniques to reveal penetration and later measurement in a POZ Warzawa microscope.

Additional tensile test were conducted on three (3) specimens taken transverse to the weld for each condition. Two (2) of them were machined to the thickness of the plate to study only the effect of the penetration and the third one to study the effect of penetration and the weld reinforcement.

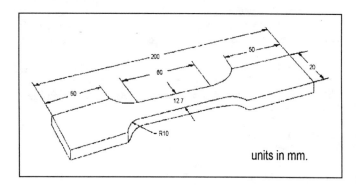

Figure 2. Tensile test specimen. (ASTM E-8, 1992)

RESULTS AND DISCUSSIONS

Once analyzed the values shown in Table 3, it can be observed that as the welding speed increases, both the penetration and the height of the weld reinforcement decreases for all the established conditions. This behavior is possible caused by the amount of alloying or mixing that occurs between the base metal and adder filler metal.

In Table 4, is shown that as voltage increases beyond 21 volts, there is a decrease in penetration, possibly due to the decrease in heat input below the limit at which maximum penetration depth is obtained, due to the combined effect of heat loss towards the extremes by radiation, the increase in arc length, and the application of low values of arc current.

Table 3. Penetration and height of the weld reinforcement values with variable welding speed (I=210 A and E=24 V)

Sample	v (mm/min.)	p ± 0.01 (mm.)	d ± 0.01 (mm.)
1	400	2.87	3.79
2	500	2.68	3.68
3	600	2.12	2.97
4	700	1.72	2.79
5	800	1.40	2.85
6	900	1.58	2.61
7	1000	1.11	2.32

Table 4. Penetration and height of the weld reinforcement values with variable arc voltage (I=150 A and v= 600 mm/min.)

Sample	E (V)	p ± 0.01 (mm.)	d ± 0.01(mm.)
10	24	0.52	1.63
12	18	0.99	2.37
13	21	1.43	2.20

The most influential variable on both the penetration and the height of the weld reinforcement is the arc current as it can be observe in Table 5. An increase in the arc current will raise the heat input. Therefore, an increase in the arc current raises both the penetration and the height of the weld reinforcement.

Table 5. Penetration and height of the weld reinforcement values with variable arc current (E=24 V and v = 600 mm/min.)

Sample	I (A)	p ± 0.01 (mm.)	d ± 0.01(mm.)
3	210	2.12	2.97
10	150	0.52	1.63
11	180	1.39	2.84

The equation for predicting the welding efficiency (μ) was developed by Datsko (1977):

$$\mu = \frac{Fu / Xweld}{Fu / Xbm} * 100 \quad (1)$$

where:

$$Fu / X\,weld = Su * p \quad (2)$$

$$Fu / X\,bm = Su * t \quad (3)$$

The equation above was used to predict the welding efficiency values for welds with and without weld reinforcement.

In Table 6, it can be observed that when the welding parameters have been changed the welding efficiency always takes different values. The welding efficiency of the welds with weld reinforcement is bigger than the welds without. This is due that the load carrying load of a weld joint increases when the area of this welded joint is being bigger.

Table 6. Welding efficiency for welds with weld reinforcement (h1) and for welds without weld reinforcement (h2).

Sample	$h_1 \pm 0.1\%$	$h_2 \pm 0.1\%$
1	101	51.3
2	107	39.1
3	85.8	31.8
4	67.6	26.2
5	63.8	20.9
6	61.5	23.7
7	49.3	16.2
8	64.4	39.1
9	42.5	31.8
10	79.3	26.2
11	51.7	20.9
12	70.6	23.7

Figure 3, shows the influence of the welding speed on the welding efficiency when maintaining both the arc voltage and arc current constant. The welding speed increases for all welding conditions and for all welds.

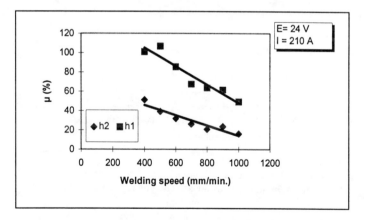

Figure 3. The effect on welding efficiency with variation of welding speed, maintaining both the arc voltage and the arc current constant.

The effect of the arc current variation on the welding efficiency when the arc voltage and the welding speed are kept constant, is shown in Figure 4. It can be observed that as the arc current increases the welding efficiency increases, since when the arc current increases both the penetration and the height of the weld reinforcement also increase because the amount of heat at the work surface is higher when the arc current is increased. On the contrary, it is observed the effect of the arc voltage on welding efficiency as can be seen in figure 5, when arc voltage increases beyond 21 V there is a decrease of welding efficiency.

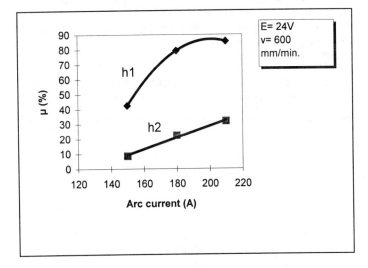

Figure 4. The effect on welding efficiency with variation of arc current, maintaining both the arc voltage and the welding speed constant.

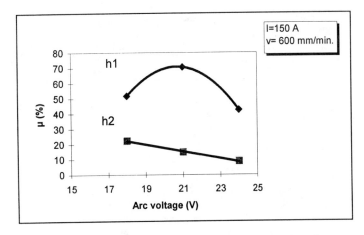

Figure 5. The effect on welding efficiency with variation of arc voltage, maintaining both the arc current and the welding speed constant.

The effect of welding speed on yield strength when the arc voltage and arc current are kept constant is shown in Fig. 6. The yield strength decreases as the welding speed is increased.

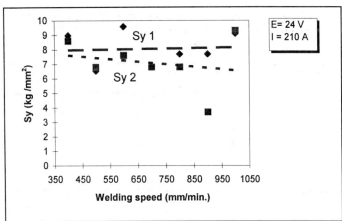

Figure 6. The effect on yield strength with variation of welding speed, maintaining both the arc voltage and the arc current constant.

Figure 7, shows the relation between yield strength and the arc voltage for all established parameters. The yield strength increases with rising the arc voltage on welds with weld reinforcement. But for welds without weld reinforcement, at low values of arc voltage the yield strength is high and for high values of arc voltage, the yield strength tend to be lower.

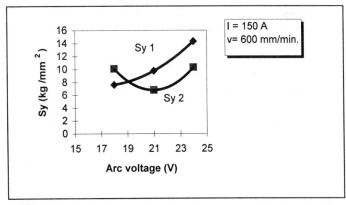

Figure 7. The effect on yield strength with variation of arc voltage, maintaining both the arc current and the welding speed constant.

As can be seen in Figure 8, the yield strength decreases as the arc current increases. Since the arc current increases the heat input increases. Consequently, the aluminum welding tend to be deformed more plastic than elastic.

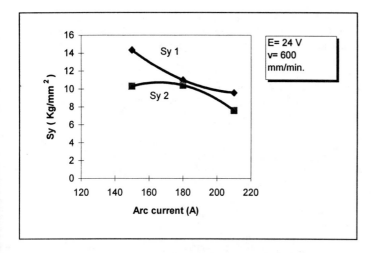

Figure 8. The effect on yield strength with variation of arc current, maintaining both the arc voltage and the welding speed constant.

As can be seen in Figure 9, the variation of the ultimate tensile strength with rising the welding speed is low, if the linear approximation is considered, even though these points do not show a defined pattern behavior, in both welds with and without reinforcement.

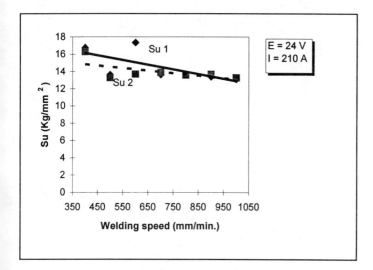

Figure 9. The effect on ultimate tensile strength with variation of welding speed, maintaining both the arc voltage and the arc current constant.

Figure 10, shows the relation between arc voltage and the ultimate tensile strength for both welds with and without weld reinforcement. At low values of arc voltage, the ultimate tensile strength is almost constant. For values of the arc voltage beyond 21 V, the ultimate tensile strength tend to be high. This result has a similar behavior as the effect of this parameter on the yield strength.

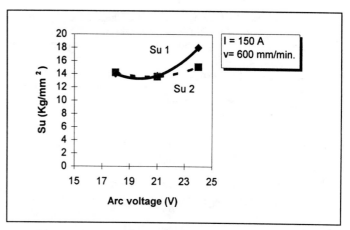

Figure 10. The effect on ultimate tensile strength with variation of arc voltage, maintaining both the welding speed and the arc current constant.

The effect of arc current on the ultimate tensile strength for welds with and without weld reinforcement is shown in Figure 11. The ultimate tensile strength remains steady as the arc current is increased from 150 A to 210 A for both curves. Therefore, the ultimate tensile strength does not change with rising arc current.

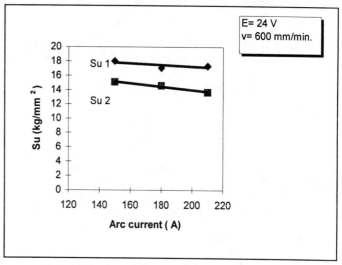

Figure 11. The effect on ultimate tensile strength with variation of arc current, maintaining both the welding speed and the arc voltage constant.

CONCLUSIONS
- The present work found that the welding efficiency of welds with weld reinforcement, is higher than welds without reinforcement for Al 6063 - T5 using ER5356 as filler metal.

- It was found that the welding efficiency of the welded joints of Alloy 6063-T5, using ER5356 as filler metal varies inversely with the welding speed, and is proportional direct to the arc current.

- For arc voltage lower than 21 volts, an increase in arc voltage causes an increase in welding efficiency. Increasing the voltages above 21 V produces a decrease in welding efficiency, when welding Al 6063-T5 using ER5356 as filler metal.

- The yield strength in aluminum 6063-T5 welds decreases as the welding speed increases.

- The yield strength in GMAW AA-6063-T5 welds decreases with increasing arc current ranging from 150 A. to 210 A.

- GMAW AA-6063-T5 Welds without a weld reinforcement using ER5356 as filler metal, the yield strength values vary, with variation of arc voltage. The trend shows that at low values of the arc voltage the yield strength is high. For high values of arc voltage, the yield strength tend to be lower.

- The ultimate tensile strength decreases low with raising the welding speed if the linear approximation is considered.

- The ultimate tensile strength in 6063-T5 with ER 5356 filler wire remains constant as arc voltage is increased until 21 V., but it increases sharply as the arc voltage is increased from 21 volts to 24 volts.

- The ultimate tensile strength observed is also constant with increasing the arc current, ranging from 150 Ampere to 210 ampere.

REFERENCES

American Society for Testing and Materials.1992 Annual Book of ASTM Standards, Section 2, Volume 02.02. ASTM, Philadelphia, Pennsylvania, 1992.

Burch, W.L.,1958; "The Effect of Weld Speed on Strength of Aluminum Joints". AWS, Annual Spring Meeting held in St. Louis, Apr.14-18. pp. 361s-367s.

Datsko J.,1977. "Materials in Design and Manufacturing", Malloy Inc. Ann Arbor, Michigan, pp. 8-1 to 8-9,

Grong, AE. and Myhr O.R., 1995; "Modeling of the Strength Distribution in the Heat Affected Zone of 6082-T6 Aluminum Weldments". Mathematical Modeling Welding Phenomena, The Institute of Materials, pp. 300-311.

Kiyohara T., Okada T., Wakino Y., Yamamoto H., 1977; "On The Stabilization of GMA Welding of Aluminum", Welding Journal, Vol. 56, No.1, pp. 21-28.

Malyn V., 1995,; "Study of Metallurgical Phenomena in the HAZ 0f 6061 T6 Aluminum Welded Joints". Welding Research Supplements" , Vol.74 , No. 9, pp. 305s-318s.

Payares, M.C. and De Barros, C., 1997; " Influence of welding Parameters on the Weld Bead Depth in Butt Joint of Aluminun 6063-T5" . Applied Mechanics in The Americas, Puerto Rico, Vol. 5, pp. 494-497.

Payares, M.C., De Barros, C., Munoz - Escalona, P., 1998; "Mathematical Model for the Prediction of Penetration in Butt Joints for Aluminum 6063 - T5". ASME 1988, Fatigue, Environmental Factors, and New Material, Vol. 374, pp. 23-27.

Shmoda T., and Doherty J., 1978; "The Relationship Between Arc Welding Parameters and Weld Bead Geometry - a Literature Survey". The Welding Institute Report, 74/1978/PE.

Tables of Miller, 1987; "Millermatic Calculator Gas Metal Arc (MIG) Welding", Miller Electric Mfg. Co.,U.S.A(1987).

Welding Engineer , 1973;" Using MIG for High Quality Aluminum Welds, Vol., 58, pp. 9-12, (1973).

Welding Handbook, "Fundamentals of Welding ", 1, AWS, U.S.A., Seventh Edition, pp. 60-62, 1975.

Yasuda K., 1991; "Welding Conditions of Aluminum Welding". J. Light Met. Weld Constr., Vol. 29, No.10, pp. 8-12.

AUTHOR INDEX

PVP-Vol. 393
Fracture, Fatigue and Weld Residual Stress

Andrade, Arnaldo H. P. 47
Broviak, Bart J. 113
Brust, Frederick W. 201, 307
Cano, V. 193
Caskey, G. R., Jr. 129
Cassier, Zulay 335
Cavallo, N. 193
Chao, Yuh-Jin 91, 113, 209
Chaouadi, Rachid 35
Chen, Chang-New 145
Chung, Yeon-Ki 155
Crespo, Germán 335
Dodds, Robert H. 11
Dong, Pingsha 177, 201
Dorta, Minerva 339
Eisele, U. 93
Gao, Xiaosheng 11
Gérard, Robert 35
Gilles, Ph. 193
Gordon, J. Robin 225
Han, Lianghao 169
He, M. Y. 17
Herter, K.-H. 121
Horii, Yukihiko 287, 297
Hour, K. 63
Jang, Jae-Il 325
Janosch, J. J. 215
Joyce, James A. 11, 53
Jullien, J. F. 193
Junek, Lubomír 179
Kim, Charlie 79
Kim, Woo-sik 325
Kirk, Mark 1, 23, 71, 79
Koppenhoefer, Kyle 225
Kubo, Shiro 105
Kwon, Dongil 325
Lam, P. S. 129, 139
Landes, John D. 47
Lawrjaniec, D. 215
Leung, Kent K. 279
Link, Rick E. 11
Liu, Cengdian 169
Liu, Shu 113
Lott, Randy 71, 79
Lucas, G. E. 17
Magula, Vladislav 179
Martin, John A. 239
McCabe, Donald E. 29, 47

Miranda, Carlos A. J. 47
Miura, Naoki 155
Mohr, William 225
Muñoz-Escalona, Patricia 339
Murakawa, Hidekazu 287, 297
Natishan, Marjorie 23
Ochodek, Vladislav 179
Odette, G. R. 17
Ohji, Kiyotsugu 105
Parker, Ronnie B. 255
Payares, María Carolina 339
Preston, Robin V. 265
Qi, Xinhai 209
Rangaswam, Partha 255
Rathbun, H. J. 17
Roos, E. 93, 121
Rosinski, Stan 71
Sato, Masanobu 287, 297
Scibetta, Marc 35
Server, William 71, 79
Sheckherd, J. W. 17
Shercliff, Hugh R. 265
Sindelar, R. L. 129, 139
Slováček, Marek 179
Smith, E. 101, 165
Smith, Simon D. 265
Sokolov, M. A. 29
Swain, Ronald L. 47
Taheri, S. 193
Taleb, L. 193
Tanaka, Jinkichi 287, 297
Tomes, Chuck 79
Tregoning, Robert L. 11, 53
Van Der Sluys, W. A. 63
van Walle, Eric 35
Vincent, Y. 193
Wallin, Kim 3
Wang, Jianhua 287, 297
Wang, Yinpei 169
Wilkowski, G. M. 153
Williams, James 79
Withers, Philip J. 265
Woo, L. Y. 129
Yang, Young-chul 325
Yoon, K. K. 1, 63
Yoshikawa, Masashige 105
Zhang, Jinmiao 201

Book Number: G01122